AF358640

Power Transmission and Control in
Power and Vehicle Machineries

# Power Transmission and Control in Power and Vehicle Machineries

Editors

**Yunhua Li**
**Nariman Sepehri**
**Kazuhisa Ito**
**Maolin Cai**
**Bing Xu**
**Yan Shi**

Basel • Beijing • Wuhan • Barcelona • Belgrade • Novi Sad • Cluj • Manchester

*Editors*

Yunhua Li
Beihang University
Beijing
China

Nariman Sepehri
University of Manitoba
Winnipeg
Canada

Kazuhisa Ito
Shibaura Institute of Technology
Saitama
Japan

Maolin Cai
Beihang University
Beijing
China

Bing Xu
Zhejiang University
Hangzhou
China

Yan Shi
Beihang University
Beijing
China


*Editorial Office*
MDPI AG
Grosspeteranlage 5
4052 Basel, Switzerland


This is a reprint of articles from the Special Issue published online in the open access journal *Applied Sciences* (ISSN 2076-3417) (available at: https://www.mdpi.com/journal/applsci/special_issues/Power_Transmission).

For citation purposes, cite each article independently as indicated on the article page online and as indicated below:

Lastname, A.A.; Lastname, B.B. Article Title. *Journal Name* **Year**, *Volume Number*, Page Range.

**ISBN 978-3-7258-2449-6 (Hbk)**
**ISBN 978-3-7258-2450-2 (PDF)**
**doi.org/10.3390/books978-3-7258-2450-2**

# Contents

*Editorial*

# Editorial of Special Issue on Power Transmission and Control in Power and Vehicle Machineries

Yunhua Li [1], Nariman Sepehri [2], Kazuhisa Ito [3], Maolin Cai [1], Bing Xu [4] and Yan Shi [1,*]

1   School of Automation Science and Electrical Engineering, Beihang University, Beijing 100191, China; yhli@buaa.edu.cn (Y.L.)
2   Department of Mechanical Engineering, University of Manitoba, Winnipeg, MB R3T 5V6, Canada
3   Department of Machinery and Control Systems, Shibaura Institute of Technology, Saitama 337-8570, Japan
4   Department of Mechanical and Electronic Engineering, Zhejiang University, Hangzhou 310058, China
*   Correspondence: shiyan@buaa.edu.cn

## 1. Introduction

Power transmission and control include the conversion, transmission, and distribution of power, and the actuating of control in the end implement block; it plays an important part in the power and the vehicle of machines [1], such as automobile and construction machinery [2], airplanes [3], and ships. Power transmission and control has also been widely used in power engineering, such as in wind power generation systems, hydroelectric power systems, solar power systems, compressed air energy storage systems, hybrid energy vehicles, geological mining machinery [4], hydrogen energy vehicles [5], and wind generators [6]. By using advanced sensors and transducers, control theory, modelling and simulation, and optimization methods, the performances of power transmission and control systems in power and vehicle machinery can be significantly improved.

This Special Issue focuses on reporting the new achievements in power and vehicle machines from 2017 to 2018.

## 2. New Theories and Technological Progresses

In this Special Issue, we have curated a selection of 18 papers that represent cutting-edge research in the fields of hydraulic and pneumatic system technologies. These papers highlight innovative approaches to design, analysis, and control strategies, emphasizing advancements in energy efficiency and precision in industrial applications. When looking back on previous Special Issues, various topics have been addressed, including hydraulic system design and control, pneumatic systems and components, and hydraulic and pneumatic analysis and optimization. In relation to hydraulic system design and control, the published papers mainly focus on novel design methodologies and control strategies for hydraulic systems. The first paper, authored by Pedro Roquet et al. (contribution 1), provides a new simplified methodology to evaluate the design specifications of hydraulic components, in which the method allows for the definition and establishment of the hydraulic cylinder design specifications while taking into account the probabilistic characterization of the load spectrum variability; this method could be extrapolated to other hydraulic or mechanical components. The second paper is authored by Che-Pin Chen and Mao-Hsiung Chiang (contribution 2), developed a novel proportional pressure control valve for an automobile hydraulic braking actuator. In this article, the proposed novel proportional pressure control valve of an automobile hydraulic braking actuator is implemented and verified experimentally. The third paper, authored by Haoling Ren et al. (contribution 3), proposes a novel automatic idle speed control system with a hydraulic accumulator and control strategy for construction machinery. In order to reduce the energy consumption and emissions of the hydraulic excavator, a two-level idle speed control system with a hydraulic accumulator for the construction machinery is proposed to reduce the energy

**Citation:** Li, Y.; Sepehri, N.; Ito, K.; Cai, M.; Xu, B.; Shi, Y. Editorial of Special Issue on Power Transmission and Control in Power and Vehicle Machineries. *Appl. Sci.* **2024**, *14*, 9762. https://doi.org/10.3390/app14219762

Received: 28 August 2024
Accepted: 24 September 2024
Published: 25 October 2024

consumption and improve the control performance of the actuator when the idle mode is cancelled. The fourth paper is authored by Qing Guo et al. (contribution 4); a prescribed performance constraint (PPC) control method is adopted in an electro-hydraulic system (EHS) to restrict the tracking position error of the cylinder position to a prescribed accuracy and to guarantee the dynamic and steady position response in a required boundedness under these uncertain nonlinearities. These papers address critical challenges in enhancing the reliability and efficiency of hydraulic systems in various applications.

The second category is pneumatic systems and components; the papers in this category mainly deal with the development and enhancement of pneumatic systems, high performance valves, and actuators. The fifth paper by by Songlin Nie et al. (contribution 5) focuses on the development of a high-pressure pneumatic on/off valve with high transient performances directly driven by a voice coil motor; this research will have a significant effect on raising the performance of the high-speed pneumatic on/off valve and the development of pneumatic precision motion control. The sixth paper in this Special Issue, which is authored by Yeming Zhang et al. (contribution 6), presents a pneumatic rotary actuator position servo system based on ADE-PD control. The flow state of the gas and the motion state of the pneumatic rotary actuator in the pneumatic rotary actuator position servo system are analyzed in this paper. The seventh paper theoretically analyses the flow ripple of a tandem crescent pump with index angles; it is authored by Hua Zhou (contribution 7) and presents a theoretical approach for lowering the outlet flow ripple of a crescent pump by applying a tandem crescent pump consisting of two gear pairs with an index angle between them. The eighth paper, authored by Fan Yang et al. (contribution 8), proposes a new method for analyzing the pressure response delay in a pneumatic brake system caused by the influence of transmission pipes; the study aims to propose an analysis method for resolving the pressure response of a pneumatic brake circuit considering the effect of a transmission pipe. The ninth paper, written by Fan Yang et al. (contribution 9), analyzed the energy efficiency of a pneumatic booster regulator with energy recovery; it discovers that a recovery chamber helps to improve the performance of the VBA-R, which included a boost ratio improvement of 15–25% and an efficiency improvement of 5–10% compared with a conventional VBA booster regulator. The tenth paper carries on an experimental study on hysteresis characteristics of a pneumatic braking system for a multi-axle heavy vehicle in emergency braking situations, and it is authored by Zhe Wang et al. (contribution 10). In the article, the hysteresis of a pneumatic brake system for an eight-axle vehicle in an emergency braking situation is studied based on a novel test bench. The eleventh paper investigates the effects of internal EGR by variable exhaust valve actuation with post injection on auto-ignited combustion and emission performance and it is authored by Insu Cho et al. (contribution 11). The paper discovers that it is possible to efficiently utilize heat to recompress retardation post injection with a downscaled specification of the exhaust valve rather than the intake valve. These contributions are pivotal in pushing forward the operational capabilities and energy efficiency of pneumatic systems.

The third category is hydraulic and pneumatic analysis and optimization, which includes research focused on analytical models and optimization techniques for both hydraulic and pneumatic systems. The 12th paper, written by Dan Jiang et al. (contribution 12) builds a pressure transient model of water–hydraulic pipelines with cavitation. Different from the traditional method of characteristics (MOCs), the present model method is advantageous in terms of its simple and convenient computation. The 13th paper studies the dimensionless energy conversion characteristics of an air-powered hydraulic vehicle, and this paper is authored by Dongkai Shen et al. (contribution 13). The research can be referred to in the performance and design optimization of the HP transformers. The 14th paper in the Special Issue conducts a performance test and internal flow field simulation of a vortex pump; it is authored by Ping Tan et al. (contribution 14). It is revealed that for the vortex pump to have advanced suction and anti-cavitation performance, the lowest pressure in the pump should be $-4 \times 10^4$ Pa and it should be located at the centre of the vortex chamber cavity. The 15th paper models and dynamically analyses the direct operating solenoid valve for improving the performance of the shifting control system and it is authored by Xiangyang Xu et al. (contribution 15). A numerical approach for solving the multi-domain physical problem of the valve is presented. Moreover, the 16th paper proposed an energy regeneration hydraulic system via a relief valve with an energy regeneration unit; the paper is authored by Tianliang Lin et al. [2] and indicates that the proposed structure of the relief valve with HERU can achieve a better performance and higher regeneration efficiency. The 17th paper, written by Wei Liu et al. (contribution 16), reviewed inlet/outlet oil coordinated control for electro-hydraulic power mechanisms under a sustained negative load and presented the existing problems and future trends in inlet/outlet coordinated control for an electro-hydraulic power mechanism under sustained negative load. The 18th paper, authored by Fangwei Ning et al. (contribution 17), reviewed the research progress of related technologies of electric–pneumatic pressure proportional valves. These studies provide foundational insights into optimizing component design and operational strategies for improved performance and sustainability.

Overall, this Special Issue not only highlights the new advances, but also anticipates the future trajectory for enhancing power transmission and control within power and vehicle machinery systems from 2017 to 2018. The curated selection of papers significantly enriches the field's knowledge base, focusing on hydraulic and pneumatic system advancements. These contributions offer both practical solutions and theoretical insights that are crucial for refining system integration, thereby fostering more sustainable and efficient technological developments in power transmission and vehicle machinery.

## 3. Emerging Technologies and Future Prospects

Although electric transmission achieved great technological progresses in the past decade [7,8], fluid power transmission and control still has unique advantages in power machines, construction machinery [1], mining machinery, aircrafts [3], hydro-electric power machines [9], agricultural machines [10], etc. Currently, hybrid power transmission, energy management, and new energy sources are being paid great attention in the field of power transmission and control [11,12]. In the meantime, cybernetic physic systems (CPSs), artificial intelligence, advanced modelling [13,14] and simulation, advanced control theory, and digital twin [15] will provide the innovation and development power for fluid power transmission and control technology.

**Acknowledgments:** This Special Issue thanks the contributions of various talented authors, hardworking and professional reviewers, and the dedicated Editorial Team of *Applied Sciences*. We believe that—no matter what the final decisions of the submitted manuscripts were—the feedback, comments, and suggestions from the reviewers and editors are valuable for the authors to improve their papers. We would like to take this opportunity to record our sincere gratefulness to all reviewers. Finally, we place on record our gratitude to the Editorial Team of *Applied Sciences*, and special thanks to the editorial team from the MDPI Branch Office in Beijing.

**Conflicts of Interest:** The authors declare no conflicts of interest.

### List of Contributions

1. Roquet, P.; Gamez-Montero, P.J.; Castilla, R.; Raush, G.; Codina, E. A Simplified Methodology to Evaluate the Design Speci-fications of Hydraulic Components. *Appl. Sci.* **2018**, *8*, 1612. https://doi.org/10.3390/app8091612.
2. Chen, C.-P.; Chiang, M.-H. Development of Proportional Pressure Control Valve for Hydraulic Braking Actuator of Automobile ABS. *Appl. Sci.* **2018**, *8*, 639. https://doi.org/10.3390/app8040639.
3. Ren, H.; Lin, T.; Zhou, S.; Huang, W.; Miao, C. Novel Automatic Idle Speed Control System with Hydraulic Accumulator and Control Strategy for Construction Machinery. *Appl. Sci.* **2018**, *8*, 496. https://doi.org/10.3390/app8040496.
4. Guo, Q.; Liu, Y.; Jiang, D.; Wang, Q.; Xiong, W.; Liu, J.; Li, X. Prescribed Performance Constraint Regulation of Electrohydraulic Control Based on Backstepping with Dynamic Surface. *Appl. Sci.* **2018**, *8*, 76. https://doi.org/10.3390/app8010076.

5.    Nie, S.; Liu, X.; Yin, F.; Ji, H.; Zhang, J. Development of a High-Pressure Pneumatic On/O
      Valve with High Transient Performances Direct-Driven by Voice Coil Motor. *Appl. Sci.* **2018**,
      611. https://doi.org/10.3390/app8040611.
6.    Zhang, Y.; Li, K.; Wei, S.; Wang, G. Pneumatic Rotary Actuator Position Servo System Based o
      ADE-PD Control. *Appl. Sci.* **2018**, *8*, 406. https://doi.org/10.3390/app8030406.
7.    Zhou, H.; Du, R.; Xie, A.; Yang, H. Theoretical Analysis for the Flow Ripple of a Tandem Crescer
      Pump with Index Angles. *Appl. Sci.* **2017**, *7*, 1148. https://doi.org/10.3390/app7111148.
8.    Yang, F.; Li, G.; Hua, J.; Li, X.; Kagawa, T. A New Method for Analysing the Pressure Respons
      Delay in a Pneumatic Brake System Caused by the Influence of Transmission Pipes. *Appl. Sc
      **2017**, *7*, 941. https://doi.org/10.3390/app7090941.
9.    Yang, F.; Tadano, K.; Li, G.; Kagawa, T. Analysis of the Energy Efficiency of a Pneumatic Booste
      Regulator with Energy Recovery. *Appl. Sci.* **2017**, *7*, 816. https://doi.org/10.3390/app70808
10.   Wang, Z.; Zhou, X.; Yang, C.; Chen, Z.; Wu, X. An Experimental Study on Hysteresis Characte
      istics of a Pneumatic Braking System for a Multi-Axle Heavy Vehicle in Emergency Brakin
      Situations. *Appl. Sci.* **2017**, *7*, 799. https://doi.org/10.3390/app7080799.
11.   Cho, I.; Lee, Y.; Lee, J. Investigation on the Effects of Internal EGR by Variable Exhaust Valv
      Actuation with Post Injection on Auto-ignited Combustion and Emission Performance. *App
      Sci.* **2018**, *8*, 597. https://doi.org/10.3390/app8040597.
12.   Jiang, D.; Ren, C.; Zhao, T.; Cao, W. Pressure Transient Model of Water-Hydraulic Pipeline
      with Cavitation. *Appl. Sci.* **2018**, *8*, 388. https://doi.org/10.3390/app8030388.
13.   Shen, D.; Chen, Q.; Wang, Y. Dimensionless Energy Conversion Characteristics of an Ai
      Powered Hydraulic Vehicle. *Appl. Sci.* **2018**, *8*, 347. https://doi.org/10.3390/app8030347.
14.   Tan, P.; Sha, Y.; Bai, X.; Tu, D.; Ma, J.; Huang, W.; Fang, Y. A Performance Test and Internal Flo
      Field Simulation of a Vortex Pump. *Appl. Sci.* **2017**, *7*, 1273. https://doi.org/10.3390/app712127
15.   Xu, X.; Han, X.; Liu, Y.; Liu, Y.; Liu, Y. Modeling and Dynamic Analysis on the Direct Operatin
      Solenoid Valve for Improving the Performance of the Shifting Control System. *Appl. Sci.* **201
      *7*, 1266. https://doi.org/10.3390/app7121266.
16.   Liu, W.; Li, Y.; Li, D. Review on Inlet/Outlet Oil Coordinated Control for Electro-Hydrauli
      Power Mechanism under Sustained Negative Load. *Appl. Sci.* **2018**, *8*, 886. https://doi.org/1
      .3390/app8060886.
17.   Ning, F.; Shi, Y.; Cai, M.; Wang, Y.; Xu, W. Research Progress of Related Technologies of Electri
      Pneumatic Pressure Proportional Valves. *Appl. Sci.* **2017**, *7*, 1074. https://doi.org/10.3390
      app7101074.

## References

1.    Li, Y.H.; Wang, Y.; Geoffrey Chase, J.; Mattila, J.; Myung, H.; Sawodny, O. Survey and Introduction to Focused Section o
      Mechatronics for Sustainable and Resilient Civil Infrastructure. *IEEE-ASME Trans. Mechatron.* **2013**, *18*, 1637–1645. [CrossRef]
2.    Lin, T.; Chen, Q.; Ren, H.; Zhao, Y.; Miao, C.; Fu, S.; Chen, Q. Energy Regeneration Hydraulic System via a Relief Valve wit
      Energy Regeneration Unit. *Appl. Sci.* **2017**, *7*, 613. [CrossRef]
3.    Li, D.; Dong, S.J.; Wang, J.; Li, Y.H. Temperature Dynamic Characteristics Analysis and Thermal Load Dissipation Assessment fo
      Airliner Hydraulic System in a Full Flight Mission Profile. *Machines* **2022**, *10*, 258. [CrossRef]
4.    Yao, J.; Zhang, X.; Sun, Z.X.; Huang, Z.Q. Numerical simulation of the heat extraction in 3D-EGS with thermal hydrauli
      mechanical coupling method based on discrete fractures model. *Geothermics* **2018**, *74*, 19–34. [CrossRef]
5.    Cristello, J.B.; Yang, J.M.; Hugo, R.; Lee, Y.S.; Park, S.S. Feasibility analysis of blending hydrogen into natural gas networks. *Int.
      Hydrogen Energy* **2023**, *48*, 17605–17629. [CrossRef]
6.    Chen, W.T.; Wang, X.Y.; Zhang, F.S.; Liu, H.W.; Lin, Y.G. Review of the application of hydraulic technology in wind turbine. *Win
      Energy* **2020**, *23*, 1495–1522. [CrossRef]
7.    Mazali, I.I.; Abu Husain, N.; Tawi, K.B.; Daud, Z.H.C.; Kob, M.S.C.; Asus, Z. Review of latest technological advancement i
      electro-mechanical continuously variable transmission. *Int. J. Veh. Des.* **2019**, *81*, 166–190. [CrossRef]
8.    Hossain, S.; Rokonuzzaman, M.; Rahman, K.S.; Habib, A.K.M.A.; Tan, W.-S.; Mahmud, M.; Chowdhury, S.; Channumsin, 
      Grid-Vehicle-Grid (G2V2G) Efficient Power Transmission: An Overview of Concept, Operations, Benefits, Concerns, and Futur
      Challenges. *Sustainability* **2023**, *15*, 5782. [CrossRef]
9.    Sharif, M.N.; Haider, H.; Farahat, A.; Hewage, K.; Sadiq, R. Water-energy nexus for water distribution systems: A literatur
      review. *Environ. Rev.* **2019**, *27*, 519–544. [CrossRef]
10.   Mocera, F.; Somà, A.; Martelli, S.; Martini, V. Trends and Future Perspective of Electrification in Agricultural Tractor-Implemen
      Applications. *Energies* **2023**, *16*, 6601. [CrossRef]
11.   Li, T.; Huang, L.; Liu, H. Energy management and economic analysis for a fuel cell supercapacitor excavator. *Energy* **2019**, *17
      840–851. [CrossRef]

2.	Li, Z.; Lin, Y.; Chen, Q.; Wu, K.; Lin, T.; Ren, H.; Gong, W. Control Strategy of Speed Segmented Variable Constant Power Powertrain of Electric Construction Machinery. *Appl. Sci.* **2022**, *12*, 9734. [CrossRef]
3.	Aletras, N.; Doulgeris, S.; Samaras, Z.; Ntziachristos, L. Comparative Assessment of Supervisory Control Algorithms for a Plug-In Hybrid Electric Vehicle. *Energies* **2023**, *16*, 1497. [CrossRef]
4.	Wang, X.; Ji, W.; Gao, Y. Optimization Strategy of the Electric Vehicle Power Battery Based on the Convex Optimization Algorithm. *Processes* **2023**, *11*, 1416. [CrossRef]
5.	Kosova, F.; Unver, H.O. A digital twin framework for aircraft hydraulic systems failure detection using machine learning techniques. *Proc. Inst. Mech. Eng. Part C-J. Eng. Mech. Eng. Sci.* **2023**, *237*, 1563–1580. [CrossRef]

 *applied sciences*

*Article*

# A Simplified Methodology to Evaluate the Design Specifications of Hydraulic Components

**Pedro Roquet [1], Pedro Javier Gamez-Montero [2], Robert Castilla [2], Gustavo Raush [2] and Esteban Codina [2,***

[1]   ROQCAR SL, Antonio Figueras 68, 08551 Tona, Spain; pereroquet@hotmail.com
[2]   Department of Fluid Mechanics, LABSON, Universitat Politecnica de Catalunya, Campus Terrassa, Colom 7, 08222 Terrassa, Spain; pedro.javier.gamez@upc.edu (P.J.G.-M.); robert.castilla@upc.edu (R.C.); gustavo.raush@upc.edu (G.R.)
*   Correspondence: esteban.codina@upc.edu; Tel.: +34-93-739-8664

Received: 29 July 2018; Accepted: 6 September 2018; Published: 11 September 2018

**Abstract:** The fatigue of a hydraulic component inherently varies due to various factors that can be divided into two categories: structural and load spectrum variability. The effects of both variabilities must be considered when determining fatigue life. Compared with the structural variability, determining the variability in the load spectrums is more difficult because the service conditions are complicated and the measurements of the load parameters are slow and expensive. The problem that arises when studying the fatigue behaviour of such components is the transferability of short data samples from real-life load histories, which are application-dependant, to laboratory test methods. Derived from the experimental background and know-how of the authors, this paper proposes a methodology that allows the definition and establishment of the hydraulic cylinder design specificactions, while taking into account the probabilistic characterisation of the load spectrum variability. This methodology could be extrapolated to other hydraulic or mechanical components.

**Keywords:** fatigue; pressure trace; pressure spectrum; damage factor; cumulative damage; off-road mobile machinery; hydraulic components; hydraulic components specifications; laboratory testing

---

## 1. Introduction

In this paper, we consider a hydraulic cylinder as a typical hydraulic component. Hydraulic cylinders obtain their power from pressurised oil. Hydraulic cylinders are frequently found in equipment and machinery, such as construction equipment (excavators, bull-dozers, and road graders) and material handling equipment (fork lift trucks, telescopic handlers, and lift gates). The cylinder is prone to structural problems, such as buckling and fatigue failure. Until recently, engineers had to choose hydraulic cylinders based on the required pressure range without any accurate life cycle data, and previous service experience was often an indirect validation of the design solution. An example of this is the DNVGL-CG-0194 guideline [1]. This class guideline provides the requirements upon which DNV bases the certification of hydraulic cylinders, including requirements for documentation, design, manufacturing, and testing.

However, these guidelines are no longer sufficient. The hydraulic components are subject to complex and random loads, which determine the reliability of the fatigue and the useful life of the machinery. Therefore, the life reliability must be analytically evaluated to full expected laboratory test specifications. This paper demonstrates how to define these particular specifications according to real load histories.

Reliability is a property expressed as the probability that the component will perform its function without breakdown when operated at a duty cycle and exposed to a specific environment for a given period of time. The reliability must be established during the design phase of the system. It is not a new concept but has created new challenge for hydraulic system designers.

The fatigue of a hydraulic component varies inherently due to various factors that can be divided into two categories: load spectrum and structural variability. The variability of the load spectrum of mobile machinery refers to the differences in the pressure history between hydraulic components of the same type and use. Variability can occur due to differences in machine operator experience, work, trajectory, and so on. Structural variability refers to the statistical variability inherent in the fatigue performance of built-up structures, which arises from the variability in manufacturing technologies and material properties, and is usually quantified by the probability distribution of the fatigue life under prescribed specifications for laboratory tests.

Generally, the hydraulic cylinder design includes basic and detailed stages. During the basic design, the principal dimensions of the rod and tube are determined by considering the working force, speed, and range in terms of yield and buckling. During the detailed design, the dimensions of the rod notch, ports, welds, tube end, gland, and cushion ring are determined by considering the fatigue specifications.

Engineering researchers have strived to obtain standard histories that can be applied efficiently to fatigue analysis as representative of the whole loading process of the target system. The automobile and aerospace industries have developed standard load-time histories (SLHs) that may be applied to different structural or mechanical parts of vehicles and airplanes. The acronym SLH is generally used for standardised load-time histories as well as for standardised load sequences, including load spectra. An extensive list of such waveforms used in industrial laboratories and research centers together with their characteristics and the range of application were presented by Heuler and Klätschke [2]. Unfortunately, this type of standardised load-time histories is not available for mobile machinery, both off-road and agricultural.

In order to assess the fatigue damage of hydraulic components, two different approaches can be used. If the load time history can be captured and recorded easily by experimental measurements or numerical simulation, the time domain fatigue damage assessment approach can be applied. In this approach, all input loading and output stress or strain responses are time-based signals [3–6]. In other situations, response stress and input loading are preferably expressed as frequency-based signals, usually in the form of power spectral density (PSD). The frequency domain analysis based on the so-called spectral approach is popular. This approach provides a solution to the random vibration fatigue problem, which, in general, is different from our case [7].

Given the aforementioned issues, this paper proposes a methodology that allows the evaluation and definition of hydraulic cylinder design specificactions, taking into account the probabilistic characterisation of the load spectrum variability. The method also provides a simple procedure for the transferability of data from real-life load histories (application-dependant) to laboratory testing methods. The approach used is based on cycle counting schemes and damage accumulation models, such as the reservoir count method and the Palmgren-Miner linear damage rule [8]. These schemes and models are efficient in terms of testing time, effectiveness, and cost.

## 2. Pressure: Load History and Damage Factor

Our main example for measuring time-load history and then performing statistical analysis is a front loader. In practice, a long-term load spectrum contains complete load information, but this is difficult to directly measure due to the restrictions of testing technology, as well as time and cost. Therefore, obtaining long-term load spectrum information based on short-term data was necessary. Until now, engineers have used traditional methods to obtain long-term load spectrum data, basically consisting of multiplying a short-term load spectrum by a constant proportionality coefficient. The disadvantages of traditional methods are that only the data measured in a finite time are repeated, extreme loads cannot be measured, and their impact on damage is ignored. Load extrapolation methods can overcome the above limitations. With the development of statistics and computer software, although difficult, it is possible to apply load extrapolation methods. Wang et al. [9] provided a selection guidance and several load extrapolation methods that can be applied.

In this paper, an approach inspired by traditional methods is presented to overcome these challenges, consisting of (i) evaluating the damage produced in the hydraulic component for each short-term spectra (damage factor); (ii) considering each short-term spectra and its damage index as statistical samples of the total damage produced by complete load information; and (iii) the laboratory specifications are estimated from the damage factors of different machine tasks, which produce the same damage the component suffers during its lifespan. For illustrative purposes, two front loader tasks were considered: loading and transport. Pressure histories (traces) for all hydraulic cylinders can be obtained by pressure transducers installed in cylinder's ports.

Current design methods use cycle counting in order to interpret the varying pressure range of the varying loads. The following two methods are the most commonly applied methods in the field of fatigue studies: reservoir and rain-flow. In our case, the obtained data (pressure vs. time) from every pressure trace were treated with the reservoir cycle counting method recommended in EN 13445-3 standard—Unfired pressure vessels Part 3: Design [10]. Also, this algorithm was standardised according to ASTM E 1049-85, Cycle Counting in Fatigue Analysis [8].

Once the cycles were identified, the damage for all the cycles in the loading history were combined to obtain the damage for the entire loading history. A number of deterministic damage accumulation models have been developed since the late 1990, that can be mainly classified into two categories: linear damage cumulative theories [11] and nonlinear damage cumulative theories [12]. The first models have some shortcomings: (i) they fail to consider load history; (ii) cumulative damage has no relationship with load sequence effects; and (iii) effects of load interaction are not taken into account. Therefore, to address the above-mentioned disadvantages, nonlinear cumulative damage theories were suggested, which were classified into six groups by Zhu [13]. Although we are aware that the linear method of mining is non-conservative, it was used because Miner's rule [14] is probably the simplest and conceptual cumulative damage model that can be used to didactically explain our approach. Miner's rule states that if there are $q$ different pressure levels (with linear damage hypothesis), the fraction of life consumed (damage) $D$ by exposure to the cycles at the different pressure levels is

$$\sum_{i=1}^{q} \frac{n_i}{N_i} = D \tag{1}$$

where $n_i$ is the number of cycles accumulated at pressure $P_i$ and $N_i$ is the number of cycles to failure at pressure $P_i$.

Fatigue damage for an individual cycle is the reciprocal of the fatigue life, $N_i$ Fatigue lives for a cycle are computed using constant amplitude methods with the appropriate pressure or stress ranges, mean stresses, and material properties. Damage is then summed for all cycles in the loading history.

What should the damage be at failure? If damage were truly a linear process, the damage at failure would be equal to 1. In simple two step block loading, a sequence of high amplitude cycles followed by low amplitude cycles has a damage sum $D < 1.0$. Similarly, a sequence of low amplitude cycles followed by high amplitude cycles ha a damage sum $D > 1.0$. In our case, as the succession of peaks of low and high amplitude were randomly distributed, we considered that $D \sim 1.0$.

From the pressure spectrum vs. percent cycles (Figure 1), the step damage for every pressure step of the spectrum$(\varepsilon_i)$ can be defined as

$$\varepsilon_i = v_i P_i^m \tag{2}$$

where $\varepsilon_i$ is the spectrum relative damage for the $i$th pressure step of a task, $v_i$ is the portion of the task in unified percentage during the $P_i$ pressure level, $m$ is the material coeficient from Basquin equation ($m = 3$ for steel for hydraulic cylinders [10]), $P_i$ is the pressure value assigned to the $i$th pressure step of a task, and $i$ is the index that identifies the pressure step number.

The spectrum relative damage ($\delta_k$) is defined as the sum of all step damages ($\varepsilon_i$) for this spectrum

$$\delta_k = \sum_1^n \varepsilon_i; \quad \phi_k = \delta_k \cdot \chi_k \tag{3}$$

where $k$ is the index that identifies the task number.

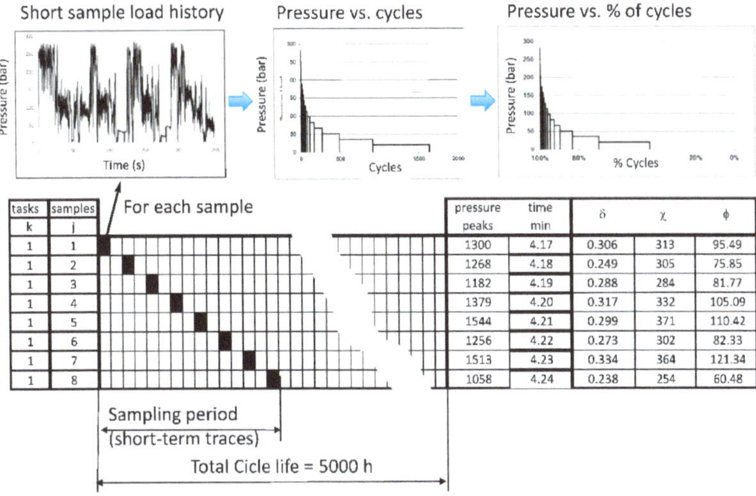

**Figure 1.** Scheme of load histories sampling and treatment.

Although the spectrum relative damage ($\delta_k$) provides a quantification of the severity of the $k$th task, it does not take into account the loading frequency. The damage factor ($\phi_k$) is defined as the multiplication of the relative damage factor ($\delta_k$) by the frequency of the application pressure cycles, ($\chi_k$). This frequency is defined as the number of pressure peaks in a time unit.

At this point, only the damage of one portion of the work performed by the machine was calculated (task $k$). Following the steps described above, it was possible to calculate the damage generated $\phi_k$ to the hydraulic components by the other tasks.

To illustrate the method, Figure 1 shows eight recorded trace samples ($j = 8$) that represent the load spectrum variability. From each trace, the pressure spectrum (200 steps are recommended) and the associated damage factor were calculated [15].

Table 1 and Figure 2 show that the statistical distribution (variability) of the damage factors for the task $k = 1$ (loading task) follows a log-normal distribution where $F = \frac{r-0.3}{n+0.4}$ is the accumulated frequency.

**Table 1.** Damage factors for the task $k = 1$ (loading task).

| $k$ | $j$ | $\phi_{kj}$ (bar$^3\cdot$min$^{-1}\cdot10^6$) | $F$ (%) | ln ($\phi_{kj}$) | $z$ |
|---|---|---|---|---|---|
| 1 | 8 | 60.48 | 8.3 | 4.10 | $-1.38$ |
| 1 | 5 | 75.85 | 20.2 | 4.33 | $-0.83$ |
| 1 | 3 | 81.77 | 32.1 | 4.40 | $-0.46$ |
| 1 | 6 | 82.33 | 44.0 | 4.41 | $-0.15$ |
| 1 | 1 | 95.49 | 56.0 | 4.56 | 0.15 |
| 1 | 4 | 105.09 | 67.9 | 4.65 | 0.46 |
| 1 | 5 | 110.42 | 79.8 | 4.70 | 0.83 |
| 1 | 7 | 121.34 | 91.7 | 4.80 | 1.38 |

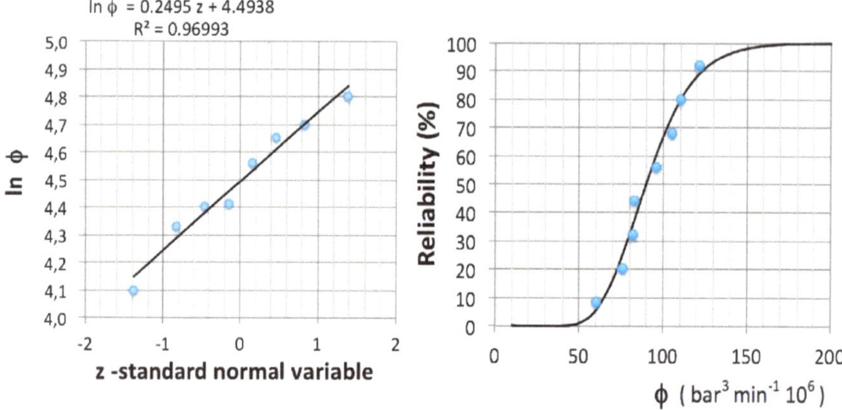



**Figure 2.** Statistical distribution of damage factor for $k = 1$ (loading task).

The damage factor statistical distribution for $k = 1$ (loading task) is

$$D(\log \phi_1) = N[4.49,\ 0.25] \tag{4}$$

Following the same process, the statistical distributions of the other tasks were calculated. For the task $k = 2$ (transport task), the following statistical distribution was considered:

$$D(\log \phi_2) = N[3.0,\ 0.31] \tag{5}$$

To estimate the equivalent damage produced in the cylinder during the full life, we worked with the following data supplied by the machine manufacturer (Table 2).

**Table 2.** Cylinder data from the machine manufacturer.

| $k$ | Task | $T$ **(Hour)** | $\pi_k$ |
|---|---|---|---|
| 1 | Loading | 3500 | 0.70 |
| 2 | Transport | 1500 | 0.30 |

The equivalent damage factor $\phi_{eq}$ can be obtained as the sum of the damage factors for all the tasks weighted by a time factor $\pi_k$ as follows:

$$\phi_{eq} = \sum_1^n \phi_k \pi_k \tag{6}$$

Figure 3 shows the statistical distributions of the damage factors of the two considered tasks, as well as the statistical distribution of the equivalent damage. This statistical distributio follows the log-normal type, given that the sum of two (or more) log-normal random variables approaches a log-normal distribution [16].

$$D(\log \phi_{eq}) = N[m_{\phi eq},\ s_{\phi eq}] = N\left[\sum_k \pi_k m_{\phi k},\ \sum_k \pi_k s_{\phi k}\right] \tag{7}$$

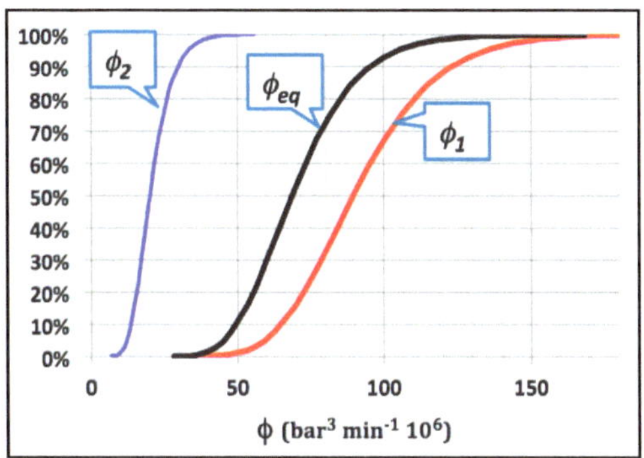


**Figure 3.** Statistical distribution of damage factors.

## 3. Pressure Specifications

Commonly, the market defines machine life expectancy using different modes, but always specifying working hours, such as 5000 h without failures. More often, statistical concepts are considered, so the life expectancy is expressed as "5000 h $\beta_{10}$" [17]. The latter means that when a sampling of 100 machines are tested over 5000 h, only 10 of them (10%) are expected to fail before reaching those working hours. Therefore, the damages generated during the useful life of the machine, when the identified tasks are performed by the machine, cannot generate a failure before the life of the objective $T$. This can be stated as

$$\phi_{eq}T = \sum_{1}^{n} \phi_k \pi_k \cdot T = \alpha \text{ and } N = \alpha P^{-m} \tag{8}$$

which is analogous to the Basquin equation. Then,

$$T\sum_{1}^{n} \phi_k \pi_k = \alpha = N_{eq} P_{eq}^{m} \tag{9}$$

where $\alpha$ is the Basquin coefficient and $N_{eq}$ is the equivalent number of cycles at a pressure level $P_{eq}$ under constant amplitude conditions.

The pair of values ($N_{eq}$, $P_{eq}$) defines a line in logarithmic coordinates with slope $m$ (for steel $m = 3$). From this curve and analogous to what is proposed in standard ISO 13445-3 [9], the curve ($N_{eq}$, $P_{eq}$) is identified by equivalent pressure value corresponding to $2 \times 10^6$ cycles, which constitutes the class curve $P_{\varnothing eq}$:

$$P_{\varnothing eq} = \left[\frac{1}{2 \cdot 10^6} T \phi_{eq}\right]^{\frac{1}{3}} \tag{10}$$

$P_{\varnothing eq}$ values also follow a log-normal distribution,

$$\begin{aligned} D\left(\log P_{\phi_{eq}}\right) &= D\left[p_\phi\right] = N\left[m_{\phi eq}, s_{\phi eq}\right] \\ D\left(\log P_{\phi_{eq}}\right) &= N[5.38, 0.085] \end{aligned} \tag{11}$$

At this step, once the statistical distribution of load is known, according to the desired or accepted reliability value, the machine manufacturer may define hydraulic cylinder pressure life specification (laboratory). As an example, if we considered a load reliability of 90%, then (see Figure 4),

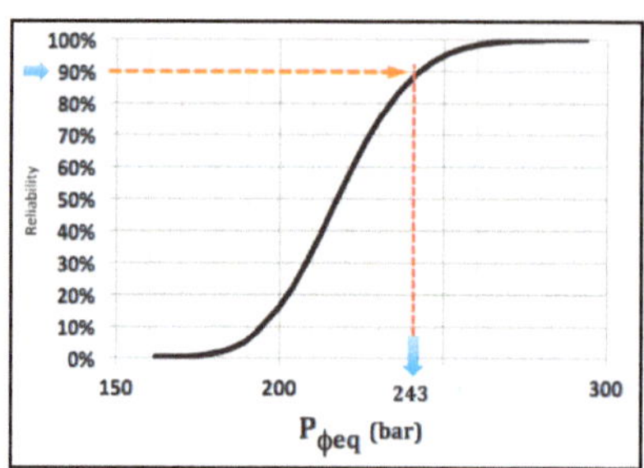

**Figure 4.** Statistical distribution of $P_{\varnothing eq}$.

$$P_S = P_{\phi_{eq(90\%)}} = 243 \text{ bar} \tag{12}$$

The specified pressure $P_S$ defines the equivalent damage produced in laboratory test. So, laboratory test parameters $(P_{eq}, N_{eq})$ that produce an equivalent damage in the cylinder are any values according to Equation (13):

$$N_{eq}P_{eq}^3 = 2{\cdot}10^6 \, P_s^3 \tag{13}$$

Figure 5 shows that any values of $(P_{eq}, N_{eq})$ fulfills the specification conditions and produces equal damage to the cylinder (i.e., 386 bar over 500,000 cycles).

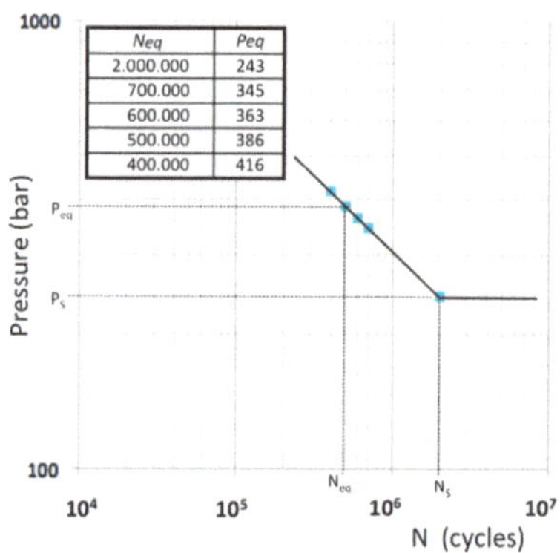



**Figure 5.** Laboratory test parameters according pressure specifications.

## 4. Reliability of Hydraulic Component

Once the load variability is defined, the structural variability of the hydraulic component is considered. Assuming that a hydraulic component manufacturer knows the S–N curves corresponding to any main possible failure modes and according to its know-how (design and manufacturing technology), determining the fatigue resistance reliability of a hydraulic component is possible. Figure 6 shows the typical S–N curves and corresponding P–N curves for a specific cylinder geometry for the same failure mode (F-01) defined by the value of the pressure $P_D$ (or stress $S_D$), which correspond to $N = 2 \times 10^6$ cycles and *nn* % of reliability (known as the pressure class *nn*, or stress class *nn* of the curve) [10].

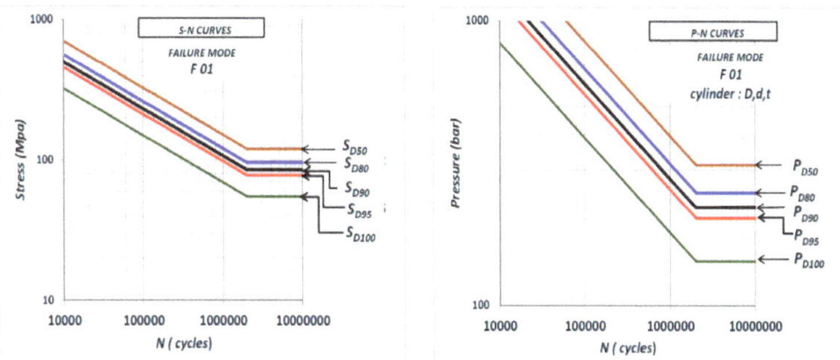

**Figure 6.** Typical (**left**) S–N curves and (**right**) P–N curves for specific hydrauklic cylinder geometry for this same failure mode (F-01).

Recalling that fatigue damage due to load conditions is represented by $P_{\phi eq}$, and damage resistance of the hydraulic cylinder is represented by $P_D$:

$$D[P_\phi] = N[m_{p\phi},\, s_{p\phi}] \tag{14}$$

$$D[P_D] = N[m_{pD},\, s_{pD}] \tag{15}$$

Both distributions have $P_S$ as a common linkage, which meets the conditions

$$p[P_{p\phi} \le P_S] = R_\phi \text{ and } p[P_D \le P_S] = R_D \tag{16}$$

Any cylinder with a resistance $P_D$ may be mounted in a machine with any of the loads $P_\phi$. The condition for the cylinder to resist is set by the resistance that is superior to the load (see Figure 7).

$$y = p_D - p_\phi > 0 \tag{17}$$

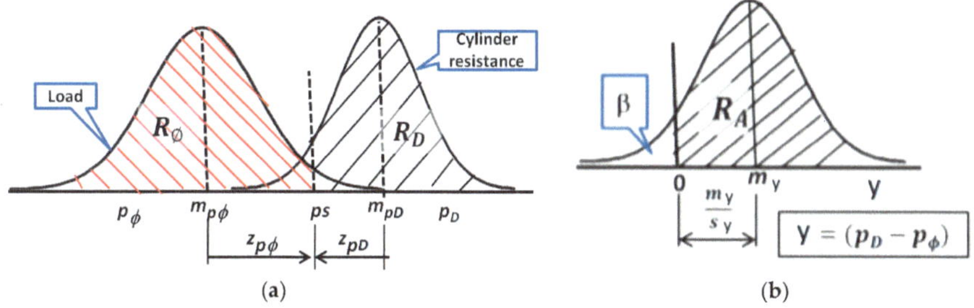


**Figure 7.** Cylinder mounted in a machine with any of the loads: (**a**) Load variability and structural variability functions; (**b**) At the end of the life will not fail more than $\beta$ % of cylinders.

The variable $y$ follows a normal distribution:

$$D[y] = N[m_y, s_y] \tag{18}$$

where

$$\begin{aligned} m_y &= m_{pD} - m_{p\phi} \\ s_y &= \sqrt{s_{pD}^2 + s_{p\phi}^2} \end{aligned} \tag{19}$$

fulfilling

$$R_A = p\left[z \geq \frac{m_y}{s_y}\right] \tag{20}$$

The design data is $\beta = (1 - R_A)$, which means "at the end of the life (desired) will not fail more than $\beta\%$ of cylinders".

If we consider

$$k_s = \frac{s_{p\varnothing}}{s_{pD}} \tag{21}$$

then,

$$\frac{m_y}{s_y} = \frac{-z_{pD} + z_{p\phi}s_{p\phi}}{\sqrt{s_{pD}^2 + s_{p\phi}^2}} = \frac{-z_{pD} + k_s z_{p\phi}}{\sqrt{1 + k_s^2}} \tag{22}$$

Equation (22) allows us to estimate the life expectancy, defined by reliability $R_A$, (or the design data $\beta$) as a function of load spectrum variability (defined by the reliability, $R_\phi$), resistance variability of the hydraulic component (defined by the reliability, $R_D$) and standard deviation ratio, $k_s$. In practice, the load standard deviation ($s_{p\phi}$) is known according the proposed approach, but the resistance standard deviation of the hydraulic cylinder ($s_{pD}$) and consequently, the value of the ratio $k_s$ cannot be known at the design stage because, sometimes, the selected cylinder manufacturer is still not known. In Table 3, $R_D$ values have been tabulated applying Equation (22), see diagram Figure 8, for values within the following ranges: $50\% \leq R_\phi \leq 95\%$, $3\% \leq \beta \leq 10\%$, and $0.25 \leq k_s \leq 2$.

**Table 3.** Values of hydraulic cylinder resistance reliability $R_D$.

| $R_\phi$ (%) | B (%) | $k_s$ 0.25 | $k_s$ 0.65 | $k_s$ 0.85 | $k_s$ 1 | $k_s$ 1.5 | $k_s$ 2 |
|---|---|---|---|---|---|---|---|
| 50 | 3 | 98 | 99 | 99 | 100 | | |
| 50 | 5 | 96 | 98 | 98 | 99 | 100 | |
| 50 | 10 | 91 | 94 | 95 | 97 | 99 | 100 |
| 80 | 3 | 96 | 96 | 97 | 97 | 99 | 100 |
| 80 | 5 | 93 | 92 | 93 | 93 | 96 | 98 |
| 80 | 10 | 87 | 84 | 83 | 83 | 85 | 88 |
| 85 | 3 | 96 | 95 | 95 | 96 | 98 | 99 |
| 85 | 5 | 92 | 90 | 90 | 90 | 92 | 95 |
| 85 | 10 | 86 | 80 | 79 | 78 | 78 | 79 |
| 90 | 3 | 96 | 93 | 93 | 93 | 95 | 97 |
| 90 | 5 | 92 | 87 | 86 | 85 | 85 | 87 |
| 90 | 10 | 84 | 76 | 72 | 70 | 65 | 62 |
| 95 | 3 | 95 | 90 | 88 | 87 | 86 | 86 |
| 95 | 5 | 90 | 81 | 78 | 75 | 69 | 65 |
| 95 | 10 | 82 | 68 | 61 | 57 | 44 | 34 |

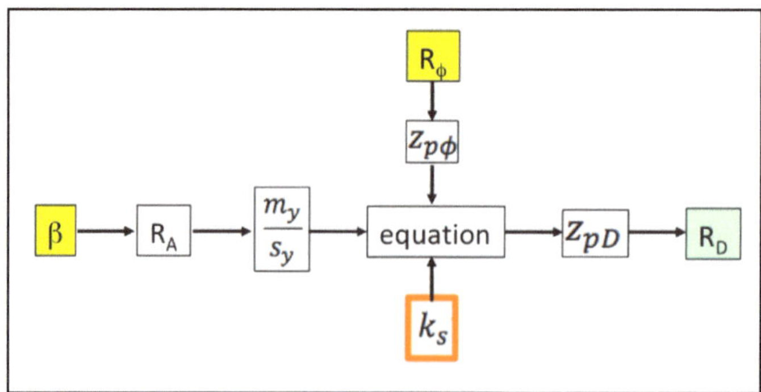


**Figure 8.** Scheme to calculate the required resistance reliability ($R_D$).

Let's see an example, to satisfy the specifications indicated in Section 3. If $\beta = 10\%$, (equivalent $R_A = 90\%$) and admitting that the load reliability is $R_\phi = 90\%$, Table 3 allows infering that it must demand a hydraulic cylinder resistance variability larger than 84% (for the assumption that $k_s$ = 0.25) or larger than 62% (for the assumption that $k_s = 2$). Adopting a conservative position, it leads to demand a minimum resistance reliability of the hydraulic cylinder $R_D = 85\%$.

## 5. Conclusions

Reliability assessment in fatigue failures is a difficult problem: the structural area due to the large scatter in fatigue life and the load history due to highly varying machine user profiles. On one hand, an assessment procedure must therefore be simple enough to be able to quantify vague input information; on the other hand, it must be sophisticated enough to be a useful engineering tool for improvements.

Earlier models of fatigue damage accumulation reported in literature focus on deterministic nature of the process whereas in practice, damage accumulation is of stochastic nature. The main differences in the present method compared to the reported ones are [3,12] that (i) it uses random variables to include the stochasticity in both external loadings and material properties and (ii) the quality of the hydraulic component is represented by the reliability values.

In literature, some similar approaches to the probabilistic damage accumulation paradigm can be found. Shen et al. [18] developed a probabilistic distribution model of stochastic fatigue damage, wherein they have considered the randomness of loading process as well as the randomness of fatigue resistance of material by introducing a random variable of single cycle fatigue damage. Liu and Mahadevan [19] proposed a general methodology for stochastic fatigue life under variable amplitude loading by combining a nonlinear fatigue damage accumulation rule and stochastic S–N curve representation technique. These models are conceptually in the vein of the approach presented in this work. Nevertheless they are more complex and none of them explained how to define the specifications of a component based on the required reliability or a specific algorithm for component reliability prediction.

The major limitation of our approach is that it treats damage accumulation as linear phenomenon (both laboratory and fields tests). It is known that the application of Miner's rule for variable amplitude life calculation is erroneous. Its effect is weak, specially for typical random loading [20]. However, as cumulative calculations with such values result only in an estimated fatigue life, they still contain a certain risk due to possible lower real damage sums. Therefore, in the case of safety-critical components, which must never fail, an experimental verification is recommended [21,22].

The methodology presented in this work allows, not only the determination of the hydraulic cylinder pressure specifications for laboratory tests, but also provides a simple and quick method to select a hydraulic cylinder since there is a continuous need to improve the durability requirements of the hydraulics components, making them more correlated to the actual customer needs. The potential here is to differentiate the requirements, for example, allowing the offer of a light-weight hydraulic component for specific demanding applications.

**Author Contributions:** Conceptualization, P.R. and E.C.; Methodology, P.R. and E.C.; Formal Analysis, P.R., E.C., P.J.G.-M., R.C. and G.R.; Investigation, P.R., E.C., P.J.G.-M., R.C. and G.R.; Writing—Original Draft Preparation, P.R. and E.C.; Writing—Review & Editing, P.J.G.-M.

**Funding:** The authors disclosed receipt of the following financial support for the research, authorship, and/or publication of this article: This research program has received funding from the European Union Seventh Framework Programme FP6 NMP-CT-2004-505466 'New design and manufacturing processes for high pressure fluid power products' (Acronym: PROHIPP).

**Conflicts of Interest:** The authors declare no conflict of interest.

## Nomenclature

| | |
|---|---|
| $D$ | fraction of life consumed (damage) |
| $F$ | accumulate frequency |
| $j$ | trace samples |
| $m$ | material coefficient from Basquin equation |
| $n_i$ | number of cycles accumulated at pressure $P_i$ |
| $N_{eq}$ | equivalent number of cycles to failure at pressure $P_{eq}$ |
| $N_i$ | number of cycles to failure at pressure $P_i$ |
| $P_D$ | damage resistance of the hydraulic cylinder |
| $P_i$ | pressure level assigned to the $i$th pressure step of a task |
| $P_S$ | equivalent damage produced in laboratory test |
| $P_\phi$ | loads |
| $P_{\phi eq}$ | fatigue damage due to load conditions |
| $R_A$ | life expectancy reliability |
| $R_D$ | resistance reliability |
| $R_\phi$ | load reliability |
| $s_{p\phi}$ | load standard deviation |
| $s_{pD}$ | resistance standard deviation of the hydraulic cylinder |
| $S_D$ | stress |
| $T$ | life objective |
| $z$ | standard normal variable |

### Greek Symbols

| | |
|---|---|
| $\alpha$ | Basquin coefficient |
| $\beta$ | live expectancy |
| $\delta_i$ | spectrum relative damage |
| $\delta_k$ | relative damage factor |
| $\varepsilon_i$ | step damage for every pressure step of the spectrum |
| $\pi_k$ | time factor |
| $\chi_k$ | frequency of the application pressure cycles |
| $f_{eq}$ | equivalent damage factor |
| $f_k$ | damage factor |

## References


1. Hydraulic Cylinders. Class Guideline DNVGL-CG-0194. Edition July 2017. Available online: https://www.dnvgl.com (accessed on 27 July 2018).
2. Heuler, P.; Klätschke, H. Generation and use of standardised load spectra and load-time histories. *Int. J. Fatigue* **2005**, *27*, 974–990. [CrossRef]
3. Johannesson, P.; Speckert, M. *Guide to Load Analysis for Durability in Vehicle Engineering*; John Wiley & Sons Ltd.: Chichester, UK, 2013; pp. 1–456, ISBN 978-1-118-64831-5.
4. Johannesson, P. Extrapolation of Load Histories and Spectra. *Fatigue Fract. Eng. Mater. Struct.* **2006**, *29*, 209–217. [CrossRef]
5. Shen, R.; Zhang, X.; Zhou, C. Dynamic Simulation of the Harvester Boom Cylinder. *Machines* **2017**, *5*, 13. [CrossRef]
6. Jerabek, I.; Horky, L.; Weigel, K. Long-term test- analysis and monitoring. In Proceedings of the EAN-55th Conference on Experimental Stress Analysis, Novy Smokovec, Slovakia, 30 May–1 June 2017. Available online: http://experimentalni-mechanika.cz/konference/2017.html?download=1339:long-term-cyclic-test-analysis-and-monitoring (accessed on 15 May 2018).
7. Qiang, R.; Hongyan, W. Frequency Domain Fatigue Assessment of Vehicle Component under Random Load Spectrum. *J. Phys. Conf. Ser.* **2011**, *305*, 012060. [CrossRef]
8. ASTM. *Standard Practices for Cycle Counting in Fatigue Analysis, Annual Book of ASTM Standards*; ASTM E1049-85; ASTM International: West Conshohocken, PA, USA, 2017; Volume 03.01, pp. 710–718.

9. Wang, J.; Chen, H.; Li, Y.; Wu, Y.; Zhang, Y. A Review of the Extrapolation Method in Load Spectrum Compiling. *J. Mech. Eng.* **2016**, *62*, 60–75. [CrossRef]

10. EN 13445. *Unfired Pressure Vessels Is a Standard that Provides Rules for the Design, Fabrication, and Inspection of Pressure Vessels*; (EN 13445 Consists of 10 Parts). Part 3: Dedicated to Design Rules; 2004. Available online: http://www.unm.fr/main/download.php?file=107_FICHIER_0.pdf (accessed on 27 July 2018).

11. Yang, X.H.; Yao, W.X.; Duan, C.M. The development of deterministic fatigue cumulative damage theory. *Eng. Sci.* **2003**, *15*, 81–87.

12. Fatemi, A.; Yang, L. Cumulative fatigue damage and life prediction theories: A survey of the state of the art for homogeneous materials. *Int. J. Fatigue* **1998**, *20*, 9–34. [CrossRef]

13. Zhu, S.P.; Huang, H.Z.; Liu, Y. A practical method for determining the Corten-Dolan exponent and its application to fatigue life prediction. *Int. J. Turbo Jet-Engines* **2012**, *29*, 79–87. [CrossRef]

14. Miner, M.A. Cumulative damage in fatigue. *J. Appl. Mech.* **1945**, *12*, 159–164.

15. Roquet, P.; Codina, E.; Perez, J. *Hydraulic Components Fatigue Assessment Based on Real Life Load Histories*; Conference TEH2017; INMA Bucharest: Bucarest, Romania, 2017.

16. Lo, C.F. The Sum and Difference of Two Log-normal Random Variables. *J. Appl. Math.* **2012**, *2012*, 1–13. [CrossRef]

17. Stimson, W.A. Appendix B: Introduction to Product Reliability. In *Forensic Systems Engineering: Evaluating Operations by Discovery*; John Wiley & Sons Ltd.: Hoboken, NJ, USA, 2018; pp. 1–368, ISBN 978-1-119-42275-4.

18. Shen, H.; Lin, J.; Mu, E. Probabilistic model on stochastic fatigue damage. *Int. J. Fatigue* **2000**, *22*, 569–572. [CrossRef]

19. Liu, Y.; Mahadevan, S. Stochastic fatigue damage model-ing under variable amplitude loading. *Int. J. Fatigue* **2007**, *29*, 1149–1161. [CrossRef]

20. Vormwalda, M. Classification of load sequence effects in metallic structures. *Procedia Eng.* **2015**, *101*, 534–542. [CrossRef]

21. Gamez-Montero, P.J.; Salazar, E.; Castilla, R.; Freire, J.; Khamashta, M.; Codina, E. Misalignment effects on the load capacity of a hydraulic cylinder. *Int. J. Mech. Sci.* **2009**, *51*, 105–113. [CrossRef]

22. Gamez-Montero, P.J.; Salazar, E.; Castilla, R.; Freire, J.; Khamashta, M.; Codina, E. Friction effects on the load capacity of a column and a hydraulic cylinder. *Int. J. Mech. Sci.* **2009**, *51*, 141–151. [CrossRef]

Article

# Development of Proportional Pressure Control Valve for Hydraulic Braking Actuator of Automobile ABS

Che-Pin Chen and Mao-Hsiung Chiang *

Department of Engineering Science and Ocean Engineering, National Taiwan University, No. 1, Sec. 4, Roosevelt Rd., Taipei City 106, Taiwan; d96525005@ntu.edu.tw
* Correspondence: mhchiang@ntu.edu.tw; Tel.: +886-2-3366-3730

Received: 25 February 2018; Accepted: 17 April 2018; Published: 20 April 2018

**Abstract:** This research developed a novel proportional pressure control valve for an automobile hydraulic braking actuator. It also analyzed and simulated solenoid force of the control valves, and the pressure relief capability test of electromagnetic thrust with the proportional valve body. Considering the high controllability and ease of production, the driver of this proportional valve was designed with a small volume and powerful solenoid force to control braking pressure and flow. Since the proportional valve can have closed-loop control, the proportional valve can replace a conventional solenoid valve in current brake actuators. With the proportional valve controlling braking and pressure relief mode, it can narrow the space of hydraulic braking actuator, and precisely control braking force to achieve safety objectives. Finally, the proposed novel proportional pressure control valve of an automobile hydraulic braking actuator was implemented and verified experimentally.

**Keywords:** hydraulic braking actuator; proportional pressure control valve; anti-lock braking system; proportional electro-hydraulic brake

---

## 1. Introduction

To provide safer and more comfortable driving conditions for motorists, the anti-lock braking system (ABS) is currently basic equipment for automobiles. This system achieves improved control capability by connecting the electro-hydraulic brake (EHB) between master cylinder and wheel cylinder. As ABS starts to work, the hydraulic pressure (braking force) within the tire calipers is controlled by four groups of solenoid valves (inlet/outlet valve). Furthermore, there are three patterns of pressure variation, namely pressure increase, pressure holding, and pressure decrease [1–4]. The hydraulic circuit of an EHB actuator is shown in Figure 1 [5].

The braking capability of automobiles relies on characteristics of the tires, which are the only contact with the ground. The main factor determining this capability is the slip ratio of the tires and ground, defined as slip.

$$S = \frac{V_v - V_w}{V_v} \tag{1}$$

where $S$ is the slip, $V_v$ is the vehicle speed, and $V_w$ is the wheel speed.

When the slip is between 0.1 and 0.3, the braking action is most effective, and, if the slip is zero, the wheels are operating with no resistance. However, when slip is more than 0.3, the braking force may be reduced. The tires are completely locked when slip is 1, and the wheels would slide along the pavement [6–11]. The results are shown in Figure 2 [12].

An ABS braking system can prevent the improper sliding of tires and it has three functions:

(1) When encountering an obstacle or any road conditions, it can be actuated by the steering wheel to obtain suitable tracking performance.

(2) It can avoid excessive braking effects that can cause uncontrolled steering (if the front wheels are locked) or drifting (if the rear wheels are locked)

(3) The electronic control braking effects can shorten the stopping distance.

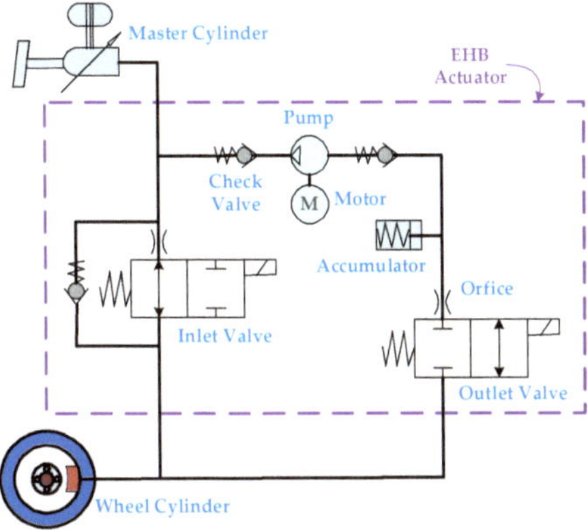


**Figure 1.** Hydraulic circuit of EHB actuator between master cylinder and wheel cylinder.

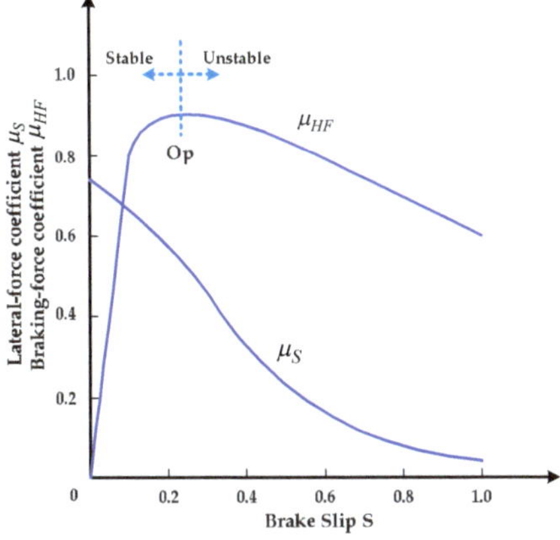


**Figure 2.** Relationship between wheels and pavement.

The general ABS drives the solenoid valve when it detects that the wheels are locked, and it operates the slip control in an on–off mode to release the calipers' pressure. Traditional solenoid valve patterns can only adjust the calipers' pressure through a fully open or closed valve port and have no accurate control for slip, so they cannot provided highly effective braking. However, an accident can occur when the ABS is actuating and high frequency vibration from the pedal make the driver panic and then release the pedal. To develop a brake actuator that uses a new pattern, and also considering the market differentiation and product performance, for the first time, a proportional pressure control valve was used for ABS to replace two solenoid valves. To improve the disadvantages of solenoid valve and achieve better slip control, the pressure to be released as determined by the corresponding input current and electromagnetic force. To improve braking quality, it can effectively control the calipers' pressure and then decide the appropriate braking force due to the stability of pressure release control, and no pressure oscillations are generated, as with a solenoid valve. A single proportional valve for each of the four wheels was substituted for the original two solenoid valves for each wheel.

This reduces the cost and space requirements compared to the traditional EHB configuration. Figure 3 shows the novel hydraulic circuit for a proportional electro-hydraulic brake (PEHB) actuator.

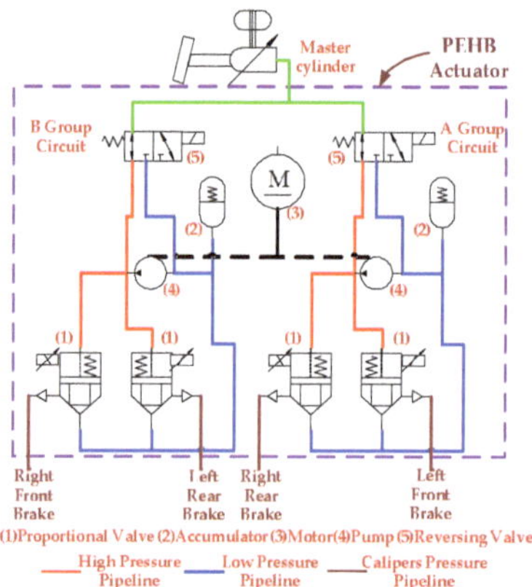


**Figure 3.** Proposed hydraulic circuit with a PEHB actuator.

This research focused on the configuration design of the proportional electromagnet within the proportional valve, and its main function is to drive the armature. The development process was divided into three stages:

(1) Firstly, using electromagnetic theory and mathematical calculations, parameters of the new cone proportional electromagnet were analyzed, and then the electromagnet force using a simulation program was estimated.
(2) Secondly, according to the configuration design (post-analysis), the production and electromagnet force test were completed.
(3) Thirdly, the proportional valve was tested, and its relief pressure control capability for ABS control applications of the PEHB actuator was confirmed.

## 2. Analysis of Proportional Solenoid Force

The principles of proportional valve operation are as follows. The proportional solenoid configuration (increase brass ring) can produce a horizontal solenoid force working zone which is not related to the stroke. When different currents are input, the solenoid force balances the oil pressure and the spring force. Base and armature attract each other by coil excitation, and the required stroke can be generated to drive the shuttle shaft to open the valve. Configuration of the proportional pressure valve is shown in Figure 4 [13]. This gives it two advantages: a closed-loop control (compared to the solenoid valve) and lower cost (compared to the servo valve). Figure 5 shows the axial force (Fa) of the base (flux path 1) and the radial force (Fr) of the flange (flux path 2) work with each other in specific section so that the solenoid force is equal to their sum. Therefore, the solenoid force is unrelated to the working stroke, which is why the linear proportional valve zone is formed. Therefore, the proportional electromagnet can be directly controlled, and the solenoid force depends on the operating current in the linear zone.

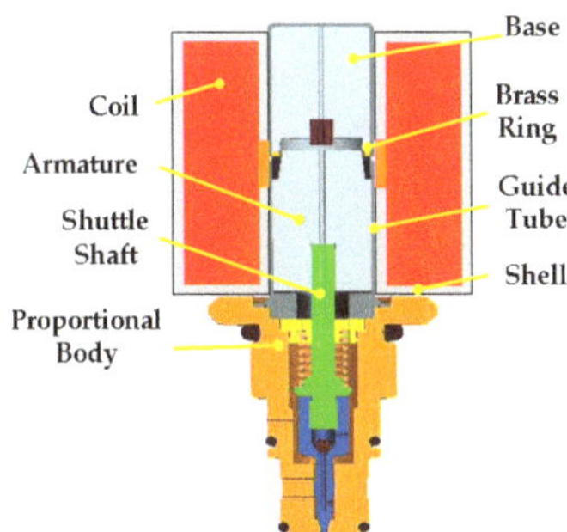

**Figure 4.** Configuration of proportional pressure control valve for PEHB actuator.

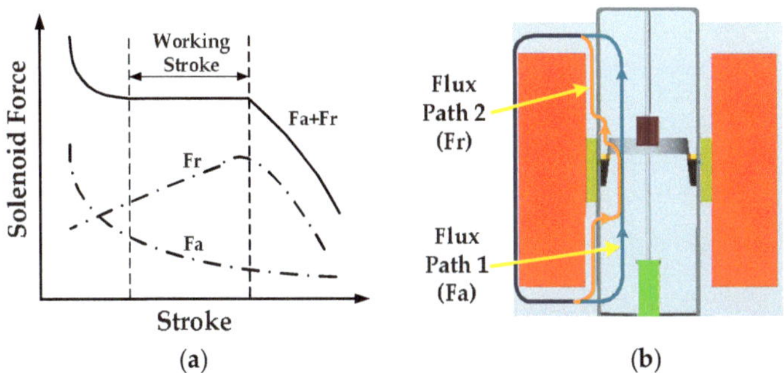


**Figure 5.** Relation between solenoid force and stroke: (**a**) form of force; and (**b**) flux path mode.

## 2.1. Permeance of Air Gaps

When the electromagnetic coil is energized, the energy can generate a magnetic field. Thus, there will be a group of closed magnetic circuits around the coils within the proportional solenoid, and the magnetic field intensity changes with the armature position. Lastly, it drives the armature to move according to the high magnetic energy. The basic proportional electromagnet is shown in Figure 6. Simplifying the six types of magnetic path can be the basis for subsequent calculations [14,15], as shown in Table 1. Here, $\mu_0$ represents the permeability of air, $g$ represents the air gap, $r$ represents the armature radius, $h$ represents the right side distance between armature and brass ring, $t$ represents the thickness of brass ring, and $w$ represents the distance from base to brass ring.

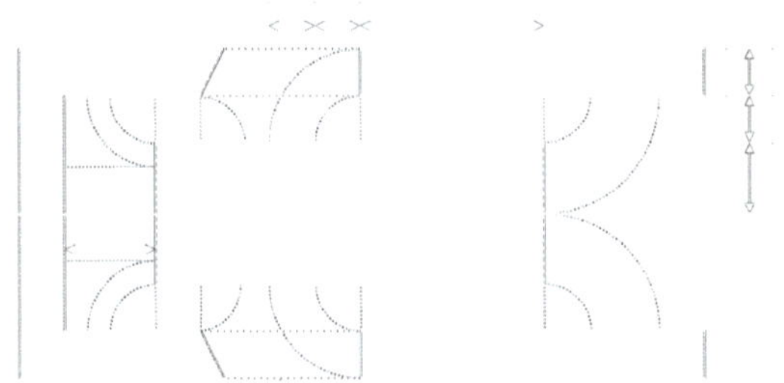

**Figure 6.** Simplification of the six types of magnetic path.

Table 1. The air gaps flux path model calculations.

| Flux Path Model | Mean Path Length | Permeance ($P$) |
| --- | --- | --- |
| I | $g$ | $\frac{2\pi\mu_0 h}{g}(r + \frac{g}{2})$ |
| II | $1.22g$ | $3.315\mu_0(r + \frac{g}{2})$ |
| III | $1.22g$ | $3.315\mu_0(r + \frac{g}{2})$ |
| IV | $\sqrt{g(g+t)}$ | $4\mu_0(r + \sqrt{g(g+t)})\ln(\frac{g+t}{g})$ |
| V | $\sqrt{g(g+r)}$ | $4\mu_0(r + g - \sqrt{g(g+r)})\ln(\frac{g+r}{g})$ |
| VI | $w$ | $\frac{\mu_0\pi}{w}(r + g - \frac{2w}{\pi})^2$ |

## 2.2. Permeability of Air of the Proportional Electromagnet

For ease of construction, the metal cone replaces the traditional brass ring in this study (when the taper angle is zero, it becomes the traditional shape, and the difference between armature and base diameter is the air gap) [16]. The magnetic field lines pass in and out of the coil along the closed path and they have the same physical characteristics with current and resistance. Then, the magnetic flux moves toward the minimum reluctance. When the metals are magnetic and in contact with each other, the magnetic field lines are distributed on the surface of the metals and the air gap permeance can be ignored. When the metals are not in contact with each other, the magnetic field lines are forced to be in and out the air gaps and the directions are both vertical to the metal surface following the shortest path and minimum reluctance.

According to the proportional electromagnetic configuration of this study, the air-gap zones located at both ends of the armature are $P_{A1}$–$P_{A6}$ and $P_{A7}$–$P_{A9}$. Around the front end gap, magnetism exists between base and armature, and the direction of flow is from base to armature. An enlarged view of this is shown in Figure 7. In Figure 7a, range CFHG represents Path Mode VI, and range DLMN represents Path Mode I. In Figure 7b, range ABCD represents Path Mode IV, and range AND represents Path Mode III. In Figure 7c, range MPQR represents Path Mode V, and range LMP represents Path Mode II.

For derivation of the rear end gap, magnetism is the same as front end gap, but the direction of flow is from armature to coil base. An enlarged view is shown in Figure 7d. Permeance $P_{A7}$ represents Path Mode I, permeance $P_{A8}$ represents Path Mode II, and permeance $P_{A9}$ represents Path Mode V.

The permeance is the reciprocal of the reluctance ($R$), and the reluctance physical characteristics are similar to the resistance. Shorter paths or larger cross-sectional areas produce smaller reluctance, easier flow and better permeability of the zone. The permeance is represented by parameters, such as path length and cross-sectional area. In Figure 7a, variable $x$ represents the armature displacement, and the definition of size of proportional electromagnet is shown in Figure 8. The results are shown in Table 2.

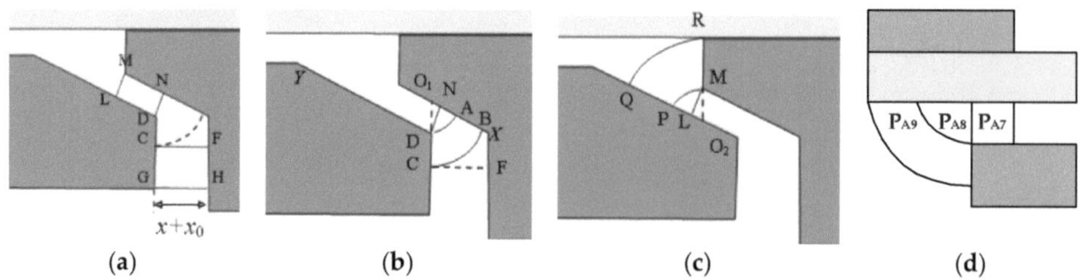


Figure 7. Air gaps magnetic flux path: (a–c) front end; and (d) rear end.

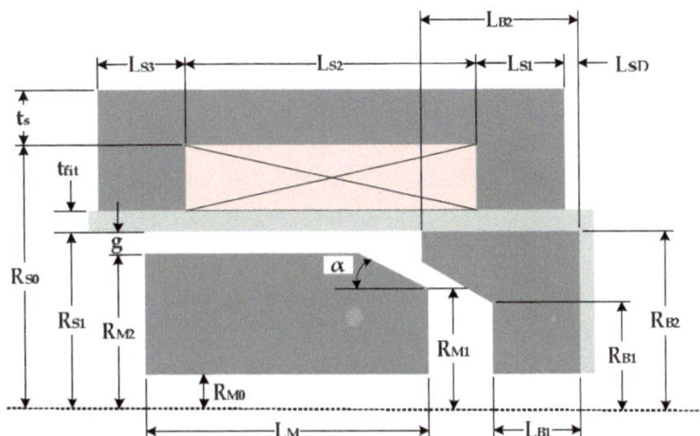

**Figure 8.** Definition of size for the proportional electromagnet.

### 2.3. Equivalent Permeance of Air Gaps

The front end equivalent permeance can be obtained after series connection, as shown in Equation (2) and Table 2. It is obtained the same way as the rear end gap, as shown in Equation (3). The proportional electromagnet in this study is designed as a modular configuration, so the non-magnetic material is used to attach the base and the armature acts as a guide tube. Because the permeability coefficient of non-magnetic material is similar to that of air, it is considered in the formula for the air gap zone. The permeance results are shown in Equation (4). The total equivalent permeance of the air gap zone can be calculated using Equations (2)–(4). The series connection is shown in Equation (5).

$$P_{air1} = P_{A1} + P_{A2} + P_{A3} + P_{A4} + P_{A5} + P_{A6} \tag{2}$$

$$P_{air2} = P_{A7} + P_{A8} + P_{A9} \tag{3}$$

$$P_{air3} = \frac{\pi u_0 (R_{B2} + R_{S1}) L_{S1} L_{S3}}{(L_{S1} + L_{S3}) t_{fit}} \tag{4}$$

$$P_{air} = \left( P_{air1}^{-1} + P_{air2}^{-1} + P_{air3}^{-1} \right)^{-1} \tag{5}$$

**Table 2.** Permeance of air gaps in the proportional electromagnet.

| Permeance | Mean Path Area | Mean Path Length ($l_i$) | Equation ($P_{Ai}$) |
|---|---|---|---|
| $P_{A1}$ | $CFHG$ | $x + x_0$ | $u A_1 / l_1$ |
| $P_{A2}$ | $DLMN$ | $x \sin \alpha$ | $u A_2 / l_2$ |
| $P_{A3}$ | $ABCD$ | $\sqrt{\overline{O_1 D} \cdot \min(\overline{O_1 C}, \overline{O_1 X})}$ | $4 u_0 P_{31} P_{32}$ |
| $P_{A4}$ | $ADN$ | — | $3.315 u_0 (R_{m1} + (\overline{O_1 D} \cdot \cos \alpha)/2)$ |
| $P_{A5}$ | $MPQR$ | $\sqrt{\overline{O_2 M} \cdot \min(\overline{O_2 R}, \overline{O_2 Y})}$ | $4 u_0 P_{51} P_{52}$ |
| $P_{A6}$ | $LMP$ | — | $3.315 u_0 [R_{O2} + (\overline{O_2 M} \cos \alpha)/2]$ |
| $P_{A7}$ | — | $g$ | $\frac{2\pi u_0}{g} (R_{M2} + \frac{g}{2}) \cdot \min(L_{S3}, L_{P7})$ |
| $P_{A8}$ | — | — | $3.315 u_0 (R_{M2} + g/2) \cdot K_8$ |
| $P_{A9}$ | — | $\sqrt{g \cdot R_{90}}$ | $4 u_0 (R_{B2} - l_9) \ln \frac{R_{90}}{g}$ |

### 2.4. Permeance of Metal Zone Calculation

As mentioned above, when the proportional electromagnet is energized, it forms a closed magnetic field and the simplified magnetic flux ($\Phi$) path of the metal zone is shown in Figure 9. Areas with different amperage ($NI$) will lead to different magnetic field intensity ($H$). Furthermore, the magnetomotive force ($F_{sum}$) and magnetic flux density ($B$) exist within the metal.

The permeability coefficient of metal is a non-constant, so there is a nonlinear correspondence between its H and B. When the magnetic flux density is higher than a specific value, it will lead to magnetic saturation, so the magnetic flux density may not increase even though the magnetic field is strengthened. The magnetic field intensity could be obtained by the correspondence with magnetic flux density through the material B-H curves, and the results are shown in Table 3.

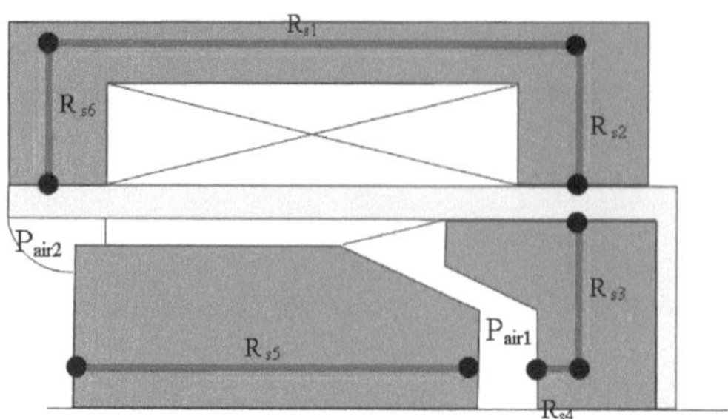


**Figure 9.** Magnetic flux path of the metal zone.

**Table 3.** Permeance parameter of metal zone.

| $i$ | Mean Path Area ($A_i$) | Mean Path Length ($l_i$) | $P_{Si}$ |
|---|---|---|---|
| 1 | $2\pi\, t_s R_{so}$ | $L_{S2} + (L_{S1} + L_{S3})/2$ | $u_S A_1 / l_1$ |
| 2 | $\pi(R_{si} + R_{so} + t_s/2) \cdot L_{S1}$ | $R_{so} + t_s/2$ | $u_S A_2 / l_2$ |
| 3 | $\pi(R_{B2} + R_{MO}) \cdot L_{B1}$ | $R_{B2} - (R_{M2} + R_{M0})/2$ | $u_S A_3 / l_3$ |
| 4 | $\pi(R_{B2}^2 - R_{MO}^2)$ | $L_{B1}/2$ | $u_S A_4 / l_4$ |
| 5 | $\pi(R_{M2}^2 - R_{MO}^2)$ | $L_M$ | $u_S A_5 / l_5$ |
| 6 | $\pi(R_{si} + R_{so} + t_s/2) \cdot L_{S3}$ | $L_{S2} + (L_{S1} + L_{S3})/2$ | $u_S A_6 / l_6$ |

When it is energized, the proportional electromagnet forms a closed magnetic circuit and the inside paths are connected in series. Therefore, the magnetic flux is the same and the magnetomotive force of each metal zone is obtained by magnetic flux multiplied by the reciprocal of permeance. The magnetic flux density of metal material is obtained by calculating the magnetic flux and then corresponding to the magnetic field intensity to allocate the magnetomotive force of metal and air gaps. The equivalent permeance of the metal zone within the proportional electromagnet is shown in Equation (6).

$$P_{steel} = R_{steel}^{-1} = \left(\sum_{i=1}^{6} R_{Si}\right)^{-1} = \left(\sum_{i=1}^{6} P_{Si}^{-1}\right)^{-1} \tag{6}$$

*2.5. Solenoid Force of Proportional Electromagnet*

The total permeance of the proportional electromagnet is the reciprocal of the total reluctance, shown in Equation (7). It can be obtained by the series connection of air gaps and metal zones. When magnetomotive force exists, the magnetomotive force of air gaps can be obtained by the permeance and magnetic flux, and it varies with the armature position, as shown in Equation (8). Due to material properties, the magnetomotive force of metal zones is determined by the magnetic field intensity, as shown in Equation (9), although the permeability of steel ($u_{Si}$) cannot be obtained in advance. When the magnetic flux is obtained, the magnetic field intensity corresponds with the B-H curves. Then, the magnetomotive force can be calculated.

$$P_{sum} = R_{sum}^{-1} = \left[P_{air}^{-1} + P_{steel}^{-1}\right]^{-1} = P_{air} P_{steel} / (P_{air} + P_{steel}) \tag{7}$$

$$F_{air} = \Phi R_{air} = \Phi P_{air}^{-1} \tag{8}$$

$$F_{steel} = \sum_{j=1}^{6} H_j l_j = \sum_{j=1}^{6} \frac{B_j}{u_{si}} l_j = \sum_{j=1}^{6} \frac{\Phi}{u_{si} A_j} l_j \tag{9}$$

$$NI = F_{sum} = F_{air} + F_{steel} \tag{10}$$

In terms of the armature solenoid force, the input energy can be converted into magnetic energy and mechanical energy according to the conservation of energy, as shown in Equation (11). The current and the metal permeance are unrelated to armature displacement. When the stable current is input, the differential value of some items is zero. It can be derived from the principle of virtual work ($W$). The armature thrust can be obtained by the magnetic flux and permeance of gap zones and its derivation, as shown in Equation (12). A negative sign indicates the opposite direction of displacement force.

$$W = \frac{1}{2}\Phi^2 R_{air} = \frac{1}{2}\Phi^2 P_{air}^{-1} \tag{11}$$

$$\begin{aligned}
F_{em} &= \frac{\partial W}{\partial x} = -\frac{\Phi^2}{2} \cdot \frac{d}{dx}\left(P_{air1}^{-1} + P_{air2}^{-1} + P_{air3}^{-1}\right) \\
&= -\frac{\Phi^2}{2} \cdot \left[P_{air1}^{-2}\frac{d}{dx}P_{air1} + P_{air2}^{-2}\frac{d}{dx}P_{air2} + P_{air3}^{-2}\frac{d}{dx}P_{air3}\right]
\end{aligned} \tag{12}$$

As mentioned above, the armature solenoid force can be obtained by the magnetic flux and air gap parameter of Equations (2)–(4). The magnetic flux is an unknown input parameter since it has a nonlinear relationship with the input current, and it is obtained by the calculation program. The database of magnetic flux and magnetomotive force should be created, the actual magnetic flux is obtained through the input current. Then, the proportional solenoid force, as derived from Equation (11), is entered into Equation (12).

## 3. Design and Simulation of Solenoid Force

### 3.1. Configuration Parameter Design

After the excitation of two materials, mutual attraction is generated and the attraction may be greatly attenuated with the separation distance. The reluctance of air gap is reduced when the two materials are close to each other, as shown in Equation (12). Most air gap reluctance increases with armature away from the base, but $P_{A3}$ is the opposite. In a particular path, the total reluctance of the proportional electromagnet is constant, so that the armature thrust must remain stable and would not change with the displacement.

Therefore, the design of parameters needs to focus on the permeance change of $P_{A3}$ and $P_{A1}$. Considering the collocation with each other, the optimized configuration of proportional electromagnet was completed. The following is a brief description of how some important parameters influence the electromagnetic force.

### 3.1.1. Base Flange

The base flange ($L_{B2} - L_{B1}$) mainly influences the parameters of the front end air gap. A shorter cone zone gives better permeance of air gap, and the permeability also increases. Therefore, the solenoid force is enhanced, and the horizontal becomes worse. The configuration boils down to the general electromagnet when there is no cone zone. The design is generally determined by the working stroke, so the stroke of the proportional valve must be within 1 mm. To avoid a strong magnetic attraction starting zone, a dimension of 1.5 mm for the product finally was taken as the design value. The simulation results are shown in Figure 10.

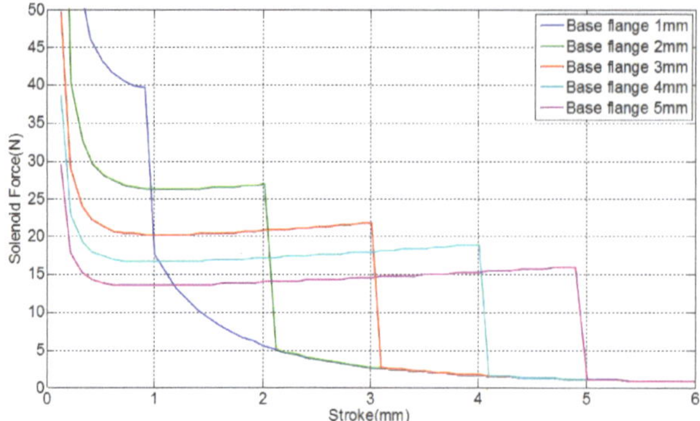

**Figure 10.** Simulation results of base flange.

### 3.1.2. Taper Angle

With the taper angle ($\alpha$), the size of configuration greatly influences the permeance of the front end air gap, and it is the most important parameter determining the electromagnetic force. When the taper angle decreases, $P_{A2}$ and $P_{A4}$ may relatively increase, and $P_{A3}$ changes with it. Therefore, the angle is reduced and the electromagnetic force clearly increases away from the armature (the trend of change is determined according to the rate of change of parameters $P_{31}$ and $P_{32}$). Using a wide angle can achieve a better horizontal dimension of the control zone, but the electromagnetic force is relatively weak. Finally, seven degrees was adopted as the design value for the product size, and simulation results are shown in Figure 11.

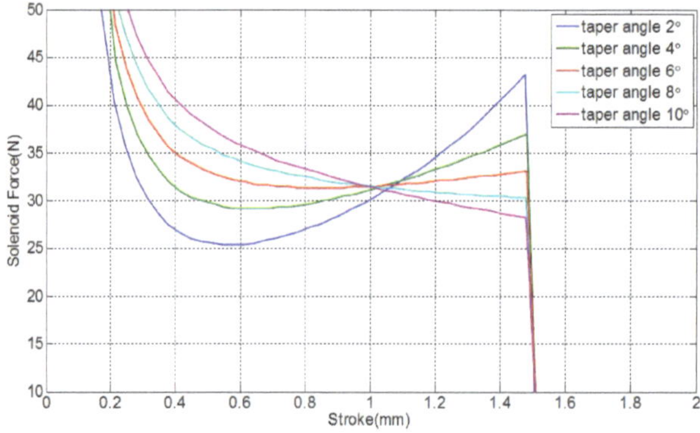

**Figure 11.** Simulation results for the taper angle.

### 3.1.3. Air Gap

To achieve a small volume but large thrust, the combination is better with smaller gap, thinner guide tube of proportional electromagnet, and smaller air gap between armature and base. Considering the actual needs of processing, the assembly was given a 0.1 mm air gap ($g$), 0.35 mm of guide tube ($t_{fit}$), and 0.1 mm of unilateral radius difference ($R_{M1} - R_{B1}$) between armature and base. The simulation results are shown in Figures 12 and 13.

For the simulation parameters, the gap between armature and base bottom diameter is the critical dimension because it directly affects the characteristic of thrust curve. When the difference between armature and base is close to zero, it can achieve better performance. To prevent the armature diameter from being bigger than the base and then generating interference, a diameter of 6.8 mm for the armature was used in the final design. The displacement cannot influence the guide tube thickness and assembly air gap, so the thrust curve will not change. The simulation results are shown in Figure 14.

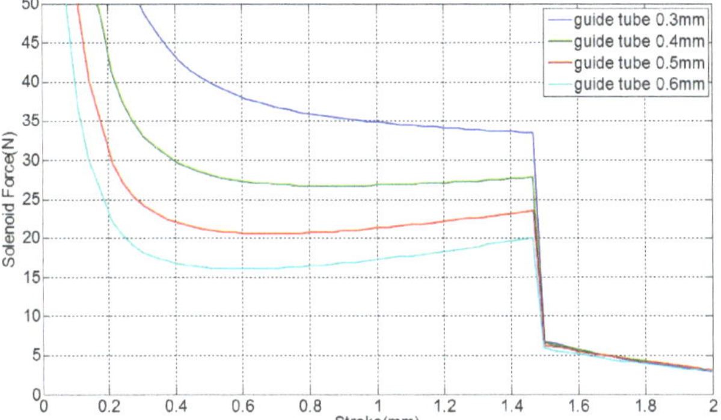

**Figure 12.** Simulation results for the guide tube.

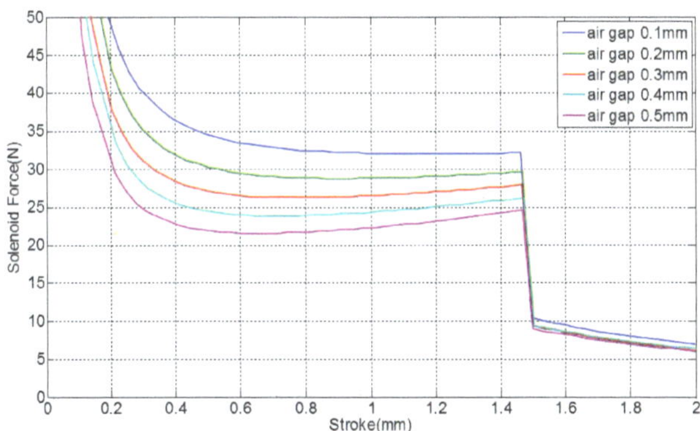

**Figure 13.** Simulation results for the air gap.

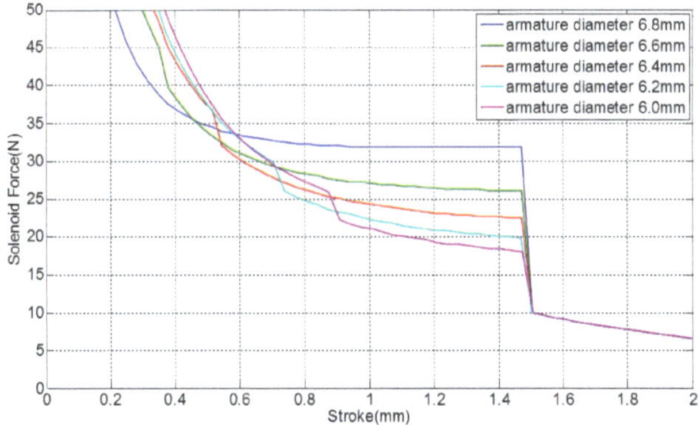

**Figure 14.** Simulation results for the armature diameter.

### 3.1.4. Parameter Design Results

Through the analysis of simulation results, detailed parameters such as base flange, taper angle, guide tube, air gap, and armature diameter were determined to obtain smooth, long-stroke, and large-thrust horizontal control zone. The relevant parameters are summarized and compared in Table 4.

**Table 4.** The relevant parameters are summarized and compared.

| Feature \ Parameter | Location (Figure 8) | Design Size | Thrust (N) | Size V.S. Thrust |
|---|---|---|---|---|
| base flange | $L_{B2} - L_{B1}$ | 1.5 mm | 31 | Size ↑, Thrust ↓ |
| taper angle | $\alpha$ | $7^0$ | 31 | Angle ↑, Thrust ↑ |
| guide tube | $t_{fit}$ | 0.35 mm | 31 | Size ↑, Thrust ↓ |
| air gap | $g$ | 0.1 mm | 32 | Size ↑, Thrust ↓ |
| armature diameter | $R_{M2} - R_{M0}$ | 6.8 mm | 32 | Size ↑, Thrust ↑ |

### 3.2. Simulation and Test Results of Armature Force

Using three different armature materials for the electromagnet and armature material, hysteresis tests are shown in Figure 15. After magnetic annealing, materials have a higher B-H. SUS 430 steel is easier to achieve magnetic saturation, and SUM 24L steel has higher magnetic flux (large slope). Thus, SUM 24L steel has a higher thrust curve and is suitable for the proportional electromagnet material.

Figure 16 shows the configuration of the proportional valve solenoid force test. The PC-based control unit is composed of the experimental software, a load cell, a digital controller, a steeper motor, and a proportional valve. Thus, the stepping motor pulls the proportional valve inside the armature to move, and the relationship between the solenoid force and the stroke of the armature at different voltages (8 V, 10 V, and 12 V) can be observed.

Figure 17 shows solenoid force simulation and test results. The solenoid force is proportional to the linear relationship by using SUM 24L steel (magnetic annealing) for the armature, magnetic base, and three different input voltages. The solenoid force required for the proportional valve is 30 N. When 12 V voltages is input, the horizontal control zone can reach 30 N for achieving the expected full pressure relief function. However, when the stroke is over 1.5 mm, the force is not completely decreased rapidly and the simulation value is different. It is possible that the test results were influenced by an electromagnetic leakage, and the future control zone should be set at a stroke from 0.7 mm to 1.5 mm to avoid non-horizontal control zone.

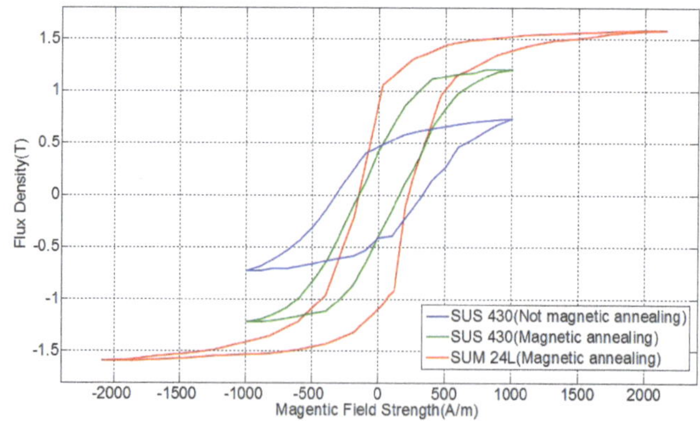


**Figure 15.** B-H curves of material.

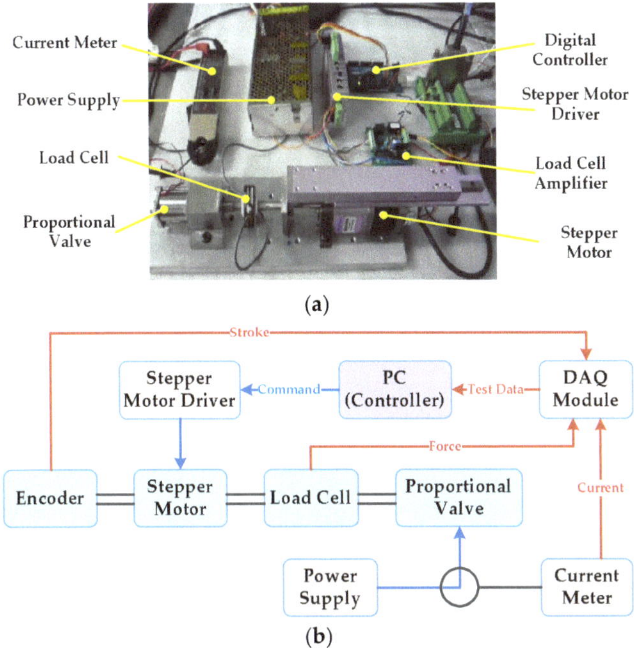


**Figure 16.** Configuration of the proportional valve solenoid force test: (**a**) test rig; and (**b**) structure.

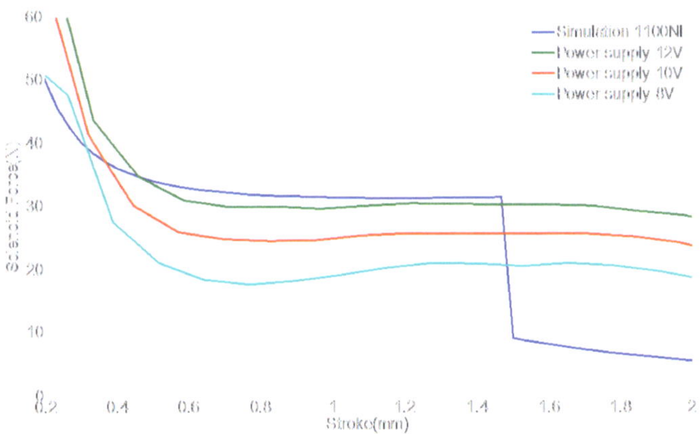


**Figure 17.** Simulation and test results for the proportional valve solenoid force.

## 4. Design and Test of Proportional Valve

### 4.1. Design of the Proportional Valve Body

A proportional pressure control valve is composed of the proportional electromagnet set and valve body set, which comprises a magnetic armature, spool valve, housing, and needle valve seat. The design is shown in Figure 18. The outlet and inlet $P_{cal}$ are connected, and they both have connection calipers. Although the valve body has four ports, it is three-port and two-position structure.

Through the shuttle shaft operation, the needle valve port must be closed, and it allows no internal leakage when there is no actuation. With the cone design, the check valve within the shuttle shaft is a safety device. If the needle valve malfunctions, the shuttle shaft cannot be returned to the original position, but the caliper pressure can still be relieved through the check valve. If the spring pre-pressuring is ignored, the shuttle shaft force equation is as follows:

$$F_m = F_{em} + F_{cal} \tag{13}$$

In addition $F_m = P_m * A_{sp}$, $F_{cal} = P_{cal} * (A_{sp} - A_{ol})$.
Substituting Equation (13),

$$P_{cal} = (F_m - F_{em})/(A_{sp} - A_{ol}) \tag{14}$$

Where $F_m$ is the master cylinder pressure acting on the shuttle shaft force; $F_{em}$ is the proportional solenoid force; $F_{cal}$ is the caliper chamber pressure acting on the shuttle shaft force; $P_m$ is the master cylinder pressure; $P_{cal}$ is the caliper pressure; $A_{sp}$ is the spool valve cross-sectional area; and $A_{ol}$ is the needle valve hole cross-sectional area.

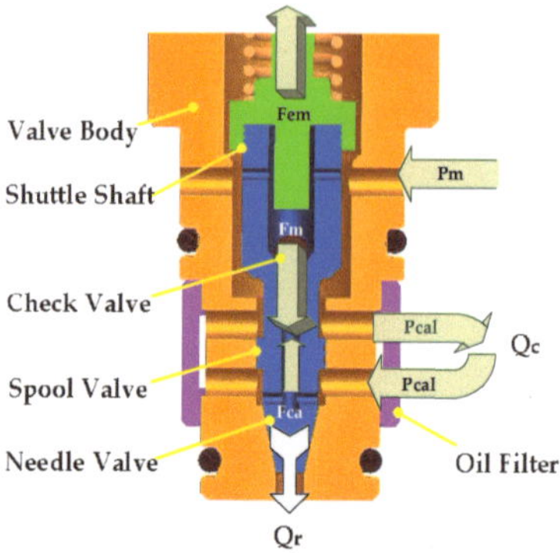


**Figure 18.** Proportional valve body construction and fluid force.

Assuming that the master cylinder pressure ($P_m$) is 172 bars, the general ABS specifications must reduce the calipers pressure ($P_{cal}$) to 69 bars or less. The spool shaft diameter is 1.8 mm, the proportional solenoid force is about 3.0 kgf, and the needle valve hole and calipers hole are 0.6 mm. The spool valve is fully closed when it moves 0.6 mm. Then, substituting values into Equation (14), the pressure relief value is calculated and the design meets requirements.

Regarding the rated flow calculation [17], the master cylinder pressure must remain constant. The pressure $P_c$ which is flowing through the spool valve is the supply pressure for the brake calipers with flow rate $Q_c$. The valve orifice area $A_c$ may vary with the displacement changes of shuttle shaft. Pressure $P_r$ flows through the needle valve with flow rate $Q_r$, and the needle valve orifice area is $A_r$. The equation which the spool valve port and the needle valve port flow through is as follows.

$$P_{cal} = \frac{\left(\frac{172 * \pi * 0.18^2}{4} - 3.0\right)}{\frac{\pi}{4}(0.18^2 - 0.06^2)} = 60.8 \; bars$$

$$Q_c = C_d A_c \sqrt{\frac{2(P_m - P_c)}{\rho}} \tag{15}$$

$$Q_r = C_d A_r \sqrt{\frac{2(P_c - P_r)}{\rho}} \tag{16}$$

The opening size of spool valve port and needle valve port are in a relation of mutual growth and decline. The maximum flow rate probably occurs when the two vale ports are in half-open position, which means the stroke is in the 0.3 mm position. The relationship of pressure is $P_m - P_c = P_c - P_r$ and $P_r$ is close to zero. Thus, $P_c$ is approximately a half of $P_m$. In the steady state, $Q_c = Q_r$, the maximum flow rate could be obtained by estimating the flow of spool valve. The half-open cross-sectional area of spool valve is as follows.

$$A_c = \frac{1}{2}\left(\frac{\pi(0.06)^2}{4}\right) = 0.001413 \text{ cm}^2$$

Substituting Equation (15),

$$Q_c = 0.7 * 0.001413 * 100 * \sqrt{\frac{2*(172-86)*10^5}{1000}}$$
$$= 13 \text{ cm}^3/\text{sec} = 0.78 \, l/\text{min}$$

*4.2. Single Proportional Valve Test*

This test is to confirm that the proportional solenoid energy is powerful enough to achieve the desired pressure relief of the proportional valve and that it could be used to control the brake calipers. When the input voltage is different, the armature can obtain a corresponding magnetic attraction and then drive the valve port to open. The single proportional valve test rig is shown in Figure 19a. To confirm the ability of relief, the high pressure of input port is compared with the pressure on the caliper port.

The experimental procedure of the proportional valve system is described and shown in Figure 19b. The system circuit is pressurized to 175 bars with a hand pump. The PC-based control unit includes an experimental software program, a pressure sensor, an AD/DA interface card and a PWM drive circuit. The control signals are computed in the PC-based control unit, and that drives the proportional valve via an AD/DA interface card and a PWM drive circuit with the sampling time of 1 ms. Thus, the solenoid force pulls the proportional valve inside the armature to move and the relationship between the command and the pressure on the calipers port at preloading (175 bars). The testing order is mainly step wave and triangular wave command-based since the step wave could confirm the reaction time, and the triangle wave could confirm the follow state and linearity.

To verify the pressure relief performance of proportional valves, the single proportional valve must be tested to confirm the pressure relief function and facilitate the ABS wheel configuration design. Because the braking force is related to the brake pedal or handle brake of drivers, the caliper chamber pressure at each brake condition is tested with dissimilar initial pressures. In addition, the proportional solenoid is applied with 10 V to drive the single proportional valve for the maximum pressure relief test.

Figure 20 shows the triangular wave command relief test results. When the power supplies 12 V (vehicle voltage), the proportional solenoid is not actuated at command 2.5 V or less, and the pressure drop is 130 bars above 10 V. The pressure drop between 2.5 V to 10 V is linear, and the relationship between voltage and pressure drop can be derived as 17.3 bars/V. The caliper pressure follows the command well and linearly, and the pressure does not have vibration when the valve is fully open. The solenoid force calculated as a result of the experiment is as follows:

$$\frac{\left(\frac{175*\pi*0.18^2}{4} - F_{em}\right)}{\frac{\pi}{4}(0.18^2 - 0.06^2)} = 45 \text{ bars} \Rightarrow F_{em} = 33.6 \text{ N}$$

The calculated test result is greater than the simulated value (30 N).

Figure 21 shows step wave command relief test results. It shows that when the voltage is lower than 3 V, the electromagnetic force is insufficient to open the valve port. The corresponding pressure drops for 3, 4.5, 6, 7.5, 9, and 10 V are, respectively, 25, 50, 75, 105, 125, and 130 bars. Figure 22 shows the time response when the step wave command is 10 V. The valve opening time is 9 ms (timing sequence from 8.821 s to 8.3 s) and the valve closing time is 5 ms (timing sequence from 9.321 s to 9.326 s).

This proportional valve test meets with application requirements, and it can be used as a reference for the internal controller of the subsequent ABS module as the basis for the slip control calculation.

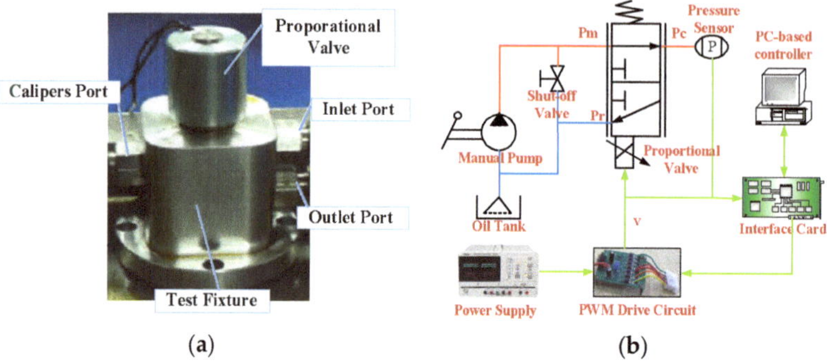


**Figure 19.** Configuration of single proportional valve test: (**a**) test rig; and (**b**) structure.

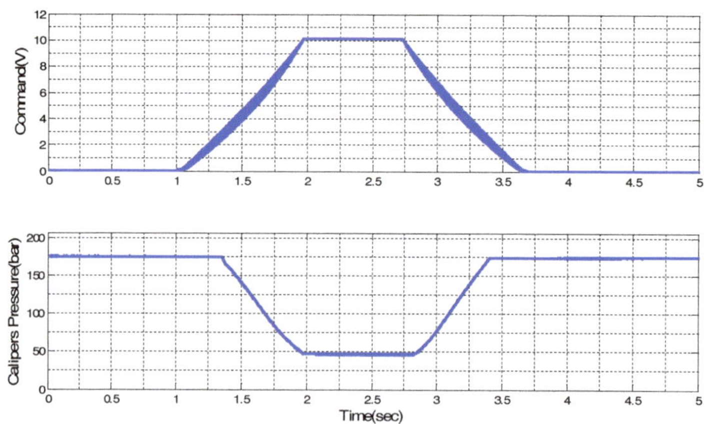

**Figure 20.** Triangular wave command at proportional valve relief test in 175 bars.

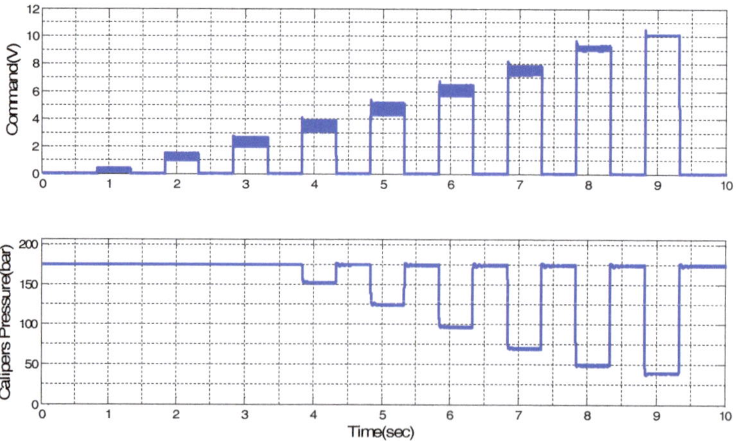

**Figure 21.** Step wave command at proportional valve relief test in 175 bars.

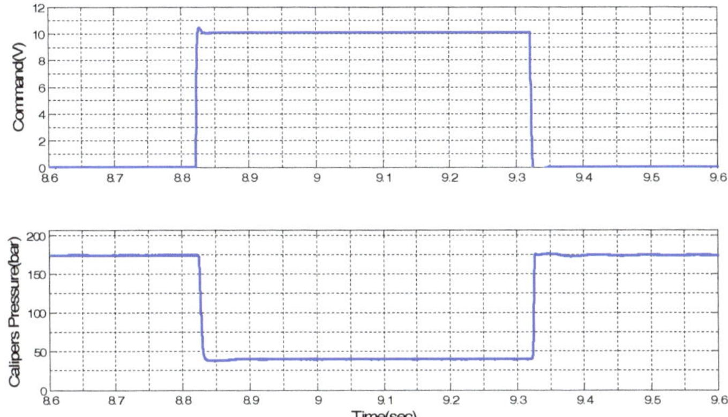

**Figure 22.** Step wave command at proportional valve relief test in 175 bars (timing sequence from 8.6 s to 9.6 s).

## 5. Conclusions

This study aimed to integrate the Inlet Valve (Inlet Valve normally open) and Outlet Valve (Outlet Valve normally closed) of an EHB, which has the same function as a three-port, two-position braking proportional pressure control valve. The assembly was composed of a cone lift lever (with electromagnet), the valve shell case, a valve body (including the top plug seats), and the proportional solenoid coil.

This new product development was tested under simulated braking force of pressing 175 bars, and its pressure drop is up to 130 bars. In terms of electromagnetic force, it can reach more than 30 N. Thus, this product can meet the demand of an ABS wheel anti-lock pressure relief and pressure relief adjustment control. Moreover, it provides ABS closed-loop control and has good linear control accuracy and repeatability. Overall, this product meets the design requirements.

**Author Contributions:** All the authors participated in the design of experiments, analysis of data and results, and writing of the paper.

**Conflicts of Interest:** The authors declare no conflicts of interest.

## References

1. Aksjonov, A.; Augsburg, K.; Vodovozov, V. Design and Simulation of the Robust ABS and ESP Fuzzy Logic Controller on the Complex Braking Maneuvers. *Appl. Sci.* **2016**, *6*, 382. [CrossRef]
2. Ivanov, V.; Savitski, D.; Augsburg, K.; Barber, P.; Knauder, B.; Zehetner, J. Wheel slip control for all-wheel drive electric vehicle with compensation of road disturbances. *J. Terramech.* **2015**, *61*, 1–10. [CrossRef]
3. Mirzaei, M.; Mirzaeinejad, H. Optimal design of a non-linear controller for anti-lock braking system. *Transp. Res. Part C* **2012**, *24*, 19–35. [CrossRef]
4. Mirzaeinejad, H.; Mirzaei, M. A novel method for non-linear control of wheel slip in anti-lock braking systems. *Control Eng. Pract.* **2010**, *18*, 918–926. [CrossRef]
5. Wang, B.; Huang, X.; Wang, J.; Guo, X.; Zhu, X. A robust wheel slip ratio control design combining hydraulic and regenerative braking systems for in-wheel-motors-driven electric Vehicles. *J. Frankl. Inst.* **2015**, *352*, 577–602. [CrossRef]
6. Wu, M.C.; Shih, M.C. Simulated and experimental study of hydraulic anti-lock braking system using sliding-mode PWM control. *Mechatronics* **2003**, *13*, 331–351. [CrossRef]
7. Qiuetal, Y.; Liang, X.; Dai, Z. Backstepping dynamic surface control for an anti-skid braking system. *Control Eng. Pract.* **2015**, *42*, 140–152.
8. Tanelli, M.; Sartori, R.; Savaresi, S.M. Combining Slip and Deceleration Control for Brake-by-wire Control Systems: A Sliding-mode Approach. *Eur. J. Control* **2007**, *6*, 593–611. [CrossRef]

9.    Lv, C.; Zhang, J.; Li, Y.; Yuan, Y. Novel control algorithm of braking energy regeneration system for an electric vehicle during safety-critical driving maneuvers. *Energy Convers. Manag.* **2015**, *106*, 520–529. [CrossRef]

10.   Patra, N.; Datta, K. Observer based road-tire friction estimation for slip control of braking system. *Procedia Eng.* **2012**, *38*, 1566–1574. [CrossRef]

11.   Bhandari, R.; Patil, S.; Singh, R.K. Surface prediction and control algorithms for anti-lock brake system. *Transp. Res. Part C* **2012**, *21*, 181–195. [CrossRef]

12.   Choa, J.R.; Choia, J.H.; Yoo, W.S.; Kim, G.J.; Woo, J.S. Estimation of dry road braking distance considering frictional energy of patterned tires. *Finite Elem. Anal. Des.* **2006**, *42*, 1248–1257. [CrossRef]

13.   Chen, C.P.; Tung, C.; Chen, C.A. A Proportional Electro-Hydraulic Breaking Control Valve, Taiwan, 2010. Patent No. I320374, 11 February 2010.

14.   Lee, C.O.; Song, C.S. Analysis of the solenoid of a hydraulic proportional compound valve. In Proceedings of the 35th National Conference on Fluid, Chicago, IL, USA, 13–15 November 1979.

15.   Chen, Y.N.; Kuo, W.H. Analysis and Design of Proportional Pressure Control Valve. Master's Thesis, National Taiwan University, Taipei, Taiwan, 1987.

16.   Chen, C.A.; Tung, C.; Chen, C.P. The Proportional Solenoid Module for a Hydraulic System, Taiwan, 2015. Patent No. I474350, 21 February 2015.

17.   Merritt, H.E. *Hydraulic Control Systems*, 3rd ed.; John Wiley & Sons: Hoboken, NJ, USA, 1985; pp. 76–113.

Article

# Development of a High-Pressure Pneumatic On/Off Valve with High Transient Performances Direct-Driven by Voice Coil Motor

Songlin Nie, Xiangyang Liu, Fanglong Yin *, Hui Ji and Jingxiu Zhang

Beijing Key Laboratory of Advanced Manufacturing Technology, Beijing University of Technology, Beijing 100124, China; niesonglin@bjut.edu.cn (S.N.); lxy6154@163.com (X.L.); jihui@bjut.edu.cn (H.J.); zhangjingxiu2016@163.com (J.Z.)
* Correspondence: yfl@bjut.edu.cn; Tel.: +86-10-6739-6362

Received: 12 February 2018; Accepted: 3 April 2018; Published: 12 April 2018

**Abstract:** The high-speed pneumatic on/off valve is one of the critical components in pneumatic systems, which has been widely investigated in the last decades. In this research, a new voice coil motor direct drive high-speed pneumatic on/off valve (VCM-DHPV) is proposed, and the mathematical model of VCM-DHPV, which consists of the fluid subsystem and electro-mechanical subsystem, is established. In addition, the key structural parameters of VCM-DHPV are optimized through the simulation analysis to improve its dynamic performance. The experiment results show that the developed VCM-DHPV has a good sealing performance by adopting the face-seal type in the valve port, and a large flow rate up to 5500 L/min, and its opening response time is 8.2 ms under the gas supply pressure of 8 MPa and exiting voltage of 240 V. With the supply pressure and the exciting voltage rising, the opening response time of VCM-DHPV is gradually increasing, and the variation tendency of the spool displacement curves is in accordance with the simulation results. This research will have significantly effects on raising the performance of the high-speed pneumatic on/off valve and the development of pneumatic precision motion control.

**Keywords:** high-speed on/off pneumatic valve; dynamic performance; AMESim simulation; voice coil motor

---

## 1. Introduction

A pneumatic high-speed on/off valve working in high pressure is one of the critical components in high-pressure pneumatic systems, which has been widely used in many applications, such as aviation, aerospace, underwater tools, drilling platform, and air-powered vehicle [1–5]. Therefore, it has gained more attention and has been investigated widely in recent years. The opening or closing of a spool orifice can control the volume flow of conventional hydraulic control valves. Compared with servo or proportional valves having complex structures and high cost, a certain number of simple on/off valves, which are known as digital hydraulic valves, have advantages of easy digital control, low power loss and compact structure, and the volume flow is controlled by on/off fluid [6–9]. However, there are some challenging issues associated with the development of pneumatic high-speed on/off valve working in high pressure, which is mainly due to the frequency response and thrust demand of the driving elements.

Recently, the research on the high-speed on/off valves concentrates upon improving the dynamic performance of the high-speed electro-mechanical actuator, developing the new structural pilot high-speed on/off valve. It has been confirmed that the conventional driving elements such as solenoid, torque motor and piezoelectric transducer are difficult to meet the requirements of high frequency response, larger thrust and low cost of the direct drive high-speed on/off valves simultaneously [10–12].

Wang et al. [13] studied the control performance of two-stage pilot high-speed switching valve driven by high-speed solenoid. It is found that the control signal, the operating frequency of high-speed switching valve, and the control chamber pressure could directly affect the response time of the main stage. However, because of the limitations of the valve stroke and magnetic force, it is difficult to achieve high pressure and large flow simultaneously. Therefore, the high-speed switching valve driven by high-speed solenoid was used as pilot valve, and the rated pressure and rated flow of the developed pilot valve are 2 MPa and 2 L/min, respectively.

Due to the limitations of the pilot valve response time, it is hard to increase dynamic response performance of the main valves. Besides, the two-stage on/off valves also have some disadvantages, for example, the machine of the nozzle is difficult and easy to be polluted, and has a high energy losses through pilot stage. Consequently, the single stage valves driven by the linear electro-mechanical actuator directly is a research hotspot at present. Voice coil motor (VCM) characterized by its fast dynamic response, ease of control, good repeatability, large thrust and compact structure is employed to direct drive the spool of high-pressure hydraulic and pneumatic direct drive valves. Li et al. [14,15] numerically and experimentally investigated a high pressure pneumatic VCM direct drive servo valve. The hybrid control scheme consisting of the disturbance observer and the proportion integration differentiation (PID) controller was established, which includes the velocity/acceleration feed-forward. The experimental results showed that the improved spool position control system has a good disturbance rejection capability and strong robustness. Guo et al. [16] designed a new voice coil motor with high acceleration by the electromagnetic finite element method and the magnetic circuit method, and developed a high-frequency response VCM direct drive valve (DDV) system. Then, the VCM-DDV system adopted a nonlinear PID control strategy and a dual closed-loop control structure. The experimental results showed that the designed VCM has an evident improved acceleration, and the static and dynamic performances of the VCM-DDV system are excellent, whose position bandwidth can achieve 350 Hz. Wu et al. [17,18] researched the influence of current–force coefficient and damping length of the VCM on the performance of the hydraulic direct drive valve, and the design parameters of VCM were optimized. The results showed that certain types of VCMs might be more appropriate for application in DDVs. Miyajima et al. [19] presented a VCM direct drive pneumatic three-port spool type servo valve, which has a high natural frequency of 300 Hz and a small leakage through the annular gap between the spool and the sleeve when the inlet pressure was 0.6 MPa.

Although some works have been presented to study the DDVs driven by VCM, very little evidence has been found in development of high pressure pneumatic on/off valve driven by VCM. The objective of this research is to propose a novel pneumatic on/off valve with a high pressure, large flow rate and high dynamic response. In Section 2, the structure of newly developed on/off valve is introduced, and then the mathematical models of electro-mechanical and fluid subsystem for the on/off valve are described. The simulations for the dynamic response of the on/off valve in the AMESim (Siemens, Munich, Germany) are presented in Section 3. In Section 4, the static and dynamic performances of the on/off valve are experimentally investigated, and the results and discussion about the experiments and simulations are analyzed. Conclusions are provided in Section 5.

## 2. Characterization and Methodology

### 2.1. Description of VCM-DHPV

Figure 1 illustrates the schematic diagram of VCM-DHPV, which consists of two parts: the pneumatic on/off valve and the VCM actuator. The on/off valve is regarded as an electro-pneumatic converter, which has the functions of the power amplified element and the energy conversion [20]. The VCM actuator proposed is shown in Figure 2. The main part of the VCM is composed of an iron yoke and a cylindrical permanent magnet, which can make a radially-oriented magnetic field in the air gap of the VCM. The moving part of the VCM is composed of the plastic bobbin and the wire winding, which can provide an Ampere force in the vertical direction perpendicular to the current flow and the magnetic flux when the coil is energized [21]. The spool connected directly with the slider is driven linearly by the VCM actuator. The permanent magnet of actuator is fixed with the valve body as the stator. To eliminate the excess pneumatic force acting on the spool, the lock nut is designed to balance the area difference of the spool, and then only the flow force and seal friction should be overcome by VCM. Therefore, the electromagnetic force of VCM is decreased and the dynamic characteristic of VCM-DHPV is improved. A high-accuracy position sensor mounted on the VCM is used to measure the displacement of the valve spool, and the position signal is fed back to the system controller, which generates the coil current. The poppet valve face-seal type is used in the valve port to reduce the leakage of VCM-DHPV when working under high-pressure condition. In addition, the face-seal in the valve port is made of graphite sleeve dipped with the phenolic resin, so the graphite valve port is softer than the valve cone made of ceramic, which is sprayed at the surface of titanium alloy spool. This kind of soft–hard contact sealing can archive zero leakage of the valve port. As a control component, the VCM-DHPV is designed to control gas flows through the pipelines and the chambers in the pneumatic actuator. Therefore, based on the position control system, the displacement of the valve spool is regulated accurately, so that it can control the gas quickly through the valve port from inlet to outlet.

When there is no exciting voltage signal, the spool is compressed tightly on the graphite sleeve by the spring, and then the VCM-DHPV is closed. As positive exciting voltage signal is given, the spool moves upward under the action of the electromagnetic force by VCM, and then the valve is opened. As a result, the inlet port and outlet port are interlinked. The high-pressure air flows from the inlet port into atmosphere through the outlet port. As negative exciting voltage signal is given, the spool moves downward under the action of the electromagnetic force by VCM, and then the VCM-DHPV is closed, and the inlet port and outlet port are disconnected.

The dynamic characteristic of VCM-DHPV is one of the main research focuses. According to ISO 5598-2008, the dynamic response time of fluid power components can be described as the elapsed time between the action initiation and the resulting reaction, which is tested under stated conditions. In this research, the opening response time is considered as the elapsed time between the initiation point when the exciting voltage is given and the point when the on/off valve is fully opened.

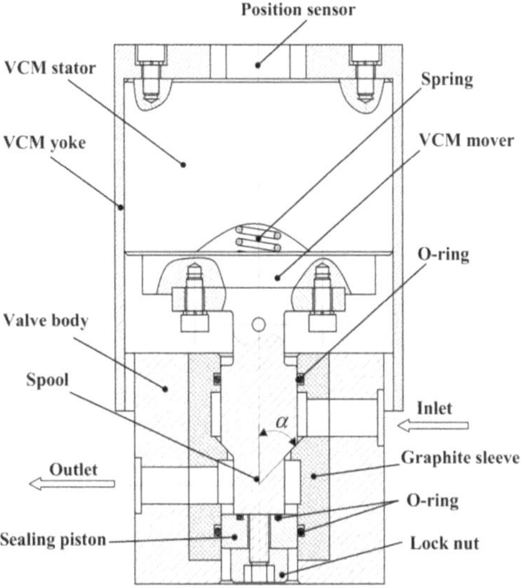


**Figure 1.** Schematic diagram of the developed high-speed pneumatic on/off valve (VCM-DHPV).

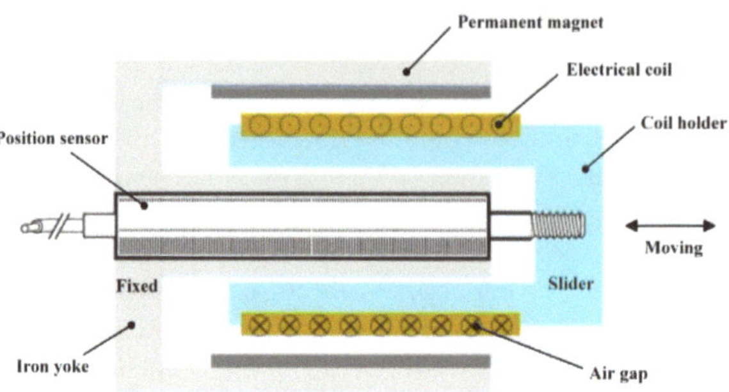


**Figure 2.** Principle of voice coil motor with position sensor.

## 2.2. Mathematical Model

As a pneumatic control component, the on/off valve was designed to provide rapid gas-flow through the cavities inside the pneumatic actuator. The characteristics of the on/off valve depends on the flow and electro-mechanical equations, whose relationships are intensively coupled with each other. Therefore, the on/off valve can be regarded as a converter between the electric and the pneumatic. Figure 3 describes the block diagram of VCM-DHPV, which illustrates the signal flow of the on/off valve model from the input voltage of the VCM to the output flow of the valve, including the intermediate interactions. As shown in Figure 3, the VCM-DHPV can be divided into pneumatic component and electromagnetic component. Then, the characteristic equations of VCM-DHPV could be obtained.

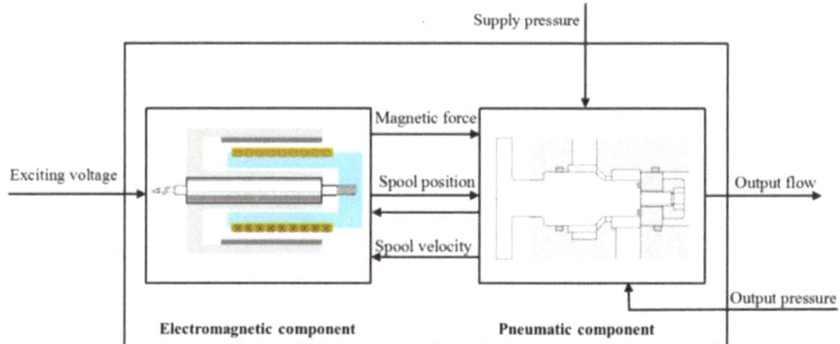


**Figure 3.** Block diagram for mathematical model of VCM-DHPV.

In the single degree of freedom (DOF) mass–spring–damper system for the moving part of VCM-DHPV, which consists of the VCM mover and the spool, Newton's second law of motion is given by:

$$m\frac{\mathrm{d}^2x(t)}{\mathrm{d}t^2} + b\frac{\mathrm{d}x(t)}{\mathrm{d}t} + k_s(x(t) + x_0) + F_s + F_f + F_t = F_e \tag{1}$$

where $x(t)$ is the displacement of the VCM mover and the spool (mm), $m$ is the mass of the coil of VCM and spool of the on/off valve assembly (kg), $b$ is damping coefficient (N/(m/s)), $k_s$ is spring constant (N/m), $x_0$ is spring pre-tension (mm), $F_s$ is steady gas flow force (N), $F_f$ is Coulomb's friction force (N), $F_t$ is transient gas flow force (N), $F_e$ is electromagnetic force of the VCM (N), and $F_f$ is Coulomb's friction force (N), which is very small compared to the electromagnetic force and can be ignored. Because the friction force and gas flow force are uncertainty, nonlinearity and time variation, the forces $F_s$ and $F_t$ can be considered as the disturbances for the valve spool position control system of VCM-DHPV [15], which will be presented as follows.

The electromagnetic component consists of a magnetic circuit and an electrical circuit [16]. According to the Lorentz force principle, the electro-mechanical conversion of the voice coil motor can be obtained. Therefore, the electromagnetic force of VCM is expressed as:

$$F_e = K_e i = BlNi \tag{2}$$

where $K_e$ is current–force coefficient, so $F_e$ is proportional to the current of the coil.

Figure 4 depicts the equivalent circuit of the voice coil motor, and its voltage balance equation can be written as:

$$Ri + L_s \frac{di}{dt} + e_m = u \tag{3}$$

where $R$ is resistance of the coil, $L_s$ is inductance of the coil, and $e_m$ is back EMF, which can be expressed as:

$$e_m = BlN \frac{dx}{dt} = K_e \frac{dx}{dt} \tag{4}$$

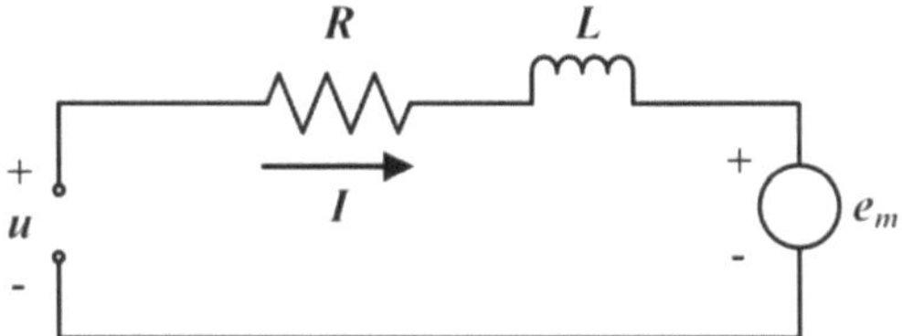


**Figure 4.** Equivalent electrical circuit of VCM.

The air-flow is a complex thermodynamic process, where it is a variable mass system in VCM-DHPV. According to the energy equation, continuity equation and dynamic equation, the mathematical model of the air-flow can be established. It should be noted that the gas supply pressure of the developed VCM-DHPV varies in the range of 0.1–8 MPa. As the pressure of the air is less than 20 MPa, the gas compressibility factor has little deviation from 1% (less than 5%), therefore the high-pressure air could be simplified to the ideal gas [14,15,22]. Because the start up process of VCM-DHPV is very short, the exhaust process of the control chamber can be simplified to an adiabatic process. The following assumptions are given as follows [23]:

1. A constant stable gas supply is considered.
2. Flowing process of gas through valve port of VCM-DHPV is isentropic.
3. The pressure and thermal fields are uniformly distributed inside every cavity of VCM-DHPV.
4. The spring of VCM-DHPV is assumed to be linear. In addition, the masses of spool, piston, and spring are integrated into one inertial parameter.
5. Dynamic flow forces, gas inertia and pressure loss in tubes can be neglected.
6. Steady state flow forces are considered only.

The electromagnetic component can be used to control gas flow through adjusting spool position of the on/off valve. For turbulent flow, the spool orifice equation can be used to represent the gas flow through VCM-DHPV, which includes the subsonic and choked flow regimes. Because the high-pressure air in pneumatic components and pneumatic system is compressible, as the ratio of the upstream pressure $p_u$ to the downstream pressure $p_d$ is larger than the critical value $\sigma_{cr}$, the flow regime can be regarded subsonic and the mass flow depends nonlinearly on both pressures. However, when the ratio is smaller than the critical value $\sigma_{cr}$, the flow achieves sonic velocity (choked flow) and depends linearly on the upstream pressure $p_u$ [8,22]. In these two cases, the air mass flow depends linearly on the spool position of the on/off valve. The standard equation for the mass flow rate through the spool orifice can be modeled as:

$$\dot{m}_t = \alpha \cdot S \cdot p_u \cdot \sqrt{\frac{k}{RT}} \cdot \phi(p_u, p_d) \tag{5}$$

where $\alpha$ is the coefficient of contraction, $S$ is the passage area of the spool orifice, $S = 4C_dDx(t)$, $C_d$ is discharge coefficient, $D$ is the orifice diameter of the valve, $x(t)$ is opening of the on/off valve, $k$ is specific heat ratio of air, $R$ is gas constant of air, and $T$ is absolute temperature. $phi(p_u,p_d)$ is given by [24,25]:

$$\phi(p_u, p_d) = \begin{cases} \sqrt{\left(\frac{2}{k+1}\right)^{\frac{k+1}{k-1}}} \; when \; \left(\frac{p_d}{p_u} \leq \sigma_{cr}\right) \\[2ex] \sqrt{\frac{2}{k-1}\left(\left(\frac{p_d}{p_u}\right)^{\frac{2}{k}} - \left(\frac{p_d}{p_u}\right)^{\frac{k+1}{k}}\right)} \; when \; \left(\frac{p_d}{p_u} > \sigma_{cr}\right) \end{cases} \tag{6}$$

where $\sigma_{cr}$ is critical pressure ratio, which is equal to 0.5283 and can be defined as:

$$\sigma_{cr} = \left(\frac{2}{k+1}\right)^{\frac{k}{k-1}} \tag{7}$$

Based on the momentum conservation principle, the steady gas flow force on valve spool can be calculated as:

$$F_s = \dot{m}_t(x(t))v\cos\theta \tag{8}$$

Besides, the transient gas flow force acting on the valve spool can be written by:

$$F_t = \pm L_v \frac{d\dot{m}_t}{dt} = \pm L_v\alpha\frac{dS}{dt}p_u \cdot \sqrt{\frac{k}{RT}} \cdot \phi(p_u, p_d) = \pm 4L_v\alpha C_dDp_u \cdot \sqrt{\frac{k}{RT}} \cdot \phi(p_u, p_d)\frac{dx(t)}{dt} \tag{9}$$

where $L_v$ is damping length of the on/off valve, $\theta$ is gas flow jet angle, and $v$ is gas flow velocity.

Therefore, after adopting the Laplace transform for Equations (1) and (3) under the boundary condition $X(0) = 0$ and the initial condition $X'(0) = 0$, they can be represented as follows:

$$ms^2X(s) + bsX(s) + k_sX(s) + F_f + F_t + F_s = K_eI(s) \tag{10}$$

$$(R + L_s)I(s) + K_esX(s) = U(s) \tag{11}$$

Consequently, with Equations (10) and (11), the transfer function between the $X(s)$ and the $U(s)$ for the on/off valve driven by the VCM directly can be derived as:

$$G_p(s) = \frac{X(s)}{U(s)} = \frac{K_e/(R + L_s)}{ms^2 + [K_e^2/(R + L_s) + b]s + k_s} \tag{12}$$

## 3. Simulation Analysis

The start up process of VCM-DHPV, which has nonlinear relation to the displacement of on/off valve spool in the case of the constant inlet pressure, is not constant stable. It is very difficult to calculate the mathematical model of pressure. To determine the dynamic and static characteristics of VCM-DHPV, the equations of the mathematical model could be solved numerically by the AMESim software, in which the relationship between supply pressure $P_s$ and $x$ could be expressed numerically. The simulation diagram was established by AMESim, as shown in Figure 5. A physical diagram was made in AMESim based on the mathematical model. The design parameters of VCM-DHPV are shown in Table 1. Then, the dynamic response curves of VCM-DHPV under different exciting voltage $u$, supply pressures $P_s$ and half cone angle $\alpha$ of the spool (as illustrated in Figure 1) can be obtained through the simulation, as shown in Figures 6–8.

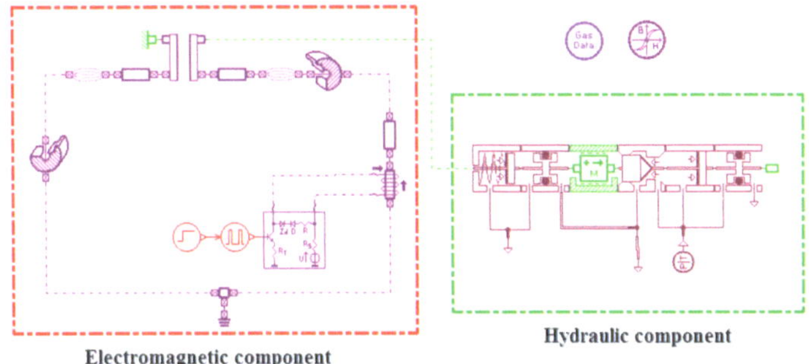



**Figure 5.** AMESim model of VCM-DHPV.

**Table 1.** Design parameters of VCM-DHPV.

| Description | Notation | Value | Unit |
| --- | --- | --- | --- |
| Mass of coil and spool | $m$ | 0.28 | kg |
| Maximum stroke of the spool | $x_{max}$ | 1.0 | mm |
| Nominal mass flow rate | $q_{tmax}$ | 0.5 | g/s |
| Gain of VCM | $K_e$ | 44.2 | N/A |
| Maximum coil current | $I$ | 8.4 | A |
| Coil resistance | $R$ | 4.6 | $\Omega$ |
| Coil inductance | $L_s$ | 220 | mH/kHz |
| Viscous friction coefficient | $k_f$ | $8 \times 10^{-3}$ | N/(m/s) |
| Spring stiffness | $k_s$ | 0.02 | N/m |

## 3.1. Influence of the Exciting Voltage

To investigate the sensitivity of dynamic response characteristic of VCM-DHPV to the different exciting voltage, the simulations are performed at the exciting voltage of 150, 180, 240 and 280 V (with the supply pressure of 8 MPa). It can be seen in Figure 6 that the opening response time of VCM-DHPV decreases while the exciting voltage $u$ gets bigger. This is because the bigger exciting voltage of the VCM coil could rapidly increase the current. Thus, the opening response time of the VCM becomes shorter. The opening response times of the on/off valve at the exciting voltage $u$ of 150, 180, 240 and 280 V are 9.1, 8.2, 6.8 and 6.2 ms, respectively. Therefore, from the increase of the dynamic characteristic purpose point of view, the exciting voltage should be appropriately improved during the process of switching for VCM-DHPV.

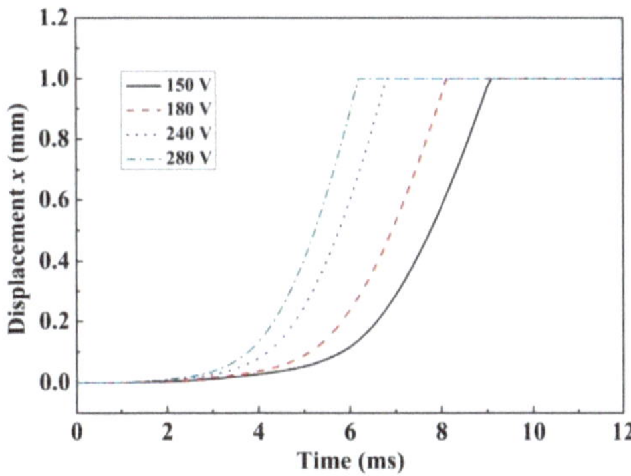



**Figure 6.** Dynamic response curves under different exciting voltage ($P_s$ = 8 MPa, $\alpha$ = 45°).

### 3.2. Influence of the Supply Pressure

Because the maximum working pressure of the developed VCM-DHPV is 8 MPa, the numerical simulation was conducted under different supply pressures which are less than 8 MPa (2, 4, 6 and 8 MPa) to investigate its dynamic characteristics. Figure 7 illustrates the displacement of spool versus simulation time of VCM-DHPV under different supply pressures (with the exciting voltage of 240 V). When only the supply pressure of VCM-DHPV is changed in the simulation, it can be seen in Figure 7 that the opening response time increased with the rise of the supply pressure. Here, the opening response times are 5.7, 5.9, 6.1 and 6.2 ms under the supply pressures of 2, 4, 6 and 8 MPa, respectively. This is attributed to the fact that the transient and steady gas flow forces gradually increased with the rise of supply pressure.

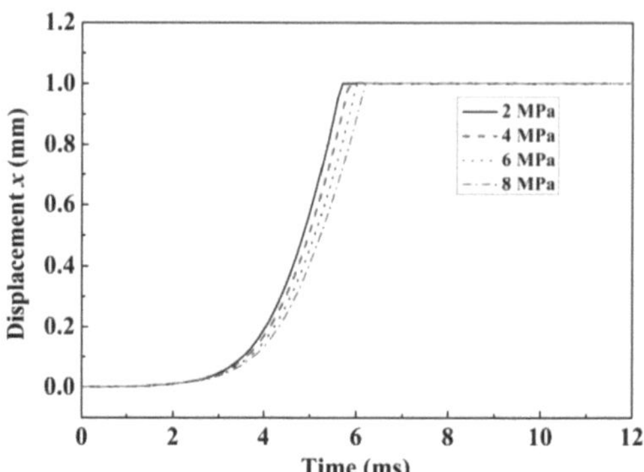


**Figure 7.** Dynamic response curves under different supply pressures ($u$ = 240 V, $\alpha$ = 45°).

### 3.3. Influence of the Half Cone Angle

Based on Equations (5) and (6), the key structural parameters of the on/off valve would also affect the dynamic response characteristics. Taking the half cone angle $\alpha$ of on/off valve spool as an example, when changing the value of the half cone angle $\alpha$ from 20° to 70° ($u$ = 240 V and $P_s$ = 8 MPa), it is achieved in Figure 8 that the bigger half cone angle $\alpha$ could slightly increase the opening response time of VCM-DHPV. The explanation of this is that as the half cone angle of the spool enlarged, the transient and steady gas flow forces would be gradually increased. However, it is also shown in Figure 8 that the maximum difference of VCM-DHPV opening response time under different half cone angle is very small (0.31 ms). This indicated that increasing the half cone angle cannot effectively improve the dynamic response of VCM-DHPV. In addition, the 45° angle is convenient for manufacturing and obtaining an appropriate sealing ability. Therefore, the half cone angle of 45° should be applied to the design of VCM-DHPV prototype.

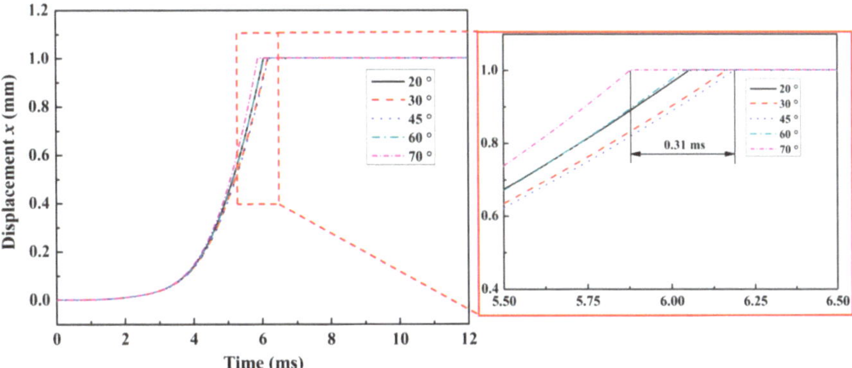

**Figure 8.** Dynamic response curves under different half cone angle ($P_s$ = 8 MPa, $u$ = 240 V).

## 4. Experiment Verification

### 4.1. Experiment Apparatus

To verify the feasibility of VCM-DHPV structure and the validity of the mathematical model, the experimental investigation was carried out. Experimental pneumatic circuit is shown in Figure 9, which is composed of a gas supply, stop valve, pressure reducing valve, flow meter, pressure sensor, pressure gauge, the on/off valve, current amplifier, position sensor, data acquisition (DAQ) and the controller. The control scheme of the on/off valve is realized by a programmable multi-axis controller (PMAC). The VCM and the position sensor were linked to a personal computer with Pewin32 Pro2 software through a DAQ, which was used to acquire position signal and control the input voltage of the VCM [26]. The rated current and rated force of the VCM used in this experiment are 8.4 A and 376 N, respectively. Therefore, a current amplifier could be adopted to amplify the current and supply to VCM. The voltage command is supplied from the computer through the DAQ, and the output voltages from the position sensor are fed back to the computer by the DAQ. Then, the spool of VCM-DHPV is actuated toward two directions. The test software is developed on the LabVIEW platform (version 11.0, National Instruments, Austin, TX, USA, 2011).

A linear variable differential transformer (LVDT) is used to detect the displacement of spool, which is transferred by a push rod connected with the spool. When the spool is promoted, it will push the rod detected by the position sensor. The developed VCM-DHPV prototype and the test bench are shown in Figure 10, and the characteristic parameters of the prototype are listed in Table 2. The experiment processes are as follows:

1. Open the stop valve and regulate the pressure reducing valve to make the compressed air as a certain pressure.
2. Use the computer to collect the pressure signal, flow signal and displacement signal when opening the on/off valve.
3. Set a new pressure and repeat Steps (1) and (2).

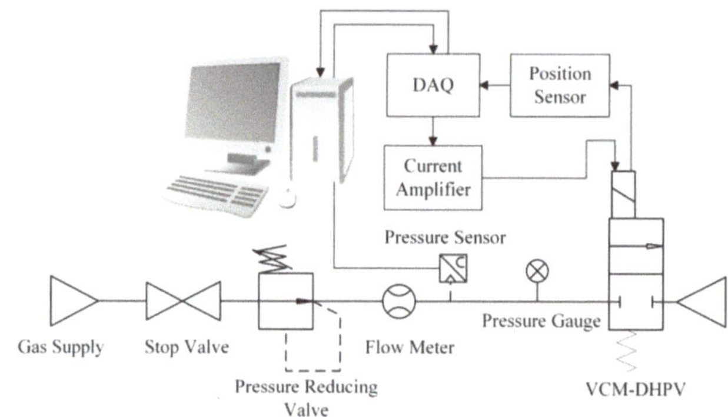



**Figure 9.** Schematic diagram of the pneumatic circuit test bench.

**Figure 10.** Pictures of VCM-DHPV prototype and test system: (**a**) VCM-DHPV prototype; and (**b**) test bench.

**Table 2.** Parameters of the proposed VCM-DHPV prototype.

| Description | Notation | Value | Unit |
| --- | --- | --- | --- |
| Orifice diameter of the valve | $D$ | 24 | mm |
| Spool diameter | $d_r$ | 16 | mm |
| Half cone angle | $\alpha$ | 45 | ° |
| Inlet diameter | $d_i$ | 11 | mm |
| Outlet diameter | $d_o$ | 11 | mm |
| VCM electromagnetic force | $F_e$ | 376 | N |
| Test pressure | $P_s$ | 8 | MPa |

*4.2. Experiment Results and Discussion*

### 4.2.1. Static Characteristics Experiments

To verify the sealing and flow characteristics of VCM-DHPV, the static characteristics of the valve is examined. Firstly, VCM-DHPV is closed by the VCM reverse power, and the supply pressures of the valve are set to 2, 4, 6 and 8 MPa, respectively. Then, the pressure sealing performance of the on/off valve is checked under the pressure holding time of 2 min. The experimental results show that the pressure sealing performances of VCM-DHPV are very good under the different supply pressures. The flow rate of VCM-DHPV under different supply pressures and is illustrated in Figure 11a. It can be seen that, with the gradual increase of supply pressure, the flow begins to increase rapidly. As the supply pressure continues to be increased to a certain value, the flow rate of VCM-DHPV could reach a stable value of about 5500 L/min, as shown in Figure 11b.

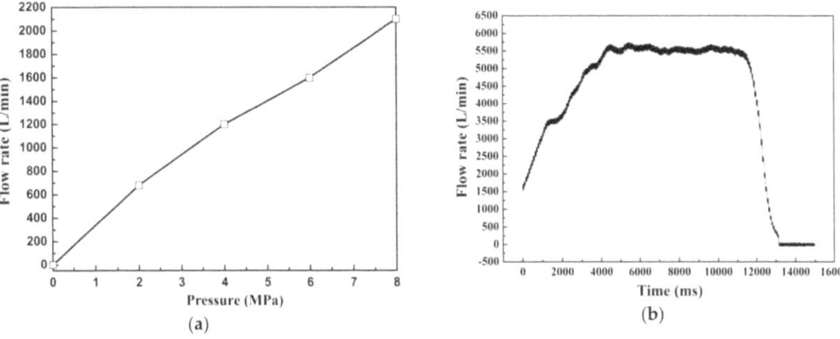

**Figure 11.** Flow curve of VCM-DHPV under different supply pressures: (**a**) VCM-DHPV prototype; and (**b**) maximum flow rate.

### 4.2.2. Dynamic Characteristics Experiments

The dynamic characteristics experiment is conducted to obtain the displacement of spool versus test time of VCM-DHPV under different supply pressures. Figure 12 shows the comparisons of the step response between the simulation results and measurements of the developed VCM-DHPV under different supple pressures. The red dashed lines stand for simulation results and the black solid lines stand for experimental results. It can be seen in Figure 12 that the curve of the simulations changes along with the trend of the experimental results, so the comparisons between the simulation results and measurements resulted in an acceptable agreement. However, as presented in Figure 12, due to the uncertainties of the sensors, some small overshoots can be observed. The opening response times of the on/off valve under different supply pressures by the simulation and experiment are listed in Table 3.

As presented in Table 3, the deviation of the simulated and measured response time under the supply pressures of 2, 4, 6 and 8 MPa is about 1.6, 1.5, 1.8 and 2.0 ms, respectively, and the deviations increase with the increasing supply pressures. This is because the friction force between the valve core and the valve body increases with the increasing supply pressure in the experiment, leading to the long response time of VCM-DHPV. However, the friction force is neglected in the numerical simulation. As a result, the deviations between the simulation and experiment increase with the increasing supply pressure. These results show that the simulated results provided a reasonable agreement within neglect of uncertainties of the sensors and the friction force between the valve core and the valve body. Therefore, it can be concluded that the numerical simulation and experimental results show the same change tendency and the presented AMESim model is acceptable in analyzing the dynamic characteristics of VCM-DHPV.

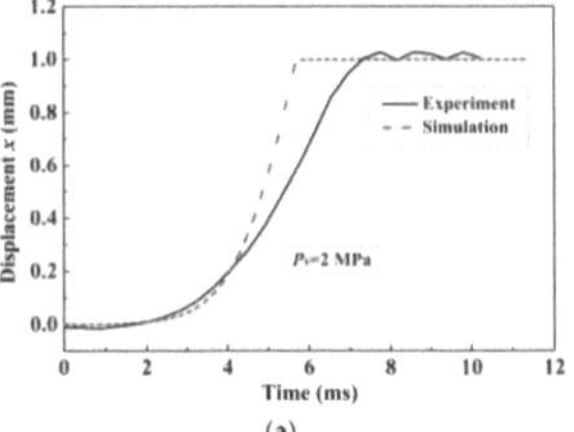
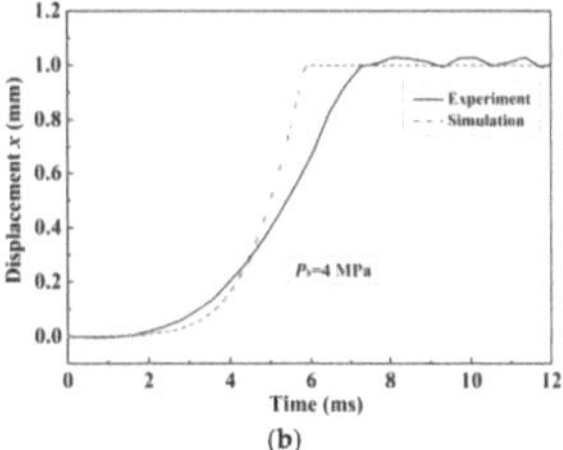

**Figure 12.** *Cont.*

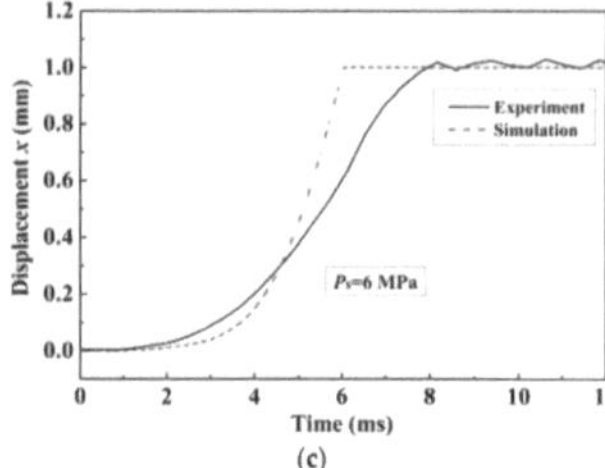 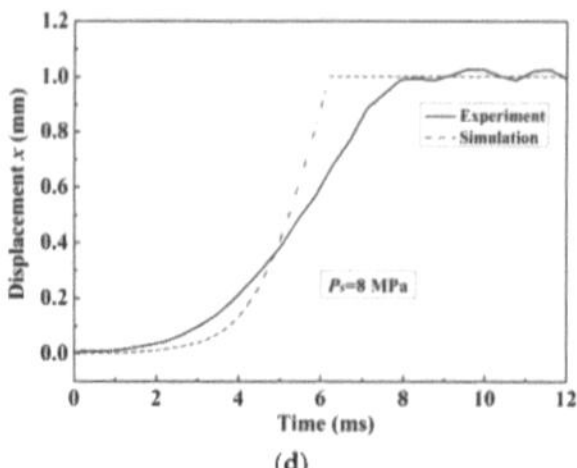

**Figure 12.** Step responses of VCM-DHPV under different supply pressures ($u$ = 240 V, $\alpha$ = 45°): (**a**) $P_s$ = 2 MPa; (**b**) $P_s$ = 4 MPa; (**c**) $P_s$ = 6 MPa; and (**d**) $P_s$ = 8 MPa.

**Table 3.** Opening response times of VCM-DHPV under different supply pressures.

| Supply Pressure | Opening Response time | | Deviation |
| :---: | :---: | :---: | :---: |
| | Simulation | Experiment | |
| 2 MPa | 5.7 ms | 7.3 ms | 1.6 ms |
| 4 MPa | 5.9 ms | 7.4 ms | 1.5 ms |
| 6 MPa | 6.1 ms | 7.9 ms | 1.8 ms |
| 8 MPa | 6.2 ms | 8.2 ms | 2.0 ms |

Meanwhile, the system identification tools in MATLAB (MathWorks, Natick, MA, USA) are employed to identify the system parameters of the developed VCM-DHPV by the simulation and experiment. The transfer function of on/off valve is acquired, and then the dynamic characteristics of the on/off valve via the frequency responses with the full stroke can be drawn in MATLAB. Figure 13 shows the Bode diagram of VCM-DHPV through the simulation and experiment under different supply pressures. The results show that the transfer function of the on/off valve could be simplified to be a second-order oscillating system. Because the damping ratio in the transfer function of the on/off valve are not considered in the simulation ($\zeta$ = 0), the bandwidth of the on/off valve is significantly influenced by the supply pressure, and its amplitude frequency characteristics of the simulations have an obvious pulse. However, due to the friction force and hydraulic leakage, the damping ratio could not be equal to zero in the actual working condition. It is worth noting in Figure 13 that the frequency responses of the on/off valve are quite different when the supply pressures are set as 2.0, 4.0, 6.0 and 8.0 MPa, respectively. Besides, it is illustrated in Figure 13 that the gain margin GM and phase margin PM of VCM-DHPV are both greater than zero. As a result, it can be concluded that VCM-DHPV is a stable minimum-phase system.

In addition, it can be seen in Figure 13 that the frequency responses of the simulations and experiments are not identical, and there is a large delay. This is attributed to the delay effect of the position sensor fixed on the spool in experiment process and the friction forces of the two O type rings ignored in the simulation model. Actually, the delay time of position sensor and friction force could disturb the dynamic characteristics of on/off valve in the experiment process, which would affect the push–pull effort of the VCM and cause the dynamic response time slowing. Consequently, further improvement of VCM-DHPV dynamic characteristics by upgrading the accuracy of position sensor and reducing the friction force between valve core and valve sleeve will be the main research concerns in the future.

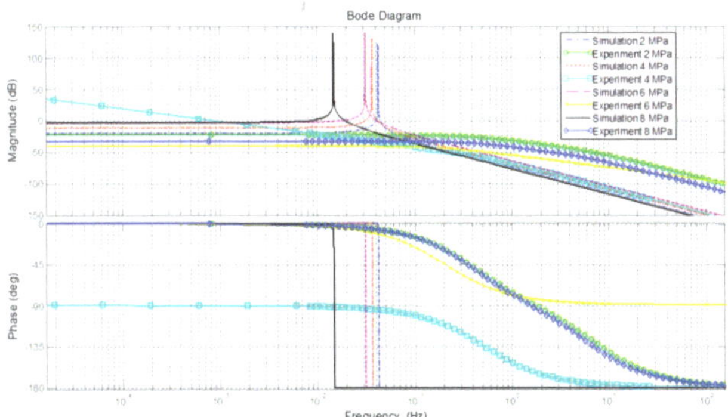

**Figure 13.** VCM-DHPV frequency response comparison between the simulated and measured results.

## 5. Conclusions

In this research, a new high-pressure large-flow pneumatic on/off valve driven directly by the voice coil motor is presented. The structure and working principle are quite different from the traditional two-stage on/off valve. The mathematical model and transfer function are derived in detail to analyze its response characteristics. The simulation results based on AMESim software show that the exciting voltage, half cone angle and supply pressure could influence the response time of the on/off valve significantly. The opening response time of the developed valve is decreased with the exciting voltage, whereas increased with the supply pressure. Besides, the half cone angle has been optimized to get the shorter opening response time. Then, an experiment is carried out to verify the feasibility of novel structure and validity of the numerical simulation, which shows that the flow rate and opening response time of the on/off valve are 5500 L/min and 8.2 ms under the gas supply pressure of 8 MPa and exiting voltage of 240 V. The experimental results indicate a reasonable match between the simulations and measurements, and both simulation analysis and experimental results exhibit that the on/off valve satisfies the initial design requirements well.

**Acknowledgments:** The authors would like to thank the National Natural Science Foundation of China (Grant Nos. 51705008 and 11572012), Beijing Natural Science Foundation (Grant Nos. 3182003, 1184012 and 3164039), Beijing Municipal Science and Technology Project (Grant No. KM201810005014), China Postdoctoral Science Foundation funded project (Grant No. 2017M620551) and Beijing Postdoctoral Research Foundation (Grant No. 2017-ZZ020) for their funding for this research. The authors are very grateful to the editors and the anonymous reviewers for their insightful comments and suggestions.

**Author Contributions:** Songlin Nie and Xiangyang Liu proposed the idea of the high-pressure pneumatic on/off valve direct-driven by voice coil motor for the purpose of increasing the frequency response; Xiangyang Liu and Fanglong Yin conceived and designed the experiments; Songlin Nie and Hui Ji contributed to the mathematical model; Fanglong Yin and Jingxiu Zhang performed the experiments and analyzed the data; Fanglong Yin wrote the Matlab code, established the AMEsim simulation model and conducted the numerical simulations; and Xiangyang Liu wrote the paper.

**Conflicts of Interest:** The authors declare no conflict of interest.

## References

1. Yang, H.; Pan, M. Engineering research in fluid power: A review. *J. Zhejiang Univ. Sci. A* **2015**, *16*, 427–442. [CrossRef]
2. Simic, M.; Herakovic, N. Reduction of the flow forces in a small hydraulic seat valve as alternative approach to improve the valve characteristics. *Energy Convers. Manag.* **2015**, *89*, 708–718. [CrossRef]
3. Richer, E.; Hurmuzlu, Y. A High Performance Pneumatic Force Actuator System: Part 1—Nonlinear Mathematical Model. *J. Dyn. Syst. Meas. Control* **2000**, *122*, 416–425. [CrossRef]
4. Shi, Y.; Wang, Y.; Cai, M.; Zhang, B.; Zhu, J. Study on the Aviation Oxygen Supply System Based on a Mechanical Ventilation Model. *Chin. J. Aeronaut.* **2018**, *31*, 197–204. [CrossRef]

5.  Ren, S.; Cai, M.L.; Shi, Y.; Xu, W.Q.; Zhang, X.D. Influence of bronchial diameter change on the airflow dynamics based on a pressure-controlled ventilation system. *Int. J. Numer. Methods Biomed. Eng.* **2018**, *34*, e2929. [CrossRef] [PubMed]

6.  Taghizadeh, M.; Ghaffari, A.; Najafi, F. Modeling and identification of a solenoid valve for PWM control applications. *C. R. Mac.* **2009**, *337*, 131–140. [CrossRef]

7.  Heikkil, M.; Linjama, M. Displacement control of a mobile crane using a digital hydraulic power management system. *Mechatronics* **2013**, *23*, 452–461. [CrossRef]

8.  Topcu, E.E.; Ibrahim, Y.; Kamis, Z. Development of electro-pneumatic fast switching valve and investigation of its characteristics. *Mechatronics* **2006**, *16*, 365–378. [CrossRef]

9.  Shi, Y.; Zhang, B.L.; Cai, M.L.; Xu, W.Q. Coupling Effect of double lungs on a VCV ventilator with automatic secretion clearance function. *IEEE/ACM Trans. Comput. Biol. Bioinform.* **2017**, *99*, 1. [CrossRef] [PubMed]

10. Kamelreiter, M.; Kemmetmüller, W.; Kugi, A. Digitally controlled electrorheological valves and their application in vehicle dampers. *Mechatronics* **2012**, *22*, 629–638. [CrossRef]

11. Lee, G.S.; Sung, H.J.; Kim, H.C. Multiphysics analysis of a linear control solenoid valve. *J. Fluid Eng.* **2013**, *135*, 011104. [CrossRef]

12. Ahn, K.; Yokota, S. Intelligent switching control of pneumatic actuator using on/off solenoid valves. *Mechatronics* **2005**, *15*, 683–702. [CrossRef]

13. Wang, S.; Zhang, B.; Zhong, Q. Study on control performance of pilot high-speed switching valve. *Adv. Mech. Eng.* **2017**, *9*, 1–8. [CrossRef]

14. Li, B.R.; Gao, L.L.; Yang, G. Modeling and control of a novel high-pressure pneumatic servo valve direct-driven by voice coil motor. *J. Dyn. Syst.* **2013**, *135*, 014507.

15. Li, B.R.; Gao, L.L.; Yang, G. Evaluation and compensation of steady gas flow force on the high-pressure electro-pneumatic servo valve direct-driven by voice coil motor. *Energy Convers. Manag.* **2013**, *67*, 92–102. [CrossRef]

16. Guo, H.; Wang, D.Y.; Xu, J.Q. Research on a high-frequency response direct drive valve system based on voice coil motor. *IEEE Trans. Power Electron.* **2013**, *28*, 2483–2492. [CrossRef]

17. Wu, S.; Jiao, Z.X.; Yan, L.; Zhang, R.; Yu, J.T.; Chen, C.Y. A fault-tolerant triple-redundant voice coil motor for direct drive valves: Design, optimization, and experiment. *Chin. J. Aeronaut.* **2013**, *26*, 1071–1079. [CrossRef]

18. Wu, S.; Jiao, Z.X.; Yan, L.; Yu, J.T.; Chen, C.Y. Development of a direct-drive servo valve with high-frequency voice coil motor and advanced digital controller. *IEEE/ASME Trans. Mechatron.* **2014**, *19*, 932–942. [CrossRef]

19. Miyajima, T.; Fujita, T.; Sakaki, K. Development of a digital control system for high-performance pneumatic servo valve. *Precis. Eng.* **2007**, *31*, 156–161. [CrossRef]

20. Yao, J.; Zhao, L.; Deng, Q. High precision locating control system based on VCM for Talbot lithography. In Proceedings of the Eighth International Symposium on Advanced Optical Manufacturing and Testing Technology, Suzhou, China, 25 October 2016.

21. Chi, W.; Cao, D.; Wang, D. Design and experimental study of a VCM-based Stewart parallel mechanism used for active vibration isolation. *Energies* **2015**, *8*, 8001–8019. [CrossRef]

22. Shi, Y.; Zhang, B.L.; Cai, M.L.; Zhang, X.D. Numerical Simulation of volume-controlled mechanical ventilated respiratory system with two different lungs. *Int. J. Numer. Methods Biomed. Eng.* **2017**, *33*, e2852. [CrossRef] [PubMed]

23. Xu, Z.P.; Wang, X.Y. Development of a Novel High Pressure Electronic Pneumatic Pressure Reducing Valve. *J. Dyn. Syst.* **2011**, *133*, 011011.

24. Shi, Y.; Wang, Y.X.; Xu, W.Q.; Liang, H.W.; Cai, M.L. Power characteristics of a new kind of air-powered vehicle. *Int. J. Energy Res.* **2016**, *40*, 1112–1121. [CrossRef]

25. Shi, Y.; Wu, T.C.; Cai, M.L.; Wang, Y.X.; Xu, W.Q. Energy conversion characteristics of a hydropneumatic transformer in a sustainable-energy vehicle. *Appl. Energy* **2016**, *171*, 77–85. [CrossRef]

26. Niu, J.L.; Shi, Y.; Cai, M.L.; Cao, Z.X.; Wang, D.D.; Zhang, Z.Z.; Zhang, X.D. Detection of Sputum by Interpreting the Time-frequency Distribution of respiratory sound signal using image processing techniques. *Bioinformatics* **2018**, *34*, 820–827. [CrossRef] [PubMed]

*applied sciences*

*Article*

# Investigation on the Effects of Internal EGR by Variable Exhaust Valve Actuation with Post Injection on Auto-ignited Combustion and Emission Performance

Insu Cho [1], Yumin Lee [2] and Jinwook Lee [3,*]

[1]  Department of Mechanical Engineering, Graduate School, Soongsil University, Seoul 06978, Korea; chois@soongsil.ac.kr
[2]  Department of Mechanical Engineering, Undergraduate Course, Soongsil University, Seoul 06978, Korea; lym6350@naver.com
[3]  Department of Mechanical Engineering, Soongsil University, Seoul 06978, Korea
*  Correspondence: immanuel@ssu.ac.kr; Tel.: +82-2-820-0929

Received: 19 February 2018; Accepted: 9 April 2018; Published: 10 April 2018

**Abstract:** Variable valve mechanisms are usually applied to a gasoline combustion engine to improve its power performance by controlling the amount of intake air according to the operating load. These mechanisms offer one possibility of resolving the conflict of objectives between a further reduction of raw emissions and an improvement in fuel efficiency. In recent years, variable valve control systems have become extremely important in the diesel combustion engine. Importantly, it has been shown that there are several potential benefits of applying variable valve timing (VVT) to a compression ignition engine. Valve train variability could offer one option to achieve the reduction goals of engine-out emissions and fuel consumption. The aim of this study was to investigate the effects on part load combustion and emission performance of internal exhaust gas recirculation (EGR) by variable exhaust valve lift actuation using a cam-in-cam system, which is an electronically variable valve device with a variable inside cam retarded to about 30 degrees. Numerical simulation based on GT-POWER has been performed to predict the $NO_x$ reduction strategy at the part load operating point of 1200 rpm in a four-valve diesel engine. A GT-POWER model of a common-rail direct injection engine with internal EGR was built and verified with experimental data. As a result, large potential for reducing $NO_x$ emissions through the use of exhaust valve control has been identified. Namely, it is possible to utilize heat efficiently as recompression of retarded post injection with downscaled specification of the exhaust valve rather than the intake valve, even if the CIC V1 condition with a reduction of the exhaust valve has a higher internal EGR rate of about 2% compared to that of the CIC V2 condition.

**Keywords:** variable exhaust valve actuation; recompression; internal EGR; GT-POWER

## 1. Introduction

In diesel engines, a robust compression-ignited combustion strongly depends on the cylinder charge temperature, composition, and cylinder pressure during valve train events [1]. Variable valve actuation (VVA) technology refers to a technique or combination of technologies that changes valve timing, valve duration, and valve lift, etc., depending on operating conditions and operating strategies, instead of opening and closing intake and exhaust valves with a single, fixed-cam profile. Especially, the compression ignition engine shows the required characteristics of the variable valve system which can control the opening period of the valve. These VVA mechanisms have limited valve lift and timing control, depending on the shape of the mechanism and method in which it is operated.

Therefore, it is desirable to design the variable valve operating mechanism in which the ability to control the range of the valve lifting and opening timing is set during the development stage of the actual technology [2].

There are several potential benefits of applying variable valve control to a diesel engine: Optimization of the charge motion, a reduced in-cylinder temperature, and pumping work only if there is no swirl valve, leading to better low-speed performance. Namely, the adoption of the variable valve in a compression ignition engine enables control of the temperature of exhaust gas and effective control of the compression ratio through intake or exhaust air quantity.

Based on these functional characteristics, a variable valve system is expected to be essential as a core technology for improving the combustion of diesel engines. The output of the compression ignition engine is affected by the fuel injection system, volume efficiency, friction loss, and pumping loss, etc. Volumetric efficiency depends on the combustion chamber geometry, engine speed, and the timing and lift of valve opening and closing. Therefore, increasing the volumetric efficiency and reducing the loss caused by friction and pumping for the maximum output according to each operating condition improves the power performance. During this optimization process, the optimum intake and exhaust flow rate and timing are adjusted according to the operating conditions without lowering the torque stability and the maximum output [3]. For instance, early exhaust valve opening (EEVO) and cylinder deactivation are implemented to raise the exhaust gas temperatures for DOC activation [4].

In a variable valve strategy known to be effective in controlling the cylinder charge temperature and composition, various exhaust valve mechanisms have been designed to control exhaust gas recirculation (EGR). The term EGR usually refers to a deliberate, external process, but there is also a level of internal EGR. This occurs when the residual combustion gas in the cylinder at the end of the exhaust stroke is mixed with the incoming charge. There is, therefore, a proportion of internal EGR which must be taken into account when planning EGR strategies. Scavenging efficiency will vary with engine load, and in an engine fitted with variable valve timing, a further parameter must be considered.

Mirko B. et al. found that internal EGR has a greater potential than external EGR to realize an accurate control of the EGR rate and $NO_x$ emission in transient operation [5]. Also, it was found that applying internal EGR moderated combustion by delaying the ignition timing, thereby suppressing in-cylinder pressure oscillations [6]. VVA for automotive diesel engines enables the pre-heating of the exhaust system to be accelerated by the effect of internal EGR [7].

Gonzalez D. et al. presented a comprehensive scheme to obtain higher exhaust gas temperatures. Internal EGR can be combined with multiple injections after the main injection event, thereby altering the heat release rate and the exothermic reactions in the exhaust stroke [8].

Furthermore, the work of Guan, W. demonstrated that the combination of the EGR and Miller cycle strategies can lead to minimum impact on the smoke emission and fuel economy, while achieving lower engine-out $NO_x$ emissions and a higher exhaust gas temperature for an efficient exhaust after treatment systems [9].

Kai D. et al. tested several VVT variants (exhaust cam phasing, late intake valve opening, Miller, second exhaust event, cylinder deactivation) as promising heating strategies. These heating strategies are finally compared to conventional heating measures to demonstrate their potentials in terms of faster after-treatment light-off with increased conversion efficiencies and benefits in $CO_2$ emissions [10].

In this study, exhaust recompression valve strategies are discussed in light of GT-POWER, 1D cycle simulation model results for a common-rail direct injection single-cylinder engine with internal EGR by using VVA and post fuel injection having an effect on the results [11]. So, the aim of this study is to improve part load fuel economy and emissions.

## 2. Numerical Model Description

### 2.1. Engine Modeling

In this study, the GT-POWER program (Gamma Technologies, Westmont, IL, USA) was used to predict the reduction of $NO_x$ at partial engine load under the engine speed of 1200 rpm, as shown in Table 1. Cam-in-cam, the variable valve scheme used in this study, is an electronically variable valve device, and the inside of the variable cam is retarded to about 30 degrees. An extensive GT-POWER model of the base R-engine with internal EGR created by variable exhaust valve actuation was built in this study.

**Table 1.** Operating points on map in R-engine.

| Low Load | Low, Middle Load |
|---|---|
| Fuel Injection Quantity: 12 mg | Fuel Injection Quantity: 12 & 19.8 mg |
| **Idle + Low Speed** | **Middle Speed** |
| Engine speed: 800 rpm | Engine speed: 1200~2000 rpm |

Among the five points on the operation map of the R-engine, as shown in Figure 1a, the standard value (hereinafter referred to as "Base" case) is selected when the engine of the basic cam follows the injection strategy according to the operating map. Figure 1b shows the prototype cam-in-cam system used in this study and variable valve profile.

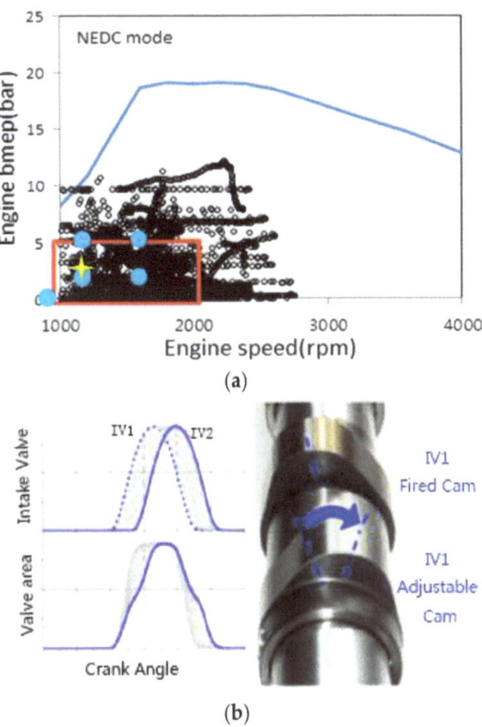


**Figure 1.** Engine map and variable valve device. (**a**) Five operating points on map in R-engine; (**b**) Prototype cam-in-cam system (MAHLE).

Commercial R-engine has 4-cylinder and 4-valve per cylinder. Cylinder pressure was measured at 1-cylinder only for combustion analysis. Engine crank shaft connected with a DC dynamometer and high-pressure fuel injected in the cylinder by signal of injection device. Table 2 shows specification of R-engine used in this study.

**Table 2.** Specification of R-engine used in this study.

| Item | Unit | Specification | |
|---|---|---|---|
| Bore x Stroke | mm | $88 \times 90$ | |
| Displacement | cc | 499 | |
| Valve per cylinder | - | 4 (2 intake and 2 exhaust) | |
| Compression ration | - | 16 | |
| Intake valve | CAD | IVO | bTDC 10 |
| | | IVC | aBDC 28 |
| Exhaust valve | CAD | EVO | bBDC 54 |
| | | EVC | aTDC 4 |

The GT-POWER program is a commercial engine analysis program widely used as an industry standard for vehicle and powertrain simulation software. It is a next-generation integrated CAE application that is becoming increasingly important in the industry due to its functions for engine evaluation, such as cooling system operations, as well as fuel consumption, noise, etc. under an arbitrary driving condition. It provides a fast solver for 1D and 3D simulations, making realistic simulations of large-scale systems possible. It also has the advantage of containing detailed information on valves, cylinders, and combustion behavior (predictive or data-based functions) for engine combustion calculations. The numerical full-circuit model developed by this study consists of the engine part (single cylinder, valves, pipes, etc.), valve train part (cam, camshaft, valve rod, etc.), and fuel injection part (piezo actuator, nozzle, rod, needle, orifice, etc.). The valve part in the engine system has received a valve profile from rotating the cam by the valve train model linked with the engine model. The injector model is built up with internal construction of a third piezo injector and linked with an engine model. The engine combustion model uses DIPULSE because $NO_x$ prediction is possible with improved accuracy. The heat transfer model uses WoschniGT.

The fuel injection model was designed with a third piezo injector and verified through comparing experimental injection data obtained by the Bosch-tube method. The Bosch-tube measurement principle is to feed the fuel into the pipe with a cross-sectional area A and select the fuel control volume in the pipe flowing at the speed of sound as c when the fuel in the pipe moves at a speed of u. By knowing the sonic velocity and fuel density in the pipe and determining the amount of change in pressure in the chamber, the injection rate can be calculated [12].

### 2.2. Four Variable Valve Train Configurations

Four different variants, as shown in Table 3, stood out as candidates for simulation testing: (1) "Base" case, for comparing with other cases; (2) CIC V1 for a recompression strategy that reduced valve lifting and duration; (3) CIC V2 for recompression that reduced valve lifting and duration about the exhaust valve; and (4) CIC V3 for added post fuel injection based on CIC V2, as shown in Table 3. With the exhaust recompression valve strategy, the cylinder charge temperature is controlled by trapping the hot exhaust gas from the previous engine cycle. This is done by closing the exhaust valve early during the exhaust stroke while opening the intake valve late in relation to the exhaust valve closing timing. The crank-angle period when both intake and exhaust valves are closed around the engine top dead center (TDC) is defined as the negative valve overlap (NVO).

**Table 3.** Four variable valve train configurations.

| Case | Valve Profile | Description |
| --- | --- | --- |
| **Base** | | All valve lift 8 mm |
| **CIC V1** | | Recompression<br>+ All valve lifting 6 mm<br>+ Negative valve overlap |
| **CIC V2** | | Recompression<br>+ Exhaust valve lifting 6 mm<br>+ Intake valve lifting 8 mm<br>+ Negative valve overlap |
| **CIC V3** | | CIC V2<br>+ Post injection<br>(Three fuel injection timing: 70°, 200°, 320°) |

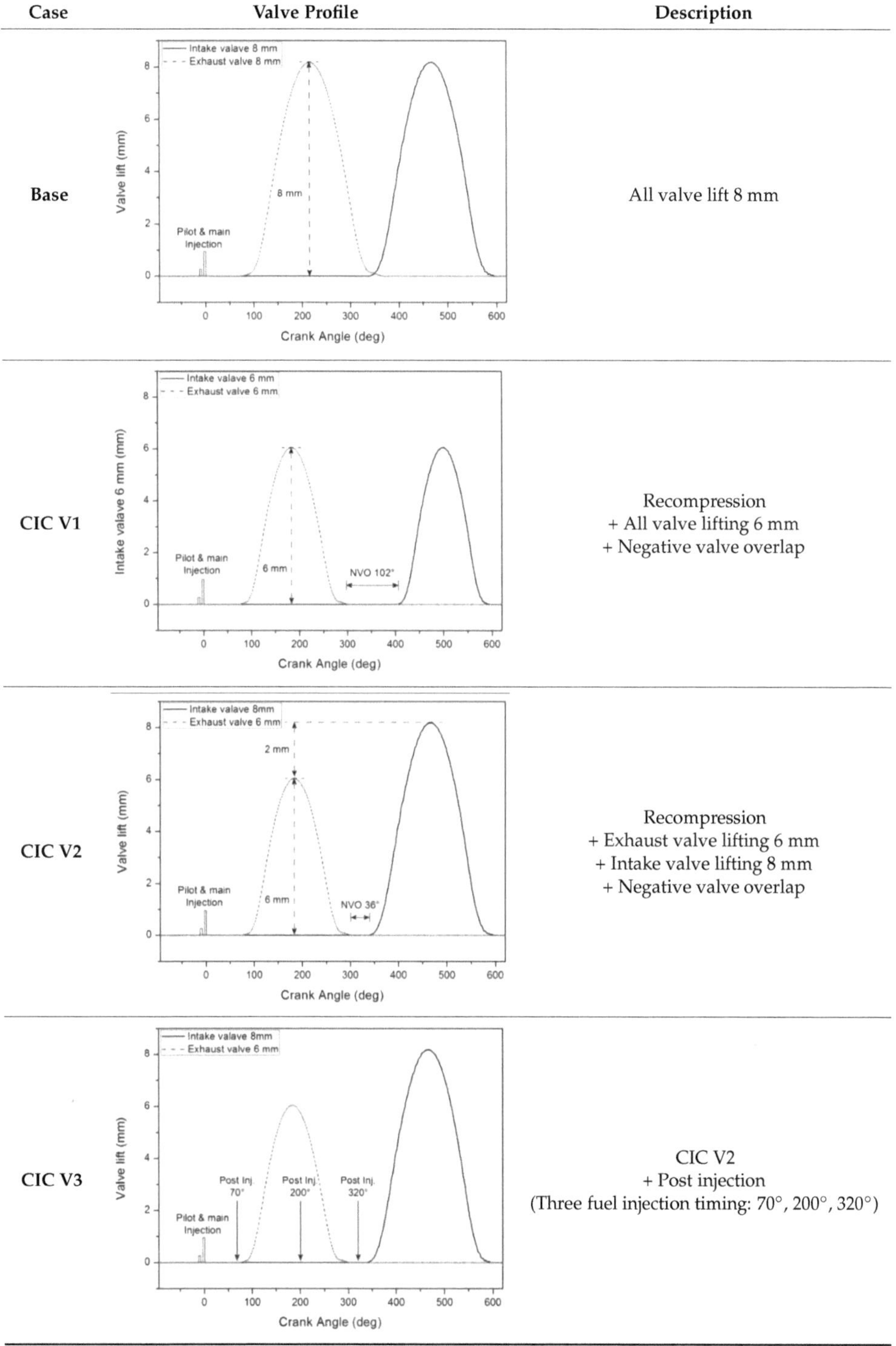

Figure 2 shows the schematic diagram of this analysis process. In order to study the effects of internal EGR by variable exhaust valve actuation, a methodology was first developed to investigate recompression internal EGR, wherein a variable exhaust valve timing mechanism is used to transfer exhaust gas from one cycle to the next, just as in the valve strategies referred to as "Base" case, CIC V1, and CIC V2. The additional step in the process begins with CIC V2 as normal and then adds three different timings of post fuel injection.

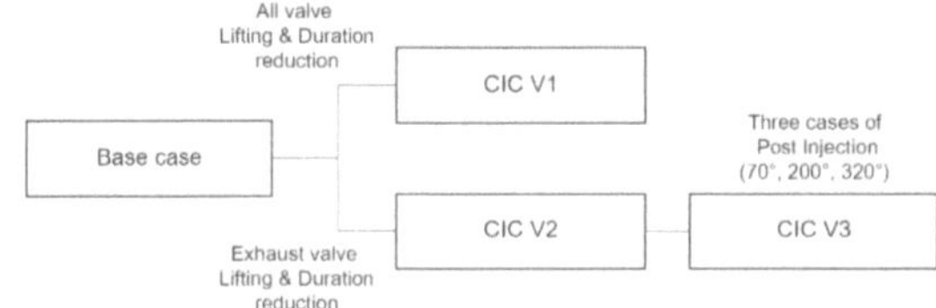


**Figure 2.** Schematic diagram of this numerical analysis process.

*2.3. Model Verification*

In order to improve the model's predictive accuracy, the engine model was developed in the 1D GT-POWER and compared with the combustion pressure and heat release of in-cylinder gas. Tables 4 and 5 show the engine operating condition in the experiment and fuel injection condition, respectively. Figure 3 shows the result of the model correlation under the test condition in Table 4. Model validation took place once the engine dynamometer data were available. It was shown that the experimental results indeed agreed with the combustion analysis.

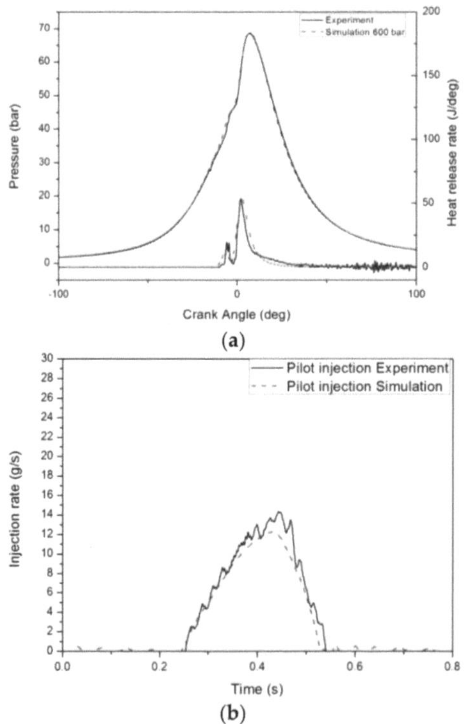


**Figure 3.** *Cont.*

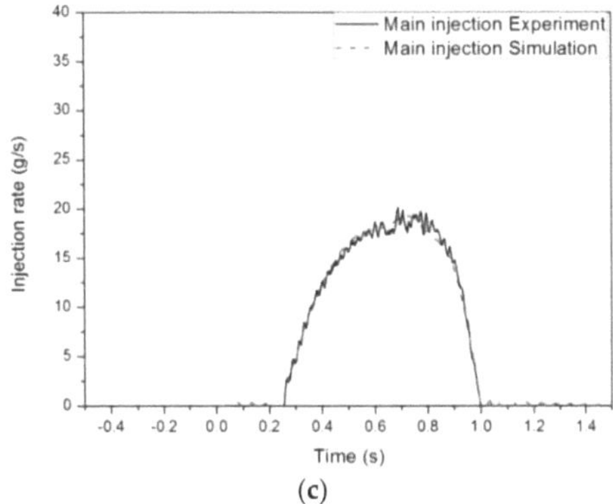


**Figure 3.** Predictive model correlation with experiment data. (**a**) Pressure and Heat Release Rate; (**b**) Pilot injection rate; (**c**) Main injection rate.

**Table 4.** Engine operating condition.

| Item | Specification |
|---|---|
| Engine speed (rpm) | 1200 |
| Fuel injection quantity (mg) | 12 |
| Fuel injection pressure (bar) | 600 |
| Intake & Exhaust valve lift (mm) | 8 |

**Table 5.** Fuel injection condition.

| Injection | Pilot | Main |
|---|---|---|
| Fuel injection timing (deg) | 15 ATDC | 5 ATDC |
| Fuel injection quantity (mg) | 1.4 | 11.1 |

## 3. Results and Discussion

### 3.1. Combustion Characteristic of Recompression Internal EGR

Figure 4 shows the temperature traces for three different recompression strategies: "Base" case, CIC V1, and CIC V2. In the recompression cases (CIC V1, CIC V2), the engine operates at a higher temperature than the "Base" case. The result is that there is a limited flow of fresh air and burned gas due to negative overlap. CIC V1 is less effective in enhancing the cylinder temperature than CIC V2, due to its relative larger intake valve lift and duration than in the exhaust valve. The maximum temperature in the case of CIC V2 is higher during the main combustion period.

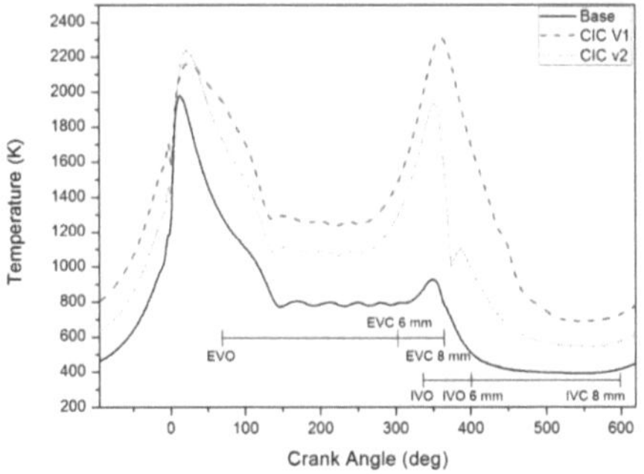


**Figure 4.** Temperature variation for different recompression strategies.

Figure 5 shows the intake flow variation for three different recompression strategies. In the case of CIC V2, there was a rapid back flow of burned gas through the intake valve. This comes from an increased flow rate through the intake valve above that of CIC V1. Additionally, the burned gas discharged to the intake valve, and the temperature of the in-cylinder gas dropped. At a crank angle of 470°, hereafter fresh air is flowing into the cylinder and then the temperature of the trapped gas is increased. Therefore, it can be concluded that back flow in the valves is one of the main factors creating combustion performance.

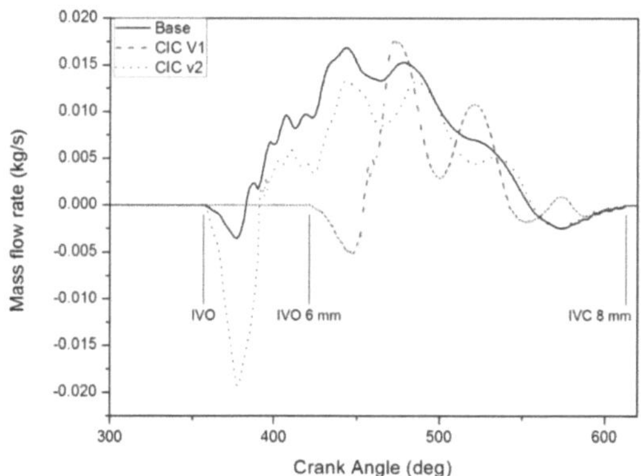


**Figure 5.** Flow traces at intake valve for different recompression strategies.

Figure 6 shows the total trapped mass traces for different recompression strategies: "Base" case, CIC V1, and CIC V2. Total trapped mass can be defined as the quantity of all chemical species (including air, fuel, EGR, residual gas, etc.) in the cylinder at the start of the cycle. In the recompression condition, the total trapped mass is reduced and the amount of the trapped mass in the case of CIC V2 is relatively more elevated before combustion, due to the increase in the intake valve lift and duration. In the case of CIC V2, when the intake valve is open at 350°, the mass quantity drops rapidly, and the negative valve overlap creates the high temperature and causes burned gas to only flow back through the intake valve at TDC.

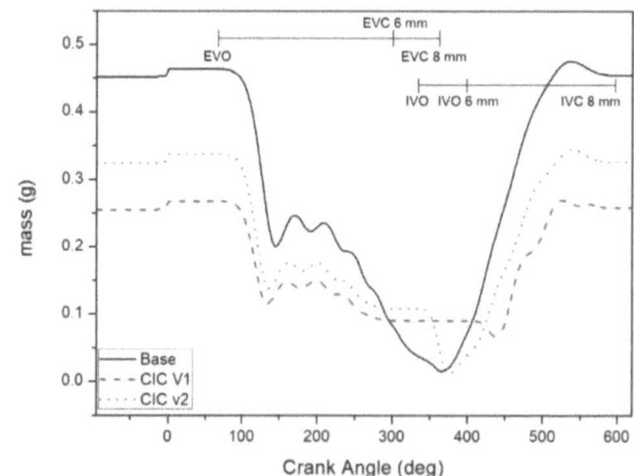


**Figure 6.** Total trapped mass traces for three different recompression strategies.

As shown in Figure 7, the increase in in-cylinder pressure during the negative valve overlap indicates its oxidation by the recompression of trapped gas. However, the in-cylinder temperature and combustion pressure in CIC V1 and CIC V2 are lower than in the "Base" case, due to a decrease in fresh air at the next ignition. This result showed that the exchange of gas was more enhanced in the case of CIC V2 because of the decreased combustion pressure in the period and peak value than the CIC V1 case during the intake stroke.

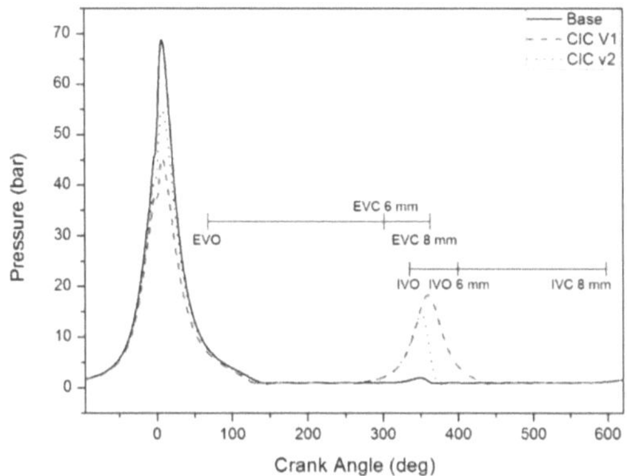

**Figure 7.** Pressure variation for three different recompression strategies.

Figure 8 shows the pressure–volume (PV) cycle for three different recompression cases. The main feature of this diagram is that the amount of energy expended or received by the system as work can be estimated as the area under the curve in the figure. In the cases of CIC V1, its pumping loss has no area on the graph due to the very low volume efficiency during the intake stroke. It indicates a higher peak pressure than the case of CIC V2. As can be seen, increased inflow of fresh air to the cylinder of CIC V2 causes pumping loss, but the maximum pumping pressure is lower than in the case of CIC V1.

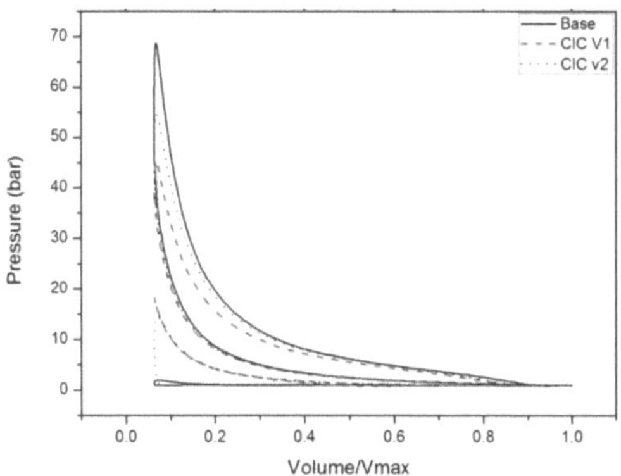

**Figure 8.** P-V diagram for three different recompression strategies.

Figure 9 shows that a decreased pressure and inflow of fresh air affects the increase in the ignition delay at the pilot injection of high-pressure fuel. Then, a decreased heat release rate and expanded diffusion combustion at main injection occur as changes in the cases of CIC V1 and V2, as characteristics of low temperature combustion.

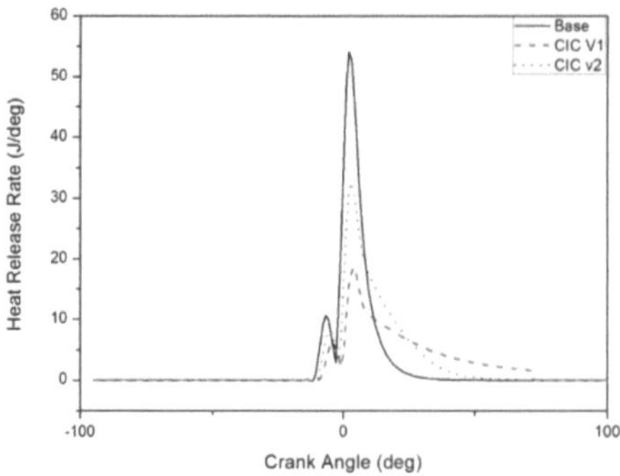

**Figure 9.** Heat Release Rate traces for three different recompression strategies.

### 3.2. Effect of Post Fuel Injection in Recompression Internal EGR

By adjusting the internal EGR applied by variable exhaust valve actuation, a great volume of heat energy occurred and was generally wasted, but it can be used to achieve lower emissions, activation of the after-treatment system, and stable operation of low temperature combustion. This study investigated the possibility of VVA that involved post fuel injection for improving thermal efficiency in the three cases. The points of post injection started (1) at 70° containing a maximum quantity according to the total trapped mass; (2) at 200°, with both the lowest pressure and temperature and the ending of the back flow through the exhaust valve during the recompression process; and (3) at 320°, with the raising of pressure and temperature by closed intake and the exhaust valve during the recompressing process. Furthermore, the quantity of post fuel injection was set to 20% of the total fuel mass in both the pilot and main injection periods. As a reference, the specific cutoff point for which the second injection is small enough relative to the main injection to constitute a post injection, rather than a generic split injection, is not well defined, but a maximum of approximately 20% of the total fuel in the post injection is consistent with most existing self-described post-injection studies [13]. Figure 10 is the result of the temperature traces for different recompression strategies with post injection. Post injection at 70° in the exhaust stroke caused an increase in temperature. In the case of post injection at 320°, it was decremented that a high temperature occurred by potential heat, which is the decreasing of temperature during NVO, which then affects gas exchange efficiently. Additionally, post injection at 320° was performed under conditions that included a lower burning mass than in other cases, so it showed the highest effect by potential heat. Figure 11 shows the result of the total trapped mass traces for different recompression strategies with post injection. In the case of post injection at 320°, it shows a higher variation in trapped mass, due to its internal gas flow activity with a high temperature.

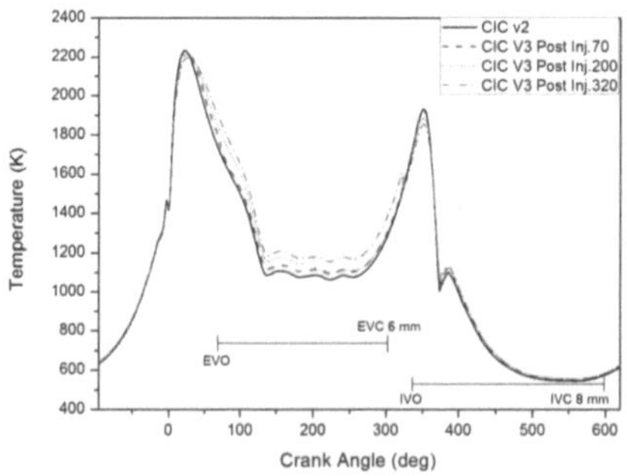

**Figure 10.** Temperature variation for different recompression strategies with three post injections.

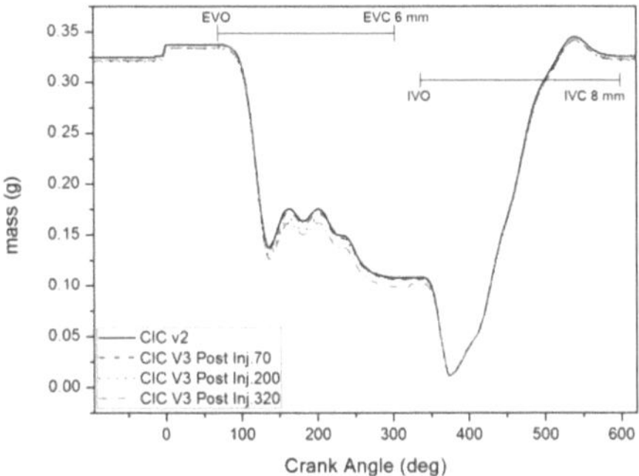

**Figure 11.** Total trapped mass traces for different recompression strategies with three post injections.

Figure 12 shows the result of the heat release rate traces for different recompression strategies with post injection. It shows the highest heat release at the pilot and main injection points in the case of post injection at 320°. This is caused by increasing the burnt fuel with post injection by decreasing the total trapped mass. In addition, it can be concluded that post injection is one of the main factors influencing thermal efficiency increased by a diffusion combustion area.

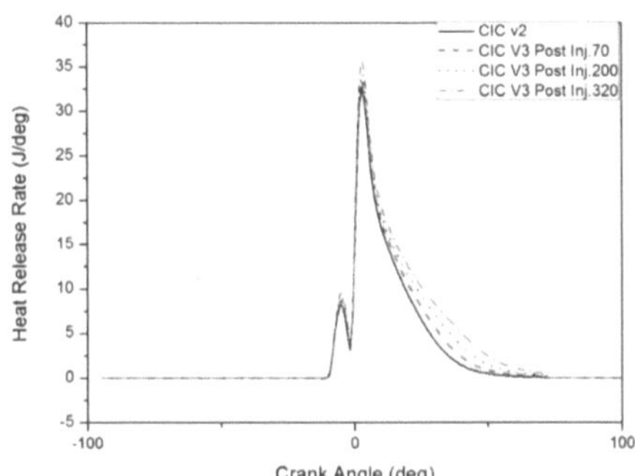

**Figure 12.** Heat release rate traces for different recompression strategies with three post injections.

Figure 13 shows the result of in-cylinder pressure variation for different recompression strategies with post injection. In the case of post injection at 320°, it shows the lowest pressure that results from reducing the pressure rate at the point of main combustion by increasing the gas exchange.

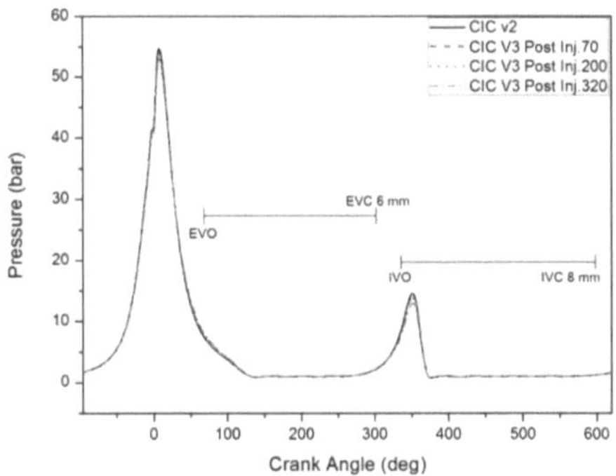

**Figure 13.** Pressure variation for different recompression strategies with three post injections.

### 3.3. Comparison of Emission Performance

Figure 14 shows comparisons of $NO_x$ emission for different variable valve control strategies. It was found that the CIC V1 case, with no-full lift and duration of intake and exhaust valve, shows the lowest $NO_x$ emission with minimum temperature, due to using recompression internal EGR. If it is difficult to control both valves, as shown in Figure 14, the CIC V3 case, with no-full lift and duration of exhaust valve only and post injection, is an effective strategy to reduce $NO_x$ emission at part load operating conditions in a compression ignition combustion. In the case of post injection at 320° in CIC V3, it shows the lowest $NO_x$ emission under variable exhaust valve control without a variable intake valve. As shown in Figure 15, the lowest volume efficiency occurred in the case of CIC V1 with full control of both the intake and exhaust valve. Whereas, it was found that CIC V2 and CIC V3 have a similar volume efficiency at each maximum combustion temperature.

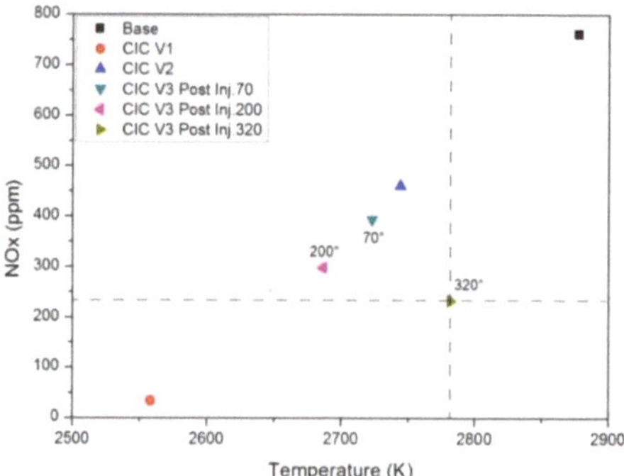


**Figure 14.** Comparison on $NO_x$ emission for different variable valve control strategies.

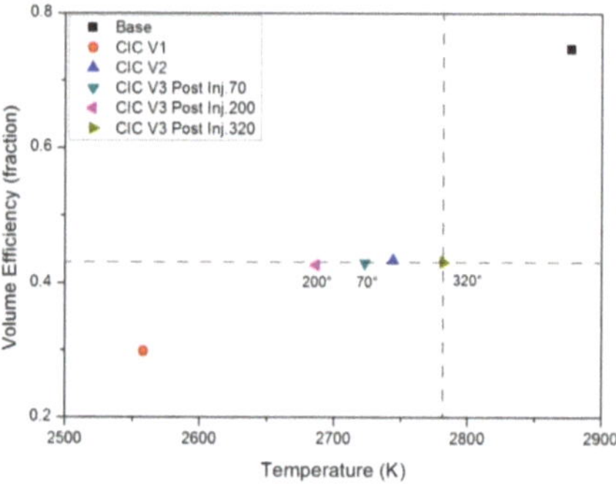


**Figure 15.** Comparison on volume efficiency for different variable valve control strategies.

Figure 16 shows the comparison results for $NO_x$ emission versus indicated mean effective pressure (IMEP) for all variable valve control strategies applied in this study. This shows a noticeable reduction in $NO_x$ emission with a reduction in IMEP. Especially, CIC V1 is a moderate case with regard to emission reduction, but this strategy was worse with regard to thermal energy. As the post injection timing in CIC V3 was retarded, it caused a decrease of $NO_x$ and an increase in IMEP linearly. The result was derived from effective work due to increased burnt fuel through post fuel injection. Figure 17 shows the comparison result for $NO_x$ emission versus indicated specific fuel consumption (ISFC) for all variable valve control strategies. This shows that the lowest fuel consumption with $NO_x$ emission was in the case of CIC V1. Furthermore, the post injection at 320° in CIC V3 is the best case with regard to the level of $NO_x$ emission under fuel penalty. Figure 18 shows the comparison of the internal EGR rate for different variable valve control strategies. The internal EGR rate is defined as a ratio between the total volume of the cylinder and the total trapped mass, with the exception of trapped air, before the beginning of combustion. It shows a difference of about 10% between the "Base" case and the recompression case. This EGR strategy is made possible by the design of valve control (lift, phase, timing, etc.), which allows for more burned gas to be trapped, without the use of any additional device for external EGR. Figure 19 shows the comparison of $NO_x$ reduction by applying different variable valve control strategies. Recompression strategy, as a bypass process to avoid the $NO_x$ emission zone in compression ignition, is beneficial to the heat release rate and volumetric efficiency at the design speed, and it also reduces $NO_x$ emission.

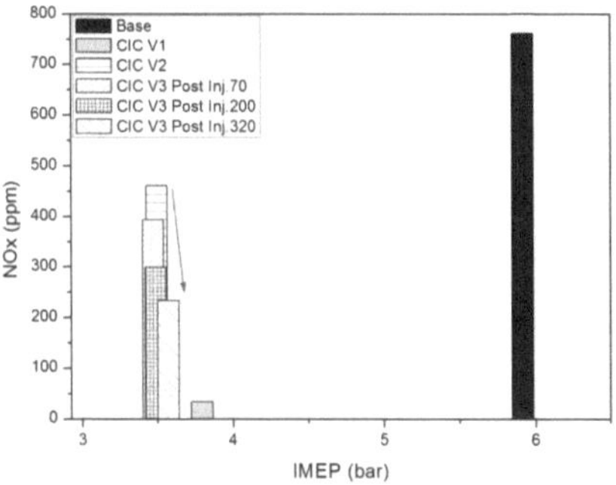

**Figure 16.** Comparison on IMEP versus $NO_x$ for different variable valve control strategies.

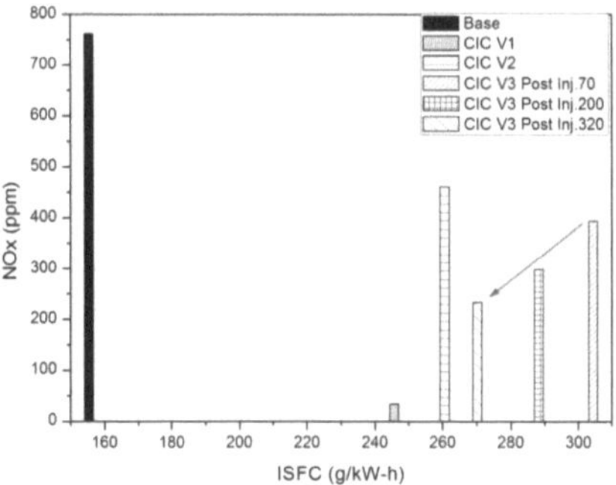

**Figure 17.** Comparison on ISFC versus $NO_x$ for different variable valve control strategies.

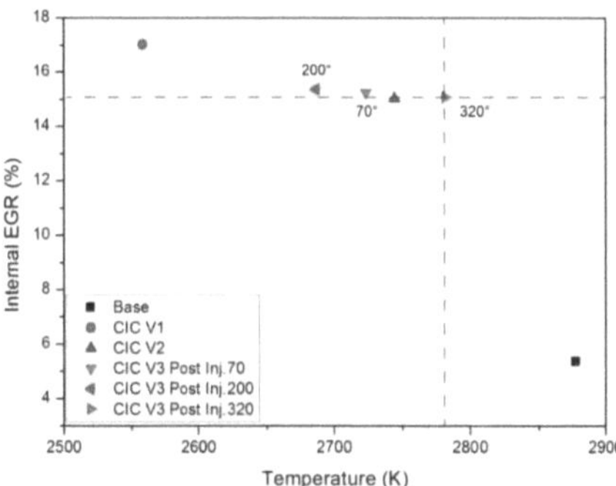

**Figure 18.** Comparison of internal EGR rate for different variable valve control strategies.

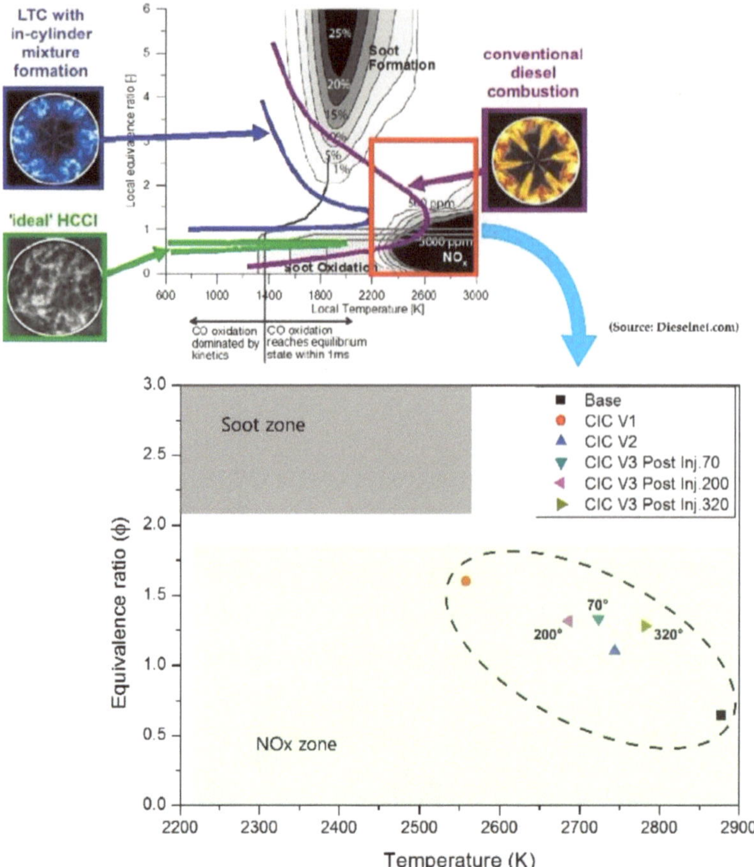



**Figure 19.** Comparison on $NO_x$ reduction by applying different variable valve control strategies.

## 4. Conclusions

In this study, the effect of internal EGR by variable valve actuation with post injection using a full-circuit GT-POWER model verified by experimental data was investigated to predict auto-ignited combustion performance that includes in-cylinder pressure and temperature, heat release, ISFC, IMEP, and $NO_x$ emission. Conclusions obtained by this study can be summarized as follows:

(1) Recompression internal EGR leads to negative valve overlap effects and a higher temperature in engine operation, due to its trapping of burnt gas and limited air flow, creating a high temperature inside the cylinder. The earlier the exhaust valve closes during the exhaust stroke, as in the case of a larger NVO, the greater the amount of the hot exhaust gas from the previous engine cycle would be trapped in the cylinder. This leaves less room in the cylinder for the in-coming fresh air mass, thereby increasing the cylinder charge temperature while decreasing the oxygen level in the cylinder charge. As a result, there is an optimal setting of NVO for each operating condition to achieve the best engine performance.

(2) The CIC V2 case is used for enhancing the internal EGR rate, which is reduced in both the lifting and duration of the exhaust valve without control of the intake valve, thus the operating temperature is decreased overall, but a higher temperature is seen at the main combustion period than in the "Base" and CIC V1 cases. This is caused by the inflow of more fresh air while keeping

internal gas in the cylinder and increasing the discharging of burnt gas through the intake valve. Therefore, effective work has risen due to a pressure drop in the intake stroke and the rising of pressure at the main combustion, compared with the CIC V1 case.

(3) For the use of thermal energy wasted when adjusting recompression by variable valve actuation, post injection with three cases was applied with total injected fuel of 20% based on CIC V2. Each post injection shows a similar volume efficiency, but retarded post injection timing caused an increase in IMEP, as well as a reduction of both ISFC and $NO_x$. In the case of post injection at 320°, when increasing the pressure and temperature for both intake and exhaust valves, which were closed in the recompression period, the highest engine operating temperature was seen, but there were lower emissions of $NO_x$ than in other post injection cases.

(4) The recompression internal EGR is effective in utilizing heat energy by enhancing the EGR rate by about 10%. Additionally, recompression with retarded post injection is beneficial to IMEP with $NO_x$ reduction, even if CIC V1 with no-full lift and duration of the intake and exhaust valve has an internal EGR rate of about 2% higher than that of CIC V2 with no-full lift and duration of the exhaust valve.

(5) Finally, the internal EGR scheme with variable exhaust valve actuation in a diesel engine is advantageous to apply post injection from the viewpoint of utilizing thermal energy. However, since the ISFC increases according to post-injection conditions, it is necessary to consider the maximum use of exhaust heat energy for $NO_x$ reduction.

**Acknowledgments:** This research was supported by the R&D project on Industrial Core Technology (2017) of MOTIE (Ministry of Trade, Industry and Energy) in Republic of Korea.

**Author Contributions:** Insu Cho, Yumin Lee and Jinwook Lee conceived and designed the numerical analysis; Insu Cho and Yumin Lee produced analytic model and performed the simulation; Insu Cho and Jinwook Lee analyzed the data and wrote the paper; Jinwook Lee supervised and advised all parts in this paper.



## References

1. Johnson, T.V. Review of Vehicular Emissions Trends. *SAE Int. J. Engines* **2015**, *8*, 1152–1167. [CrossRef]
2. Tran, T.H.T.; Enomoto, H.; Nishioka, K.; Simizu, K. *Development of Small Gasoline Engine with Electronic Variable Valve Timing Unit*; SAE Technical Paper 2011-32-0579; SAE International: Warrendale, PA, USA, 2011.
3. Johnson, T.; Joshi, A. *Review of Vehicle Engine Efficiency and Emissions*; SAE Technical Paper 2017-01-0907; SAE International: Warrendale, PA, USA, 2017.
4. Bharath, A.N.; Yang, Y.; Reitz, R.D.; Rutland, C. *Comparison of Variable Valve Actuation, Cylinder Deactivation and Injection Strategies for Low-Load RCCI Operation of a Light Duty Engine*; SAE International 2015-01-0843; SAE International: Warrendale, PA, USA, 2015.
5. Baratta, M.; Finesso, R.; Misul, D.; Spessa, E. *Comparison between Internal and External EGR Performance on a Heavy Duty Diesel Engine by Means of a Refined 1D Fluid-Dynamic Engine Model*; SAE International 2015-24-2389; SAE International: Warrendale, PA, USA, 2015.
6. Shimada, T.; Yoshida, Y.; Rin, C.; Yamada, M.; Ito, N.; Iijima, A.; Yoshida, K.; Shoji, H. Influence of Internal EGR on Knocking in an HCCI Engine. In Proceedings of the JSAE/SAE 2015 Small Engine Technologies Conference, Osaka, Japan, 17–19 November 2015.
7. Piano, A.; Millo, F.; Di Nunno, D.; Gallone, A. *Numerical Analysis on the Potential of Different Variable Valve Actuation Strategies on a Light Duty Diesel Engine for Improving Exhaust System Warm Up*; SAE Technical Paper 2017-24-0024; SAE International: Warrendale, PA, USA, 2017.
8. Gonzalez, D.M.; Di Nunno, D. *Internal Exhaust Gas Recirculation for Efficiency and Emissions in a 4-Cylinder Diesel Engine*; SAE Technical Paper 2016-01-2184; SAE International: Warrendale, PA, USA, 2016.
9. Guan, W.; Pedrozo, V.; Zhao, H.; Ban, Z.; Lin, T. *Investigation of EGR and Miller Cycle for $NO_x$ Emissions and Exhaust Temperature Control of a Heavy-Duty Diesel Engine*; SAE Technical Paper 2017-01-2227; SAE International: Warrendale, PA, USA, 2017.

10. Deppenkemper, K.; Özyalcin, C.; Ehrly, M.; Pischinger, S. 1D engine simulation approach for optimizing engine and exhaust aftertreatment thermal management for passenger car diesel engines by means of variable valve train (VVT) applications. In Proceedings of the 2018 SAE World Congress Experience Conference, Detroit, MI, USA, 10–12 April 2018.

11. Martin, J.; Sun, C.; Boehman, A.; O'Connor, J. *Experimental Study of Post Injection Scheduling for Soot Reduction in a Light-Duty Turbodiesel Engine*; SAE Technical Paper 2016-01-0726; SAE International: Warrendale, PA, USA, 2016.

12. Han, S.; Kim, J.; Lee, J. A Study on the Optimal Actuation Structure Design of a Direct Needle-Driven Piezo Injector for a CRDi Engine. *Appl. Sci.* **2017**, *7*, 320. [CrossRef]

13. O'Connor, J.; Musculus, M. *Post Injections for Soot Reduction in Diesel Engines: A Review of Current Understanding*; 2013 SAE International 2013-01-0917; SAE International: Warrendale, PA, USA, 2013.

*Article*

# Novel Automatic Idle Speed Control System with Hydraulic Accumulator and Control Strategy for Construction Machinery

Haoling Ren, Tianliang Lin *, Shengyan Zhou, Weiping Huang and Cheng Miao

College of Mechanical Engineering and Automation, Huaqiao University, Xiamen 361021, China; happyrhlly@126.com (H.R.); 1611303048@hqu.edu.cn (S.Z.); 1400203036@hqu.edu.cn (W.H.); 1400403007@hqu.edu.cn (C.M.)
* Correspondence: ltlkxl@163.com

Received: 23 January 2018; Accepted: 19 March 2018; Published: 26 March 2018

**Abstract:** To reduce the energy consumption and emissions of the hydraulic excavator, a two-level idle speed control system with a hydraulic accumulator for the construction machinery is proposed to reduce the energy consumption and improve the control performance of the actuator when the idle mode is cancelled. The structure and working principle are analyzed. The hydraulic accumulator (HA) is used to store the energy, which can provide backup pressured fluid when the idle mode is cancelled. Then, a method of how to set the pressure differential between the hydraulic accumulator and the load is proposed and the control law is discussed. The test rig is built. The experimental result shows that the idle speed can be switched among the first idle speed, the second idle speed and the normal speed automatically. Though the idle speed in the novel system can be reduced more than that in the conventional automatic idle speed control system (AISCS), the proposed system can still build the actuator pressure more quickly when the idle mode is cancelled. When compared to the system without the idle speed control, the energy saving of the proposed system is about 67%. The proposed two-level idle speed control system with a HA can achieve a high energy efficiency and a good control performance.

**Keywords:** construction machinery; hydraulic excavator; energy saving; idle speed control; hydraulic accumulator; control strategy

---

## 1. Introduction

Energy saving and environment friendly become the primary demands for the construction machinery under the environment of stringent emission regulations. Currently, there are several energy saving methods for construction machinery, for example, positive and negative flow control system [1,2], load sensing control [3,4], hybrid power system [5–8], energy regeneration [9–12] and automatic idle speed control system (AISCS) [13], are proposed and utilized in some models. Whether it is positive and negative flow control system or load sensing control, the purpose of them is to balance the output of the pump and the requirement of the load to reduce the energy loss. They do reduce the throttle loss and improve the efficiency of the machine. In hybrid power systems, the engine works at an optimal power consumption area to reduce fuel consumption and emissions via more than one power train [6–8]. Because there is more than one power train, the structure is complex and the cost is high, which limits its wide application. However, the above energy saving technologies mainly concentrate on the working time of the construction machinery. In fact, construction machines consume about 30% of the total energy in the idle state. AISCS is used to reduce the fuel consumption and emissions in the idle time. Commonly, the traditional AISCS is utilized to adjust the engine speed switching between the first idle speed, sometimes there is second idle speed and the normal working speed.

A lot of researches have been reported in the field of AISCS of automobiles. He et al. investigated a CNG engine and reduced the idle speed from original 800 rpm to 700 rpm [14]. Li et al. set the desired idle speed to 611 rpm instead of the normal (production) idle speed of 740 rpm, and used the sliding mode control to AISCS [15]. Though a low idle speed could reduce the fuel consumption, the cost was to increase the risk of missing fire or even stalling. That is to say, the transition to and from idle speed should be smooth and well controlled [16]. Many research devoted themselves to solving the AISCS problems with different control strategies [17–21].

Though some useful results have been achieved on the AISCS of automobiles and the technology is relatively mature, only a few research on the AISCS have been developed for construction machinery. Due to the different working conditions, the AISCS used in automobiles cannot be adopted to construction machinery directly. Xiong et al. analyzed the working principle and the implementation method of the AISCS in a rotary drilling rig [22]. They tested the fuel consumption under different speed and found that the speed when the minimum fuel consumption achieved was the preset idle speed. Liu designed an AISCS according to the working conditions, speed sensing, and engine power matching. The adjusting time of the engine from idle speed to the rated speed was about 2 s [23]. Hao optimized the duty ratio of the PWM, minimum ramping time, and idle speed through adapting the adaptive control method and achieved a better energy saving effect [24]. Those researches were still based on the traditional engine and the idle speed cannot be too low to avoid the misfire. Meantime, the hydraulic system still utilized the traditional pumps system. Therefore, the adjusting time of the transition from the idle speed to the normal speed of the engine cannot be very short and the pumps cannot build the pressure quickly to drive the actuator when the idle mode is cancelled. This leads to the instability of the actuator movement and a slow response to the working signal. Hydraulic accumulator (HA) is widely used in the hydraulic systems because it can be used as an auxiliary power source to supply pressured oil in a short time. When the pump cannot supply enough oil to the actuator at the moment that the AISCS is cancelled, HA is a best choice to supply the oil that the actuator needed, despite its short supply time. Though the engine using the AISCS can consume less fuel and reduce emissions, there are still emissions, especially when the idle speed cannot be set too low. While the electric motor (EM) is true zero emissions and it has a high efficiency within a wider speed range. Therefore, if the EM and HA are used in the AISCS of the construction machinery, the above problems may be avoided. This paper is to verify the feasibility of this idea.

In this research, a prototype of a two-ton hydraulic excavator that is (HE) driven by the EM for experiment has been built. The key object of this research is the novel AISCS with the HA for a two-ton HE. The remainder of this paper is organized as follows: Section 2 gives the structure and working principle of the AISCS with HAs. The control strategy of the proposed AISCS is discussed in detail in Section 3. Then, the experiment results are analyzed in Section 4. Conclusions are presented in Section 5.

## 2. Structure and Working Principle of the Novel AISCS

Figures 1 and 2 show the configuration of the traditional and the proposed AISCS, respectively. The multi-way valves that are used in Figures 1 and 2 are controlled by the pilot pressure produced by the joystick and the valve stroke is proportional to the pilot pressure. The working principle of the traditional AISCS is when the joystick returns to the middle position, the multi-way valve works at the middle position too. The output of the pump flows to the tank through the multi-way valve. When the controller detects the time that the joystick stays at the middle position is larger than a set value, the engine speed drops to a low value to reduce the fuel consumption. When the joystick leaves the middle position, the multi-way valve moves to the corresponding position and the engine accelerates to the normal speed. There is a delay when the actuator starts to move and due to the low pressure in the working chamber of the actuator, the movement of the actuator is unstable. When compared to the traditional AISCS, the proposed AISCS has the following advantages:

1.  A HA is connected to the outlet of the pump via a solenoid directional valve 1. The HA is used to store energy during the first level idle mode, which can be serve as an auxiliary energy source to drive the actuator when the idle mode is cancelled.
2.  There is a pressure loading unit at the bypass way of the multi-way valve. When the multi-way valve is at middle working position and the solenoid directional valve 2 is powered off, it can separate the pump from the tank by the solenoid directional valve 2 and charge the HA.
3.  The EM has a faster response than the traditional engine and a high efficiency within a wider speed range. Thanks to the utilization of the EM, there is no pollution at all.

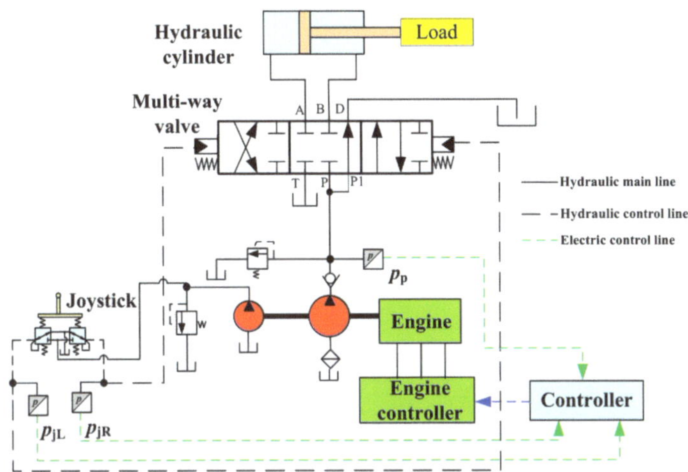


**Figure 1.** Schematic of the traditional automatic idle speed control system (AISCS).

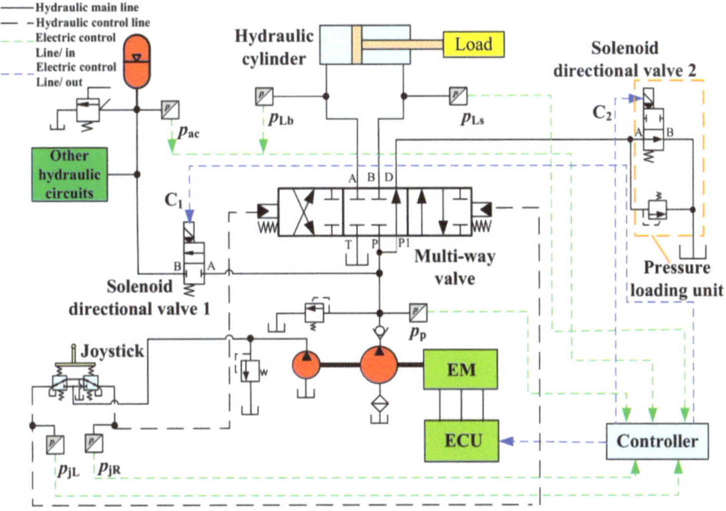


**Figure 2.** Schematic of the novel AISCS.

Therefore, the pressure in the working chamber of the actuator can be built quickly to drive the load with the assistance of the HA. Meanwhile, the idle speed can be set to a low value to obtain good energy saving and decrease noise, even if the response of the EM cannot be guaranteed.

## 3. Control Strategy

There are two goals of the control strategy. One is to make EM speed be as low as possible to reduce energy consumption. However, the low speed should not be at the cost of control performance and pump suction capacity. As the pump suction capacity cannot be improved easily, the other goal of control strategy is to guarantee the control performance when the idle mode is cancelled, which creates a sense of the hydraulic cylinder moving at the same speed as the target velocity, which is decided by the joystick. In other words, the input pressure $p_p$ of the multi-way valve should be built up as quickly as possible. However, the EM speed cannot be increased rapidly from the idle speed to its normal speed, especially when the idle speed is low.

### 3.1. Optimization of Pressure Differential $\Delta p$ between the HA and the Maximum Load Pressure

One case should be considered: as long as the EM starts to run, the joystick returns to its maximum position in an instant and the EM speed cannot reach to its target speed as quickly as possible. Hence, the flow rate of the pump will be too low to drive the load in time and the velocity of the load cannot be guaranteed. Meanwhile, when the HE works in a crane mode and pilot-oil discloses the multi-way valve, the output pressure of the pump is at a rather low level and the pressure in the hydraulic cylinder will drop suddenly with a big oscillation, especially in the case that the load is heavy and the load had not been lowered to the ground. As a matter of fact, one of the disadvantage of the HA is that only when the pressure of the HA is larger than the load pressure can it release the oil to drive the load. If the pressure differential $\Delta p$ between the HA and the maximum load pressure is too high, though the actuator will start immediately, excess loss will be produced across the hydraulic control valve. While if the pressure differential $\Delta p$ between the HA and the maximum load pressure is too low, it cannot build up the pressure quickly to drive the load. Therefore, the pressure differential $\Delta p$ between the HA and the maximum load pressure should be optimized to ensure the control performance and the energy saving.

The maximum load pressure $p_{Lmax}$ is determined from the expression

$$p_{Lmax} = \max(p_{Lb}, p_{Ls}) \tag{1}$$

where, $p_{Lb}$ is the non-rod side chamber of the hydraulic cylinder, MPa. $p_{Ls}$ is the rod side chamber of the hydraulic cylinder, MPa.

The target velocity of the actuator is given by

$$v_t = k \cdot (p_{Lb} - p_{Ls}) \tag{2}$$

where, $k$ is the proportionality coefficient between the target velocity and the joystick pressure.

The actual velocity of the load is calculated as

$$v = \frac{q_p \cdot n_p - (C_{ip} + C_{ep})p_p - Q_{bp} + Q_{ac}}{A_L} \tag{3}$$

where, $q_p$ is the displacement of the pump, m$^3$/rad. $n_p$ is the speed of the pump, rad/s. $C_{ip}$ and $C_{ep}$ are the internal and external leakage coefficient of the pump, respectively, m$^3$/(Pa·s). $p_p$ is the output port pressure of the pump, Pa. $Q_{bp}$ is the bypass flow rate of the pump, m$^3$/s. $Q_{ac}$ is the flow rate of the HA, and m$^3$/s. $A_L$ is the effective area of driving chamber of the cylinder, m$^2$.

The Boyle's law expression for this case is

$$p_{a0} V_0^n = p_{ax}(V_0 \pm \Delta V)^n \tag{4}$$

where, $p_{a0}$ is the pre-charge pressure of the HA, MPa. $V_0$ is the gas volume under the pressure $p_{a0}$, $m^3$. $p_{ax}$ is the pressure of the HA, MPa. $\Delta V$ is the volume change of the HA, $m^3$. $n$ is the polytrophic exponent, in this research, and $n$ can be set at 1.4 because the time for releasing the flow to the multi-way valve is short, which is less than 5 s. When the HA is charged, the minus sign is chosen in Equation (4).

The flow rate of the HA can be achieved by differentiating the Equation (4) with respect to time, as follows:

$$Q_{ac} = \frac{d\Delta V}{dt} = \pm \frac{V_0 p_{a0}^{\frac{1}{n}}}{n} (p_{ax})^{\frac{-1-n}{n}} \frac{dp_{ax}}{dt} \tag{5}$$

When the idle mode is cancelled, the HA is the main power device to drive the load in the initial stage, namely,

$$Q_{ac} \approx v A_L \tag{6}$$

When considering that the pressures of HA satisfies,

$$0.25 p_{a2} < p_{a0} < 0.9 p_{a1} \tag{7}$$

where $p_{a1}$ and $p_{a2}$ are the minimum and maximum working pressure of HA, respectively.

Then, the pressure change rate of the HA in Equation (5) can be obtained by

$$\frac{dp_{ax}}{dt} = \pm \frac{Q_{ac} n}{V_0 p_{a0}^{\frac{1}{n}} (p_{ax})^{\frac{-1-n}{n}}} \approx \pm \frac{v \cdot A_L \cdot n}{V_0 \cdot p_{a0}^{\frac{1}{n}} \cdot (p_{ax})^{\frac{-1-n}{n}}} \approx \pm \frac{v \cdot A_L \cdot n}{V_0 \cdot p_{a0}^{\frac{1}{n}} \cdot (1.2 p_{a0})^{\frac{-1-n}{n}}} \tag{8}$$

When the polytrophic exponent $n$ is 1.4, the pressure change rate of the HA can be rewritten as,

$$\frac{dp_{ax}}{dt} \approx \pm \frac{v \cdot A_L \cdot 1.4 \cdot p_{a0}}{V_0 \cdot 1.2^{\frac{-2.4}{1.4}}} \approx \pm \frac{v \cdot A_L \cdot p_{a0}}{V_0} \tag{9}$$

Seen from Figure 2, the pressure differential $\Delta p$ between the HA and the load is equal to the sum of the pressure change of the HA and the pressure drop across the solenoid directional valve 1, multi-way valve, and oil pipes. Then, the pressure differential $\Delta p$ between the HA and the load is developed as

$$\Delta p = p_{ax} - p_L \approx \int \frac{v A_L p_{a0}}{V_0} dt + \Delta p_{valve} \tag{10}$$

where $p_L$ is the load pressure, MPa. $\Delta p_{valve}$ is pressure drop across the solenoid directional valve 1, multi-way valve, and oil pipes, MPa. However, the actual velocity of the load, which is decided by the joystick and the load on the next working cycle is uncertain. Hence, the maximum target velocity of the actuator is used to substitute the actual velocity. Therefore, the pressure differential $\Delta p$ between the HA and the load is developed as

$$\Delta p = p_{ax} - p_L = \Delta p_{valve} + \int \frac{A_L p_{a0} v_{tmax}}{V_0} dt \tag{11}$$

where $v_{tmax}$ is the maximum target velocity of the actuator, m/s.

According to Equation (11), the pressure differential $\Delta p$ can be divided into two parts: the first part is the drop pressure in directional valve 1, multi-way valve, and oil pipes. It can be considered as a constant. The second part is the pressure change of the HA. It can be deduced that the pressure change of the HA depends on the rated volume $V_0$ of the HA, the pre-charge pressure $p_{a0}$ of the HA, and the maximum velocity $v_{tmax}$ of the actuator.

### 3.2. Control Law

The overall control process of the AISCS of Figure 2 is illustrated in Figure 3. When the joystick returns to the middle position, the controller detects that the pressure differential of the joystick is under the preset small positive value and sends the pilot control oil to the multi-way valve to make it work at the middle position. The proposed AISCS is to control EM speed to be switched between the different values, including the switch from the normal working speed to the first idle speed, the second idle speed, and that from the two idle speeds to its normal working speed. When reducing EM speed, the main consideration is the energy saving of the system. While the set value of the idle speed is the primary factor that influences the energy consumption. When the idle mode is cancelled, the control performance is the main consideration. Therefore, the time for building up the pressure of the pump should be smooth and short. Taking the self-suction capacity of the pump into account, the first level idle speed is set to 800 rpm and the second level idle speed is set to 500 rpm, because the HA can help the pump to build up the pressure quickly. During the different process, the control strategy changes accordingly. The control strategy is based on the following principles.

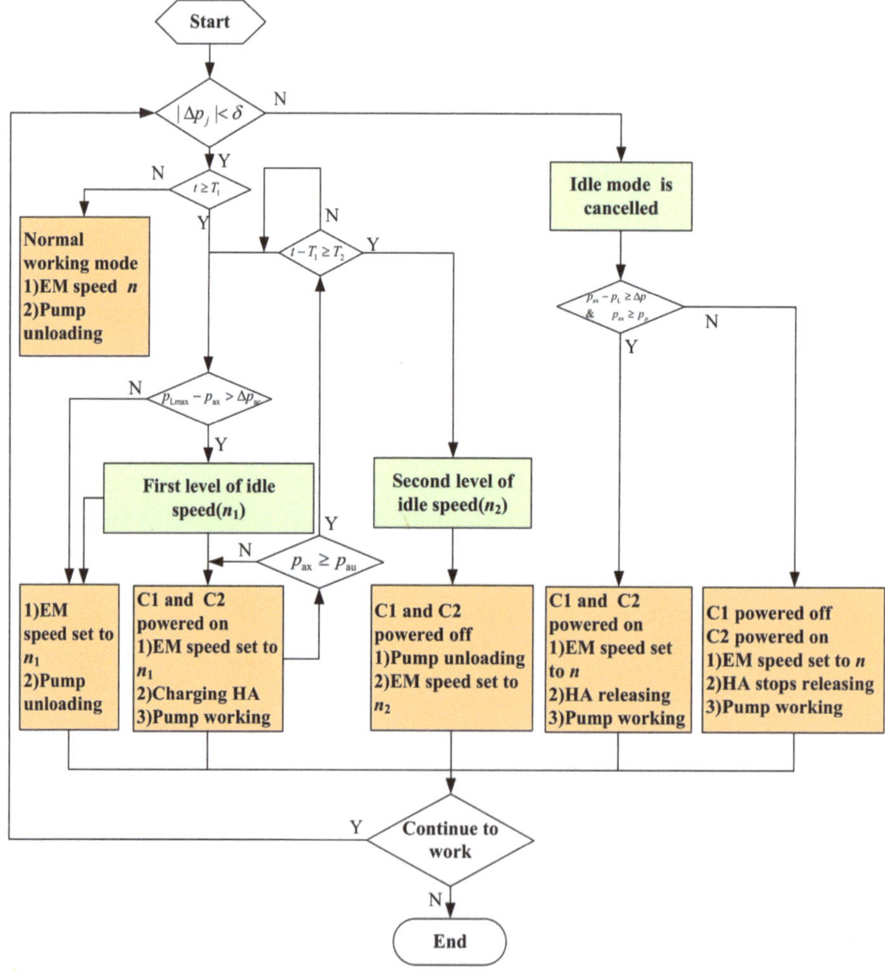


**Figure 3.** Overall control diagram of the AISCS.

- Mode 1: First level idle mode

When the time of the joystick backs to the middle position is longer than time $T_1$, the controller makes the EM speed reduce to the first level idle speed $n_1$ to reduce the energy loss. Meanwhile, the maximum load pressure $p_{Lmax}$ is adopted to decide to charge the HA by the pump or not. When the pressure differential between the maximum load pressure and the HA is larger than the preset value $\Delta p_{ac}$, then the solenoid directional valves 1 and 2 are powered on. Then the pump charges the HA via solenoid directional valve 1. After the pressure of the HA increases to the upper limited value, the solenoid directional valves 1 and 2 are powered off. The pump is connected to the tank through solenoid directional valve 2 and is unloaded.

- Mode 2: Second level idle mode

With an HA in AISCS, the HA had store a certain amount of energy in the first level idle mode. When the time of the joystick backs to middle position is longer than time $T_1 + T_2$ and the pressure of HA is under the threshold values, then the controller makes the EM work at the second level idle speed $n_2$, which is lower than the first level idle speed $n_1$ to further reduce the energy consumption.

- Mode 3: Idle mode is cancelled

When the joystick leaves its middle position, the AISCS is cancelled. In a HE, the joysticks are the most common interfaces between the drivers and the hydraulic manipulators. The target velocity of the actuator is characterized by the status of the joystick. The joystick signal gives a pressure differential. The target velocity of the actuator is the control object, which is compared with the joystick command through a conversion gain, as shown in Equation (2). When considering the dynamic response of EM, it cannot quickly reach to the normal speed which is usually about 1800 rpm. That is to say, the pump cannot build up the pressure quickly to drive the actuator. Hence, the actuator should be driven by the pump and the HA. In this case, both the solenoid directional valves 1 and 2 are powered on. The stored oil in the HA is released through solenoid directional valve 1 to the inlet of the multi-way valve to help the pump build up the pressure quickly to drive the actuator. The pressure of HA decreases with the discharging process and the load pressure changing, according to the pressure of HA. When the pump pressure is larger than the HA pressure, the solenoid directional valve 1 is powered off, and the actuator is driven by the pump only. The proportion of the flow rate supplied by the HA and the pump is scheduled to achieve a smooth and quick movement of the actuator.

During this control process, the HA pressure is the control object, which is achieved by controlling the on or off of the solenoid directional valves 1 and 2. Further, when the AISCS is cancelled, the pressure differential $\Delta p$ between the HA and the load is detected and controlled by the sensor signals shown in Figure 2. Due to the pressure, the differential is proportional to the target velocity of the actuator, which is characterized by the status of the joystick shown in Equation (2), therefore, the pressures of the joysticks are the key to monitor.

## 4. Experiment Research

### 4.1. Test Rig

To investigate the effects of the control strategy of AISCS, a test rig was developed, which is shown in Figure 4. The EM consists of two coaxial motors. The pump is a triple pump, including the main pump and the pilot pump. The hydraulic cylinder of the arm is chosen as the actuator to control. A 1.6 L HA is installed and the solenoid directional valves are integrated into one block. The sensors are installed to detect the status of the key pressures. The basic parameters of the test rig are listed in Table 1. To compare and analyse conveniently, the control signal of the joystick is the same and the boom cylinder has the same start/end points and displacements when the same type experiment is carried out. Because the idle speed period and the performance when the idle speed is cancelled are the main considerations of this research, the working period of the construction machinery, including digging, swing, and other performance, which is the same to the traditional construction machinery, is not discussed. The following part discusses the period when the joystick backs to the middle position and the system works in the idle mode.

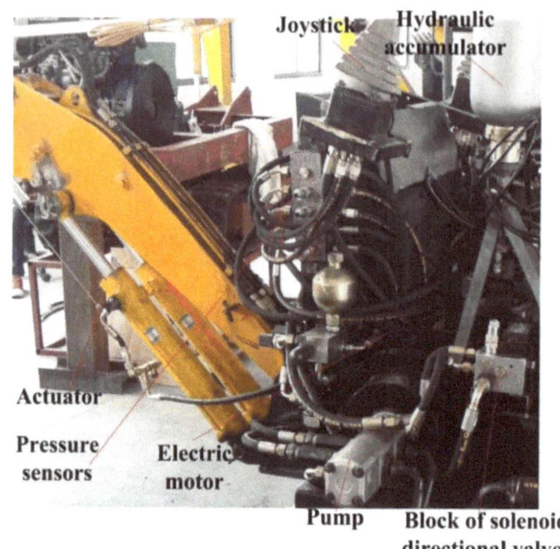


**Figure 4.** Layout of the test rig.

**Table 1.** Key parameters used in the test rig.

| Key Components | Parameter | Value |
| --- | --- | --- |
| Actuator | Rod diameter/mm | 35 |
| | Cylinder diameter/mm | 63 |
| | Maximum stroke/mm | 100 |
| Electric motor | Power/kW | 8 |
| | Rated speed/(rpm) | 1800 |
| Pump | Displacement/(mL·r$^{-1}$) | 16 |
| Hydraulic accumulator | Volume/L | 1.6 |
| | Pre-charge pressure/MPa | 2 |

*4.2. Control Performance*

The control performance is used to evaluate the responsiveness of the pressure of the actuator's cylinder to follow the signal that is produced by the joystick. If the pressure of the actuator cylinder can reach its target value quickly and smoothly, it is considered as a good control performance.

Figure 5 shows the EM speed of the AISCS and joystick pressure according to the state of the joystick. It can be seen that the EM speed changes according to the state of the joystick. The joystick backs to the middle position and the actuator stops to work at time 7 s. When the controller detects the joystick staying at middle position for 8 s, which means that the actuator stops working for 8 s, and then the controller makes the EM work at the first level idle speed (800 rpm) from time 15 s. After 20 s and if the joystick is still at the middle position, then the controller makes the EM work at the second level idle speed (500 rpm) from time 35 s to further reduce the energy consumption. When the controller detects the joystick leaves the middle position, it makes the EM accelerate to its normal working speed (1800 rpm). Hence, the EM can switch the speed among the normal speed

(1800 rpm), first level idle speed (800 rpm), and the second level idle speed (500 rpm), judging by the control strategy.

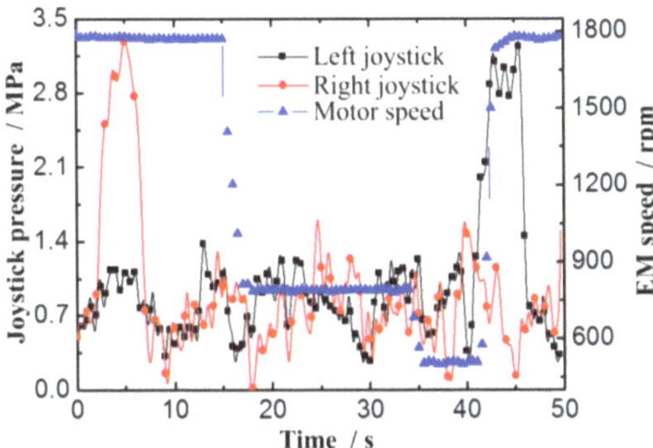

**Figure 5.** Experimental curves of the electric motor (EM) speed and joystick pressure.

Figure 6 describes the pressures of the HA, pump, joystick, and load during the proposed AISCS process. When the joystick backs to the middle position at time 7 s, the actuator stops, therefore, the velocity and the pressure of the actuator are close to zero and the pump is unloading. When the system works at the first level idle mode at time 15 s, the actuator still does not work, while the pump supplies the oil to the HA and both the pressures of the HA and pump rise until the pressure of the HA reaches the upper limited value. Then, the pump is unloading through the solenoid directional valve 2. When the joystick leaves the middle position at time 40 s, both the HA and the pump supply the oil to the actuator to make it work quickly. As the EM speed cannot reach its target speed as quickly as possible, the outlet pressure of the pump is less than the load pressure, while the pressure of the HA is above the load pressure. Accordingly, only the HA can drive the actuator in this working mode at the time from 40 s to 44.6 s. The pressure differential between the HA and the load pressure almost keeps constant to ensure that the HA have the capability to release the oil to the actuator. This indicates that the HA can be an auxiliary energy to build up the pressure of the actuator quickly. When the outlet pressure of the pumps reaches the pressure of the HA, the HA stops to work and only the pumps drive the actuator. Even though it takes 2–5 s for the EM's acceleration process, the system can still build up the pressure quickly. Hence, the pressure differential control strategy can guarantee the control performance for the actuator. Both Figures 5 and 6 indicate that the proposed AISCS can realize the preset goals and the control strategy works well.

Figure 7 shows the pressures of the non-rod chamber of the actuator cylinder under different pressure differential in the proposed AISCS process. The pressure differential $\Delta p$ may influence the dynamic response of the actuator movement. The larger the pressure differential, the more quickly the actuator pressure is built up. When the pressure differential is 0 MPa, the pressure of HA is not larger than the load pressure and HA cannot release oil to the actuator. The pressure is only built up by the pump, so the load pressure drops at the beginning when AISCS is cancelled. This can conclude that the proposed AISCS can build up the pressure quickly to drive the actuator with the suitable pressure differential. The release characteristics of the HA is the key factor to decide the pressure differential $\Delta p$, while the dynamic response of the EM is not rigorous in the proposed system.

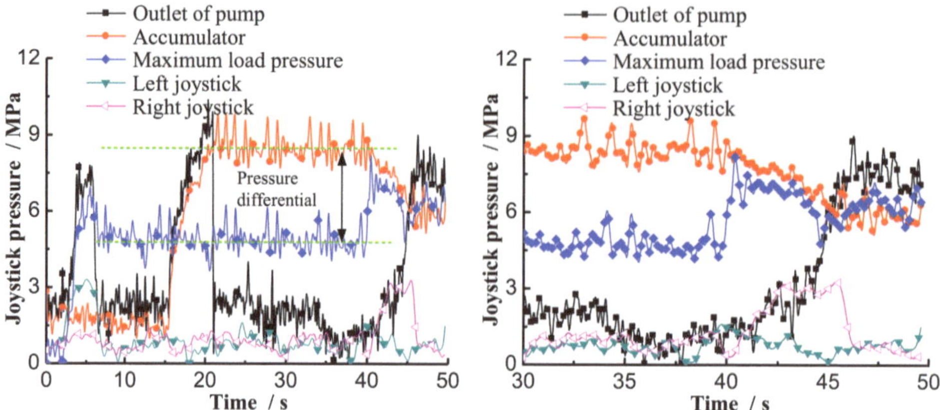

**Figure 6.** Pressures during the proposed AISCS process.

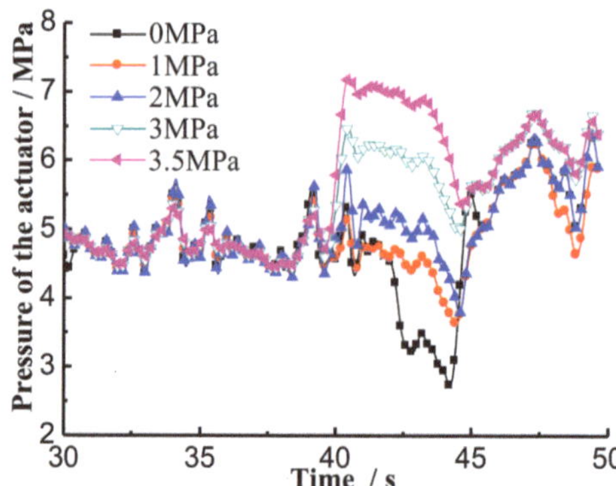

**Figure 7.** Pressures of the non-rod chamber of the actuator cylinder under different pressure differential.

### 4.3. Energy Saving

Energy saving is the main consideration of the modern construction machinery. It can be calculated by the energy that is consumed by the traditional construction machinery and that of the proposed construction machinery. Though the working time, which is about 20 s, is periodic, the idle time is irregular and the actual idle time depends on the driver. Hence, the idle mode for testing the energy saving in the novel automatic idle speed control system with hydraulic accumulator works only once.

The energy consumed of the system can be characterized by the output energy of the pump, as follows,

$$E_{\mathrm{p}} = \int p_{\mathrm{p}} Q_{\mathrm{p}} dt = q_{\mathrm{p}} n_{\mathrm{p}} \eta_{\mathrm{c}} \int p_{\mathrm{p}} dt \tag{12}$$

where, $E_{\mathrm{p}}$ is the output energy of the pump, J. $\eta_{\mathrm{c}}$ is the volumetric efficiency of the pump.

The AISCS with and without the HA has some influence on the energy consumption. Seen from Figures 8 and 9, only at the time from 15 s to 17.6 s, the output pressure of the pump in AISCS with HA is more high than that in the traditional AISCS, because the pump consumes more energy to charge the HA at this process. However, both the EM speed and the output pressure of the pump in the idle time is less than the other systems, which means that the energy consumption of the pump can be reduced in the proposed system. It can be seen from Table 2 and Figure 10 that during the idle mode, including the first level and second level idle mode, the system without the AISCS consumes 48.23 kJ energy, the

AISCS without the HA consumes 30.07 kJ energy, and the proposed AISCS with the HA consumes 25.46 kJ energy. When compared with the system without AISCS, the energy-saving efficiency of the system with the HA is 67% and that without the HA is 47%, which means that the proposed AISCS can both improve the energy saving and control performance than the traditional AISCS.

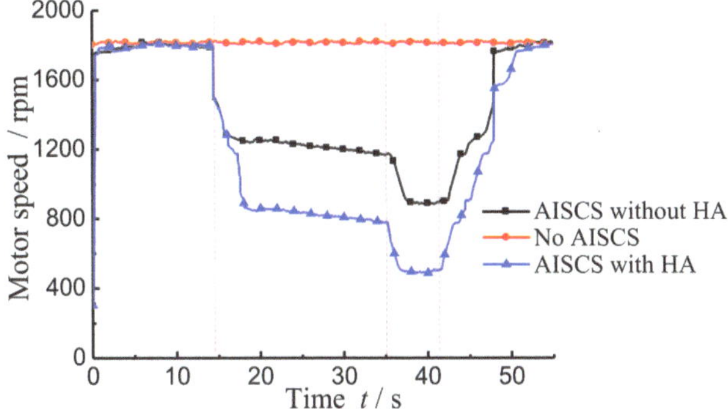

**Figure 8.** EM speed comparison under different control system.

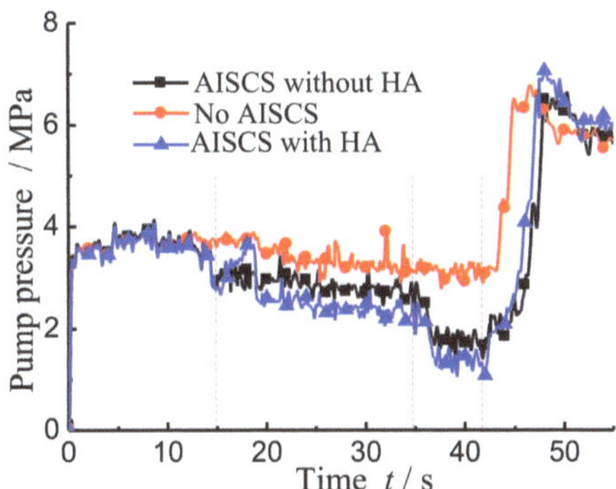

**Figure 9.** Pump pressure comparison under different control system.

**Table 2.** Energy saving in different system.

| System Type | Energy Consumed | Energy Efficiency |
|---|---|---|
| No idle speed control | 48,233 J | - |
| AISCS with HA | 15,950 J | 67% |
| AISCS without HA | 25,459 J | 47% |

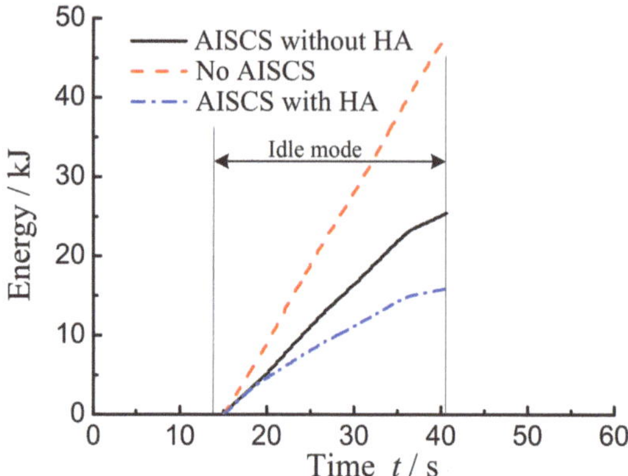


**Figure 10.** Energy consumption comparison in idle time under different control system.

## 5. Summary and Conclusions

According to the working cycle and the characteristics of the traditional AISCS of construction machinery, a novel two-level AISCS with the HA is proposed. Some useful conclusions are obtained, as follows.

1.  The proposed AISCS can realize the EM speed switching among the first level idle speed, second level idle speed, and normal speed based on the control strategy. The idle speed can set to a low value, which is 500 rpm in the proposed AISCS to reduce the energy consumption.
2.  The proposed AISCS with the HA can build up the load pressure quickly and make the load follow the action of the joystick rapidly when the idle mode is cancelled. This can make the actuator move more stably and smoothly.

The proposed AISCS with the HA can save 67% energy than the system without AISCS and can improve the energy saving and control performance than the traditional AISCS.

**Acknowledgments:** The authors acknowledge the support of National Natural Science Foundation of China (51505160), China Postdoctoral Science Foundation (2017M612139), Industry Cooperation of Major Science and Technology Project (2017H6013), Natural Science Foundation of Fujian Province (2017J01087) and Open Foundation of the State Key Laboratory of Fluid Power and Mechatronic Systems (GZKF-201611).

**Author Contributions:** H.R. proposed the idea of the proposed AISCS. T.L. improved the schematic of the novel AISCS. S.Z. set up the model and tested the motor used in the test. W.H. did the experiment and made the control strategy. C.M. analyzed the data. T.L. and H.L. wrote the paper.

**Conflicts of Interest:** The authors declare no conflicts of interest.

## Abbreviations

| | |
|---|---|
| EM | Electric motor |
| HA | Hydraulic accumulator |
| HE | Hydraulic excavator |
| AISCS | Automatic idle speed control system |
| CNG | Compressed natural gas |
| PWM | Pulse-width modulation |

## Nomenclature

| | |
|---|---|
| $A_L$ | effective area of drive chamber of the cylinder |
| $C_{ep}$ | external leakage coefficient of the pump |
| $C_{ip}$ | internal leakage coefficient of the pump |
| $k$ | proportionality coefficient between the target velocity and the joystick pressure |
| $n_p$ | speed of the pump |
| $n$ | polytrophic exponent |
| $p_{a0}$ | pre-charge pressure of the HA |
| $p_{a1}$ | minimum working pressure of HA |
| $p_{a2}$ | maximum working pressure of HA |
| $p_{au}$ | upper limited pressure of the HA |
| $p_{ax}$ | working pressure of the HA |
| $p_{jL}$ | left side output pressure of the joystick |
| $p_{jR}$ | right side output pressure of the joystick |
| $\Delta p_j$ | pressure differential of the two sides of the joystick |
| $p_L$ | load pressure |
| $p_{Lb}$ | non-rod chamber pressure of the actuator |
| $p_{Ls}$ | rod chamber pressure of the actuator |
| $p_{Lmax}$ | maximum pressure of the load |
| $\Delta p_{ac}$ | Preset pressure differential of HA |
| $p_p$ | output port pressure of the pump |
| $\Delta p$ | the pressure differential between the HA and the load |
| $\Delta p_{valve}$ | pressure drop across the solenoid directional valve 1, multi-way valve and oil pipes |
| $Q_{ac}$ | flow rate of the HA |
| $q_p$ | displacement of the pump |
| $t$ | time |
| $T_1$ | time that the system stay at the first level idle speed |
| $T_2$ | time that the system stay at the second level idle speed |
| $V_0$ | gas volume under the pressure $p_{a0}$ of the HA |
| $\Delta V$ | volume change of the HA |
| $v$ | actual velocity of the load |
| $v_t$ | target velocity of the load |
| $v_{tmax}$ | maximum target velocity of the actuator |
| $\delta$ | a small positive value |
| $\eta_c$ | volumetric efficiency of the pump |

## References


1. Chen, Q.; Wu, W.; Liu, W.; Zhang, X. Research on pump control signal of excavator positive control system. *J. Hefei Univ. Technol.* **2014**, *37*, 645–649.
2. Cheng, M.; Xu, B.; Yang, H. Efficiency Improvement for Electrohydraulic Flow Sharing Systems. In Proceedings of the 9th International Fluid Power Conference, RWTH Aachen University, Aachen, Germany, 24–26 March 2014.
3. Sugimura, K.; Murrenhoff, H. Hybrid Load Sensing—Displacement Controlled Architecture for Excavators. In Proceedings of the 14th Scandinavian International Conference on Fluid Power (SICFP'15), Tampere University of Technology, Tampere, Finland, 20–22 May 2015; pp. 20–22.
4. Jackson, R.S.; Clanton, R.R.; Pfaff, J.L. Hydraulic Control Valve System with Electronic Load Sense Control, U.S. Patent 7,089,733, 15 August 2006.
5. Lin, T.; Wang, Q.; Hu, B.; Gong, W. Development of hybrid powered hydraulic construction machinery. *Autom. Constr.* **2010**, *19*, 11–19. [CrossRef]
6. Hippalgaonkar, R.; Ivantysynova, M.; Zimmerman, J. Fuel savings of a mini-excavator through a hydraulic hybrid displacement controlled system. In Proceedings of the 8th International Fluid Power Conference, Dresden, Germany, 26–28 March 2012; pp. 139–153.

7.    Montazeri-Gh, M.; Poursamad, A.; Ghalichi, B. Application of genetic algorithm for optimization of control strategy in parallel hybrid electric vehicles. *J. Frankl. Inst.* **2006**, *343*, 420–435. [CrossRef]

8.    Cipek, M.; Pavković, D.; Petrić, J. A control-oriented simulation model of a power-split hybrid electric vehicle. *Appl. Energy* **2013**, *101*, 121–133. [CrossRef]

9.    Wang, T.; Wang, Q. Optimization design of a permanent magnet synchronous generator for a potential energy recovery system. *IEEE Trans. Energy Convers.* **2012**, *27*, 856–863. [CrossRef]

10.   Wang, T.; Wang, Q.; Lin, T. Improvement of boom control performance for hybrid hydraulic excavator with potential energy recovery. *Autom. Constr.* **2013**, *30*, 161–169. [CrossRef]

11.   Minav, T.A.; Virtanen, A.; Laurila, L.; Pyrhönena, J. Storage of energy recovered from an industrial forklift. *Autom. Constr.* **2012**, *22*, 506–515. [CrossRef]

12.   Lin, T.; Wang, Q.; Hu, B.; Gong, W. Research on the energy regeneration systems for hybrid hydraulic excavators. *Autom. Constr.* **2010**, *19*, 1016–1026. [CrossRef]

13.   Cairano, S.D.; Yanakiev, D.; Bemporad, A.; Kolmanovsky, L.V.; Hrovat, D. Model predictive idle speed control: Design, analysis, and experimental evaluation. *IEEE Trans. Control Syst. Technol.* **2012**, *20*, 84–97.

14.   He, Y.; Ma, F.; Deng, J.; Shaoa, Y.; Jian, X. Reducing the idle speed of an SI CNG engine fueled by HCNG with high hydrogen ratio. *Int. J. Hydrog. Energy* **2012**, *37*, 8698–8703. [CrossRef]

15.   Li, X.; Yurkovich, S. Sliding mode control of delayed systems with application to engine idle speed control. *IEEE Trans. Control Syst. Technol.* **2001**, *9*, 802–810.

16.   Hrovat, D.; Sun, J. Models and control methodologies for IC engine idle speed control design. *Control Eng. Pract.* **1997**, *5*, 1093–1100. [CrossRef]

17.   Huang, X.; Xie, H.; Song, K. Idle speed control of FSAE racing engines based on Mid-ranging ADRC. In Proceedings of the 33rd Chinese Control Conference (CCC), Nanjing, China, 28–30 July 2014; pp. 201–206.

18.   Chien, T.L.; Chen, C.C.; Hsu, C.Y. Tracking control of nonlinear automobile idle-speed time-delay system via differential geometry approach. *J. Frankl. Inst.* **2005**, *342*, 760–775. [CrossRef]

19.   Kandler, C.; Koenings, T.; Ding, S.X.; Weinhold, N.; Schultalbers, M. Stability Investigation of an Idle Speed Control Loop for a Hybrid Electric Vehicle. *IEEE Trans. Control Syst. Technol.* **2015**, *23*, 1189–1196. [CrossRef]

20.   Laurain, T.; Lauber, J.; Palhares, R.M. Observer design to control individual cylinder spark advance for idle speed management of a SI engine. In Proceedings of the IEEE International Conference on Industrial Electronics and Applications (ICIEA), Auckland, New Zealand, 15–17 June 2015; pp. 262–267.

21.   Chen, T.; Xie, H.; Li, L.; Zhang, L.; Wang, X.; Zhao, H. Methods to achieve HCCI/CAI combustion at idle operation in a 4VVAS gasoline engine. *Appl. Energy* **2014**, *116*, 41–51. [CrossRef]

22.   Xiong, Y.; Zhang, P.; Guo, S. Control and experiment research of XR220D rotary drilling Rig. *Equip. Manuf. Technol.* **2015**, *81*, 240–241.

23.   Liu, R. Research on the power matching control system for the hydraulic excavators. *Mach. Tool Hydraul.* **2015**, *43*, 111–115.

24.   Hao, P.; He, Q.; Zhang, X.; Xie, S. Study on Load and Operating Mode Identification of Excavator. *Hydraul. Pneum. Seals* **2008**, *5*. Available online: http://en.cnki.com.cn/Article_en/CJFDTotal-YYQD200805005.htm (accessed on 23 March 2018).

*Article*

# Pneumatic Rotary Actuator Position Servo System Based on ADE-PD Control

**Yeming Zhang *, Ke Li *, Shaoliang Wei and Geng Wang**

School of Mechanical and Power Engineering, Henan Polytechnic University, Jiaozuo 454000, China;
wsl_ify@163.com (S.W.); wgmouse@163.com (G.W.)

* Correspondence: tazhangyeming@163.com (Y.Z.); leeketech@163.com (K.L.)

Received: 17 January 2018; Accepted: 5 March 2018; Published: 9 March 2018

**Abstract:** In order to accurately control the rotation position of a pneumatic rotary actuator, the flow state of the gas and the motion state of the pneumatic rotary actuator in the pneumatic rotary actuator position servo system are analyzed in this paper. The mathematical model of the system and the experiment platform are established after that. An Adaptive Differential Evolution (ADE) algorithm which adaptively ameliorates the scaling factor and crossover probability in the process of individual evolution is proposed and applied to the parameter optimization of PD controller. The experimental platform is used to compare the controller with Differential Evolution (DE) algorithm and NCD-PID controller. Finally, the characteristics of the system are tested by increasing the inertial load. The experimental results illustrate that system using ADE-PD control strategy has greater position precision and faster response than using DE-PD and NCD-PID strategies, and shows great robustness.

**Keywords:** pneumatic rotary actuator; rotation position; mathematical model; adaptive differential evolution; parameter optimization

---

## 1. Introduction

Pneumatic systems have been rapidly developed in industrial production because of the advantages of low cost, quick response, and great energy-saving [1–3]. The pneumatic rotary actuator can realize the change of rotation angle, so it is widely used in the rotation of the mechanical arm, the rotation of the platform, the opening and closing of the valve, and so on [4]. At present, the "two points" control method is the most common method to control the rotation position of a pneumatic rotary actuator, but it is difficult to realize the pinpoint at any position [5]. The pneumatic servo control system can control the motion trajectory of the cylinder and make the positioning accurate and efficient, which has become a hot research topic in pneumatic technology [6].

Short working strokes and great friction of the pneumatic rotary actuator make it difficult to apply the control strategy of ordinary pneumatic actuator to the pneumatic rotary actuator [7]. Scholars in various countries have made thorough studies of the servo control system of ordinary pneumatic actuator, but the research on rotation position control of the pneumatic rotary actuator is rare. Therefore, it is necessary to study the model establishment [8–10], parameter identification, and control strategy of the pneumatic rotary actuator deeply [11].

Many scholars utilized the feedback of the actuator's position signal or the two chamber pressure to control the position accuracy. Rad et al. [12] studied the method that the position precision was well controlled by position and pressure feedbacks of the actuator. Ren et al. [13] only used the feedback of the position of the actuator to realize the tracking control of the position by an Adaptive Backstepping Sliding Mode Control (ABSMC) method, which reduced the cost of the pneumatic system.

Adding the feedback of the pressure can solve the phenomenon of position slip caused by the dead zone of the proportional valve. By further testing the performance of widely-used proportional directional control valves, we conclude that the valve MPYE-5-M5-010-B has very high sensitivity with virtually no dead zone. Therefore, in this text, only the position signal is fed back to the controller, and then the result is output to the proportional valve, the fast and accurate response of the position can be realized.

The establishment of the model can lay a good foundation for the research [14–17], and the application of control algorithm is also particularly important. Proportional-Integral-Differential (PID) control is a typical representative of the classical control theory, which has the characteristics of a simple algorithm, good robustness, high reliability, and relatively easy adjustment. However, due to the serious nonlinearity of the pneumatic servo system, the simple classical PID cannot effectively control it. The scholars, based on PID control, added some intelligent algorithms to improve the control accuracy. BAI [18] established a mathematical model for the servo system of a vane rotary actuator, and three parameters of PID control are optimized by fuzzy controller. Ahn et al. [19] applied a pneumatic servo system to the PAM manipulator with artificial muscles and combined the traditional PID controller with the Neural Network (NN) to design a nonlinear Neural Network PID controller which did not need training in advance and adaptively adjusted the parameters. But, Fuzzy control has low control accuracy, and the structure of the neural network is too complex. These limit the application in practice.

At present, the development trend of pneumatic servo control systems is high efficiency and energy saving [20–23]. Simplifying the structure of the system and controller and maintaining good control accuracy and stability [24] have broad prospects in industrial applications. Kaitwanidvilai et al. [25] applied Genetic Algorithm (GA) to the design of robust controller. By searching and evolving the parameters, the optimal solution was obtained, which made the system control effect better and more robust while simplifying the structure of controller.

To sum up, it is an effective method to optimize the three parameters of PID by using the intelligent algorithm. The Nonlinear Control Design (NCD) toolbox in MATLAB/Simulink has been able to optimize the three control parameters of the classic PID, which can be described as NCD-PID. However, there will still be greater errors and fluctuations. Therefore, this paper proposes an Adaptive Differential Evolution (ADE) algorithm to optimize the two parameters proportional and differential (PD), which can be expressed as ADE-PD. ADE-PD can adjust parameters adaptively only with the approximate range of the parameters. While using Differential Evolution (DE) algorithm to optimize proportional and differential, DE-PD, needs the appropriate parameters to ensure a relatively good result. The results of experiment show that system using ADE-PD control strategy has greater position precision and faster response than using DE-PD and NCD-PID strategies, and shows great robustness.

## 2. Experimental Set-Up

The rack and pinion pneumatic rotary actuator is the actuator of pneumatic servo control system. The composition and control principle of the system are described in Figure 1.

In Figure 1, the pneumatic connections are shown in solid lines and the electrical connections are expressed in dashed lines. The air compressor generates compressed air, and the gas flows through air filter, air regulator and air lubricator to become pure gas with stable pressure, which provides power for the pneumatic system. Rack and pinion pneumatic rotary actuator is placed horizontally, and the load is installed on the rotary table of it. Proportional directional control valve dominates flow direction and flow of the gas at the inlet and outlet to control the rotation angle of the rotary table. The rotary encoder transmits the rotation angle of actuator to the data acquisition card in Industrial Personal Computer (IPC) by transistor-to-transistor logic (TTL) levels. The TTL levels are calculated by the counter in data acquisition card. Therefore, the rotation angle of the actuator can be obtained. The host computer established by Labview calls the control strategy compiled by MATLAB/Simulink to deal with the difference between actual rotation angle and set rotation angle, and converts the result into 0–10 V voltage signals. Then the signals are transmitted to the proportional directional control valve to control actuator to reduce the deviation of rotation angle. The position of pneumatic rotary actuator is controlled accurately in the above process of real time control.

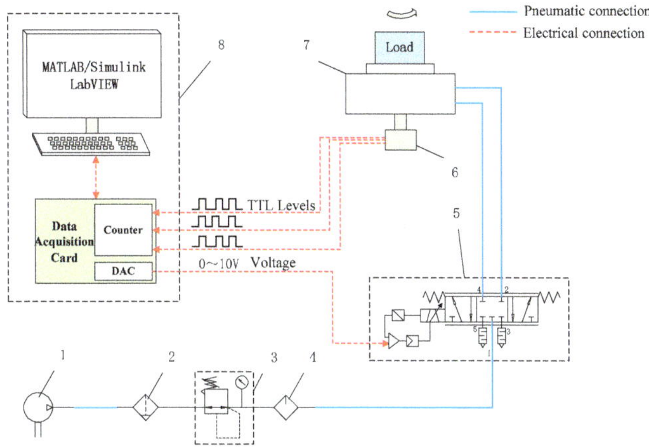


**Figure 1.** Schematic diagram of pneumatic rotary actuator position servo system (1—Air compressor; 2—Air filter; 3—Air regulator; 4—Air lubricator; 5—Proportional directional control valve; 6—Rotary encoder; 7—Rack and pinion pneumatic rotary actuator; and 8—IPC).

The experimental components are selected reasonably according to the principle of the system, and the experimental platform shown in Figure 2 is set up. Models and parameters of the main components are shown in Table 1.

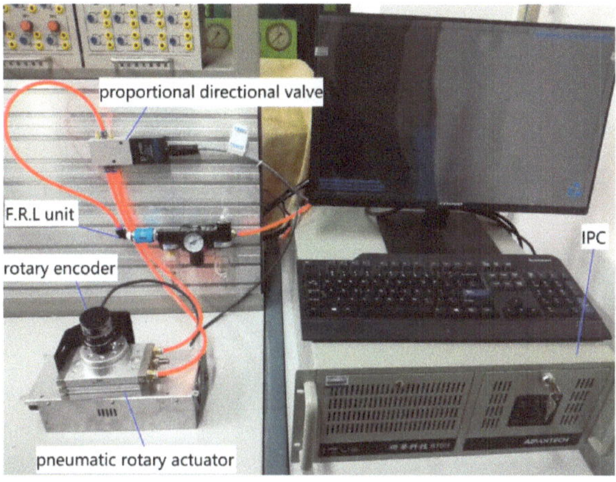


**Figure 2.** Experimental platform of pneumatic rotary actuator position servo system.

**Table 1.** Models and parameters of main components.

| Component | Model | Parameters |
| --- | --- | --- |
| Pneumatic rotary actuator | SMC MSQA30A | bore: 30 mm, maximum angle: 190° |
| Proportional directional control valve | FESTO MPYE-5-M5-010-B | control voltage: 0~10 V, rated flow:100 L/min |
| Data acquisition card | NI PCI-6229 | 32 bit counter, output voltage: −10~10 V |
| Rotary encoder | OMRON E6B2-CWZ3E | resolution: 1800 P/R |
| IPC | IPC-610H | standard configuration |
| F.R.L Unit | AirTAC AC2000 | maximum pressure: 1 MPa |

### 3. Modeling the System

Figure 3 is the schematic diagram of pneumatic rotary actuator controlled by proportional directional control valve. Chamber **a** of the rack and pinion pneumatic rotary actuator has two chambers in series, and so does chamber **b**. The proportional directional control valve is a 5/3-way valve with five symmetrical valve ports. The direction of the proportional valve spool moving toward the right is positive direction, in Figure 3.

It is assumed that the working medium is an ideal gas that satisfies the ideal gas state equation. The gas in the actuator is uniform and the parameters of each point in the chamber are equal. There is no leakage between the actuator and the environment or between the two chambers. Charge and discharge processes from the outside to the chambers are fast, and the flow of gas in the system has no friction loss and has no heat exchange with environment. Therefore, the flow process can be considered as a reversible adiabatic process, that is, the isentropic process [26].

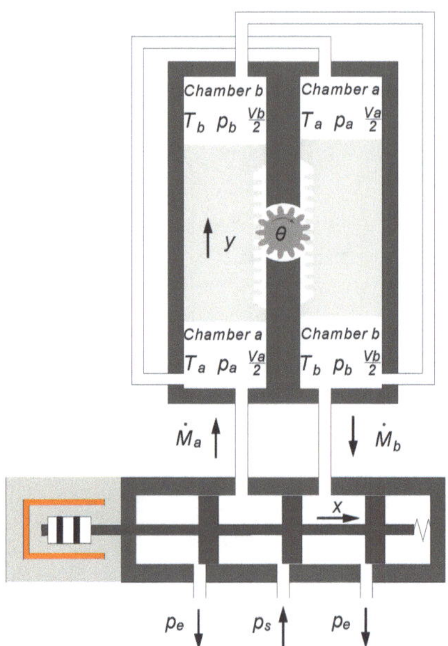


**Figure 3.** Schematic diagram of the valve-controlled actuator.

*3.1. Mass Flow Throttling Equation*

According to the basic theories of gas dynamics and thermodynamics, the mass flow passing through the port of proportional directional control valve is a function of independent variables which are the displacement $x$ of proportional valve spool and the gas pressures $p_a$, $p_b$ of two chambers of the actuator [27]. The function can be written as:

$$
\begin{array}{ll}
x \geq 0, & \left\{ \begin{array}{l} \dot{M}_a = f_a(x, p_a) \\ \dot{M}_b = f_b(x, p_b) \end{array} \right. \\
x < 0, & \left\{ \begin{array}{l} \dot{M}_a = f_a\prime(x, p_a) \\ \dot{M}_b = f_b\prime(x, p_b) \end{array} \right.
\end{array}
\tag{1}
$$

where $\dot{M}_a$ denotes mass flow flowing into chamber **a**, $\dot{M}_b$ is mass flow flowing out from chamber **b**, $f_a$, $f_a\prime$, $f_b$, and $f_b\prime$ are symbols of different equations.

When the proportional valve spool moves in the positive direction, $x \geq 0$, and chamber **a** is in charge state. The air supply is upflow, and chamber **a** is downflow. While chamber **b** is in discharge state, in which chamber **b** is upflow, and the environment is downflow. The mass flow flowing through chamber **a** and chamber **b** can be expressed by Equations (2) and (3) [28,29].

$$
\dot{M}_a = \begin{cases} \dfrac{CWxp_s}{\sqrt{RT_s}}\sqrt{\dfrac{2k}{k-1}\left[\left(\dfrac{p_a}{p_s}\right)^{\frac{2}{k}} - \left(\dfrac{p_a}{p_s}\right)^{\frac{k+1}{k}}\right]} & \left(\dfrac{p_a}{p_s} \geq c_0\right) \\[2em] \dfrac{CWxp_s}{\sqrt{RT_s}}\sqrt{\dfrac{2k}{k+1}\left(\dfrac{2}{k+1}\right)^{\frac{2}{k-1}}} & \left(\dfrac{p_a}{p_s} < c_0\right) \end{cases}
\tag{2}
$$

$$
\dot{M}_b = \begin{cases} \dfrac{CWxp_b}{\sqrt{RT_b}}\sqrt{\dfrac{2k}{k-1}\left[\left(\dfrac{p_e}{p_b}\right)^{\frac{2}{k}} - \left(\dfrac{p_e}{p_b}\right)^{\frac{k+1}{k}}\right]} & \left(\dfrac{p_e}{p_b} \geq c_0\right) \\[2em] \dfrac{CWxp_b}{\sqrt{RT_b}}\sqrt{\dfrac{2k}{k+1}\left(\dfrac{2}{k+1}\right)^{\frac{2}{k-1}}} & \left(\dfrac{p_e}{p_b} < c_0\right) \end{cases}
\tag{3}
$$

where $C$ represents flow coefficient of orifice in proportional valve; $W$ is geometric cross-sectional area gradient of orifice in proportional valve; and $p_a$ and $p_b$ are pressures of chamber **a** and **b** respectively; $p_s$ is supply pressure; $p_e$ is atmospheric pressure; $R$ is gas constant; $T_s$ is ambient temperature; $T_b$ is gas temperature in chamber **b**; $k$ is adiabatic index of ideal gas; and $c_0$ is critical pressure ratio.

According to Shearer's conclusion that the dynamic response of the system is independent of the initial position [30], the position $i$ of the piston in proportional valve can be set as the initial position. At this point, the Equations (2) and (3) are linearized as Equation (4) by Taylor Formula.

$$
\begin{cases} \Delta \dot{M}_a = K_{ma}\Delta x - K_{ca}\Delta p_a \\ \Delta \dot{M}_b = K_{mb}\Delta x - K_{cb}\Delta p_b \end{cases}
\tag{4}
$$

where $K_{ma}$ and $K_{mb}$ are flow gains flowing through **a** port and **b** port of proportional valve, while $K_{ca}$ and $K_{cb}$ are flow-pressure coefficients correspondingly.

$$
K_{ma} = \left.\dfrac{\partial \dot{M}_a}{\partial x}\right|_i = \begin{cases} \dfrac{CWp_s}{\sqrt{RT_s}}\sqrt{\dfrac{2k}{k-1}\left[\left(\dfrac{p_{ai}}{p_s}\right)^{\frac{2}{k}} - \left(\dfrac{p_{ai}}{p_s}\right)^{\frac{k+1}{k}}\right]} & \left(\dfrac{p_a}{p_s} \geq c_0\right) \\[2em] \dfrac{CWp_s}{\sqrt{RT_s}}\sqrt{\dfrac{2k}{k+1}\left(\dfrac{2}{k+1}\right)^{\frac{2}{k-1}}} & \left(\dfrac{p_a}{p_s} < c_0\right) \end{cases}
\tag{5}
$$

$$
K_{ca} = -\left.\dfrac{\partial \dot{M}_a}{\partial p_a}\right|_i = \begin{cases} -\dfrac{CWx_i}{\sqrt{RT_s}}\sqrt{\dfrac{2k}{k-1}}\dfrac{\frac{2}{k}\left(\frac{p_{ai}}{p_s}\right)^{\frac{2-k}{k}} - \frac{k+1}{k}\left(\frac{p_{ai}}{p_s}\right)^{\frac{1}{k}}}{2\sqrt{\left(\frac{p_{ai}}{p_s}\right)^{\frac{2}{k}} - \left(\frac{p_{ai}}{p_s}\right)^{\frac{k+1}{k}}}} & \left(\dfrac{p_a}{p_s} \geq c_0\right) \\[2em] 0 & \left(\dfrac{p_a}{p_s} < c_0\right) \end{cases}
\tag{6}
$$

$$
K_{mb} = \left.\dfrac{\partial \dot{M}_b}{\partial x}\right|_i = \begin{cases} \dfrac{CWp_{bi}}{\sqrt{RT_{bi}}}\sqrt{\dfrac{2k}{k-1}\left[\left(\dfrac{p_e}{p_{bi}}\right)^{\frac{2}{k}} - \left(\dfrac{p_e}{p_{bi}}\right)^{\frac{k+1}{k}}\right]} & \left(\dfrac{p_e}{p_b} \geq c_0\right) \\[2em] \dfrac{CWp_{bi}}{\sqrt{RT_{bi}}}\sqrt{\dfrac{2k}{k+1}\left(\dfrac{2}{k+1}\right)^{\frac{2}{k-1}}} & \left(\dfrac{p_e}{p_b} < c_0\right) \end{cases}
\tag{7}
$$

$$
K_{cb} = -\left.\dfrac{\partial \dot{M}_b}{\partial p_b}\right|_i = \begin{cases} -\dfrac{CWx_i}{\sqrt{RT_{bi}}}\sqrt{\dfrac{2k}{k-1}\left[\left(\dfrac{p_e}{p_{bi}}\right)^{\frac{2}{k}} - \left(\dfrac{p_e}{p_{bi}}\right)^{\frac{k+1}{k}}\right]}\left\{1 - \dfrac{\frac{2}{k}\left(\frac{p_e}{p_{bi}}\right)^{\frac{2-k}{k}} - \frac{k+1}{k}\left(\frac{p_e}{p_{bi}}\right)^{\frac{1}{k}}}{2\left[\left(\frac{p_e}{p_{bi}}\right)^{\frac{2}{k}} - \left(\frac{p_e}{p_{bi}}\right)^{\frac{k+1}{k}}\right]}\left(\dfrac{p_e}{p_{bi}}\right)\right\} & \left(\dfrac{p_e}{p_b} \geq c_0\right) \\[2em] -\dfrac{CWx_i}{\sqrt{RT_{bi}}}\sqrt{\dfrac{2k}{k+1}\left(\dfrac{2}{k+1}\right)^{\frac{2}{k-1}}} & \left(\dfrac{p_e}{p_b} < c_0\right) \end{cases}
\tag{8}
$$

where $p_{ai}$ and $p_{bi}$ stand for pressures of chamber **a** and **b** in working position $i$; $T_{bi}$ stands for temperature of chamber **b** in working position $i$; $x_i$ is displacement of proportional valve spool in working position $i$.

When the system is in the initial state, valve spool is in the middle position. The displacement of valve spool at this time is the same as the gap between spool and inwall of the valve, that is $x_i = x_0$. And equations $T_{ai} = T_{bi} = T_s$ and $p_{ai} = p_{bi}$ are established. If the supply pressure $p_s$ and its temperature $T_s$ are given, $T_{ai}$, $T_{bi}$, $p_{ai}$ and $p_{bi}$ can be measured, and $x_0$ can also be tested experimentally. Therefore, $K_{ma}$, $K_{mb}$, $K_{ca}$ and $K_{cb}$ can be calculated by Equations (5)–(8), respectively.

When the proportional valve spool moves in the negative direction, $x < 0$, and chamber **a** is in discharge state. Chamber **a** is upflow, and the environment is downflow. While chamber **b** is in charge state, in which air supply is upflow, and chamber **b** is downflow. The mass flow flowing through chamber **a** and chamber **b** can be replaced by Equations (9) and (10).

$$\dot{M}_a = \begin{cases} \dfrac{CWxp_a}{\sqrt{RT_a}}\sqrt{\dfrac{2k}{k-1}\left[\left(\dfrac{p_e}{p_a}\right)^{\frac{2}{k}} - \left(\dfrac{p_e}{p_a}\right)^{\frac{k+1}{k}}\right]} & \left(\dfrac{p_e}{p_a} \geq c_0\right) \\[3ex] \dfrac{CWxp_a}{\sqrt{RT_a}}\sqrt{\dfrac{2k}{k+1}\left(\dfrac{2}{k+1}\right)^{\frac{2}{k-1}}} & \left(\dfrac{p_e}{p_a} < c_0\right) \end{cases} \tag{9}$$

$$\dot{M}_b = \begin{cases} \dfrac{CWxp_s}{\sqrt{RT_s}}\sqrt{\dfrac{2k}{k-1}\left[\left(\dfrac{p_b}{p_s}\right)^{\frac{2}{k}} - \left(\dfrac{p_b}{p_s}\right)^{\frac{k+1}{k}}\right]} & \left(\dfrac{p_b}{p_s} \geq c_0\right) \\[3ex] \dfrac{CWxp_s}{\sqrt{RT_s}}\sqrt{\dfrac{2k}{k+1}\left(\dfrac{2}{k+1}\right)^{\frac{2}{k-1}}} & \left(\dfrac{p_b}{p_s} < c_0\right) \end{cases} \tag{10}$$

where $T_a$ is the gas temperature in chamber **a**.

Equation (4) is also applicable, but the values of $K_{ma}$, $K_{mb}$, $K_{ca}$ and $K_{cb}$ are changed. Equation (4) can be rewritten as the form of load flow, which is expressed by Equation (11).

$$\Delta \dot{M}_a + \Delta \dot{M}_b = (K_{ma} + K_{mb})\Delta x - K_{ca}\Delta p_a - K_{cb}\Delta p_b \tag{11}$$

### 3.2. Mass Flow Continuity Equation

According to the law of mass flow, the mass flow of the gas in chamber a and b is equal to the mass change of inflow and outflow of gas per unit time, respectively, which can be written as

$$\begin{cases} \dot{M}_a = \dfrac{dM_a}{dt} = \dfrac{d(\rho_a V_a)}{dt} = \rho_a \dfrac{dV_a}{dt} + V_a \dfrac{d\rho_a}{dt} \\[2ex] -\dot{M}_b = \dfrac{dM_b}{dt} = \dfrac{d(\rho_b V_b)}{dt} = \rho_b \dfrac{dV_b}{dt} + V_b \dfrac{d\rho_b}{dt} \end{cases} \tag{12}$$

where $M_a$ and $M_b$ are masses of gas in chamber **a** and **b**; $\rho_a$ and $\rho_b$ are densities of gas in chamber **a** and **b**; and $V_a$ and $V_b$ are volumes of gas in chamber **a** and **b**.

If gas state equation $\rho = \frac{p}{RT}$ is substituted into Equation (12), we can obtain Equation (13).

$$\begin{cases} \dot{M}_a = \dfrac{1}{RT_a}\left(p_a \dfrac{dV_a}{dt} + V_a \dfrac{dp_a}{dt} - \dfrac{p_a V_a}{T_a}\dfrac{dT_a}{dt}\right) \\[2ex] -\dot{M}_b = \dfrac{1}{RT_b}\left(p_b \dfrac{dV_b}{dt} + V_b \dfrac{dp_b}{dt} - \dfrac{p_b V_b}{T_b}\dfrac{dT_b}{dt}\right) \end{cases} \tag{13}$$

According to the previous hypotheses, the temperature $T$ and initial temperature $T_0$ in the whole process of the system satisfy the isentropic condition [31]:

$$T = T_0 \left(\frac{p}{p_0}\right)^{\frac{k-1}{k}} \tag{14}$$

where $p_0$ is the initial pressure. We can get the derivative of Equation (14).

$$\frac{\mathrm{d}T}{\mathrm{d}t} = \frac{k-1}{k}\frac{T_0}{p_0}\left(\frac{p}{p_0}\right)^{-\frac{1}{k}}\frac{\mathrm{d}p}{\mathrm{d}t} = \frac{k-1}{k}\frac{T_0}{p}\left(\frac{p}{p_0}\right)^{\frac{k-1}{k}}\frac{\mathrm{d}p}{\mathrm{d}t} = \frac{k-1}{k}\frac{T}{p}\frac{\mathrm{d}p}{\mathrm{d}t} \tag{15}$$

Substituting Equation (15) into (13) yields

$$\begin{cases} \dot{M}_a = \frac{1}{RT_ak}\left(V_a\frac{\mathrm{d}p_a}{\mathrm{d}t} + kp_a\frac{\mathrm{d}V_a}{\mathrm{d}t}\right) \\ -\dot{M}_b = \frac{1}{RT_bk}\left(V_b\frac{\mathrm{d}p_b}{\mathrm{d}t} + kp_b\frac{\mathrm{d}V_b}{\mathrm{d}t}\right) \end{cases} \tag{16}$$

When the piston of pneumatic rotary actuator works near position $i$, it can be considered to make small changes near the position [27]. By linearizing Equation (16), we obtain the increment of mass flow:

$$\begin{cases} \Delta\dot{M}_a = \frac{1}{RT_{ai}k}\left[(V_{ai})\frac{\mathrm{d}(\Delta p_a)}{\mathrm{d}t} + (kp_{ai})\frac{\mathrm{d}(\Delta V_a)}{\mathrm{d}t}\right] \\ -\Delta\dot{M}_b = \frac{1}{RT_{bi}k}\left[(V_{bi})\frac{\mathrm{d}(\Delta p_b)}{\mathrm{d}t} + (kp_{bi})\frac{\mathrm{d}(\Delta V_b)}{\mathrm{d}t}\right] \end{cases} \tag{17}$$

where $T_{ai}$ stands for temperature of chamber **a** in working position $i$, $V_{ai}$ and $V_{bi}$ are volumes of gas of chamber **a** and **b** in working position $i$. After setting the initial position as an intermediate position, we take the total working volume of two chambers as $2V_0$, then the initial values are as follows:

$$T_{ai} = T_{bi} = T_s, V_{ai} = V_{bi} = V_0,\ p_{ai} = p_{bi} = p_i$$

where $p_i$ stands for pressure of chambers in initial position $i$.

The displacement of the actuator piston is represented by $y$, and the effective area of the piston is represented by $A$. The rotation angle of the actuator is represented by $\theta$, and the pitch diameter of the pinion is $d_f$. Then there is the following relationship:

$$-\frac{\mathrm{d}(\Delta V_a)}{\mathrm{d}t} = \frac{\mathrm{d}(\Delta V_b)}{\mathrm{d}t} = A\frac{\mathrm{d}(\Delta y)}{\mathrm{d}t} = \frac{1}{2}Ad_f\frac{\mathrm{d}(\Delta\theta)}{\mathrm{d}t} \tag{18}$$

We combine Equation (17) with Equation (18), and obtain

$$\Delta\dot{M}_a + \Delta\dot{M}_b = \frac{1}{RT_sk}\left[V_0\left(\frac{\mathrm{d}(\Delta p_a)}{\mathrm{d}t} - \frac{\mathrm{d}(\Delta p_b)}{\mathrm{d}t}\right) + Akp_id_f\frac{\mathrm{d}(\Delta\theta)}{\mathrm{d}t}\right] \tag{19}$$

*3.3. Mechanical Equation of the Pneumatic Rotary Actuator*

From Newton's second law, we can get the moment balance equation of rack and pinion pneumatic rotary actuator [32]:

$$Ad_f(p_a - p_b) = \frac{1}{2}md_f^2\frac{\mathrm{d}^2\theta}{\mathrm{d}t^2} + J\frac{\mathrm{d}^2\theta}{\mathrm{d}t^2} + \frac{1}{2}Bd_f^2\frac{\mathrm{d}\theta}{\mathrm{d}t} + M_L + M_f \tag{20}$$

where: $m$ is mass of one piston of the actuator; $J$ is moment of inertia of pinion; $B$ is viscosity damping coefficient; $M_L$ is load moment; and $M_f$ is friction moment.

*3.4. Block Diagram and Transfer Function of the Pneumatic Rotary Actuator*

We set two parameters $p_L$ and $K_m$, and $p_L = p_a - p_b$, $K_m = K_{ma} + K_{mb}$.
By Laplace transform, Equations (11), (19), and (20) can be rewritten as:

$$\Delta\dot{M}_a(s) + \Delta\dot{M}_b(s) = K_mx(s) - K_{ca}p_L(s) - (K_{ca} + K_{cb})\Delta p_b(s) \tag{21}$$

$$\Delta\dot{M}_a(s) + \Delta\dot{M}_b(s) = \frac{1}{RT_sk}\left\{V_0sp_L(s) + Akp_id_fs\theta(s)\right\} \tag{22}$$

$$\left[\left(\frac{1}{2}md_f^2 + J\right)s^2 + \frac{1}{2}Bd_f^2 s\right]\theta(s) = Ad_f p_L(s) - M_L(s) - M_f(s). \tag{23}$$

The block diagram of the pneumatic rotary actuator system, shown in Figure 4, can be obtained by Equations (21)–(23).

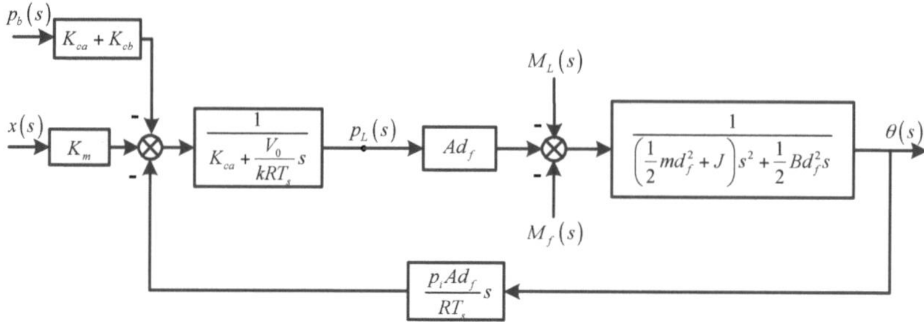


**Figure 4.** Block diagram of the pneumatic rotary actuator system.

The transfer function of the pneumatic rotary actuator system can be derived directly by the block diagram:

(1) After setting load moment $M_L$, friction moment $M_f$, and $p_b$ to zero, we obtain the transfer function from the displacement $x$ of proportional valve spool to the output angle $\theta$ of pneumatic rotary actuator, which can be simplified as:

$$\frac{\theta_x(s)}{x(s)} = \frac{K_T}{s\left(\frac{s^2}{\omega_0^2} + \frac{2\zeta}{\omega_0}s + 1\right)} \tag{24}$$

where: gain

$$K_T = \frac{2K_m ART_s}{Bd_f K_{ca} RT_s + 2A^2 d_f p_i}, \tag{25}$$

natural frequency

$$\omega_0 = \sqrt{\frac{Bd_f^2 K_{ca}kRT_s + 2kA^2 d_f^2 p_i}{\left(md_f^2 + 2J\right)V_0}}, \tag{26}$$

and damping ratio

$$\zeta = \frac{K_{ca}kRT_s\left(md_f^2 + 2J\right) + Bd_f^2 V_0}{2\sqrt{\left(md_f^2 + 2J\right)\left(Bd_f^2 K_{ca}kRT_s + 2kA^2 d_f^2 p_i\right)V_0}} \tag{27}$$

(2) After setting displacement $x$ of proportional valve spool and $p_b$ to zero, we obtain the transfer function from $M_L + M_f$ to the output angle $\theta$ of pneumatic rotary actuator, which can be predigested as:

$$\frac{\theta_M(s)}{M_L(s) + M_f(s)} = -\frac{\frac{K_T}{K_m Ad_f}\left(K_{ca} + \frac{V_0}{kRT_s}s\right)}{s\left(\frac{s^2}{\omega_0^2} + \frac{2\zeta}{\omega_0}s + 1\right)}. \tag{28}$$

(3)   After setting displacement $x$ of proportional valve spool, load moment $M_L$ and friction moment $M_f$ to zero, we obtain the transfer function from the friction $p_b$ to the output angle $\theta$ of pneumatic rotary actuator, which can be simplified as

$$\frac{\theta_p(s)}{p_b(s)} = -\frac{\frac{K_T}{K_m}(K_{ca} + K_{cb})}{s\left(\frac{s^2}{\omega_0^2} + \frac{2\zeta}{\omega_0}s + 1\right)} \tag{29}$$

(4)   Total output of the pneumatic rotary actuator system is

$$\theta(s) = \theta_x(s) + \theta_M(s) + \theta_p(s) = \frac{K_T x(s) - \frac{K_T}{K_m A d_f}\left(K_{ca} + \frac{V_0}{kRT_s}s\right)\left[M_L(s) + M_f(s)\right] - \frac{K_T}{K_m}(K_{ca} + K_{cb})p_b(s)}{s\left(\frac{s^2}{\omega_0^2} + \frac{2\zeta}{\omega_0}s + 1\right)}. \tag{30}$$

### 3.5. Transfer Function of Proportional Directional Control Valve and Servo Amplifier

The natural frequency of the proportional directional control valve is 5~7 times larger than that of pneumatic rotary actuator. Hence, the proportional directional control valve responds quickly. Generally, its impact on the system can be ignored, and its transfer function can be regarded as a proportional link [33], which can be expressed by Equation (31).

$$\frac{x(s)}{U(s)} = K_v \tag{31}$$

where $U(s)$ is the Laplace transform of the output voltage of servo amplifier; $K_v$ is gain of proportional directional control valve.

The transfer function of the servo amplifier can be described as follows:

$$\frac{U(s)}{\theta_r(s) - \theta(s)} = K_a \tag{32}$$

where $\theta_r(s)$ is the Laplace transform of the system input angle; $K_a$ is gain of servo amplifier.

### 3.6. Transfer Function of the System

From Equations (30)–(32), the block diagram of the servo system shown in Figure 5 can be obtained.

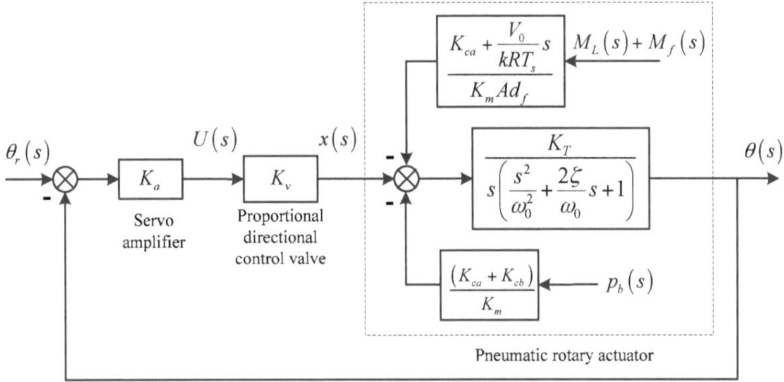


**Figure 5.** Block diagram of the pneumatic rotary actuator servo system.

Then the open-loop transfer function of the system can be obtained by the block diagram of the pneumatic rotary actuator servo system.

$$G(s) = \frac{K_a K_v K_T}{s\left(\frac{s^2}{\omega_0^2} + \frac{2\zeta}{\omega_0}s + 1\right)} \tag{33}$$

The initial parameter values of the model can be obtained, and shown in Table 2.

**Table 2.** Values of model parameters.

| Parameter | Value | Parameter | Value |
|---|---|---|---|
| $A$ | $3.4636 \times 10^{-4}$ [m$^2$] | $m$ | 0.21 [kg] |
| $B$ | 228 [N·s/m] | $p_e$ | $1.013 \times 10^5$ [Pa] |
| $C$ | 0.68 | $R$ | 287 [J/(kg·K)] |
| $c_0$ | 0.21 | $T_s$ | 293 [K] |
| $d_f$ | 0.014 [m] | $V_0$ | $1.6767 \times 10^{-5}$ [m$^3$] |
| $J$ | $1.678 \times 10^{-4}$ [kg·m$^2$] | $W$ | $3.1415 \times 10^{-2}$ [m] |
| $k$ | 1.4 | $x_0$ | $5 \times 10^{-6}$ [m] |

## 4. Adaptive Differential Evolution (ADE) Algorithm

Differential Evolution (DE) algorithm is a global optimization method based on swarm intelligence, which intelligently optimizes the evolution of the population through mutation, crossover and selection among individuals [34]. The idea of Differential Evolution (DE) algorithm is [35]: the initial individuals are randomly generated in the population, and the vector difference between any two individuals in the population is calculated; the vector difference is weighted according to a certain rule and summed up with the third individual to generate a new individual; if the fitness of the new individual is better than the target individual, the new individual will be used instead of the target individual to enter the next generation; otherwise, the old individual will continue the calculation of the next generation, so that the result will approximate to the optimal solution through the above iterative operation.

Adaptive Differential Evolution (ADE) algorithm ameliorates the scaling factor and crossover probability so that they can be adaptively changed during the iterative process, which improves the convergence precision and speeds up the convergence [36]. Since Adaptive Differential Evolution (ADE) algorithm has fewer parameters, strong global convergence and rapid convergence [37], it is suitable for the parameters tuning of pneumatic rotary actuator servo system.

Adaptive Differential Evolution (ADE) algorithm contains the following four steps.

### 4.1. Generating the Initial Population

The population contains $N_p$ $D$-dimensional vectors, whose parameters are real numbers. The $i$-th vector of the $g$-th generation can be marked as $x_{i,g}$. $i = 1, 2, 3 \ldots N_p$, $g = 1, 2, 3 \ldots G$, where $G$ is the maximum generation. $x_{i,g} = \left(x_{j,i,g}\right)$, $j = 1, 2, 3 \ldots D$, where $x_{j,i,g}$ is $j$-th parameter of $i$-th vector.

The parent vector parameters $x_{j,i,0}$ are randomly generated and can be described by Equation (34).

$$x_{j,i,0} = rand_j(0, 1) \cdot \left(b_j^U - b_j^L\right) + b_j^L \tag{34}$$

where $rand_j(0, 1)$ is the random number generated in the interval [0, 1]; $b_j^U$ and $b_j^L$ are two $D$-dimensional initial vectors. The superscripts $U$ and $L$ denote the upper bound and lower bound, respectively.

### 4.2. Mutation Operation

After initialization, a population consisting of $N_p$ test vectors is produced by mutating the parent population. The mutation operation is to add a scalable difference between two randomly selected vectors to a third vector. Equation (35) shows how to combine three different and randomly selected vectors to produce a mutant vector $v_{i,g}$.

$$v_{i,g} = v_{r0,g} + F \cdot \left(v_{r1,g} - v_{r2,g}\right) \tag{35}$$

where random indexes $r_0, r_1, r_2 \in \{1, 2, 3...N_p\}$.

We adjust the value of the scaling factor $F$ with a diminishing concave function. In the early stage of the algorithm, the value of $F$ is larger, and there is a larger population space, so the local extremum is not easy to appear and the convergence precision is improved. In the latter part of the algorithm, the value of $F$ is smaller, which helps to converge rapidly in the best range and increase the convergence rate [38]. The scaling factor $F$ can be given as:

$$F = (F_{\max} - F_{\min})\left(\frac{G - g}{G}\right)^2 + F_{\min} \tag{36}$$

where $F_{\max}$ is the maximum value of $F$; $F_{\min}$ is the minimum value of $F$.

Merging Equations (36) and (35) yields:

$$v_{i,g} = v_{r0,g} + \left[(F_{\max} - F_{\min})\left(\frac{G - g}{G}\right)^2 + F_{\min}\right](v_{r1,g} - v_{r2,g}). \tag{37}$$

### 4.3. Crossover Operation

Cross operation makes use of the parameters copied from two different vectors to form the test vector $u_{i,g}$. Cross operation compares the results of crossover probability $CR$ to a uniform random number generator $rand_j(0, 1)$. If the random number is less than or equal to $CR$, the test parameter will inherit the mutant vector $v_{i,g}$, otherwise the parameter will be copied from the vector $x_{i,g}$. Cross operation can be described by:

$$u_{i,g} = u_{j,i,g} = \begin{cases} v_{j,i,g}, & rand_j(0,1) \leq CR \\ x_{j,i,g}, & rand_j(0,1) > CR \end{cases} \tag{38}$$

where $v_{j,i,g}$ is $j$-th parameter of $i$-th mutant vector, $u_{j,i,g}$ is $j$-th parameter of $i$-th test vector.

We use an increasing convex function to adjust the crossover probability, $CR$. In the initial stage of the algorithm, $CR$ has a smaller value, and there is a larger population space, which improves the global convergence capability. In the latter part of the algorithm, $CR$ takes a larger value and improves the convergence speed. The crossover probability, $CR$, can be given as:

$$CR = CR_{\min} + (CR_{\max} - CR_{\min})\left(\frac{g}{G}\right)^2 \tag{39}$$

where $CR_{\max}$ is the maximum value of $CR$; $CR_{\min}$ is the minimum value of $CR$.

### 4.4. Selection Operation

If the target function value of the test vector is less than or equal to the target function value of the target vector, the test vector will enter the next generation, otherwise the target vector will enter the next generation. The target function can be represented by the symbol $f_c$. So the number of vectors remains unchanged, and the fitness will be better or remain the same. The selection operation process can be expressed as follows:

$$x_{i,g+1} = \begin{cases} u_{i,g}, & f_c(u_{i,g}) < f_c(x_{i,g}) \\ x_{i,g}, & \text{else} \end{cases}. \tag{40}$$

Operations of mutation, crossover, and selection were circulated in turn until the maximum number of evolutionary iterations, $G$, was reached.

## 5. ADE-PD Controller Design

The PD controller can make the system respond quickly and have little lag. So, it is used to optimize the control of the pneumatic rotary actuator servo system, and the parameters $k_p$ and $k_d$ of PD controller are tuned by Adaptive Differential Evolution (ADE) algorithm [39]. The first order low pass filter is also used in the system in order to reduce other interferences of the measurement channel and ensure the stability of the deviation $e(n)$. The schematic diagram of the control system with the ADE-PD controller is shown in Figure 6. The controller can be expressed in a time-discrete way [40].

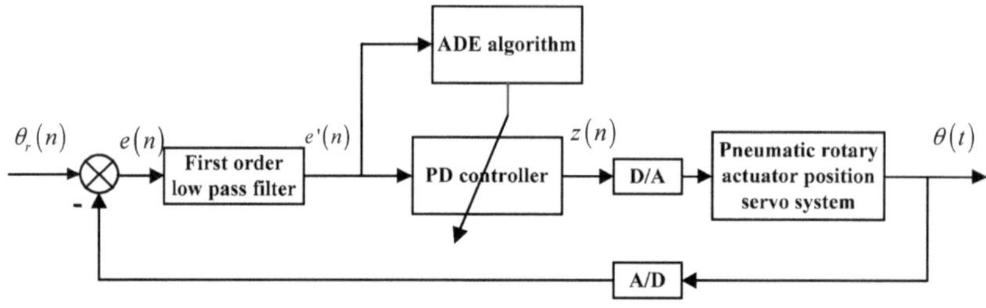


**Figure 6.** Schematic diagram of control system with ADE-PD controller.

The relationship between the input and output signals of the first order low pass filter is:

$$e\prime(n) = (1 - \alpha)e\prime(n - 1) + \alpha e(n) \tag{41}$$

where $e(n)$ and $e\prime(n)$ are input and output of the filter at the $n$-th sampling, $n$ = 1, 2, 3 ... ; filter coefficient $\alpha = 1 - e^{-T_P/\tau}$, where $T_P$ is sampling period, and $\tau$ is time constant.

The discrete PD algorithm can be expressed as:

$$z(n) = k_p\left\{e\prime(n) + \frac{T_d}{T_P}[e\prime(n) - e\prime(n - 1)]\right\} = k_p e\prime(n) + k_d[e\prime(n) - e\prime(n - 1)] \tag{42}$$

where $z(n)$ is output of the controller, $T_d$ is differential time constant, $k_p$ is proportional coefficient, and $k_d$ is differential coefficient, $k_d = k_p T_d / T_P$.

In order to obtain the satisfactory dynamic characteristics of the transient process, the discrete integral of time-weighted absolute value of the error (ITAE) is used as the minimum target function of the parameter selection. The square of the input is added to the target function, so as to prevent excessive control. The optimal index in the process of optimizing parameters $k_p$ and $k_d$ is given as:

$$I = \sum_{n=1}^{Z}\left[w_1|e\prime(n)| + w_2\theta_r^2(n)\right]T_P \tag{43}$$

where $Z$ is total points of sampling; $w_1$ and $w_2$ are weights.

In order to avoid overshoot, the penalty function is used, that is, once the overshoot is produced, the overshoot is added as one part of the optimal index. At this time, the optimal index is:

$$\text{when } e\prime(n) < 0, I = \sum_{n=1}^{Z}\left[w_1|e\prime(n)| + w_2\theta_r^2(n) + w_3|e\prime(n)|\right]T_P \tag{44}$$

where $w_3$ is weight, $w_3 \gg w_1$.

The flowchart of optimizing parameters with Adaptive Differential Evolution (ADE) algorithm is described in Figure 7.

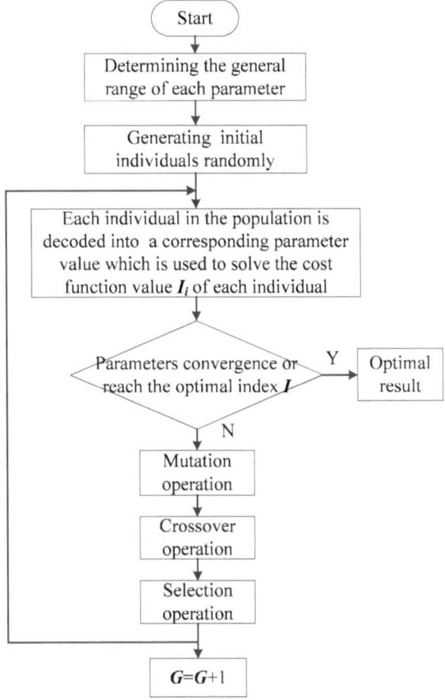

**Figure 7.** Flow chart of parameter tuning of ADE-PD controller.

## 6. Experiment and Analysis

The supply pressure of the experiment is 0.6 MPa, and the gas temperature is 20 °C. We select some parameters which are suitable for this system. The number of individuals is 30, the proportional coefficient $k_p$ is in the interval $[-5, 20]$, the range of the differential coefficient $k_d$ is $[-1, 1]$, where adding ranges of negative numbers can lead to better iterative processes, and cannot affect final results that are positive. The weights $w_1 = 0.999$, $w_2 = 0.001$, and $w_3 = 10$, and the evolutionary iterations $G = 50$. Sampling period $T_P$ is 1 ms.

### 6.1. Parameters Selection and Comparison with DE Algorithms

The DE-PD controller is used to control the system in the experiment firstly. When crossover probability $CR$ is set to 0.6, we can obtain different trajectory curves of pneumatic rotary actuator by changing the value of scaling factor $F$ to analyze the influence of parameter $F$ on the system. Figure 8 shows different results of rotation positions of the actuator and cost functions $I_i$ when $F = 0.04, 0.05$–$1.4$, 1.6, and 1.8. It can be observed from Figure 8a that the overshoot and fluctuation of position curve are the least and almost unchanged when $F$ is 0.05~1.4. Overly high or low $F$ can affect the stability of the trajectory. Figure 8b reveals different convergent processes when $F$ changes, and it verifies that the higher $F$ is, the slower the convergence rate is.

In the same way, different trajectory curves are shown in Figure 9a when $CR = 0.1, 0.3, 0.5$~$1.5$, and 1.7 after $F$ is set to 1.0. In the interval of $CR$ $[0.5,1.5]$, the overshoot and fluctuation of position curve are the least, and there are large fluctuations of rotation with other values of $CR$. From Figure 9b, we can also conclude that larger $CR$ can make the convergence faster.

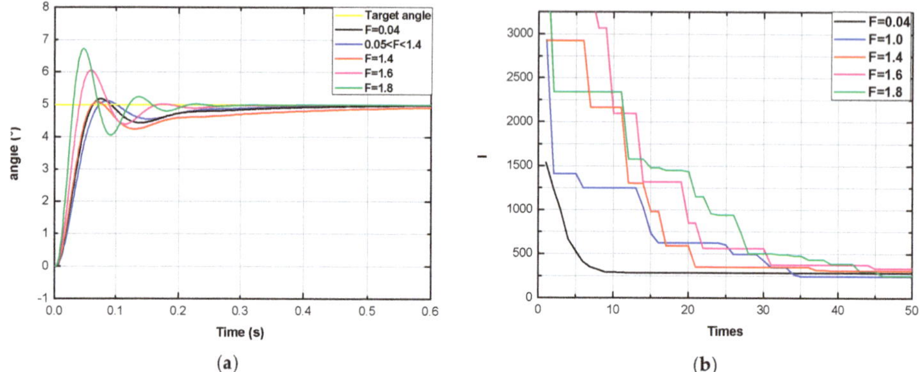

**Figure 8.** Comparison of experimental results with changed F, (**a**) Rotation positions; (**b**) Cost functions $I_i$.

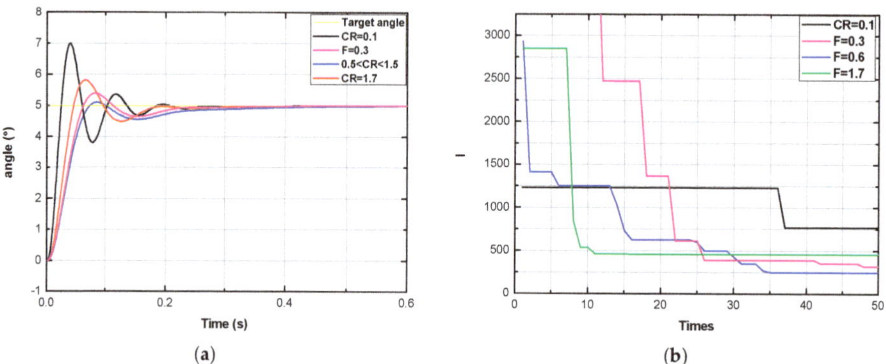

**Figure 9.** Comparison of experimental results with changed CR, (**a**) Rotation positions; (**b**) Cost functions $I_i$.

Then another experiment is done with ADE-PD control by changing $F_{max}$, $F_{min}$, $CR_{max}$, and $CR_{min}$. Figure 10 shows different rotation positions with ADE-PD control and the trajectory of the actuator performs best when $F_{max} = 1.5$, $F_{min} = 1.0$, $CR_{max} = 0.9$, and $CR_{min} = 0.6$. We compare the curves controlled by ADE and DE in Figure 11, and come to a conclusion that position curve controlled by ADE-PD has smaller overshoot and fluctuation.

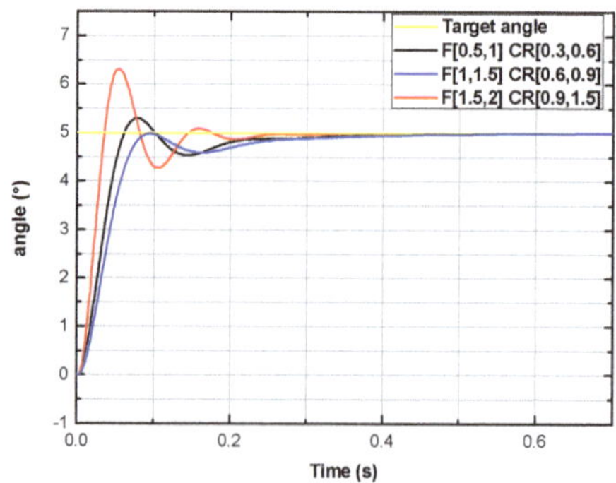

**Figure 10.** Different rotation positions with ADE-PD control.

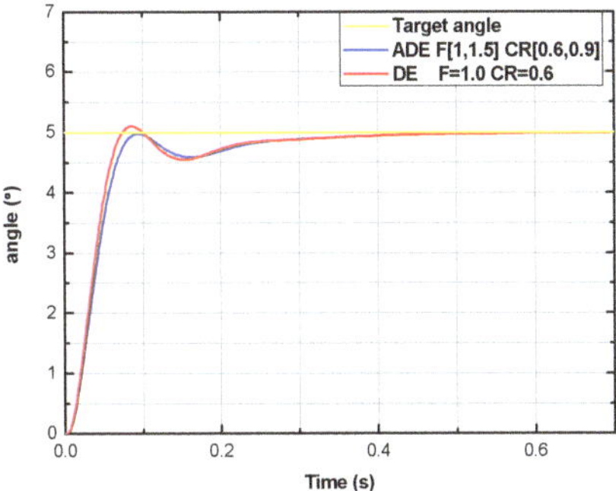

**Figure 11.** Comparison of curves controlled by ADE-PD and DE-PD.

## 6.2. Comparison with NCD-PID Control

In the experiment, the PID controller optimized by NCD is also used to control rotation position and is compared with the ADE-PD controller. Firstly, the target angle is set as $10°$. As shown in Figure 12, the response curve controlled by ADE-PD controller reaches peak in 0.101 s, and reaches the steady state in 0.355 s, and the steady state error is controlled within $0.1°$. However, the peak time of the response curve controlled by NCD-PID controller is 0.95 s, and at the same time, the steady state error is $0.15°$, and there are slight fluctuations throughout the process. So system based on ADE-PD control has greater position precision and faster response.

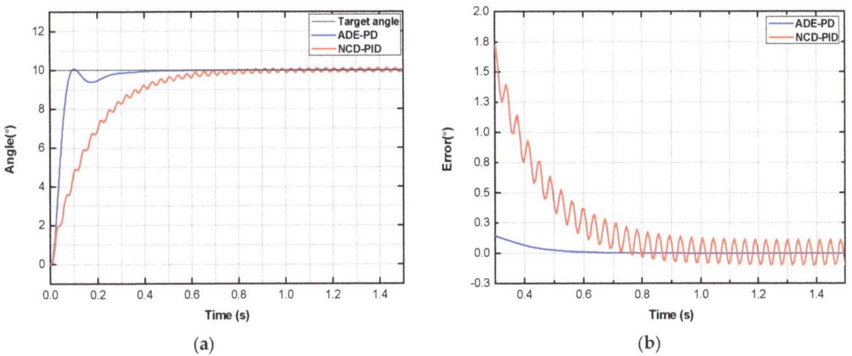

**Figure 12.** Comparison of curves controlled by ADE-PD and NCD-PID: (**a**) Rotation positions and (**b**) Errors.

If the input signals are changed to sinusoidal waves with different frequencies, we can obtain the trackings of the position. Figures 13–15 show comparisons of tracking curves when the frequency of sinusoidal signal is 5 rad/s, 10 rad/s, and 3 rad/s, respectively. As shown in Figure 13c, steady state error controlled by ADE-PD is $0.06°$, while the error controlled by NCD-PID is $0.25°$. In Figure 14c, the steady state error controlled by ADE-PD is $0.22°$, and the error controlled by NCD-PID is $0.41°$. Finally, the steady state error controlled by ADE-PD is $0.04°$, while the error controlled by NCD-PID is $0.17°$ in Figure 15c.

By contrasting the errors in Figure 13c, Figure 14c andFigure 15c, we come to a conclusion that the control precision of the system becomes worse with the increase of the frequency of sinusoidal wave. And in Figures 12a, 13b, 14b and 15b, there are persistent fluctuations in steady state because of stickiness in the actuator when changing the direction of motion. The system based on ADE-PD control can reduce the occurrence of stickiness and has less errors and fluctuations as shown in Figure 12b, Figure 13c, Figure 14c andFigure 15c.

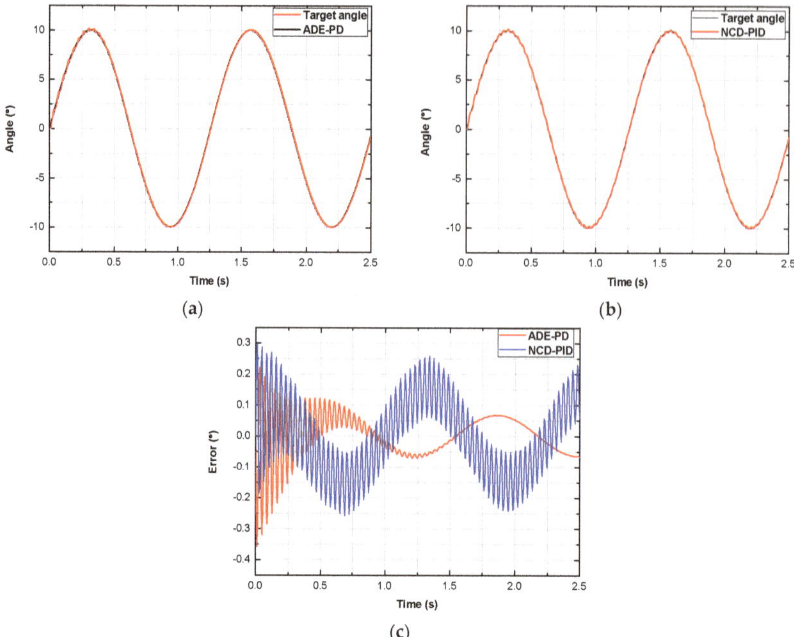

**Figure 13.** Comparison of tracking curves when the frequency of sinusoidal signal is 5 rad/s: (**a**) Tracking curve controlled by ADE-PD; (**b**) Tracking curve controlled by NCD-PID; and (**c**) Errors.

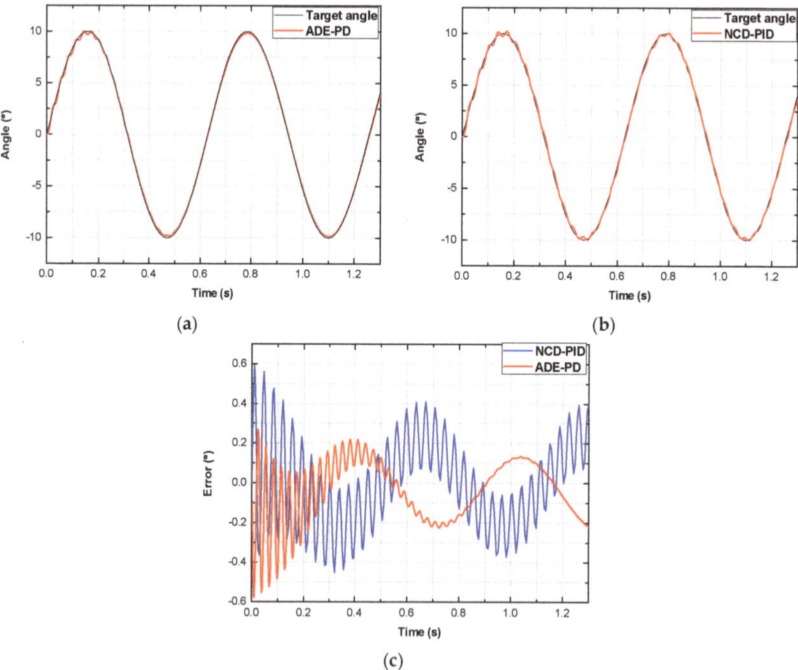

**Figure 14.** Comparison of tracking curves when the frequency of sinusoidal signal is 10 rad/s: (**a**) Tracking curve controlled by ADE-PD; (**b**) Tracking curve controlled by NCD-PID; and (**c**) Errors.

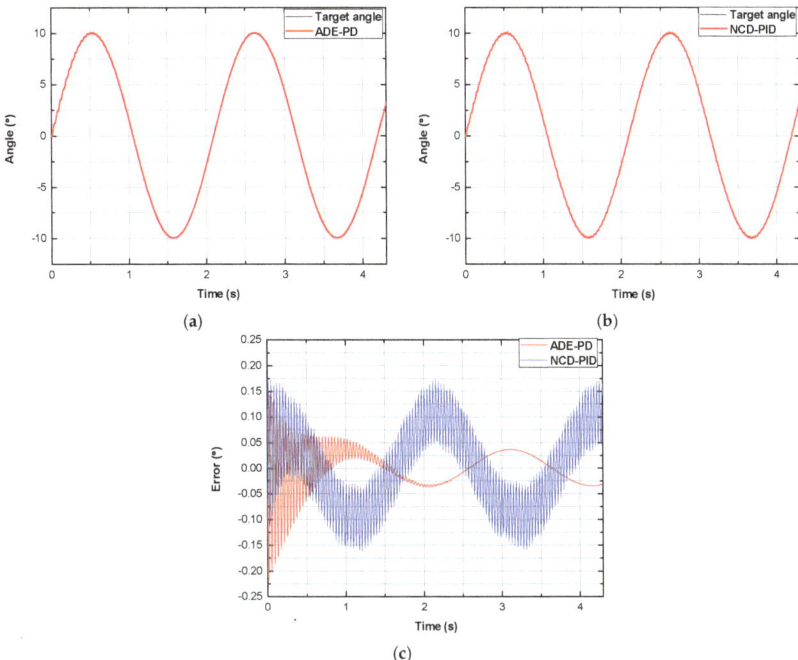

**Figure 15.** Comparison of tracking curves when the frequency of sinusoidal signal is 10 rad/s:
(**a**) Tracking curve controlled by ADE-PD; (**b**) Tracking curve controlled by NCD-PID; and (**c**) Errors.

### 6.3. Discussion on Increasing Inertia Load

The moment of inertia of the pneumatic rotary actuator itself is $1.678 \times 10^{-4}$ kg·m$^2$, which can be increased by loading block on rotating platform. Tracking curves are shown in Figure 16 when moment of inertia is $3.2 \times 10^{-4}$ kg·m$^2$, and the curves are shown in Figure 17 when moment of inertia is $4.5 \times 10^{-4}$ kg·m$^2$. Figure 16a,b describe rotation positions tracking fixed value, while Figure 17a,b provide rotation positions tracking sinusoidal signal whose frequency is 5 rad/s.

From the results in Figure 12b, Figure 16b, and Figure 17b, we summarize that increasing the inertia load can increase the rotation fluctuation and error. And it can be observed from Equations (25) and (26) that increasing the moment of inertia will reduce natural frequency and damping ratio. Thereby it makes the error increase and even deviate from center, which is obtained from Figure 13c, Figure 16d, and Figure 17d. Even so, the system has relatively stable trajectories and tiny errors which are within $0.27°$, and shows good robustness.

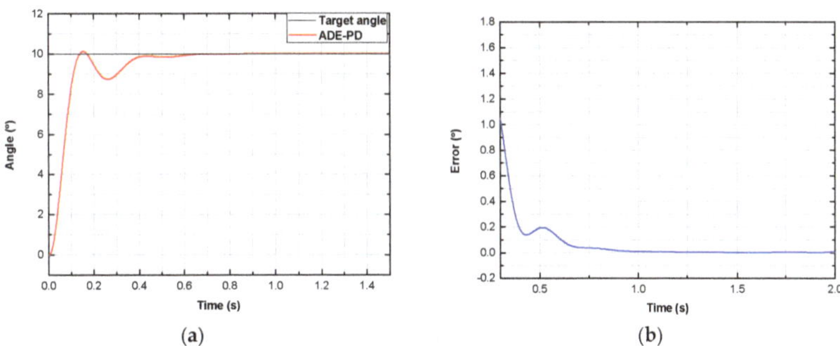

**Figure 16.** *Cont.*

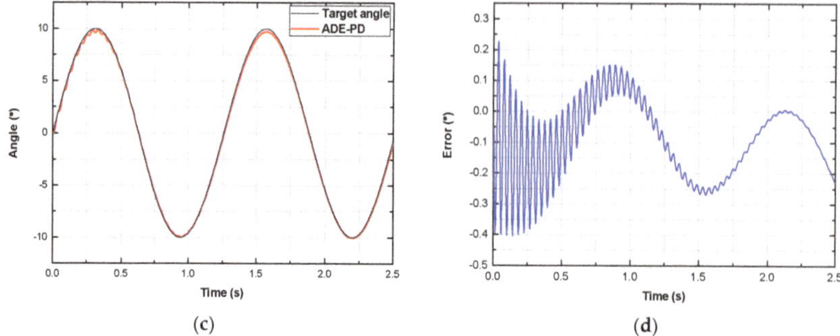

**Figure 16.** Tracking curves when moment of inertia is $3.2 \times 10^{-4}$ kg·m$^2$: (**a**) Rotation position tracking fixed value; (**b**) Error tracking fixed value; (**c**) Rotation position tracking sinusoidal signal; and (**d**) Error tracking sinusoidal signal.

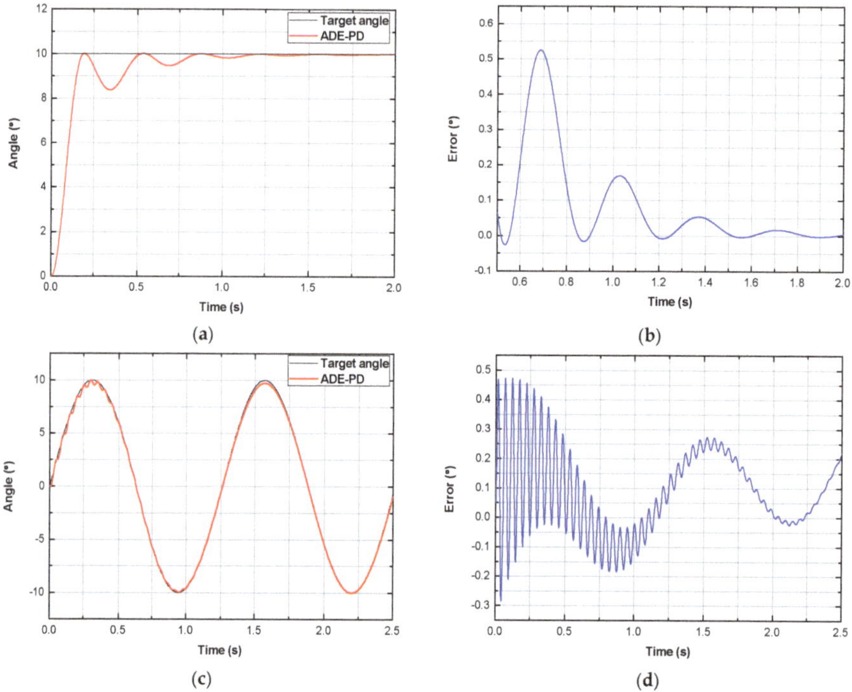

**Figure 17.** Tracking curves when moment of inertia is $4.5 \times 10^{-4}$ kg·m$^2$: (**a**) Rotation position tracking fixed value; (**b**) Error tracking fixed value; (**c**) Rotation position tracking sinusoidal signal; and (**d**) Error tracking sinusoidal signal.

## 7. Conclusions

In this paper, the pneumatic rotary actuator position servo system is studied and analyzed, and an experimental platform is set up. The motion laws of the rack and pinion pneumatic rotary actuator and the proportional directional control valve are analyzed. Starting from the flow of gas flowing through the proportional directional control valve and the motion state of the pneumatic rotary actuator, the basic equations of the actuator controlled by the valve are obtained and linearized near the middle position, and the mathematical model of the pneumatic rotary actuator is deduced.

Then the mathematical model of the whole system is obtained. The establishment of this model lays a foundation for the research of theory and experiment.

An Adaptive Differential Evolution (ADE) algorithm is proposed in this paper. In the process of individual variation, the values of scaling factor and crossover probability are adjusted by a decreasing concave function and an increasing convex function, respectively, which improves the accuracy and the speed of algorithm optimization. The ADE-PD controller is designed to control the pneumatic rotary actuator position servo system, and the experimental platform is used to select parameters, compare the controller with DE algorithm, and compare with PID controller by changing input signals. Finally, the characteristics of the system are discussed by increasing the inertia load. The conclusions obtained from experiments can be summarized as follows:

(a) Different values of $F$ and $CR$ can change the stability of the control. Rotation position controlled by ADE-PD has smaller overshoot and fluctuation than others controlled by DE-PD when $F_{max} = 1.5$, $F_{min} = 1.0$, $CR_{max} = 0.9$, $CR_{min} = 0.6$.

(b) The control precision of the system becomes worse with the increase of the frequency of sinusoidal wave. And there are persistent fluctuations in steady state because of stickiness in the actuator when changing the direction of motion. The system based on ADE-PD control can reduce the occurrence of stickiness, and has greater position precision and faster response than NCD-PID control.

(c) Increasing the moment of inertia will reduce natural frequency and damping ratio, and make the error increase and even deviate from center. Even so, the system has relatively stable trajectories and tiny errors, which shows great robustness.

(d) Although a resonance is expected by the compressibility and the friction could cause a small displacement instability, a reasonable control can be obtained.

**Acknowledgments:** The research reported in the article was funded by the Fundamental Research Funds for Henan Province Colleges and Universities (Grant No. NSFRF140120); Henan Province Science and Technology Key Project (Grant No. 172102310674); the Doctor Foundation of Henan Polytechnic University (Grant No. B2012-101); the Science and Technology Research Projects of Education Department of Henan province (Grant No. 14B460033); The Construction Project of the Case Library Course for Graduate Students of Henan Polytechnic University (Grant No. 2016YAL09).

**Author Contributions:** Yeming Zhang and Ke Li conceived and designed the experiments; Shaoliang Wei performed the experiments and Geng Wang analyzed the data; Ke Li wrote the paper and Yeming Zhang modified the manuscript.

**Conflicts of Interest:** The authors declare no conflict of interest.

## Nomenclature

| | |
|---|---|
| $A$ | effective area of the piston of actuator [m$^2$] |
| $b_j^U$; $b_j^L$ | two $D$-dimensional initial vectors |
| $B$ | viscosity damping coefficient [N·s/m] |
| $c_0$ | critical pressure ratio = 0.21 |
| $C$ | flow coefficient of orifice of proportional valve = 0.68 |
| $CR$; $CR_{max}$; $CR_{min}$ | crossover probability |
| $d_f$ | pitch diameter of the pinion [m] |
| $D$ | the number of dimensions of vectors |
| $e(n)$; $e\prime(n)$ | input and output of the filter at the $n$-th sampling |
| $f_{a,b}$; $f_{a,b}\prime$ | function of mass flow passing through the port of proportional valve |
| $f_c$ | target function |
| $F$ | scaling factor |
| $g$ | the number of evolutionary iterations |
| $G$ | the maximum generation |
| $I$; $I_i$ | optimal index; cost function value |
| $J$ | moment of inertia of pinion [kg·m$^2$] |

| | |
|---|---|
| $k$ | adiabatic index of ideal gas = 1.4 |
| $k_d$ | differential coefficient |
| $k_p$ | proportional coefficient |
| $K_a$ | gain of servo amplifier [V/rad] |
| $K_{ca,cb}$ | flow-pressure coefficients |
| $K_m$ | sum of flow-pressure coefficient and flow gain |
| $K_{ma,mb}$ | flow gains flowing through **a** port and **b** port of proportional valve |
| $K_T$ | gain of pneumatic rotary actuator [rad/m] |
| $K_v$ | gain of proportional directional control valve [m/V] |
| $m$ | mass of one piston in the actuator [kg] |
| $M_{a,b}$ | mass of gas in chamber **a** or **b** [kg] |
| $M_f$ | friction moment [N·m] |
| $M_L$ | load moment [N·m] |
| $\dot{M}_{a,b}$ | mass flow flowing into chamber **a** or flowing out from chamber **b** [kg/s] |
| $n$ | sampling times |
| $N_p$ | the number of $D$-dimensional vectors |
| $p_{a,b}$ | pressure of chamber **a** or **b** [Pa] |
| $p_{ai,bi}$ | pressure of chamber **a** or **b** in working position $i$ [Pa] |
| $p_e$ | atmospheric pressure [Pa] |
| $p_L$ | difference between $p_a$ and $p_b$ [Pa] |
| $p_s$ | supply pressure [Pa] |
| $rand_j(0,1)$ | random number generated in the interval [0,1] |
| $R$ | gas constant = 287 [J/(kg·K)] |
| $t$ | time [s] |
| $T_0$ | initial temperature [K] |
| $T_{a,b}$ | gas temperature in chamber **a** or **b** [K] |
| $T_{ai,bi}$ | temperature of chamber **a** or **b** in working position $i$ [K] |
| $T_d$ | differential time constant |
| $T_P$ | sampling period [ms] |
| $T_s$ | ambient temperature = 293 [K] |
| $u_{i,g}; u_{j,i,g}$ | test vector; $j$-th parameter of $i$-th test vector |
| $U(s)$ | Laplace transform of the output voltage of servo amplifier |
| $v_{i,g}; v_{j,i,g}$ | mutant vector; $j$-th parameter of $i$-th mutant vector |
| $V_0$ | value of the volume of gas [m³] |
| $V_{a,b}$ | volume of gas in chamber **a** or **b** [m³] |
| $V_{ai,bi}$ | volume of gas of chamber **a** or **b** in working position $i$ [m³] |
| $w_{1,2,3}$ | weights |
| $W$ | geometric cross-sectional area gradient of orifice in proportional valve [m] |
| $x$ | displacement of proportional valve spool [m] |
| $x_0$ | the gap between valve spool and inwall of the valve [m] |
| $x_i$ | displacement of proportional valve spool in working position $i$ [m] |
| $x_{i,g}; x_{j,i,g}$ | the $i$-th vector of the $g$-th generation; $j$-th parameter of $i$-th vector |
| $y$ | displacement of the actuator piston [m] |
| $z(n)$ | output of the controller |
| $Z$ | total points of sampling |
| $\alpha$ | filter coefficient |
| $\zeta$ | damping ratio |
| $\theta$ | rotation angle of the actuator [rad] |
| $\theta_{x,M,p}(s)$ | Laplace transform of rotation angle of the actuator when input is $x$, $M_L + M_f$ and $p_b$ |
| $\theta_r(s)$ | Laplace transform of the system input angle |
| $\rho_{a,b}$ | density of gas in chamber **a** or **b** [kg/m³] |
| $\tau$ | time constant |
| $\omega_0$ | natural frequency [Hz] |

## References

1. Zhang, Y.; Cai, M. Whole life-cycle costing analysis of pneumatic actuators. *J. Beijing Univ. Aeronaut. Astronaut.* **2011**, *37*, 1006–1010.
2. Yang, F.; Tadano, K.; Li, G.; Kagawa, T. Analysis of the energy efficiency of a pneumatic booster regulator with energy recovery. *Appl. Sci.* **2017**, *7*, 816. [CrossRef]
3. Zhang, Y.M.; Cai, M.L. Overall life cycle comprehensive assessment of pneumatic and electric actuator. *Chin. J. Mech. Eng. (Engl. Ed.)* **2014**, *27*, 584–594. [CrossRef]
4. Zang, X.; Liu, Y.; Li, W.; Lin, Z.; Zhao, J. Design and experimental development of a pneumatic stiffness adjustable foot system for biped robots adaptable to bumps on the ground. *Appl. Sci.* **2017**, *7*, 1005. [CrossRef]
5. Jiang, G.; Luo, M.; Bai, K.; Chen, S. A precise positioning method for a puncture robot based on a PSO-optimized BP neural network algorithm. *Appl. Sci.* **2017**, *7*, 969. [CrossRef]
6. Sheng, Z.; Li, Y. Hybrid robust control law with disturbance observer for high-frequency response electro-hydraulic servo loading system. *Appl. Sci.* **2016**, *6*, 98. [CrossRef]
7. Pujana-Arrese, A.; Mendizabal, A.; Arenas, J.; Prestamero, R.; Landaluze, J. Modelling in modelica and position control of a 1-DoF set-up powered by pneumatic muscles. *Mechatronics* **2010**, *20*, 535–552. [CrossRef]
8. Niu, J.; Shi, Y.; Cao, Z.; Cai, M.; Chen, W.; Zhu, J.; Xu, W. Study on air flow dynamic characteristic of mechanical ventilation of a lung simulator. *Sci. China Technol. Sci.* **2017**, *60*, 243–250. [CrossRef]
9. Ren, S.; Shi, Y.; Cai, M.; Xu, W. Influence of secretion on airflow dynamics of mechanical ventilated respiratory system. *IEEE/ACM Trans. Comput. Biol. Bioinform.* **2017**, *99*, 1. [CrossRef] [PubMed]
10. Ren, S.; Shi, Y.; Cai, M.; Xu, W.; Zhang, X.D. Influence of bronchial diameter change on the airflow dynamics based on a pressure-controlled ventilation system. *Int. J. Numer. Method Biomed. Eng.* **2017**. [CrossRef] [PubMed]
11. Saravanakumar, D.; Mohan, B.; Muthuramalingam, T. A review on recent research trends in servo pneumatic positioning systems. *Precis. Eng. J. Int. Soc. Precis. Eng. Nanotechnol.* **2017**, *49*, 481–492. [CrossRef]
12. Rad, C.-R.; Hancu, O. An improved nonlinear modelling and identification methodology of a servo-pneumatic actuating system with complex internal design for high-accuracy motion control applications. *Simul. Model. Pract. Theory* **2017**, *75*, 29–47. [CrossRef]
13. Ren, H.; Fan, J. Adaptive backstepping slide mode control of pneumatic position servo system. *Chin. J. Mech. Eng. (Engl. Ed.)* **2016**, *29*, 1003–1009. [CrossRef]
14. Shi, Y.; Wang, Y.; Cai, M.; Zhang, B.; Zhu, J. An aviation oxygen supply system based on a mechanical ventilation model. *Chin. J. Aeronaut.* **2017**, *31*, 197–204. [CrossRef]
15. Shi, Y.; Zhang, B.; Cai, M.; Xu, W. Coupling effect of double lungs on a VCV ventilator with automatic secretion clearance function. *IEEE/ACM Trans. Comput. Biol. Bioinform.* **2017**, *99*, 1. [CrossRef] [PubMed]
16. Shi, Y.; Wu, T.; Cai, M.; Wang, Y.; Xu, W. Energy conversion characteristics of a hydropneumatic transformer in a sustainable-energy vehicle. *Appl. Energy* **2016**, *171*, 77–85. [CrossRef]
17. Shi, Y.; Wang, Y.; Liang, H.; Cai, M. Power characteristics of a new kind of air-powered vehicle. *Int. J. Energy Res.* **2016**, *40*, 1112–1121. [CrossRef]
18. Bai, Y.H. Research on Pneumatic Rotation Position Servo Control Technology. Ph.D. Thesis, Nanjing University of Science and Technology, Nanjing, China, 2006.
19. Thanh, T.U.D.C.; Ahn, K.K. Nonlinear PID control to improve the control performance of 2 axes pneumatic artificial muscle manipulator using neural network. *Mechatronics* **2006**, *16*, 577–587. [CrossRef]
20. Zhang, Y.; Cai, M.; Kong, D. Overall energy efficiency of lubricant-injected rotary screw compressors and aftercoolers. In Proceedings of the IEEE Computer Society 2009 Asia-Pacific Power and Energy Engineering Conference (APPEEC 2009), Wuhan, China, 27–31 March 2009; Chinese Society for Electrical Engineering; IEEE Power and Energy Society; Scientific Research Publishing; Wuhan University: Wuhan, China, 2009.
21. Lin, T.; Chen, Q.; Ren, H.; Zhao, Y.; Miao, C.; Fu, S.; Chen, Q. Energy regeneration hydraulic system via a relief valve with energy regeneration unit. *Appl. Sci.* **2017**, *7*, 613. [CrossRef]
22. Kagawa, T.; Cai, M.; Kameya, H. Overall efficiency consideration of pneumatic systems including compressor, dryer, pipe and actuator. In Proceedings of the JFPS International Symposium on Fluid Power 2002, Nara, Japan, 12–15 November 2002; pp. 383–388.

*Appl. Sci.* **2018**, *8*, 406

23. Shi, Y.; Zhang, B.; Cai, M.; Zhang, X.D. Numerical simulation of volume-controlled mechanical ventilated respiratory system with 2 different lungs. *Int. J. Numer. Method Biomed. Eng.* **2016**, *33*, 2852. [CrossRef] [PubMed]

24. Gai, Y.; Cai, M.; Shi, Y. Analytical and experimental study on complex compressed air pipe network. *Chin. J. Mech. Eng. (Engl. Ed.)* **2015**, *28*, 1023–1029. [CrossRef]

25. Kaitwanidvilai, S.; Parnichkun, M. Position control of a pneumatic servo system by genetic algorithm based fixed-structure robust h, loop shaping control. In Proceedings of the IECON 2004—30th Annual Conference of IEEE Industrial Electronics Society, Busan, Korea, 2–6 November 2004; IEEE Computer Society: Busan, Korea, 2004; pp. 2246–2251.

26. Li, J.P. *Pneumatic Transmission System Dynamics*; South China University of Technology Press: Guangzhou, China, 1991; pp. 32–35. ISBN 7562302677.

27. Yin, Y.B. *High Speed Pneumatic Theory and Technology*; Shanghai Science and Technology Press: Shanghai, China, 2014; pp. 165–171. ISBN 978-7-5478-2068-1.

28. Ning, F.; Shi, Y.; Cai, M.; Wang, Y.; Xu, W. Research progress of related technologies of electric-pneumatic pressure proportional valves. *Appl. Sci.* **2017**, *7*, 1074. [CrossRef]

29. Shi, Y.; Ren, S.; Cai, M.; Xu, W.; Deng, Q. Pressure dynamic characteristics of pressure controlled ventilation system of a lung simulator. *Comput. Math. Methods Med.* **2014**, 761712–761722. [CrossRef] [PubMed]

30. Shearer, J.L. Study of pneumatic processes in the continuous control of motion with compressed air-I. *Trans. ASME* **1956**, *2*, 233–242.

31. Wu, Z.S. *Pneumatic Transmission and Control*, 2nd ed.; Harbin Institute of Technology Press: Harbin, China, 2009; pp. 197–199. ISBN 7-5603-0989-5.

32. Cai, M.; Wang, Y.; Shi, Y.; Liang, H. Output dynamic control of a late model sustainable energy automobile system with nonlinearity. *Adv. Mech. Eng.* **2016**, *8*. [CrossRef]

33. Yang, Z.R.; Hua, K.Q.; Xu, Y. *Electro-Hydraulic Ratio and Servo Control*; Metallurgical Industry Press: Beijing, China, 2015; pp. 108–120. ISBN 978-7-5024-4622-2.

34. Storn, R.; Price, K. Differential evolution—A simple and efficient heuristic for global optimization over continuous spaces. *J. Glob. Optim.* **1997**, *11*, 341–359. [CrossRef]

35. Wu, L.; Wang, Y.; Zhou, S.; Tan, W. Design of PID controller with incomplete derivation based on differential evolution algorithm. *J. Syst. Eng. Electron.* **2008**, *19*, 578–583.

36. Zhang, X.; Zhang, X. Shift based adaptive differential evolution for PID controller designs using swarm intelligence algorithm. *Clust. Comput.* **2017**, *20*, 291–299. [CrossRef]

37. Liu, J.K. *Advanced Pid Control Matlab Simulation*, 4th ed.; Publishing House of Electronics Industry: Beijing, China, 2016; pp. 331–332. ISBN 978-7-121-28846-3.

38. Cheng, L.; Chen, J.; Chen, M.; Xu, J.; Wang, W.; Wang, T.; Guo, J. Adaptive differential evolution algorithm identification of photoelectric tracking servo system. *Hongwai Yu Jiguang Gongcheng Infrared Laser Eng.* **2016**, *45*. [CrossRef]

39. Andromeda, T.; Yahya, A.; Samion, S.; Baharom, A.; Hashim, N.L. Differential evolution for optimization of PID gain in electrical discharge machining control system. *Trans. Can. Soc. Mech. Eng.* **2013**, *37*, 293–301.

40. Niu, J.; Shi, Y.; Cai, M.; Cao, Z.; Wang, D.; Zhang, Z.; Zhang, X.D. Detection of sputum by interpreting the time-frequency distribution of respiratory sound signal using image processing techniques. *Bioinformatics* **2018**, *34*, 820–827. [CrossRef] [PubMed]

*Article*

# Pressure Transient Model of Water-Hydraulic Pipelines with Cavitation

**Dan Jiang *, Cong Ren, Tianyang Zhao and Wenzhi Cao**

School of Mechanical and Electrical Engineering, University of Electronic Science and Technology of China, Chengdu 611731, China; congR0420@163.com (C.R.); zty1990321@gmail.com (T.Z.); wenzhicao2014@gmail.com (W.C.)

* Correspondence: jdan2002@uestc.edu.cn

Received: 15 February 2018; Accepted: 1 March 2018; Published: 7 March 2018

**Abstract:** Transient pressure investigation of water-hydraulic pipelines is a challenge in the fluid transmission field, since the flow continuity equation and momentum equation are partial differential, and the vaporous cavitation has high dynamics; the frictional force caused by fluid viscosity is especially uncertain. In this study, due to the different transient pressure dynamics in upstream and downstream pipelines, the finite difference method (FDM) is adopted to handle pressure transients with and without cavitation, as well as steady friction and frequency-dependent unsteady friction. Different from the traditional method of characteristics (MOC), the FDM is advantageous in terms of the simple and convenient computation. Furthermore, the mechanism of cavitation growth and collapse are captured both upstream and downstream of the water-hydraulic pipeline, i.e., the cavitation start time, the end time, the duration, the maximum volume, and the corresponding time points. By referring to the experimental results of two previous works, the comparative simulation results of two computation methods are verified in experimental water-hydraulic pipelines, which indicates that the finite difference method shows better data consistency than the MOC.

**Keywords:** water-hydraulic pipelines; pressure transients; cavitation; finite difference method; method of characteristics

## 1. Introduction

Transient pressure pulsations generated by the rapid closure of a valve would easily cause hydraulic pipeline systems to burst because the pressure pulsations exceed the safe operating range of such pipelines. Violent pressure pulsations result in cavitation growth and collapse in these systems. To study the mechanism of cavitation during pressure transient pulsations, it is necessary to investigate the cavitation appearance, its volume evaluation, and the effect on the pipeline and hydraulic systems.

Pressure transients with cavitation in pipelines have been investigated by many researchers. Kojima et al. [1] presented the gas-nonbubbly flow model to predict pressure increments, which involved cavitation on the downstream side of the pipeline as a valve was instantaneously closed. They used a water–glycol mixture and an oil/water emulsion fluid including mineral oil as working fluids and compared the computed pressure pulsations with experimental results. Chaudhry et al. [2,3] then proposed a MacCormack scheme and a Gabutti scheme for pressure transient analysis, which was verified both in computed simulation and experimental studies. Although modeling accuracy was achieved, discrepancies in the pressure magnitudes between simulations and experiments were found. Transient pressure pulsations often lead to unexpected chatter, overshooting, and a zero bias of tracking error in the electrohydraulic control system [4–7]. Shu et al. [8–10] developed a vaporous cavitation model that used a two-phase homogeneous equilibrium to simulate pipeline pressure transients with upstream, midstream, and downstream cavitation. Bergant et al. [11,12] discussed three cavitation models: the discrete vapor cavity model (DVCM), the discrete gas cavity model (DGCM), and the generalized interface vaporous cavitation model (GIVCM). The comparative results of the three cavitation models indicated that the GIVCM was able to directly obtain the regions of vaporous cavitation occurrence. Jiang et al. [13–16], via genetic algorithms, developed the parametric identification of the gas bubble model and the frequency-dependent friction model. Parametric identification and noise suppression are also addressed in mechanical ventilation [17–20]. Sadafi et al. [21] recently studied water hammers with cavitation in a simple reservoir-pipeline-valve system and a pumping station. Karadžić et al. [22] verified the robustness of the DGCM via analysis of the experimental results. Iglesias-Rey et al. [23] performed a detailed study of the actual behavior of different valves (both air intake and exhaust) and described the mathematical characterization of different commercial valves. Fuertes-Miquel et al. [24] presented a numerical modeling of pipelines with air pockets and air valves to study the behavior of the air inside pipes as the air was expelled through air valves. Majd et al. [25] investigated the unsteady flow of a non-Newtonian fluid due to the instantaneous valve closure in a pipeline. Comparison revealed a remarkable deviation in pressure history and velocity profile with respect to the water hammer in Newtonian fluids. Zhou et al. [26] adopted a second-order finite volume method for cavitation in the water column separation of pipelines to capture vapor cavities and predict their growth and collapse. Wang et al. [27] adopted a two-dimensional CFD model to characterize liquid column separation. The simulation results revealed the formation of an intermediate cavity and both the location and shape of the region undergoing distributed vaporous cavitation. Himr [28] also studied water hammers with column separation as a one-dimensional flow. The volume of the cavity was determined by Gibson's method, and the air bubbles were considered to affect the speed of sound.

Thanks to the research development of transient pressure in the fluid transmission field, the main contributions of this paper are as follows:

(i) Different from [10], a pressure transient model of water-hydraulic pipelines is constructed to reveal both the transient pressure magnitude and the dynamic characteristics of cavitation volume. To the authors' best knowledge, there have been few attempts to predict cavitation volume changes and illustrate its influence on pressure transients in hydraulic pipelines, as described in [14,15]. Although the simplified cavitation model is based on a flow continuity principle [29], the frictional force in pipelines involving the steady friction force and the frequency-dependent unsteady friction should be a primary consideration in pressure transient analysis.

(ii) Different from the MOC, the FDM is adopted to estimate the magnitudes of the pressure peaks and the changes in cavitation volume to adapt the transient pressure both with and without cavitation. Simultaneously, the respective boundary conditions of both the upstream and downstream sides of the valve are also considered. A comparison with the MOC is made; results are verified by the percentage of the integral of the absolute difference (IAD) between simulation and experimental reference results.

## 2. Mathematical Models

### 2.1. Basic Equations without Cavitation

The simple water-hydraulic pipe is shown in Figure 1. To illustrate the propagation and reflection of the pressure transients, the sequence of events, which is caused by a valve closure in the middle of the pipe connected with the downstream and upstream tanks, will be discussed. Without a loss of generality, the pipe diameter is assumed to be constant, and the released gas is negligible.

The general model of pressure transients in the pipeline involves the continuity equation and the momentum equation, which are mentioned in Wylie et al. [29]. The continuity equation is derived from the mass conservation law as follows:

$$\frac{1}{c_0^2}\frac{\partial p}{\partial t} + \frac{\rho}{\pi r_0^2}\frac{\partial q}{\partial x} = 0, \tag{1}$$

where $p$ is the pressure in pipeline, $q$ is the flow rate, $\rho$ is the density of fluid, $c_0$ is the acoustic velocity in the fluid, $r_0$ is the radius of the pipeline, $x$ is the spatial variable, and $t$ is the time variable.

Meanwhile, the equation of momentum is constructed by Newton's law of motion as follows:

$$\frac{\rho}{\pi r_0^2}\frac{\partial q}{\partial t} + \frac{\partial p}{\partial x} + F(q) = 0. \tag{2}$$

In Equation (1), $c_0$ can be given by

$$c_0 = \sqrt{B_{eff}/\rho}, \tag{3}$$

where $B_{eff}$ is the effective bulk modulus.

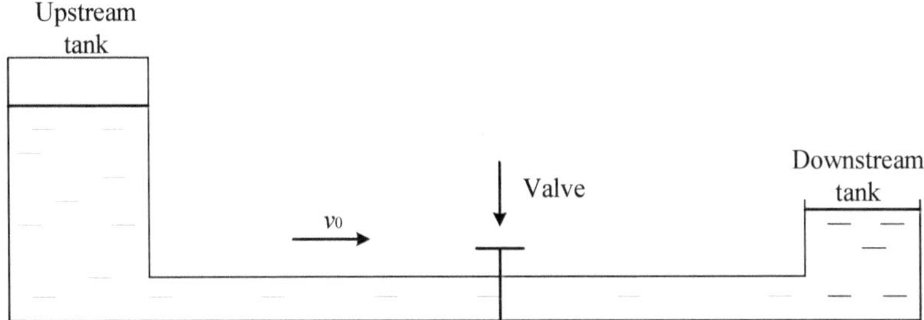


**Figure 1.** Tank-pipeline-valve system.

### 2.2. Continuity Equation under Vaporous Cavitation Condition

The cavitation normally arises when the pressure transients in pipelines are closed to the vapor pressure. If the pressure in the pipeline drops to or below the vapor pressure, vapor cavitation will form. Furthermore, if the pressure stays at the level of the vapor pressure and the cavitation size has approached a critical diameter, the cavitation will continue to grow rapidly. However, the cavitation will be unstable and collapse until the pressure is greater than the vapor pressure.

Pettersson et al. [30] and Harris et al. [31] presented a relatively simple cavitation model that illustrates the dynamic characteristics of piston pumps and can capture the key aspects of cavitation. Since the pressure in the pipeline element is assumed to be the vapor pressure under a vaporous cavitation condition, according to the flow continuity principle, the dynamics of the cavitation volume $V_{cav}$ is given by

$$\frac{dV_{cav}}{dt} = q_{out} - q_{in}, \tag{4}$$

where $q_{out}$ and $q_{in}$ are the outflow rate and inflow rate of an element in the pipe, respectively.

### 2.3. Frictional Items

Due to the fluid viscosity, the frictional force $F(q)$ in Equation (2) can be described as the sum of the steady friction item and the frequency-dependent unsteady friction item. Zielke [32] considered the frequency-dependent friction item as some weighting functions described in the frequency domain via Laplace transform. Subsequently, Trikha [33] proposed three exponent function items to estimate the frequency-dependent friction. Then, Taylor et al. [34] optimized the coefficients of the Trikha model and proposed an approximate model with four exponent function items as follows:

$$F(q) = F_0 + \frac{1}{2}\sum_{i=1}^{4} Y_i, \tag{5}$$

where the first item $F_0$ is the steady friction, and the second item is the frequency-dependent unsteady friction. Based on the Darcy–Weisbach equation, $F_0$ can be expressed as

$$F_0 = \frac{\rho f v |v|}{4 r_0}.$$
(6)

The four items of frequency-dependent friction $Y_i$ can be computed by

$$\begin{cases} \dfrac{\partial Y_i}{\partial t} = -\dfrac{n_i \mu}{\rho r_0^2} Y_i + m_i \dfrac{\partial F_0}{\partial t} \\ Y_i(0) = 0 \end{cases} , \quad i = 1, \ldots, 4$$
(7)

where the constants $n_i$ and $m_i$ are listed in Table 1.

**Table 1.** $n_i$ and $m_i$.

| i | 1 | 2 | 3 | 4 |
|---|---|---|---|---|
| $n_i$ | $3.9479 \times 10^1$ | $2.9829 \times 10^2$ | $2.2279 \times 10^3$ | $8.8782 \times 10^4$ |
| $m_i$ | $2.0141$ | $5.3946$ | $1.6259 \times 10^1$ | $3.2048 \times 10^2$ |

## 3. Simulation Methods

Two predictive methods, i.e., MOC and FDM, are presented. In order to solve the two partial differential equations in terms of pressure and flow rate, the pipeline is divided into $n$ elements of equal length $\Delta x = L/n$, where $L$ is the pipeline length. It should be noted that different test values of the pipeline length $L$ can be selected in simulation. The FDM is implemented to describe pressure transients on the downstream and upstream sides of the valve, respectively.

### 3.1. Method of Characteristics

The continuity equation (Equation (1)) should be solved together with the momentum equation (Equation (2)) since they are partial differential forms about the two unknown parameters $p$ and $q$. However, via the MOC, they can be transformed into ordinary differential equations (Equations (8) and (9)) along the characteristic lines $C^+$ and $C^-$.

$$C^+ : \begin{cases} \dfrac{\rho c_0}{\pi r_0^2} \dfrac{dq}{dt} + \dfrac{dp}{dt} + c_0 F(q) = 0 \\ \dfrac{dx}{dt} = c_0 \end{cases}$$
(8)

$$C^- : \begin{cases} \dfrac{\rho c_0}{\pi r_0^2} \dfrac{dq}{dt} - \dfrac{dp}{dt} + c_0 F(q) = 0 \\ \dfrac{dx}{dt} = -c_0 \end{cases}$$
(9)

As shown in Figure 2, the pressure and flow rate values at points $A$ ($p_A$ and $q_A$) and $B$ ($p_B$ and $q_B$) are known. Integrating Equations (8) and (9) along the characteristic lines $C^+$ and $C^-$, the following Equation (10) is obtained to further derive the pressure and flow rate at point $P$ ($p_P$ and $q_P$).

$$\begin{cases} \dfrac{\rho}{\pi r_0^2} q_P + \dfrac{1}{c_0} p_P = \dfrac{\rho}{\pi r_0^2} q_A + \dfrac{1}{c_0} p_A - \dfrac{\Delta x}{c_0} F(q_A) \\ \dfrac{\rho}{\pi r_0^2} q_P - \dfrac{1}{c_0} p_P = \dfrac{\rho}{\pi r_0^2} q_B - \dfrac{1}{c_0} p_B - \dfrac{\Delta x}{c_0} F(q_B) \end{cases}$$
(10)

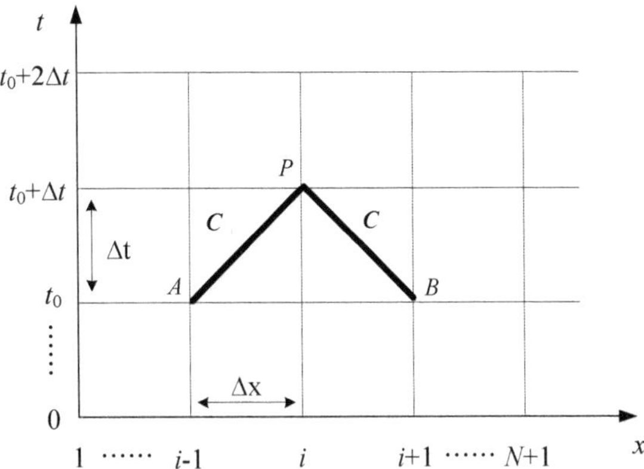

**Figure 2.** Method of characteristics.

Then, from Equation (10), the values of $p_P$ and $q_P$ can be obtained by

$$\begin{cases} p_P = \dfrac{c_0}{2}C_L - C_R \\[2mm] q_P = \dfrac{\pi r_0^2}{2\rho}C_L + C_R \end{cases}, \tag{11}$$

where

$$\begin{cases} C_L = \dfrac{\rho}{A}q_A + \dfrac{1}{c_0}p_A - \dfrac{\Delta x}{c_0}F(q_A) \\[2mm] C_R = \dfrac{\rho}{A}q_B - \dfrac{1}{c_0}p_B - \dfrac{\Delta x}{c_0}F(q_B) \end{cases}. \tag{12}$$

Details of the MOC are given by Wylie et al. [29] and Chaudhry et al. [3]. Incorporated with Equation (4), this method determines the time at which cavitation first arises in respective elements and the volume of cavitation. Furthermore, the method also determines whether cavitation has already collapsed at each time step $\Delta t = \Delta x / c_0$ and the time at which cavitation occurs again.

### 3.2. Finite Difference Method

The flow rate and pressure inside the pipeline are constructed as n-dimensional vectors as follows: $q = (q_1, q_2, \ldots, q_n)^{\mathrm{T}}$, and $p = (p_1, p_2, \ldots, p_n)^{\mathrm{T}}$. Here, the first element is close to the valve and the last element approaches the upstream or downstream tank. This simulation scheme has been implemented by the Matlab/Simulink platform and the partial derivative terms in time domain $\partial/\partial t$ can be readily calculated by the integral block of Simulink.

#### 3.2.1. The Downstream Side of the Valve

Pressure transients on the downstream side of the valve were investigated. The sketch is illustrated in Figure 3. Here, the initial velocity $v_0$ is constant and the valve is suddenly shut off. For the boundary condition, the flow rate in the element close to the valve is set to zero, and the pressure in the element close to downstream tank is constant. If the boundary condition $q_{valve} = 0$ is assumed to be the first element together with the other $n - 1$ elements, the Selector Block is used to re-order specified elements of the vector. A new flow rate vector $q'$ for the case of the flow rate is formed such that

$$q' = (q_{valve}, q_1, \ldots, q_{n-1})^{\mathrm{T}}. \tag{13}$$

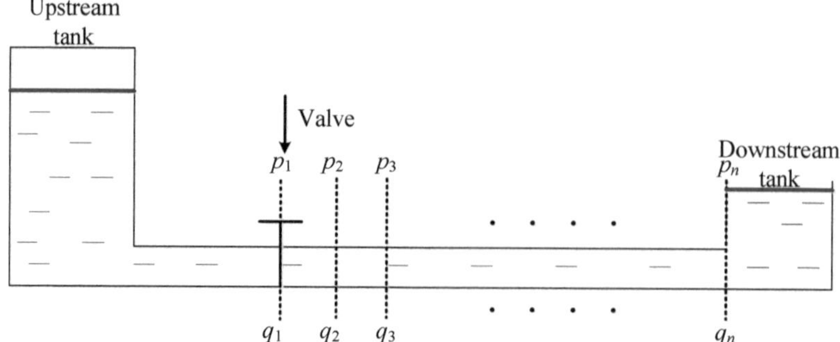


**Figure 3.** On the downstream side of a valve.

Based on the first-order upwind difference scheme, the derivative item $\partial q/\partial x$ can be expressed as follows:

$$\frac{\partial q}{\partial x} = \frac{q - q'}{\Delta x}. \tag{14}$$

For the pressure vector, the Selector Block is used to construct a new pressure vector $p'$ as follows:

$$p' = (p_2, \ldots, p_n, p_{resd})^{\mathrm{T}}, \tag{15}$$

where $p_{resd}$ is the boundary condition, which is equal to the pressure in the downstream tank.

$\partial p/\partial x$ is also given by

$$\frac{\partial p}{\partial x} = \frac{p' - p}{\Delta x}. \tag{16}$$

### 3.2.2. The Upstream Side of the Valve

Similar to the downstream side of the valve, an upstream pipeline model can be constructed. The variables of flow rate and pressure in n elements along the pipeline are considered as the vectors as shown in Figure 4. The first element is close to the valve, and the initial velocity is positive constant. For the case of the flow rate, a new flow rate vector $q'$ is formed with the corresponding boundary condition, such that

$$q' = (q_{valve}, q_1, \ldots, q_{n-1})^{\mathrm{T}}. \tag{17}$$

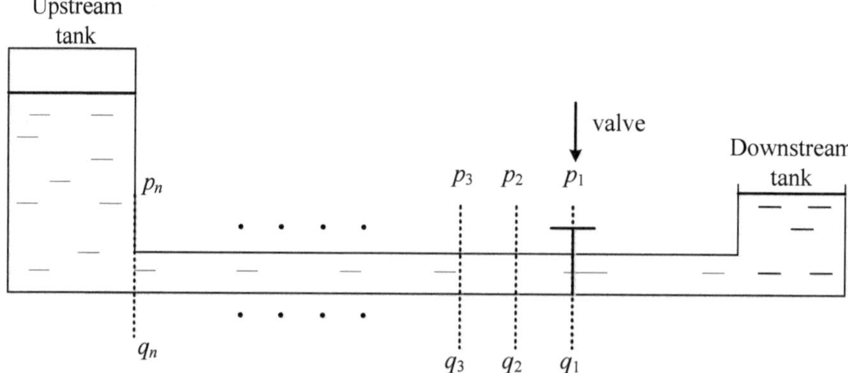


**Figure 4.** On the upstream side of a valve.

Different from the condition of the downstream side of the valve, according to the first-order upwind difference scheme, $\partial q/\partial x$ can be expressed as follows:

$$\frac{\partial q}{\partial x} = \frac{q' - q}{\Delta x}.\tag{18}$$

For the pressure vector, the Selector Block is used to create a new pressure vector $p'$:

$$p' = (p_2, \ldots, p_n, p_{resu})^{\mathrm{T}},\tag{19}$$

where $p_{resu}$ is the boundary condition of the pressure in the upstream tank. Thus, the $\partial p / \partial x$ can be described as follows:

$$\frac{\partial p}{\partial x} = \frac{p - p'}{\Delta x}.\tag{20}$$

## 4. Simulation Results

### 4.1. Case 1: Pressure Transients without Cavitation on the Downstream Side of Valve

The experimental results of the transient pressure pulsations close to the valve in the horizontal downstream pipeline are given by Vitkovsky et al. [35] and the related parameters of the tested pipeline are listed in Table 2. Here, the element number $n$ is selected as 30 in the simulation. The sensitivity of this element number $n$ has been discussed in [13]. The corresponding experimental results of transient pressure pulsations close to the valve are shown as the solid line in Figure 5.

**Table 2.** Parameters of pressure transients without cavitation on the downstream side of the valve.

| Parameter | Value |
| --- | --- |
| Upstream tank pressure $p_{resu}$ (bar) | 4.25 |
| Downstream tank pressure $p_{resd}$ (bar) | 4.22 |
| Pipe radius $r_0$ (mm) | 11.05 |
| Pipe length $L$ (m) | 37.2 |
| Water density $\rho$ (kg/m$^3$) | 1000 |
| Initial velocity $v_0$ (m/s) | 0.3 |
| Acoustic velocity in the fluid $c_0$ (m/s) | 1319 |

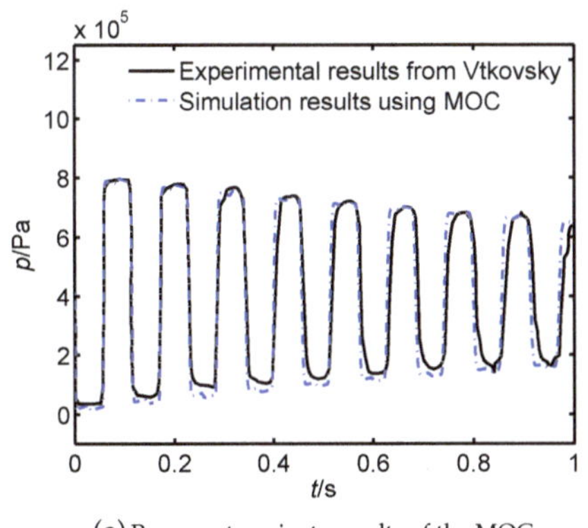


(**a**) Pressure transients results of the MOC

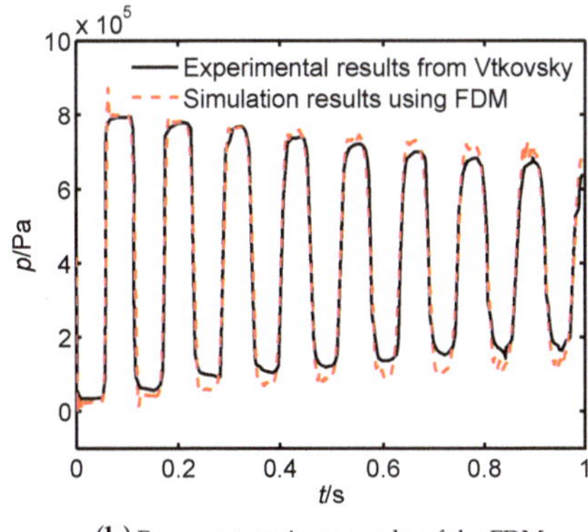


(**b**) Pressure transients results of the FDM

**Figure 5.** Simulation and experimental pressure transients without cavitation on the downstream side of the valve.

Figure 5 denotes that, in the downstream pipeline, the pressure at the vicinity of the valve is reduced quickly when the valve is closed. At the same time, the negative pressure wave propagates to the downstream tank. Then, the positive pressure wave is reflected from the downstream tank and travels back to the valve, which leads to the first positive pressure peak. This process may be repeated several times before the fluid energy is dissipated due to the frictional force of the pipeline.

As shown in the experimental results, from 0 to 1 s, the attenuated peaks of the pressure pulsations are decreased very slowly. Because of the lower frictional force, the magnitudes of the pressure peaks decay slower in water, which is different from the corresponding pressure transient results with the working fluid as hydraulic oil (Jiang et al. [13]). Thus, if the pressure pulsations are always greater than the saturated vapor pressure, no cavitation forms.

For comparison, the simulation results of MOC are illustrated as the dash-dotted line in Figure 5a. The simulation results of the FDM platform are presented as the dashed line in Figure 5b. It can be seen that, from 0.4 to 1 s, the phase difference between the MOC simulation and the experimental results is more obvious. However, the simulation results of the FDM are still consistent with the experimental pressure results.

To further compare the two predictive methods, the error between the simulation and the experimental results was evaluated by the percentage of the integral of absolute difference (IAD) as follows (Rabie et al. [36]):

$$IAD = \frac{\int_0^T \left| p_{Lth} - p_{L\exp} \right| dt}{p_{Lss}T} \times 100\%. \tag{21}$$

The IAD results of the two predictive methods in the three cases are listed in Table 3. In Case 1, the final steady-state pressure at the valve is equal to the downstream tank pressure (4.22 bar), and the IAD of the FDM is about 0.05%. Thus, the simulation of the FDM is consistent with the experimental result, which is superior to the MOC (IAD = 0.91%).

**Table 3.** The integral of absolute difference (IAD) of the method of characteristics (MOC) and the finite difference method (FDM).

| Case | $p_{Lss}$ (bar) | IAD of MOC | IAD of FDM |
|:---:|:---:|:---:|:---:|
| 1 | 4.22 | 0.91% | 0.05% |
| 2 | 0.98065 | 2.75% | 2.47% |
| 3 | 4.90325 | 12.39% | 10.84% |

*4.2. Case 2: Pressure Transients with Cavitation on the Downstream Side of Valve*

The case of transient pressure pulsations with cavitation in the horizontal downstream pipeline was also investigated. Some experimental parameters from Sanada et al. [37] are listed in Table 4. The corresponding experimental results are shown as the solid line in Figure 6. As the valve is quickly closed, the pressure reduces and stays at vapor pressure for about 3 s. The pressure then drops again and stays at vapor pressure for about 1.5 s. For the third time, the pressure falls and stays at vapor pressure for about 1 s.

**Table 4.** Parameters of pressure transients with cavitation on the downstream side of the valve.

| Parameter | Value |
| --- | --- |
| Upstream pressure $p_{resu}$ (bar) | 6.55164 |
| Downstream pressure $p_{resd}$ (bar) | 0.98065 |
| Pipe radius $r_0$ (mm) | 7.6 |
| Pipe length $L$ (m) | 200 |
| Water density $\rho$ (kg/m$^3$) | 1000 |
| Initial velocity $v_0$ (m/s) | 1.5 |
| Viscosity of the fluid $c_0$ (cP) | 1 |

The results obtained from the MOC and the FDM are also shown in Figure 6. It is clear that obvious differences exist between the MOC simulation and the experimental results, especially in terms of the phase differences of the subsequent peaks. However, the results of the FDM via Matlab/Simulink Platform are consistent with the experimental results.

As listed in Table 3, for Case 2, the final steady-state pressure at the valve is 0.98065 bar, which is equal to the downstream tank pressure. Different from Case 1, the two predictive methods have similar effects (the IADs of the two predictive methods are 2.75% and 2.47%).

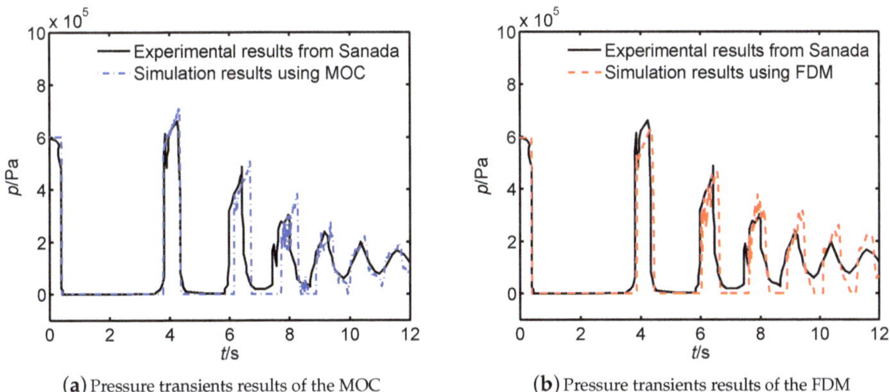



(**a**) Pressure transients results of the MOC        (**b**) Pressure transients results of the FDM

**Figure 6.** Simulation results and experimental data of pressure transients with cavitation on the downstream side of the valve.

The corresponding cavitation volumes in the element close to the valve predicted by the FDM and the MOC are shown in Figure 7. The trends of vaporous cavitation volume under three cases are listed in Table 5, which includes the cavitation start time, the end time, the duration, the maximum volume, and the corresponding time points.

Similar to the MOC, the FDM is also able to track the trends of cavitation volume. The results indicate that the computed pressure peak declines to the saturated vapor pressure after the valve is rapidly closed and after cavitation forms. However, this new cavitation collapses at 3.87 s (FDM) and 3.77 s (MOC). The maximum volumes of cavitation first are $1.372 \times 10^{-4}$ m$^3$ (FDM) and $1.401 \times 10^{-4}$ m$^3$ (MOC). When the pressure declines again, cavitation is generated again, but it is much smaller ($1.892 \times 10^{-5}$ m$^3$ from the FDM and $4.155 \times 10^{-5}$ m$^3$ from the MOC) than the first instance. Once again, cavitation collapses at the arrival of the third pressure peak. The durations of the third cavitation is 1.08 s (FDM) and 0.99 s (MOC). Thus, over this short period (12 s), the cavitation demonstrates generation and collapse three times.

**Table 5.** Trends of cavitation volume.

| Times | Method | MOC (case 1) | FDM (case 1) | MOC (case 2) | FDM (case 2) | MOC (case 3) | FDM (case 3) |
|---|---|---|---|---|---|---|---|
| 1st time | start time (s) | | | 0.40 | 0.38 | 0.60 | 0.65 |
| | end time (s) | | | 3.77 | 3.87 | 1.21 | 1.24 |
| | duration (s) | – | – | 3.37 | 3.49 | 0.61 | 0.59 |
| | maximum volume time (s) | | | 2.32 | 2.11 | 1.08 | 1.21 |
| | maximum volume (m$^3$) | | | $1.401 \times 10^{-4}$ | $1.372 \times 10^{-4}$ | $1.305 \times 10^{-5}$ | $3.955 \times 10^{-6}$ |
| 2nd time | start time (s) | | | 4.34 | 4.37 | 2.05 | 1.98 |
| | end time (s) | | | 6.14 | 6.03 | 2.17 | 2.23 |
| | duration (s) | – | – | 1.80 | 1.66 | 0.12 | 0.25 |
| | maximum volume time (s) | | | 5.12 | 5.73 | 2.15 | 2.21 |
| | maximum volume (m$^3$) | | | $4.155 \times 10^{-5}$ | $1.892 \times 10^{-5}$ | $1.281 \times 10^{-6}$ | $6.785 \times 10^{-7}$ |
| 3rd time | start time (s) | | | 6.72 | 6.54 | | |
| | end time (s) | | | 7.71 | 7.62 | | |
| | duration (s) | – | – | 0.99 | 1.08 | – | – |
| | maximum volume time (s) | | | 7.46 | 6.76 | | |
| | maximum volume (m$^3$) | | | $1.009 \times 10^{-5}$ | $1.535 \times 10^{-6}$ | | |

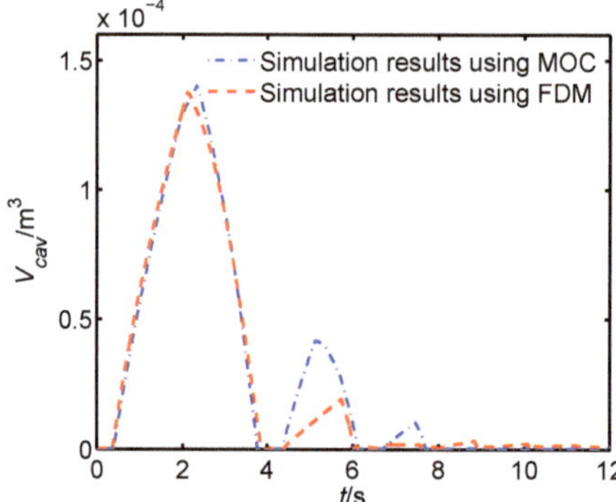


**Figure 7.** Simulation results of the cavitation volume in the first element on the downstream side of the valve using the MOC and the FDM.

### 4.3. Case 3: Pressure Transients with Cavitation on the Upstream Side of the Valve

Sanada et al. [37] also provided parameters of tested pipelines in the case of transient pressure pulsations in a horizontal upstream pipeline, as listed in Table 6. Compared with Case 2 listed in Table 4, the parameters of the test pipeline are the same, except for the values of the upstream pressure and the initial velocity.

**Table 6.** Parameters of pressure transients with cavitation on the upstream side of the valve.

| Parameter | Value |
|---|---|
| Upstream pressure $p_{resu}$ (bar) | 4.90325 |
| Downstream pressure $p_{resd}$ (bar) | 0.98065 |
| Pipe radius $r_0$ (mm) | 7.6 |
| Pipe length $L$ (m) | 200 |
| Water density $\rho$ (kg/m$^3$) | 1000 |
| Initial velocity $v_0$ (m/s) | 1.45 |
| Viscosity of fluid $c_0$ (cP) | 1 |

Figure 8 demonstrates the sequence of pressures with cavitation caused by instant valve closure. Compared with the pressure pulsations in the downstream pipeline, after a sudden valve closure, the fluid is brought to rest, firstly causing a high pressure peak at the upstream side of the valve.

Experimental results from Sanada (the solid line in Figure 8) show that the initial pressure at the valve is about $16 \times 10^5$ Pa when the valve is closed. It then reduces to the vapor pressure and keeps a steady state until about 0.5 s. Upon collapse of the cavitation, another pressure wave is generated at the valve. The subsequent pressure peak is reduced because of the friction force in the pipeline.

The figure also shows the simulation results from the MOC and the FDM. For the case of the upstream side of the valve, the final steady-state pressure at the valve is equal to the upstream tank pressure (4.90 bar). As listed in Table 3, the IADs of the MOC and the FDM are 12.39% and 10.84%. It is clear that the results of the FDM has much better consistency with the experimental results than those using the MOC.

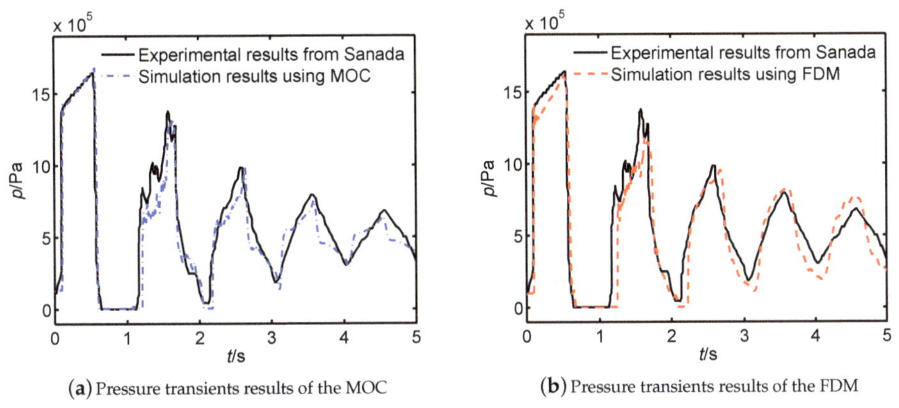


(**a**) Pressure transients results of the MOC       (**b**) Pressure transients results of the FDM

**Figure 8.** Simulation results and experimental data of pressure pulsations on the upstream side of the valve.

The corresponding cavitation volumes in the element close to the valve are shown in Figure 9. The maximum size of the vaporous cavity is $3.955 \times 10^{-6}$ m$^3$ (the duration is about 0.59 s) using the FDM and $1.305 \times 10^{-5}$ m$^3$ (the duration is about 0.61 s) using the MOC. When the pressure reduces to vapor pressure again, using the FDM, the second cavity has a volume of $6.785 \times 10^{-7}$ m$^3$ and a duration of 0.25 s; however, using the MOC, the cavity has a volume of $1.281 \times 10^{-6}$ m$^3$ and a duration of 0.12 s.

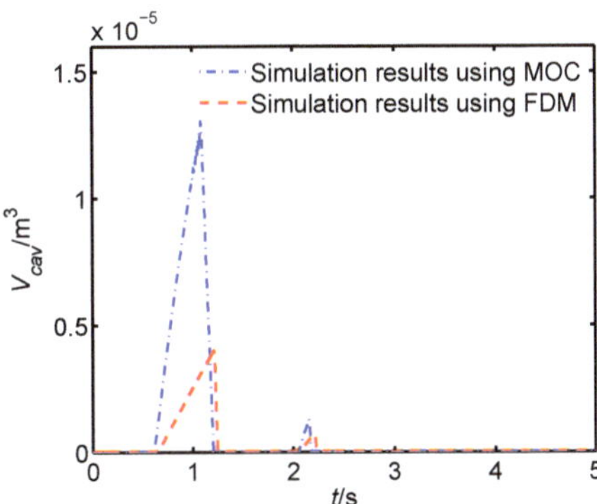


**Figure 9.** Simulation results of the cavitation volume in the first element between the MOC and the FDM.

As listed in Table 5, based on a comparison between Case 2 and Case 3, the durations of cavitation are much longer and the maximum cavitation volumes are larger on the downstream side of the valve. It is clear that cavitation is more likely to occur on the downstream side of the valve.

## 5. Conclusions

To reveal the mechanism of cavitation growth and collapse both on upstream and downstream of the water-hydraulic pipeline, this paper proposes the finite difference method (FDM) for determining the transient pressure to estimate pipeline pressure transients caused by sudden changes in fluid velocity. Firstly, the dynamic model of cavitation volume was derived during pressure transients. Then, the cavitation appearance durations and volume changes were analyzed. Furthermore, the frictional force model with the steady and frequency-dependent unsteady items were constructed in the proposed dynamic model. By referring to experimental results in [35,37], the simulation results of two computation methods were verified to indicate that the proposed FDM for transient pressure estimation has the following two advantages:

(i) The FDM is consistent with experimental results, which is improvement over the MOC in terms of the phase differences and magnitudes of the pressure peaks.

(ii) The FDM estimates not only the magnitudes of the pressure peaks but also the changes in cavitation volume to adapt the transient pressure both with and without cavitation. By statistical results, the IAD values of the FDM are much more favorable than that of the MOC.

However, the aforementioned discussion assumes that no air is released during cavitation. In fact, water usually contains some dissolved air or gas. If the pressure declines under the saturation pressure, especially under vapor pressure with agitation, then a certain amount of air will be released as gas bubbles. Thus, these effects of gas bubbles on pressure transients with cavitation will be investigated in the near future.

**Acknowledgments:** The authors are grateful to the anonymous reviewers and the editor for their constructive comments. This research was supported by the National Natural Science Foundation of China (51205045, 61305092, 51775089) and the Open Foundation of the State Key Laboratory of Fluid Power & Mechatronic Systems (GZKF-201515).

**Author Contributions:** Dan Jiang conceived and designed the structure of this paper; Tianyang Zhao and Wenzhi Cao performed the simulation; Cong Ren reviewed the literature.

**Conflicts of Interest:** The authors declare no conflict of interest.

## Abbreviation

| | |
|---|---|
| IAD | Integral of absolute difference |
| $B_{eff}$ (Pa) | Effective bulk modulus |
| $c_0$ (m/s) | Acoustic velocity in the fluid |
| $f$ | Coefficient of Darcy–Weisbach |
| $F_0$ (N) | Steady friction |
| $F(q)$ (N) | Friction |
| $m_i$ | Weighting constant |
| $n_i$ | Weighting constant |
| $p$ | Vector of pressures at nodes |
| $p_A, p_B, p_P$ (Pa) | Pressure at points $A$, $B$, and $P$ |
| $p_{resu}$ (Pa) | Pressure in the upstream tank |
| $p_{resd}$ (Pa) | Pressure in the downstream tank |
| $p'$ | New vector of pressures at nodes |
| $p_{Lexp}$ | Experimental results of pressure transients at the valve |
| $p_{Lss}$ | Steady-state pressure at the valve |
| $p_{Lth}$ | Simulation results of pressures transients at the valve |
| $q$ | Vector of flow rate at nodes |

$q_A, q_B, q_P$ (m$^3$/s)    Flow rate at points $A$, $B$, and $P$
$q'$    New vector of flow rate at nodes
$q_{in}$ (m$^3$/s)    Inflow rate
$q_{out}$ (m$^3$/s)    Outflow rate
$r_0$ (m)    Radius of the pipeline
$v$ (m/s)    Velocity in the fluid
$v_0$ (m/s)    Initial velocity in the fluid
$V_{cav}$ (m$^3$)    Cavitation volume
$Y_i$ (N)    Weighting function
$\rho$ (kg/m$^3$)    Density of fluid
$\mu$ (Pa $\cdot$ s)    Viscosity of fluid

## References


1. Kojima, E.; Shinada, M.; Shindo, K. Fluid transient phenomena accompanied with column separation in fluid power pipeline. *Bull. JSME* **1984**, *27*, 2421–2429.
2. Chaudhry, M.H.; Bhallamudi, S.M.; Martin, C.S.; Naghash, M. Analysis of transient pressures in bubbly, homogeneous, gas-liquid mixtures. *J. Fluids Eng.-Trans. ASME* **1990**, *112*, 225–231.
3. Chaudhry, M.H. *Applied Hydraulic Transients*, 3rd ed.; Springer: New York, NY, USA, 2014.
4. Guo, Q.; Zhang, Y.; Celler, B.G.; Su, S.W. Backstepping control of electro-hydraulic system based on extended-state-observer with plant dynamics largely unknown. *IEEE Trans. Ind. Electron.* **2016**, *63*, 6909–6920.
5. Guo, Q.; Zhang, Y.; Celler, B.G.; Su, S.W. State-constrained control of single-rod electrohydraulic actuator with parametric uncertainty and load disturbance. *IEEE Trans. Control Syst. Technol.* **2017**, doi:10.1109/TCST.2017.2753167.
6. Guo, Q.; Wang, Q.; Liu, Y. Anti-windup control of electro-hydraulic system with load disturbance and modeling uncertainty. *IEEE Trans. Ind. Inform.* **2017**, doi:10.1109/TII.2017.2768106.
7. Guo, Q.; Yu, T.; Jiang, D. Robust $H_\infty$ positional control of 2-DOF robotic arm driven by electro-hydraulic servo system. *ISA Trans.* **2015**, *59*, 55–64.
8. Shu, J.J.; Burrows, C.R.; Edge, K.A. Pressure pulsation in reciprocating pump piping systems Part 1: Modelling. *Proc. Inst. Mech. Eng. Part I* **1997**, *211*, 229–237.
9. Shu, J.J. A finite element model and electronic analogue of pipeline pressure transients with frequency-dependent friction. *J. Fluids Eng.* **2003**, *125*, 194–199.
10. Shu, J.J. Modeling vaporous cavitation on fluid transient. *Int. J. Press. Vessels Pip.* **2003**, *80*, 187–195.
11. Bergant, A.; Simpson, A.R. Pipeline column separation flow regimes. *J. Hydraul. Eng.* **1999**, *125*, 835–848.
12. Bergant, A.; Simpson, A.R.; Tijsseling, A.S. Waterhammer with column separation: A historical review. *J. Fluids Struct.* **2006**, *22*, 135–171.
13. Jiang, D.; Li, S.J. Simulation of hydraulic pipeline pressure transients accompanying cavitation and gas bubbles using Matlab/Simulink. In Proceedings of the 2006 ASME Joint U.S.-European Fluids Engineering Summer Meeting, Miami, FL, USA, 17–20 July 2006; pp. 657–665.
14. Jiang, D.; Li, S.J.; Bao, G. Parameter identification of gas bubble model in pressure pulsations using genetic algorithms. *Acta Phys. Sin.* **2008**, *57*, 5072–5080.
15. Jiang, D.; Li, S.J.; Edge, K.A.; Zeng, W. Modeling and simulation of low pressure oil-hydraulic pipeline transients. *Comput. Fluids* **2012**, *67*, 79–86.
16. Jiang, D.; Li, S.J.; Yang, P.; Zhao, T.Y. Frequency-dependent friction in pipelines. *Chin. Phys. B* **2015**, *24*, 034701.
17. Shi, Y.; Wang, Y.; Cai, M.; Zhang, B.; Zhu, J. An aviation oxygen supply system based on a mechanical ventilation model. *Chin. J. Aeronaut.* **2018**, *31*, 197–204, doi:10.1016/j.cja.2017.10.008.
18. Shi, Y.; Zhang, B.; Cai, M.; Zhang, D. Numerical Simulation of volume-controlled mechanical ventilated respiratory system with two different lungs. *Int. J. Numer. Methods Biomed. Eng.* **2016**, *33*, 2852, doi:10.1002/cnm.2852.
19. Shi, Y.; Zhang, B.; Cai, M.; Xu, W. Coupling Effect of Double Lungs on a VCV Ventilator with Automatic Secretion Clearance Function. *IEEE/ACM Trans. Comput. Biol. Bioinform.* **2017**, doi:10.1109/TCBB.2017.2670079.
20. Shi, Y.; Wu, T.; Cai, M.; Wang, Y.; Xu, W. Energy conversion characteristics of a hydropneumatic transformer in a sustainable-energy vehicle. *Appl. Energy* **2016**, *171*, 77–85.

21. Sadafi, M.; Riasi, A.; Nourbakhsh, S.A. Cavitating flow during water hammer using a generalized interface vaporous cavitation model. *J. Fluids Struct.* **2012**, *34*, 190–201.
22. Karadžić, U.; Bulatović, V.; Bergant, A. Valve-Induced Water Hammer and Column Separation in a Pipeline Apparatus. *J. Mech. Eng.* **2014**, *60*, 742–754.
23. Iglesias-Rey, P.L.; Fuertes-Miquel, V.S.; Garcia-Mares, F.J.; Martínez-Solano, F.J. Characterization of Commercial Air Intake and Exhaust Valves. *Tecnol. Cienc. Agua* **2016**, *7*, 57–69.
24. Fuertes-Miquel, V.S.; López-Jiménez, P.A.; Martínez-Solano, F.J.; López-Patiño, G. Numerical modelling of pipelines with air pockets and air valves. *Can. J. Civ. Eng.* **2016**, *43*, 1052–1061.
25. Majd, A.; Ahmadi, A.; Keramat, A. Investigation of Non-Newtonian Fluid Effects during Transient Flows in a Pipeline. *J. Mech. Eng.* **2016**, *62*, 105–115.
26. Zhou, L.; Wang, H.; Liu, D.Y.; Ma, J.J.; Wang, P.; Xia, L. A second-order Finite Volume Method for pipe flow with water column separation. *J. Hydro-Environ. Res.* **2017**, *17*, 47–55.
27. Wang, H.; Zhou, L.; Liu, D.Y.; Karney, B.; Wang, P.; Xia, L.; Ma, J.J.; Xu, C. CFD Approach for column separation in water pipelines. *J. Hydraul. Eng.* **2016**, *142*, 04016036.
28. Himr, D. Investigation and numerical simulation of a water hammer with column separation. *J. Hydraul. Eng.* **2016**, *141*, 04014080
29. Wylie, E.B.; Streeter, V.L.; Suo, L.S. *Fluid Transients in Systems*; Prentice-Hall: Englewood Cliffs, NJ, USA, 1993.
30. Pettersson, M.; Weddfelt, K.; Palmberg, J.O. Modelling and measurement of cavitation and air release in a fluid power piston pump. In Proceedings of the Third Scandinavian International Conference on Fluid Power, Linköping, Sweden, 25–26 May 1993; p. 113.
31. Harris, R.M.; Edge, K.A.; Tillley, D.G. The suction dynamics of positive displacement axial piston pumps. *J. Dyn. Syst. Meas. Control* **1994**, *116*, 281–287.
32. Zielke, W. Frequency-dependent Friction in Transient Liquid Flow. *J. Basic Eng.* **1968**, *90*, 109–115.
33. Trikha, A.K. Efficient Method for Simulation Frequency-dependent Friction in Transient Liquid Flow. *J. Fluids Eng.* **1975**, *97*, 97–105.
34. Taylor, S.E.M.; Johnston, D.N.; Longmore, D.K. Modeling of transient flow in hydraulic pipelines. *Proc. Inst. Mech. Eng. Part I* **1997**, *211*, 447–456.
35. Vitkovsky, J.P.; Bergant, A.; Simpson, A.R.; Martin M.A.; Lambert, F. Systematic evaluation of one-dimensional unsteady friction models in simple pipelines. *J. Hydraul. Eng.* **2006**, *132*, 696–708.
36. Rabie, M.G.; Rabie, M. *Fluid Power Engineering*; McGraw-Hill Education: New York, NY, USA, 2009.
37. Sanada, K.; Kitagawa, A.; Takenaka, T. A study on analytical methods by classification of column separations in a water pipeline. *Bull. JSME* **1990**, *56*, 585–593.

 *applied sciences*

*Article*

# Dimensionless Energy Conversion Characteristics of an Air-Powered Hydraulic Vehicle

**Dongkai Shen, Qilong Chen and Yixuan Wang ***

School of Automation Science and Electrical Engineering, Beihang University, Beijing 100191, China;
shen_dk@buaa.edu.cn (D.S.); chenqilong@buaa.edu.cn (Q.C.)
* Correspondence: magic_wyx@163.com; Tel.: +10-156-5292-8692

Received: 11 February 2018; Accepted: 23 February 2018; Published: 28 February 2018

**Abstract:** Due to the advantages of resource conservation and less exhaust emissions, compressed air-powered vehicle has attracted more and more attention. To improve the power and efficiency of air-powered vehicle, an air-powered hydraulic vehicle was proposed. As the main part of the air-powered hydraulic vehicles, HP transformer (short for Hydropneumatic transformer) is used to convert the pneumatic power to higher hydraulic power. In this study, to illustrate the energy conversion characteristics of air-powered hydraulic vehicle, dimensionless mathematical model of the vehicle's working process was set up. Through experimental study on the vehicle, the dimensionless model was verified. Through simulation study on the vehicle, the following can be obtained: firstly, the increase of the hydraulic chamber orifice and the area ratio of the pistons can lead to a higher output power, while output pressure is just the opposite. Moreover, the increase of the output pressure and the aperture of the hydraulic chamber can lead to a higher efficiency, while area ratio of the pistons played the opposite role. This research can be referred to in the performance and design optimization of the HP transformers.

**Keywords:** hydropneumatic transformer; air-powered vehicle; working characteristic; modelling simulation; experimental study

---

## 1. Introduction

In recent years, to reducing emissions, new energy vehicles has been widely promoted. As a kind of new energy vehicle, air-powered vehicle has some advantages, such as simple structure, lightweight, explosion prevention, no pollution emissions, and so on. Hence, it has important potentials in mines, chemical plants and airports, as well as having attracted more and more engineers' attention [1–4].

In 2013, Shaw et al. proposed an air-oil converter using an equal-area cylinder [5]. The converter was driven by the compressed air. In this article, the system efficiency was shown to be nearly 50% at 200 rpm according to the calculations and experimental study. However, because of the areas of air cylinder and the oil cylinder are equal, so the pressure of the compressed air must be higher than the output oil pressure, so the residual air pressure may lead to a lower efficiency. Researchers of Yijuan Zang in the same year has identified a new pressurizing system which can achieve a high pressure output with the use of a new type bidirectional Pneumatic-hydraulic converter. Because of the saving of the hydraulic station, it can save the space and can reduce the energy consumption. However, one of the key issues in this study is that the two hydraulic chamber may have an effect on the efficiency [6]. So in 2016, Yan Shi, Maolin Cai et al. proposed an expansion energy used Hydropneumatic transformer (short for EEUHP transformer). Compared with before, it has a compact size and excellent flexibility, and can improve the system efficiency using the expansion energy of the compressed air. However, one regret is the demand of electricity, so the safety performance need improvement [7].

In this study, firstly, the working principle of the HP transformer has been analyzed. In addition, according to the analysis, power system of the air-powered vehicle used HP transformer has been presented. Furthermore, the mathematic model of the HP transformer was built. By selecting the appropriate parameter values, a dimensionless model was set up. To confirm the dimensionless mathematic model, an archetype was built and studied. Moreover, the output characteristics of the HP transformer can be studied through simulation study. Finally, the influence of the key parameters was showed to get the most suitable value.

## 2. Structure of the Power System of the Air-Powered Vehicle

Structure of the power system of the air-powered vehicle was showed as Figure 1. From the diagram, we can easily get that the power system is primarily comprised by a compressed air tank, the HP transformer and a hydraulic motor. As the most important component of the system, the HP transformer consists of a pressure regulator, pneumatic and hydraulic chambers and an accumulator. Compressed air is stored in the air tank to provide the proper input pressure. The accumulator is used to accumulate and release energy [8,9].

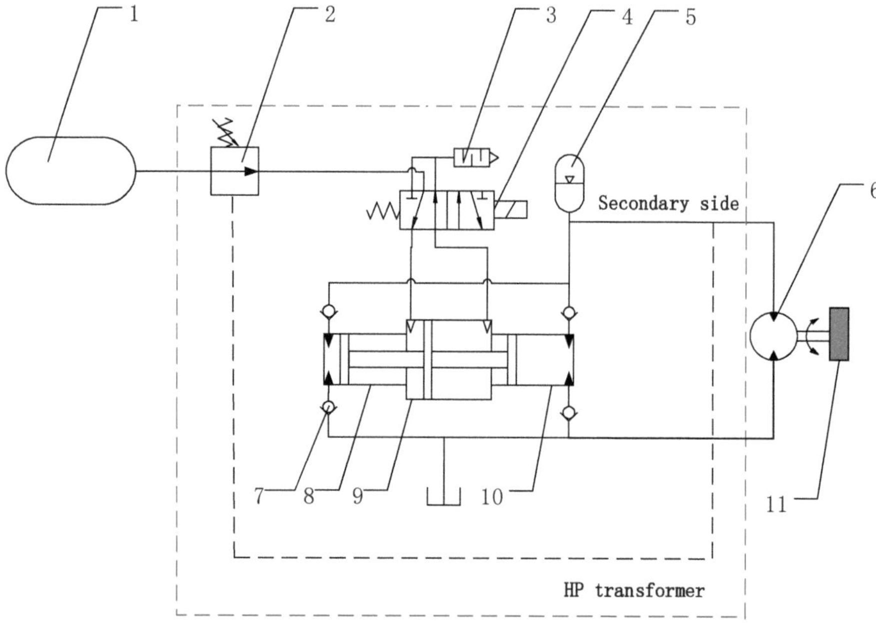


**Figure 1.** Composition of the compressed air-powered vehicle: 1 Air source; 2 Pressure governor; 3 Silencer; 4 Electromagnetic directional valve; 5 Accumulator; 6 Hydraulic motor; 7 Retaining valve; 8 Left hydraulic cylinder; 9 Pneumatic cylinder; 10 Right hydraulic cylinder; 11 Mechanical part.

Using compressed air, the air source can easily provide high enough input pressure. When the electromagnetic directional valve change to its left position, the left pneumatic chamber is connected with the air source. The compressed air flows into the left pneumatic chamber from the air source and drive the piston toward right. The piston then pushes the oil out of the chamber from the right hydraulic chamber to the hydraulic motor. When the electromagnetic directional valve change to its right position, the compressed air charged into the right chamber and drive the position towards left. Oil in the left hydraulic chamber get the force from the piston and was pressurized out to the hydraulic motor [10].

### 3. Mathematical Modeling and Experimental Verification

According to the careful study of the principles of the HP transformer, the basic mathematical was set up as follows [11].

#### 3.1. Pneumatic System Energy Equations

In this study, all the ideal gas formulas can be used because of the hypothesis that the compressed air can be deemed to be an ideal gas. Ignore the impact of the air leakage, the energy equations for the charge and discharge side of the pneumatic chamber can be expressed by the following equations [8]:

$$C_v m \frac{dT}{dt} = S \times h_d (T_a - T) + RqT - pAu \tag{1}$$

$$C_v m \frac{dT}{dt} = (S \times h_c + C_v \times q)(T_a - T) + RqT_a - pAu \tag{2}$$

where

$C_v$: specific thermal capacity at constant volume, 718 J/(kg·K);
$m$: air mass;
$T$: temperature in pneumatic chamber;
$t$: time;
$S$: effective air heat exchange area in pneumatic chamber;
$q$: air mass flow in pneumatic chamber;
$R$: gas constant, 287 J/(kg·K);
$T_a$: ambient temperature;
$p$: pneumatic pressure;
$A$: pneumatic piston area;
$u$: piston velocity;
$h_c$: heat exchange coefficient in the charge side;
$h_d$: heat exchange coefficient in the discharge side.

#### 3.2. Continuity Equations of Pneumatic System

On the basis of the pressure ratio $P_d/P_u$, equations describing the air mass flow while through a throttle can be given as follows (here 0.528 is the critical pressure ratio) [7]:

$$q = \begin{cases} \dfrac{A_{ep}p_u}{\sqrt{T_u}} \sqrt{\dfrac{2\kappa}{R(\kappa-1)}} \left[ \left(\dfrac{p_d}{p_u}\right)^{\frac{2}{\kappa}} - \left(\dfrac{p_d}{p_u}\right)^{\frac{\kappa+1}{\kappa}} \right] & \dfrac{p_d}{p_u} > 0.528 \\[2em] \dfrac{A_{ep}p_u}{\sqrt{T_u}} \left(\dfrac{2}{\kappa+1}\right)^{\frac{1}{\kappa-1}} \sqrt{\dfrac{2\kappa}{R(\kappa+1)}} & \dfrac{p_d}{p_u} \leq 0.528 \end{cases} \tag{3}$$

where

$A_{ep}$: effective pneumatic area in intake and exhaust ports;
$p_u$: pressure in upstream side;
$p_d$: pressure in downstream side;
$\kappa$: specific thermal ratio;
$T_u$: upstream side temperature.

### 3.3. State Equation of Pneumatic System

By deriving the state equation of ideal gases, we can get the equation which describe the air pressure changes in the two-pneumatic chamber [12,13]:

$$\frac{dp}{dt} = \frac{1}{V}\left[\frac{pV}{T} \times \frac{dT}{dt} + RTq - pAu\right]$$

(4)

where $V$ represents air volume.

### 3.4. Motion Equations

The velocity of the piston is calculated from Newton's second law of motion. In this study, the friction force model is considered to be the sum of the Coulomb friction and viscous friction. The viscous friction force is considered to be a linear function of piston velocity. The forces on the piston of the HP transformer are shown in Figure 2.

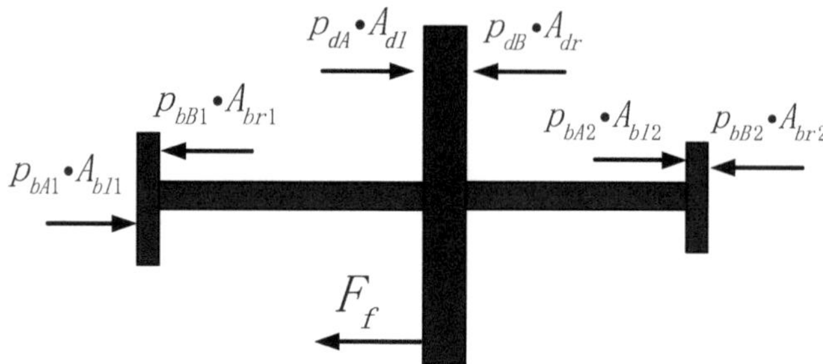


**Figure 2.** The forces on the piston of the HP (hydropneumatic) transformer.

The right side was considered to be the positive direction of the vector. The motion equations of the piston can be given by the following equations:

$$\frac{d^2x}{dt^2} = \begin{cases} \frac{1}{M}\left(p_{dA} \times A_{dl} - p_{dB} \times A_{dr} + p_{bA1} \times A_{bl1} - p_{bB1} \times A_{br1} + p_{bA2} \times A_{bl2} - p_{bB2} \times A_{br2} - F_f\right) & x \neq 0, L \\ 0 & x = 0, L \end{cases}$$

(5)

$$F_f = \begin{cases} F_s & v = 0 \\ F_c + Cu & v \neq 0 \end{cases}$$

(6)

where

$x$: piston displacement;

$M$: piston mass;

$F_f$: friction force;

$F_s$: maximum static friction force;

$F_c$: Coulomb friction force;

$C$: viscosity friction coefficient;

$p_{dA}$: pressure of driving chamber A;

$p_{dB}$: pressure of driving chamber B;

$A_{dl}$: left piston area of driving chamber A;

$A_{dr}$: right piston area of driving chamber B;

$p_{bA1}$: pressure of the first pumping chamber A;

$p_{bB1}$: pressure of the first pumping chamber B;

$A_{bl1}$: left piston area of the first pumping chamber A;
$A_{br1}$: right piston area of the first pumping chamber B;
$p_{bA2}$: pressure of the second pumping chamber A;
$p_{bB2}$: pressure of the second pumping chamber B;
$A_{bl2}$: left piston area of the second pumping chamber A;
$A_{br2}$: right piston area of the second pumping chamber B;
$L$: motion stroke.

### 3.5. Pressure Equations of Hydraulic System

The continuous equations of the driving chamber and can be written as [14]:

$$\frac{dp_{lh}}{dt} = \frac{\beta}{V_{lh}}(Q_{lhin} - Q_{lhout} - A_h u) \tag{7}$$

$$\frac{dp_{rh}}{dt} = \frac{\beta}{V_{rh}}(Q_{rhin} - Q_{rhout} + A_h u) \tag{8}$$

where

$\beta$: effective bulk modulus;
$V_{lh}$: hydraulic volume in left chamber;
$Q_{lhin}$: input hydraulic volume flow in left chamber;
$Q_{lhout}$: output hydraulic volume flow in left chamber;
$A_h$: hydraulic chamber area;
$V_{rh}$: hydraulic volume in right chamber;
$Q_{rhin}$: input hydraulic volume flow in right chamber;
$Q_{rhout}$: output hydraulic volume flow in right chamber;
$p_{lp}$: pneumatic pressure in left chamber;
$p_{rp}$: pneumatic pressure in right chamber;
$p_{lh}$: hydraulic pressure in left chamber;
$p_{rh}$: hydraulic pressure in right chamber.

### 3.6. Flow Equation of Hydraulic System

The volume flow of oil through a throttle can be described as:

$$Q_h = C_d A_{eh} \sqrt{\frac{2(p_{uh} - p_{dh})}{\rho_h}} \tag{9}$$

where

$C_d$: pore throttling coefficient;
$A_{eh}$: effective throttle area;
$\rho_h$: flow density.

According to Equation (9), when the effective area of intake and exhaust port was fixed, output flow of oil can be increased through increasing the pressure in the working boosting chambers and decreasing output oil pressure [15].

### 3.7. Power of Pneumatic System

In this study, an energy consumption evaluation criterion of pneumatic system, namely air power, is adopted. This energy consumption evaluation criterion of compressed air has been formulated as a National Standard of China (GB/T 30833-2014). According to the references [16–18], the air power of compressed air is expressed as:

$$P_p = p_a Q_p \left[ \ln \frac{p_p}{p_a} + \frac{k}{k-1} \left( \frac{T_p - T_a}{T_a} - \ln \frac{T_p}{T_a} \right) \right] \tag{10}$$

where

$p_a$: ambient air pressure.

$\theta_a$: ambient air temperature.

*3.8. Power of Hydraulic System*

The power of pressurized oil is given as [19–22]:

$$P_h = p_h Q_h \tag{11}$$

## 4. Dimensionless Modelling of HP Transformer

By selecting the appropriate reference values, in this part, the basic mathematical model is transformed to a dimensionless expression for simulation. Therefore, based on the reference values, the dynamic characteristics of the HP transformer can be obtained through checking diagrams of dimensionless parameters. Moreover, through the analysis of the dimensionless model and reference values, the influences of independent parameters on the dynamic characteristics of the HP transformer can be obtained. In addition, that is significant to grasp the essential characteristics of the HP transformer [17].

*4.1. Reference Values*

When the pressure in the pneumatic cylinder is as the same as the atmosphere pressure, the air mass flow from air source to the pneumatic cylinder, named the maximum air mass flow, can be obtained:

$$q_{pmax} = \frac{A_{ep} p_s}{\sqrt{T_s}} \left( \frac{2}{\kappa + 1} \right)^{\frac{1}{\kappa - 1}} \sqrt{\frac{2\kappa}{R(\kappa + 1)}} \tag{12}$$

The maximum charged air mass of a pneumatic chamber is named as the maximum air mass, $m_{max}$. The time to discharge $m_{max}$ of air at $q_{max}$ of air mass flow is named as the minimum time, which can be calculated as:

$$t_{min} = \frac{m_{pmax}}{q_{pmax}} \tag{13}$$

To expel a full pneumatic chamber of oil to the oil tank with the minimum time, $T_{min}$, the hydraulic oil volume flow and the pressure in the hydraulic chamber are named as the maximum hydraulic volume flow, $Q_{hmax}$, and the highest hydraulic pressure, $p_f$, which can be gotten as:

$$Q_{hmax} = \frac{V_h}{t_{min}} = C_d A_{ep} \sqrt{\frac{2\left(p_f - p_a\right)}{\rho_h}} = A_h \overline{u} \tag{14}$$

$$p_f = \frac{\rho_h V_h^2}{2 C_d^2 A_{ep}^2 t_{min}^2} + p_a \tag{15}$$

The reference values and the dimensionless variables are shown in Table 1. The basic mathematical model can be made dimensionless as described in the following section.

The ratio of the maximum power of compressed air ($P_{pmax}$) and the maximum power of hydraulic oil ($P_{hmax}$) is defined as the efficiency coefficient ($\eta_0$) of HP transformer, which is used for calculation of the efficiency of HP transformer. The efficiency coefficient ($\eta_0$) of HP transformer is given as

$$\eta_0 = \frac{P_{pmax}}{P_{hmax}} \tag{16}$$

**Table 1.** Reference values and dimensionless variables.

| Variable | Reference Value | | Dimensionless Variable |
|---|---|---|---|
| Affective area | $A_p$ | Area of e piston of the pneumatic chamber | $A_e{}^* = \frac{A_e}{A_p}$ |
| Time | $t_{min}$ | Time to totally exhaust $W_b$ of air at $G_{max}$ of air mass flow | $t^* = \frac{t}{t_{min}}$ |
| Velocity | $\bar{u} = \frac{L}{t_{min}}$ | Average velocity | $u^* = \frac{u}{\bar{u}}$ |
| Pressure | $P_s$ | Supply pressure | $P^* = \frac{P}{P_s}$ |
| Temperature | $T_a$ | Atmosphere temperature | $T^* = \frac{T}{T_a}$ |
| Air mass flow | $q_{max}$ | Maximum air mass flow | $q^* = \frac{q}{q_{max}}$ |
| Volume flow | $Q_{max} = \frac{V_h}{t_{min}} = A_h\bar{u}$ | Maximum oil volume flow | $Q^* = \frac{Q}{Q_{max}}$ |
| Air mass | $m_{pmax} = roupn * Vp$ | Maximum air mass | $m^* = \frac{m}{m_{pmax}}$ |
| Displacement | $L$ | Piston stroke | $x^* = \frac{x}{L}$ |
| Volume | $V_{pmax} = L \times A_p$ | Maximum volume of pneumatic chamber | $V^* = \frac{V}{V_{pmax}}$ |
| Power of compressed air | $P_{pmax} = \frac{p_a q_{max}}{\rho_a} \ln \frac{p_s}{p_a}$ | Maximum power of compressed air | $P_p^* = \frac{P_p}{P_{pmax}}$ |
| Power of hydraulic oil | $P_{hmax} = p_s Q_{max}$ | Maximum power of hydraulic oil when its pressure equals the supply pressure ($p_s$) | $P_h^* = \frac{P_h}{P_{hmax}}$ |

### 4.2. Dimensionless Energy Equations of Pneumatic System

The dimensionless energy equations for the discharge side and the charge side can be written as follows:

$$m^* \frac{dT^*}{dt^*} = \frac{S^*}{S_p^* t_d^*}(1 - T^*) + (k - 1)(p^* u^* - q^* T^*) \tag{17}$$

$$m^* \frac{dT^*}{dt^*} = \left(\frac{S^*}{S_p^* t_c^*} + q^*\right)(1 - T^*) + (k - 1)(q^* - p^* u^*) \tag{18}$$

where, the parameter $T_d{}^*$, which is the dimensionless temperature settling time of the discharge side, is the ratio of the temperature settling time constant, $T_{hd}$, and the isothermal pressure time constant, $T_p$. The dimensionless and dimensional time constant can be written as follows:

$$t_d^* = \frac{t_d}{t_{min}} \tag{19}$$

$$t_d = \frac{C_v m_{pmax}}{S_p h_d} \tag{20}$$

$$S_p = 2A_p + 2L\sqrt{\pi A_p} \tag{21}$$

The dimensionless maximum heat transfer area can be given as:

$$S_p^* = \frac{2A_p + 2L\sqrt{\pi A_p}}{A_p} = 2 + 2L\sqrt{\frac{\pi}{A_p}} \tag{22}$$

For the charge side:

$$t_c^* = \frac{t_c}{t_{min}} \tag{23}$$

$$t_c = \frac{C_v m_{pmax}}{S_p h_c} \tag{24}$$

### 4.3. Dimensionless Continuity Equations of Pneumatic System

Dimensionless Continuity Equations of Pneumatic System can be given as:

$$q^* = \begin{cases} \dfrac{p_{pu}{}^*}{\sqrt{T_u{}^*}} \left[ \left( \dfrac{p_d^*}{p_u{}^*} \right)^{\frac{2}{\kappa}} - \left( \dfrac{p_d^*}{p_u{}^*} \right)^{\frac{\kappa+1}{\kappa}} \right] & \dfrac{p_d^*}{p_u{}^*} > 0.528 \\[4ex] \dfrac{p_{pu}{}^*}{\sqrt{T_u{}^*}} & \dfrac{p_d^*}{p_u{}^*} \leq 0.528 \end{cases} \tag{25}$$

### 4.4. Dimensionless State Equation of Pneumatic System

Dimensionless State Equation of Pneumatic System can be given as:

$$\frac{dp_p{}^*}{dt^*} = \frac{p_p{}^*}{T_p{}^*} \frac{dT_p{}^*}{dt^*} + \frac{T_p{}^* q^*}{V_p{}^*} - \frac{p_p{}^* u^*}{V_p{}^*} \tag{26}$$

### 4.5. Motion Equations

Motion Equations can be given as:

$$\frac{d^2 x^*}{d(t^*)^2} = \begin{cases} \left( \dfrac{1}{t_f^*} \right)^2 \left( p_{lh}{}^* \times A_h{}^* - p_{rh}{}^* \times A_h{}^* + p_{lp}{}^* - p_{rp}{}^* - F_f{}^* \right) & x^* \neq 0,1 \\[2ex] 0 & x^* = 0,1 \end{cases} \tag{27}$$

$$F_f = \begin{cases} F_s{}^* & u^* = 0 \\ F_c{}^* + C^* u^* & u^* \neq 0 \end{cases} \tag{28}$$

where

$F_s{}^*$: dimensionless maximum static friction force;

$F_c{}^*$: dimensionless Coulomb friction force;

$C^*$: dimensional viscous friction force coefficient.

All of the parameters can be given as follows:

$$F_s{}^* = \frac{F_s}{P_s A_p} \tag{29}$$

$$F_c{}^* = \frac{F_c}{P_s A_p} \tag{30}$$

$$C^* = \frac{C \times u_0}{P_s A_p} \tag{31}$$

Dimensionless parameter, $T_f{}^*$ corresponds to the $J$-parameter that is used in the current selection method of a pneumatic cylinder. The $J$-parameter is given by Equation (35).

$$t_f{}^* = \frac{t_f}{t_{min}} \tag{32}$$

$$t_f = \sqrt{\frac{ML}{A_p P_s}} \tag{33}$$

$$J = \frac{t_p^2 P_s A_p}{LM} \tag{34}$$

$$J = \left(\frac{1}{t_f{}^*}\right)^2 \tag{35}$$

As the *J*-parameter represents a coefficient of acceleration, it was considered that this parameter related to the inertia of the HP transformer, and it is known as the inertia coefficient. The dimensionless parameter, $T_f{}^*$, is regarded as the dimensionless natural period of the HP transformer.

*4.6. Dimensionless Pressure Equations of Hydraulic System*

Dimensionless Pressure Equations of Hydraulic System can be given as:

$$\frac{dp_{lh}{}^*}{dt^*} = \beta^*(Ah^* * Q_{lhin}^* - Ah^* * Q_{lhout}^* - u^*) \tag{36}$$

$$\frac{dp_{rh}{}^*}{dt^*} = \beta^*(Q_{rhin}^* - Q_{rhout}^* + u^*) \tag{37}$$

where

$$\beta^* = \frac{\beta}{p_s} \tag{38}$$

*4.7. Dimensionless Flow Equations of Hydraulic System*

Dimensionless Flow Equations of Hydraulic System can be given as:

$$Q_{hin}{}^* = A_{ehin}^* \sqrt{\frac{p_{uh}^* - p_{dh}^*}{p_f^* - p_a^*}} \tag{39}$$

$$Q_{hout}{}^* = A_{ehout}^* \sqrt{\frac{p_{uh}^* - p_{dh}^*}{p_f^* - p_a^*}} \tag{40}$$

$$A_{ehin}^* = \frac{A_{ehin}}{A_{ep}} \tag{41}$$

$$A_{ehout}^* = \frac{A_{ehout}}{A_{ep}} \tag{42}$$

*4.8. Power of Pneumatic System*

Power of Pneumatic System can be calculated as:

$$P_p{}^* = \frac{q_p{}^*\left[\ln\frac{p_p{}^*}{p_a{}^*} + \frac{k}{k-1}\left(T_p{}^* - 1 - \ln T_p{}^*\right)\right]}{\ln\frac{1}{p_a{}^*}} \tag{43}$$

*4.9. Power of Hydraulic System*

Power of hydraulic System can be calculated as:

$$P_h{}^* = p_h{}^* Q_h{}^* \tag{44}$$

## 5. Simulation and Experimental Study on a HP Transformer

### 5.1. Experimental Verification of the Mathematical Model

To verify the dimensionless mathematical model, a compressed air-powered hydraulic system was set up. Its schematic structure can refer to Figure 3. The compressed air, charged from the air source, flowed through a regulator (IR3010-03BG, SMC, Tokyo, Japan), and its pressure was reduced to a fixed value (about 0.6 MPa). When the compressed air was supplied to the Hydropneumatic transformer (SWB-100D-5, SIWELL Supercharger Technology, Suzhou, China), pressurized hydraulic oil was output from the HP transformer. In order to stabilize the output pressure of the HP transformer, an accumulator (GXQA-0.35/25-L-A, AOQI, Guangdong, China) and a relief valve (DBDS6P1X/315, Rexroth, Lohr am Main, Germany) were installed in the downstream of the compressed air-powered hydraulic system. A pressure sensor (AK-4B, 701, Beijing, China), a data acquisition card (USB4711, Advantech, Taipei, Taiwan) and a computer (X430, Lenovo, Beijing, China) were utilized to measure the output pressure of the HP transformer.

A dedicated test bench for the HP transformer, as shown in Figure 4, was designed and built to measure the output pressure of the HP transformer.

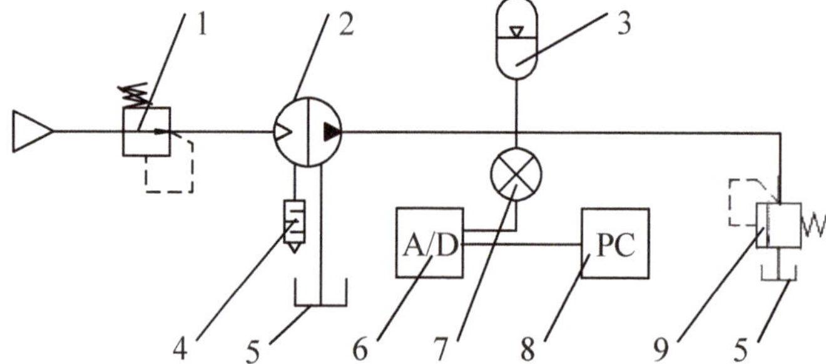


**Figure 3.** Schematic structure of the HP transformer: (1) Regulator; (2) HP transformer; (3) Accumulator; (4) Silencer; (5) Tank; (6) Data acquisition card; (7) Pressure sensor; (8) Computer; (9) Relief valve.

**Figure 4.** Dedicated test bench for the HP transformer.

According to previous study [17], because the temperature of the compressed air is almost same as the atmospheric temperature, so the influence of the temperature on the working characteristics of a pneumatic booster can be neglected. As the structure and working principle of the pneumatic booster like the structure and working principle of the HP transformer, in this study, the working process of the HP transformer is considered to be an isothermal process.

In simulation and experimental study on the HP transformer, the values of the dimensionless area of the hydraulic chamber and dimensionless output oil pressure were set to 0.25 and 2.92. As the output pressure of the HP transformer can be acquired with a pressure sensor, which is more precise and cheaper than a hydraulic flow sensor. Figure 5 depicts the simulation and experimental output hydraulic pressure characteristics.

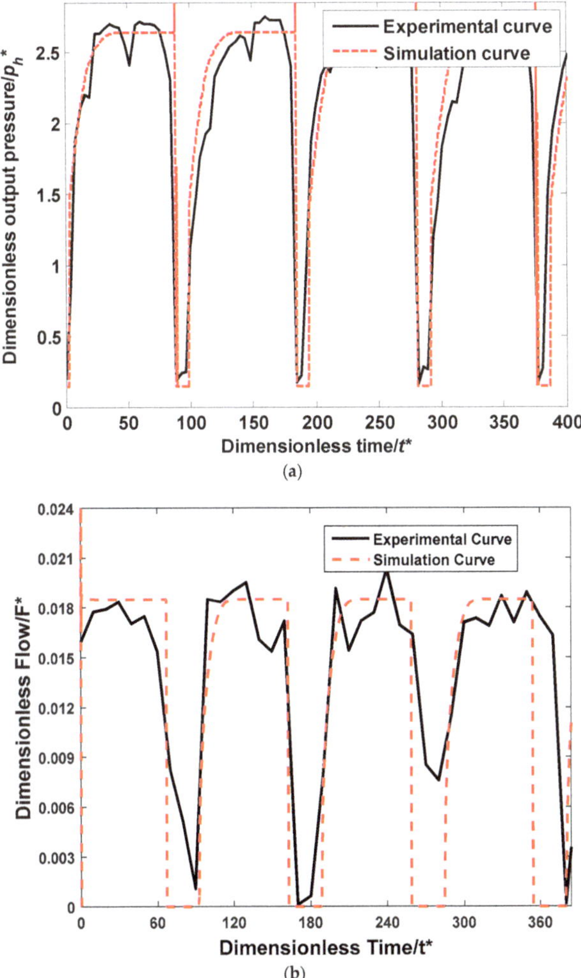


**Figure 5.** Simulation and experimental contrast curves. (**a**) Simulation and experimental curves of the dimensionless output hydraulic pressure; (**b**) Simulation and experimental curves of the dimensionless flow.

As shown in Figure 5, it is clear that the simulation results are consistent with the experimental results, and that verifies the mathematical model above. However, there are two differences between the simulation results and the experimental results: (1) The dimensionless output pressure in the experiment results increases more sluggishly than the dimensionless output pressure in the simulation results; (2) There is a fluctuation in the dimensionless output pressure of the HP transformer, when the dimensionless output pressure gets to its top.

The main reasons for the differences are listed as follows. In the experiment, the accumulator absorbed the fluctuation of the dimensionless output pressure, which slowed down the increase of the dimensionless output pressure of the HP transformer. Moreover, the opening size of the relief valve was regulated according to the output pressure of the HP transformer. With a decrease in the output pressure of the HP transformer, the opening size of the relief valve got smaller, then the output pressure of the HP transformer rose accordingly.

### 5.2. Dynamic Power Characteristics of the HP Transformer

The simulation study on the HP transformer was proceeded according to the study above, and the input power and output power dynamics of the HP transformer can be obtained, which are shown in Figure 6.

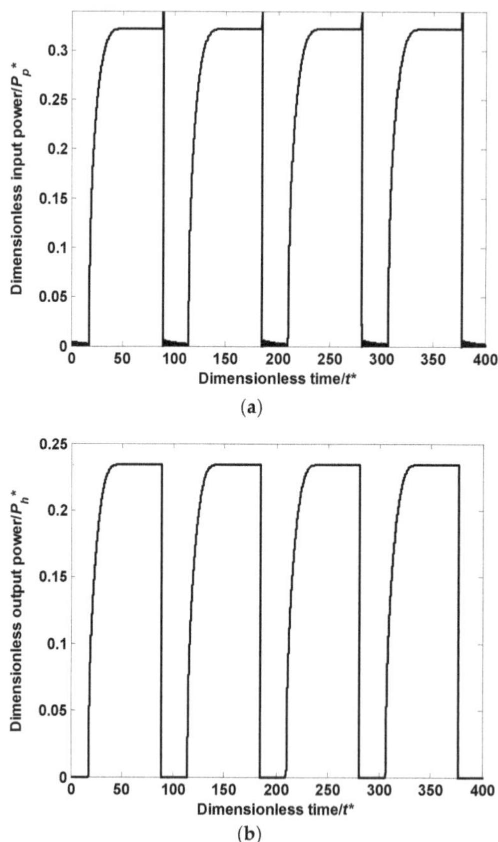

(a)

(b)

**Figure 6.** Dimensionless input power output power characteristics of the HP transformer. (a) Dimensionless input power of the HP transformer; (b) Dimensionless output power of the HP transformer.

As shown in Figure 6, it is clear that, firstly, the dimensionless input power and output power fluctuate regularly. Furthermore, when the HP transformer starts to work when the piston moves from its destination, after a pause, the dimensionless input power and output power increase rapidly and get their platforms. However, when the piston reaches another destination, the dimensionless input power and output power decrease to zero immediately. Finally, as the efficiency coefficient ($\eta_0$) of HP transformer is about 0.6522, when the dimensionless input power and output power reach their platforms, the efficiency of the HP transformer is 47.52% approximately.

## 6. Study on the Power and Efficiency of the HP Transformer

Based on the structure and principle of HP transformer, in order to get the parameters which can make the HP transformer better, we carried out the study about the efficiency and output power. After that, the parameters affecting these two properties were clarified.

To illustrate the influence of the dimensionless parameters on the output power and efficiency of HP transformer, each parameter changed for comparison when the other parameters are constant. The values of the dimensionless parameters are shown in Table 2.

**Table 2.** Values of the dimensionless parameters.

| Parameters | $T_f{}^*$ | $F_s{}^*$ | $C^*$ | $F_c{}^*$ | $A_h{}^*$ | $P_o{}^*$ | $A_{eh}{}^*$ |
|---|---|---|---|---|---|---|---|
| Values | 0.1425 | 0.00018 | 0.0004 | 0.00036 | 0.25 | 2.857 | 0.000625 |

### 6.1. Influence of the Output Pressure

In this part, for getting the change rules of the output power and efficiency when change the output oil pressure, we keep other parameters such as the dimensionless aperture of the hydraulic chamber orifice (set to 0.028), the area ratio of the pistons (set to 4) stay the same when setting the dimensionless output oil pressure to 2.42, 2.57, 2.71, 2.85 and 3.00. After careful analysis and professional simulation study, we get the results shown in Figures 7 and 8. As we can see, curves in Figure 1 depicts the dynamic characteristics of the dimensionless output power, and curve in Figure 8a shows the variation relationship of the dimensionless output power and the dimensionless output oil pressure. Curve in Figure 8b shows the corresponding efficiency of the system with different output oil pressure.

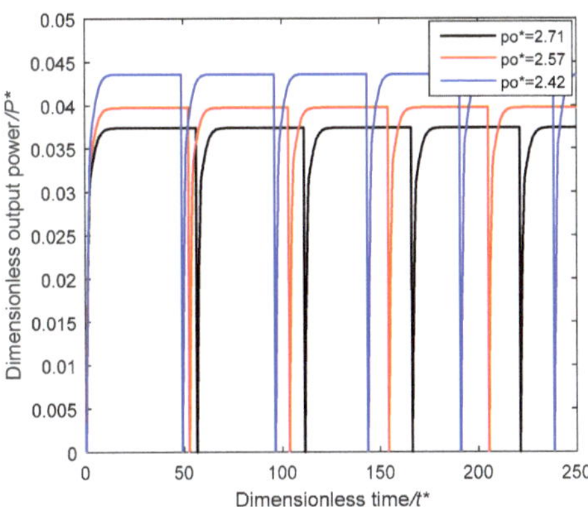


**Figure 7.** Dimensionless output power–time curves of the power system.

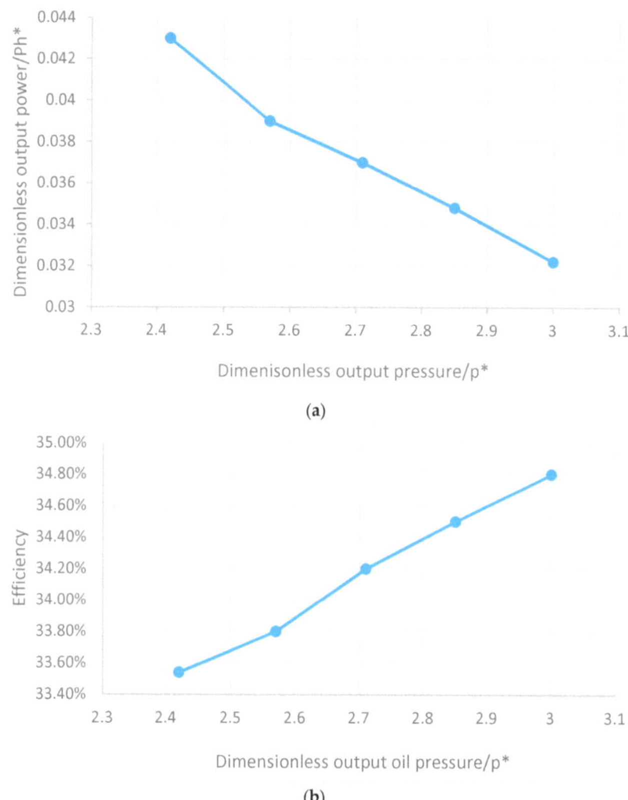

**Figure 8.** Relationships of the output power, efficiency and the output pressure. (**a**) Output power trend curve influenced by the output pressure; (**b**) Efficiency trend curve influenced by the output pressure.

Through careful study on curves in Figures 7 and 8, two results can be obtained as follows:

(1)  When the dimensionless output oil pressure increases with a proper difference from 2.42 to 3.00, the output power decreases from 0.043 to 0.0322 at the same time. This phenomenon can be explained as the rise of the output oil pressure may result in the decrease of the output flow, so the output oil pressure which is calculated by output flow and output oil pressure decreases.
(2)  As we can see in Figure 8, the curves of the dimensionless output power, system efficiency and dimensionless output oil pressure are similar to the trend line.

### 6.2. Influence of the Aperture of the Orifice of the Hydraulic Chamber

In this part, for getting the change rules of the dimensionless output power and efficiency when change the aperture of the orifice of the hydraulic chamber, we keep other parameters such as the dimensionless output oil pressure (set to 2.71), the area ratio of the pistons (set to 4) stay the same when setting the dimensionless aperture of the orifice of the hydraulic chamber to 0.0256, 0.0267, 0.0278, 0.0289 and 0.030. After careful analysis and professional simulation study, we get the results shown in Figures 9 and 10. As we can see, curves in Figure 9 depicts the dynamic characteristics of the dimensionless output power, and curve in Figure 10a shows the variation relationship of the dimensionless output power and the dimensionless aperture of the orifice of the hydraulic chamber. Curve in Figure 10b shows the corresponding efficiency of the system with different aperture.

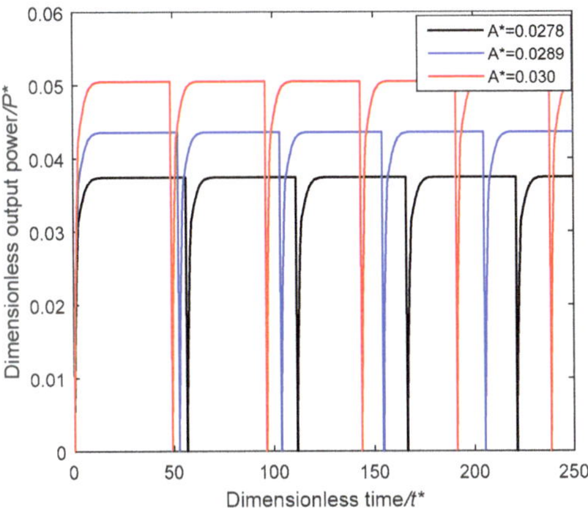

**Figure 9.** Dimensionless output power–time curves of the power system.

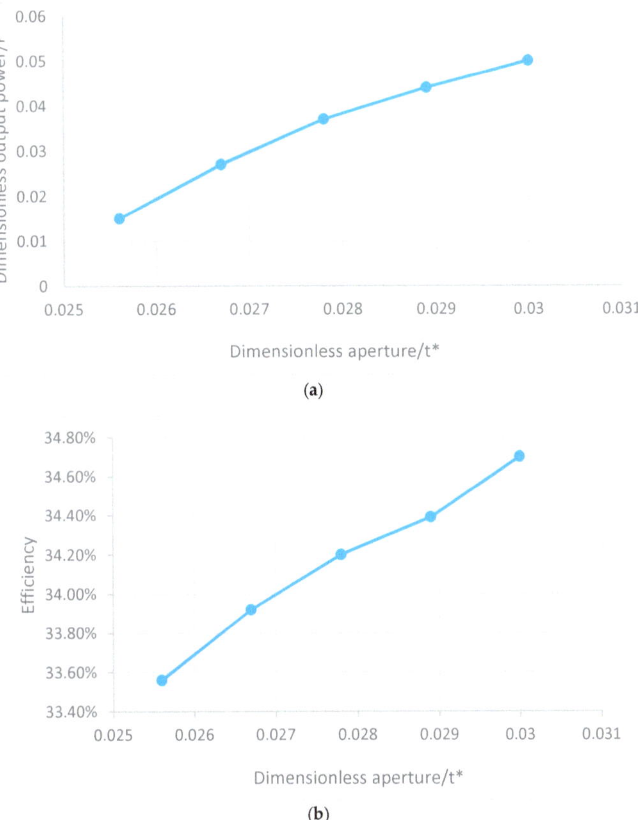

**(a)**

**(b)**

**Figure 10.** Relationships of dimensionless output power, efficiency and the aperture of the orifice of the hydraulic chamber. (**a**) Output power trend curve influenced by the aperture of the orifice of the hydraulic chamber; (**b**) Efficiency trend curve influenced by the aperture of the orifice of the hydraulic chamber.

Through careful study on curves in Figures 9 and 10, two results can be obtained as follows:

(1)  When the dimensionless aperture increases with a proper difference from 0.0256 to 0.030, the output power increase from 0.015 to 0.05 and the efficiency increase from 33.56% to 34.7% properly. This phenomenon can be explained as follows: when the aperture of the orifice of the hydraulic increase, the output flow is sure increase, so the output power increase while the output pressure keep constant

(2)  As we can see in Figure 10, the curves of the dimensionless output power, system efficiency and dimensionless aperture of the hydraulic chamber orifice are similar to the trendline.

*6.3. Influence of the Area Ratio of the Pistons*

In this part, for getting the change rules of the dimensionless output power and efficiency when change the area ratio of the pistons, we keep other parameters such as the dimensionless output oil pressure (set to 2.71), the dimensionless aperture of the hydraulic chamber orifice (set to 0.028) stay the same when setting the dimensionless area ratio of the pistons to 3.6, 3.8, 4.0, 4.2 and 4.4. After careful analysis and professional simulation study, we get the results shown in Figures 11 and 12. As we can see, curves in Figure 11 depicts the dynamic characteristics of the dimensionless output power, and curve in Figure 12a shows the variation relationship of the dimensionless output power and the area ratio of the pistons. Curve in Figure 12b shows the corresponding efficiency of the system with different area ratio of the pistons.

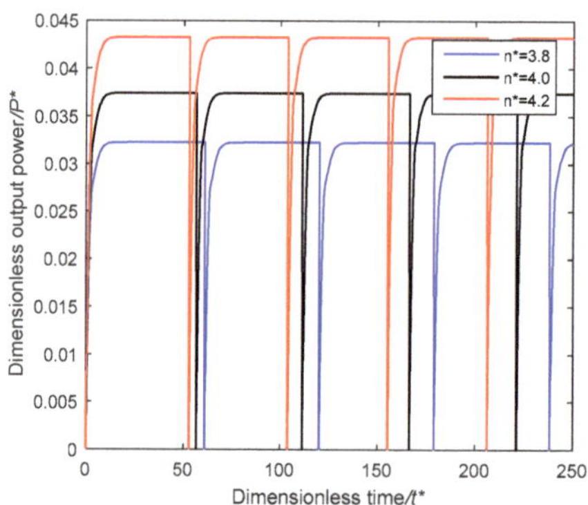

**Figure 11.** Dimensionless output power–time curves of the power system.

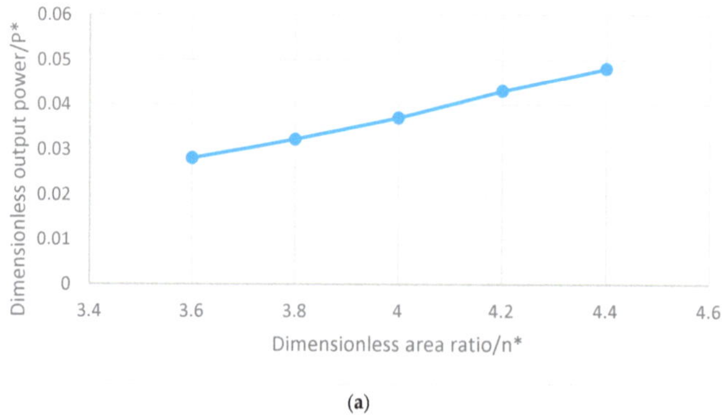

(a)

**Figure 12.** *Cont.*

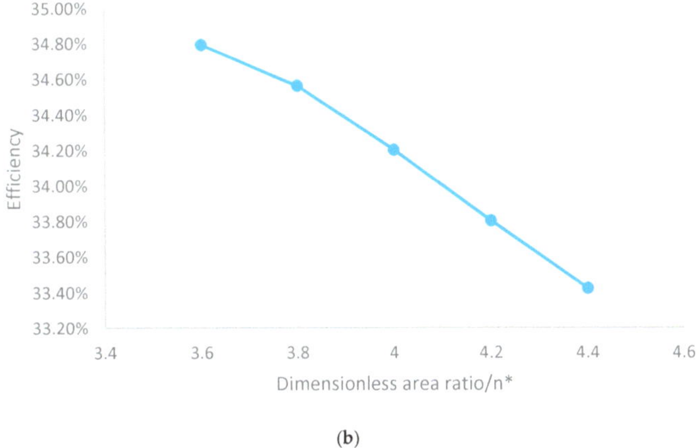

(b)

**Figure 12.** Relationships of the output power, efficiency and the area ratio of the pistons. (**a**) Output power trend curve influenced by the area ratio of the pistons; (**b**) Efficiency trend curve influenced by the area ratio of the pistons.

After the detailed analysis of Figures 11 and 12, two results can be obtained.

(1)    With the increase of the dimensionless area ratio of the pistons, the output power increases from 0.028 to 0.048 while the efficiency decreases from 34.79% to 33.42%. This is because when the area of the pneumatic piston is constant and the area of the hydraulic piston increase, the output pressure increases correspondingly. So the output power, which is calculated by output pressure and output flow, will increase.

(2)    As we can see in Figures 8, 10 and 12, the curves of the dimensionless output power, system efficiency and dimensionless area ratio of the pistons are almost linear.

## 7. Conclusions

In this study, a dimensionless mathematic model of the power systems of air-powered hydraulic vehicle was set up. The dimensionless model, which is different from the original model, has many advantages: firstly, unit conversion needs to be considered when the original model is calculated, but the dimensionless model is clearly easy; secondly, the dimensionless model is convenient to make the comparison of parameters, but the original model plays opposite.

To confirm the dimensionless mathematic model, a protocol was built and studied. Through experimental and simulation studies on the power system of air-powered hydraulic vehicle, it can be obtained as follows:

(1)    The experimental curve of the archetype has a good match with the simulation curve of the dimensionless model, so it can be derived that the mathematical model is effective and accurate.

(2)    When the dimensionless output pressure increase from 2.42 to 3.00, the dimensionless output power decrease from 0.043 to 0.0322, while the system efficiency increase from 33.54% to 34.8%.

(3)    While the dimensionless aperture increases from 0.0256 to 0.030, the dimensionless output power increase from 0.015 to 0.05, the system efficiency increase from 33.56% to 34.7% at the same time.

(4)    When the area ratio of the pistons increase from 3.6 to 4.4, the dimensionless output power increase from 0.028 to 0.048, and the system efficiency decrease from 34.79% to 33.42% properly.

From the simulation and analysis, we can get the conclusion that the increase of the aperture of the hydraulic chamber orifice and the area of the pistons can lead to a higher output power, and the increase of the output pressure and the aperture of the hydraulic chamber orifice can lead to a higher efficiency.

This research can be referred to in the performance and design optimization of the HP transformer.

**Author Contributions:** Dongkai Shen carried on all the simulation and experiment works, and he drew the conclusion of this article. Qilong Chen writes the manuscript, and in addition, he drew the pictures and tables of this paper. Yixuan Wang comes up with the method of this project, and he is responsible for the handling work of this article as the corresponding autho.

**Conflicts of Interest:** The authors declare no conflict of interest.

## References


1. Yu, Q.; Shi, Y.; Cai, M.; Xu, W. Fuzzy logic speed control for the engine of an air-powered vehicle. *Adv. Mech. Eng.* **2016**, *8*. [CrossRef]
2. Hung, Y.H.; Chen, J.H.; Wu, C.H.; Chen, S.Y. System design and mechatronics of an air supply station for air-powered scooters. *Comput. Electr. Eng.* **2016**, *54*, 185–194. [CrossRef]
3. Wijngaarden, L.V. On the equations of motion for mixtures of liquid and gas bubbles. *J. Fluid Mech.* **2006**, *33*, 465–474. [CrossRef]
4. Zhao, P.; Dai, Y.; Wang, J. Design and thermodynamic analysis of a hybrid energy storage system based on A-CAES (adiabatic compressed air energy storage) and FESS (flywheel energy storage system) for wind power application. *Energy* **2014**, *70*, 674–684. [CrossRef]
5. Naranjo, J.; Kussul, E.; Ascanio, G. A new pneumatic vanes motor. *Mechatronics* **2010**, *20*, 424–427. [CrossRef]
6. Zang, Y.J.; Polytechnic, S. Design and Analysis of a New Type Bidirectional Pneumatic-hydraulic Pressurization System. *Hydraul. Pneum. Seals* **2013**, *9*, 57–58.
7. Shi, Y.; Wu, T.; Cai, M.; Liu, C. Modelling and study on the output flow characteristics of expansion energy used hydropneumatic transformer. *J. Mech. Sci. Technol.* **2016**, *30*, 1163–1170. [CrossRef]
8. Niu, J.L.; Shi, Y.; Cao, Z.X.; Cai, M.L.; Wei, C.; Zhu, J. Study on air flow dynamic characteristic of mechanical ventilation of a lung simulator. *China Technol. Sci.* **2017**, *60*, 1–8.
9. Pimm, A.J.; Garvey, S.D.; Jong, M.D. Design and testing of Energy Bags for underwater compressed air energy storage. *Energy* **2014**, *66*, 496–508. [CrossRef]
10. Cai, M.; Kawashima, K.; Kagawa, T. Power Assessment of Flowing Compressed Air. *J. Fluids Eng.* **2006**, *128*, 40–405. [CrossRef]
11. Li, J.; Li, C.F.; Zhang, Y.X.; Yue, H.G. Compressed air energy storage system exergy analysis and its combined operation with nuclear power plants. *Appl. Mech. Mater.* **2013**, *448*, 2786–2789. [CrossRef]
12. Zheng, X.; Xie, L.; Liu, L. *Stability Analysis of Pneumatic Cabin Pressure Regulating System with Complex Nonlinear Characteristics*; Hindawi Publishing Corp: Cairo, Egypt, 2015.
13. Lafmejani, A.S.; Masouleh, M.T.; Kalhor, A. An experimental study on friction identification of a pneumatic actuator and dynamic modeling of a proportional valve. In Proceedings of the IEEE International Conference on Robotics and Mechatronics, Hefei/Tai'an, China, 27–31 August 2017.
14. Yang, F.; Li, G.; Hua, J.; Li, X.; Kagawa, T. A New Method for Analysing the Pressure Response Delay in a Pneumatic Brake System Caused by the Influence of Transmission Pipes. *Appl. Sci.* **2017**, *7*, 941. [CrossRef]
15. Shi, Y.; Zhang, B.; Cai, M.; Zhang, D. Numerical Simulation of volume-controlled mechanical ventilated respiratory system with two different lungs. *Int. J. Numer. Methods Biomed. Eng.* **2016**, *33*, 2852. [CrossRef] [PubMed]
16. Shi, Y.; Cai, M.L. Dimensionless study on outlet flow characteristics of an air-driven booster. *J. Zhejiang Univ. Sci. A* **2012**, *13*, 481–490. [CrossRef]
17. Vaezi, M.; Izadian, A. Control of a hydraulic wind power transfer system under disturbances. In Proceedings of the IEEE International Conference on Renewable Energy Research and Application, Bandung, Indonesia, 5–7 October 2015; pp. 886–890.
18. Shi, Y.; Wang, Y.X.; Cai, M.L.; Zhang, B.L.; Zhu, J. An aviation oxygen supply system based on a mechanical ventilation model. *Chin. J. Aeronaut.* **2018**, *31*, 197–204. [CrossRef]
19. Ren, S.; Cai, M.; Shi, Y.; Xu, W.; Zhang, X.D. Influence of Bronchial Diameter Change on the airflow dynamics Based on a Pressure-controlled Ventilation System. *Int. J. Numer. Methods Biomed. Eng.* **2017**. [CrossRef] [PubMed]
20. Ren, S.; Shi, Y.; Cai, M.; Xu, W. Influence of secretion on airflow dynamics of mechanical ventilated respiratory system. *IEEE/ACM Trans. Comput. Biol. Bioinform.* **2017**. [CrossRef] [PubMed]

21. Shi, Y.; Zhang, B.; Cai, M.; Xu, W. Coupling Effect of Double Lungs on a VCV Ventilator with Automatic Secretion Clearance Function. *IEEE/ACM Trans. Comput. Biol. Bioinform.* **2017**. [CrossRef] [PubMed]
22. Shi, Y.; Wu, T.; Cai, M.; Wang, Y.; Xu, W. Energy conversion characteristics of a hydropneumatic transformer in a sustainable-energy vehicle. *Appl. Energy* **2016**, *171*, 77–85. [CrossRef]

*Article*

# Prescribed Performance Constraint Regulation of Electrohydraulic Control Based on Backstepping with Dynamic Surface

Qing Guo [1,2,*], Yili Liu [1], Dan Jiang [3,*], Qiang Wang [1], Wenying Xiong [1], Jie Liu [1] and Xiaochai Li [1]

[1] School of Aeronautics and Astronautics, University of Electronic Science and Technology of China, Chengdu 611731, China; ylliu_123@163.com (Y.L.); qwang@std.uestc.edu.cn (Q.W.); xiongwy57@163.com (W.X.); LiuJieLLLJ@163.com (J.L.); xiaochai_li@163.com (X.L.)

[2] State Key Laboratory of Fluid Power and Mechatronic Systems, Zhejiang University, Hangzhou 310027, China

[3] School of Mechatronics Engineering, University of Electronic Science and Technology of China, Chengdu 611731, China

* Correspondence: guoqinguestc@uestc.edu.cn (Q.G.); jdan2002@uestc.edu.cn (D.J.)

Received: 16 November 2017; Accepted: 25 December 2017; Published: 8 January 2018

**Abstract:** In electro-hydraulic system (EHS), uncertain nonlinearities such as some hydraulic parametric uncertainties and external load disturbance often degrade the output dynamic performance. To address this problem, a prescribed performance constraint (PPC) control method is adopted in EHS to restrict the tracking position error of the cylinder position to a prescribed accuracy and guarantee the dynamic and steady position response in a required boundedness under these uncertain nonlinearities. Furthermore, a dynamic surface is designed to avoid the explosion of complexity due to the repeatedly calculated differentiations of the virtual control variables derived in backstepping. The effectiveness of the proposed controller has been verified by a comparative results.

**Keywords:** electro-hydraulic system; uncertain nonlinearity; prescribed performance constraint; backstepping

## 1. Introduction

Electro-hydraulic systems are currently widely used in mechatronic control engineering as they have a superior load efficiency. It was found that EHS starts to be commonly applied for large power systems such as wheel loaders [1], fatigue test devices [2], load simulators [3] and exoskeletons [4]. However, there exist uncertain nonlinearities including parametric uncertainty and external load disturbance in EHS. The former is caused by unknown viscous damping, load stiffness, variations in control fluid volumes, physical characteristics of valve, bulk modulus and oil temperature variations existed in EHS [5,6]. Thus, the high-quality dynamic performance of EHS cannot be always maintained. While the latter is often presented as the driven force or torque of mechanical plant and bias the load pressure of EHS [7]. Thus, the performance holding of EHS under these uncertain nonlinearities is still a challenge problem in EHS control loop. By the way, the parametric uncertainty and noise disturbance also obviously exist in pneumatic system such as mechanical ventilation [8–17], network distributed control plant [18,19], multiple-input single-output processes [20].

The output-constrained control is welcomed in practice, since the required dynamic behavior can be maintained in the case of different disturbance and uncertainty. Tee and Ge [21,22] originated the barrier Lyapunov function (BLF) to describe the dissipative energy instead of the quadratic Lyapunov function. Then He [23–26] and Ren [27] employed BLF in general nonlinear system, manipulator and rehabilitation robot. Subsequently, Won [28] proposed backstepping based on BLF with disturbance

*Appl. Sci.* **2018**, *8*, 76; doi:10.3390/app8010076                     www.mdpi.com/journal/applsci

observer in EHS. Qiu [29] presented backstepping control with dynamic surface for anti-skid braking system. Guo [30] presented a state-constrained controlled by BLF to restrict the position tracking error to a prescribed accuracy and guarantee the load pressure in the maximal power boundary. The merit of BLF is to constrain the system output in the satisfactory boundary by the logarithm transformation of the equivalent output error. However, since the output constraint boundary by BLF is often a constant not a time-varying constraint, the control saturation and chatter output response will emerge in initial time due to the initial large state error, as the boundary is selected very small. Thus, to relax this problem, the prescribed performance constraint (PPC) is initiated by Bechlioulis [31] to guarantee the satisfactory error response and overcome the controllability loss due to the input saturation. Then, Zhang [32,33] used PPC to restrict the attack-of-angle of hypersonic aircraft and the electromechanical system position. In fact, the servo valve control in EHS has limited throttle constraint, which indicates the oversized control will degrade the performance and the stable margin of EHS. The PPC technique transform the original constrained system into a free-constraint model by a designed weighted performance function, which can address both static and time-varying constraints by the regulation of the parameters of weighted performance function.

There exists a potential problem in the common backstepping method, i.e., the explosion of complexity of high-order nonlinear system [34,35] due to the repeatedly calculated differentiations of the virtual control variables emerged in backstepping iteration. These high-order derivatives will magnify noise and uncertainty in the actual control signals which results into violent control and chatter response [32,36]. To solve this problem, the dynamic surface control (DSC) has been proposed to design a stabilizing function instead of the repeatedly calculated derivative of virtual control. The purpose of DSC is to not only eliminate the severe proliferation and system singularity and but also guarantee fast convergence and satisfactory dynamic behavior [29].

In this study, to refuse the negative effect of the external load and hydraulic parametric uncertainty, a novel prescribed performance constraint control is proposed in the position control loop of EHS to constrain the position tracking error to a desirable performance. Different from the constraint holding technique of BLF, the PPC employed a weighted performance function to design an adjustable time-varying output-constraint and improve the system stable margin and dynamic performance. By this controller, all the signals of the single-rod EHS are uniformly bounded and the tracking error of the cylinder position can converge to a small compact set without violating the constraints. Furthermore, the dynamic surface is used to design a stabilizing functions instead of the virtual control derivative in backstepping iteration to avoid violent control and chatter response. Both theoretical proof and comparative results have been provided to verify the effectiveness of the proposed method.

The remainder of this paper is organized as follows. The plant is described in Section 2. The output-constrained controller is given in Section 3 including PPC technique and dynamic surface design. The comparative results of two controllers are given in Section 4. Finally, the conclusion is drawn in Section 5.

## 2. Plant Description

The EHS is composed of a servo valve, a symmetrical cylinder, a fixed displacement pump, a motor, and a relief valve as shown in Figure 1. The external load on this EHS is a driven force or torque of any mechatronic plant. The pump outputs the supply pressure $p_s$, which is also the pressure threshold of the relief valve.

**Hypothesis 1.** *Since the cut-off frequency of servo valve is far greater than the control system bandwidth, the valve dynamics can be neglected in EHS model construction as $x_v = K_{sv}u$, where $x_v$ is the spool position of servo valve, $u$ is the control voltage of servo valve, $K_{sv}$ is the gain of the servo valve [37].*

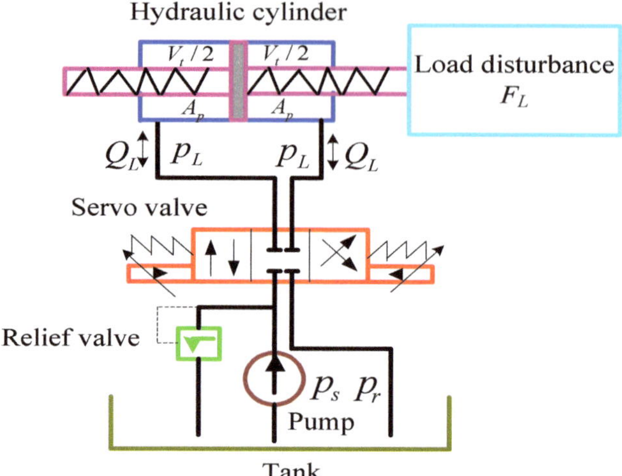



**Figure 1.** The EHS control mechanism.

According to Hypothesis 1, if the three state variables are defined as $[x_1, x_2, x_3]^T = [y, \dot{y}, p_L]^T$ where $y$ and $\dot{y}$ are the cylinder position and velocity, $p_L$ is the load pressure of the hydraulic cylinder, then the state space model of the EHS is given by

$$\begin{cases} \dot{x}_1 = x_2 \\ \dot{x}_2 = \dfrac{1}{m}(A_p x_3 - K x_1 - b x_2 - F_L) \\ \dot{x}_3 = -\dfrac{4\beta_e A_p}{V_t} x_2 - \dfrac{4\beta_e C_{tl}}{V_t} x_3 + \dfrac{4\beta_e C_d w K_{sv} u}{V_t \sqrt{\rho}} \sqrt{p_s - \mathrm{sgn}(u) x_3} \end{cases} \tag{1}$$

where $C_d$ is the discharge coefficient, $w$ is the area gradient of the servo valve, $\rho$ is the density of the hydraulic oil, $C_{tl}$ is the coefficient of the total leakage of the cylinder, $\beta_e$ is the effective bulk modulus, $A_p$ is the annulus area of the cylinder chamber, $V_t$ is the half-volume of cylinder, $m$ is the load mass, $K$ is load spring constant, $b$ is the viscous damping coefficient of the hydraulic oil, $F_L$ is the external load on the EHS, $\mathrm{sgn}(\cdot)$ is the sign function.

**Remark 1.** *In practice, the hydraulic parameters $C_d$, $\rho$, $w$, $b$, $\beta_e$, $C_{tl}$ are usually uncertain constants, but the other parameters are known [38,39].*

**Remark 2.** *The external load $F_L$ is unknown dynamic variable, which is caused by the driving force of someone mechatronic plant. Although the dynamic value of $F_L$ depends on the variables $y, \dot{y}, \ddot{y}$, $F_L$ is bounded by $|F_L(t)| \leq F_{L\max}$, where $F_{L\max}$ is an unknown bounded constant [40,41].*

Thus from Remarks 1 and 2, the state space model (1) is rewritten as follow

$$\begin{cases} \dot{x}_1 = x_2 \\ \dot{x}_2 = \bar{f}_2(x_1, x_2) + \bar{g}_2 x_3 + \Delta_2(x_1, x_2) \\ \dot{x}_3 = \bar{f}_3(x_2, x_3) + \bar{g}_3(x_3, u)u + \Delta_3(x_1, x_2, x_3) \end{cases}, \tag{2}$$

where $\bar{C}_d$, $\bar{\rho}$, $\bar{K}$, $\bar{b}$, $\bar{\beta}_e$, $\bar{C}_{tl}$ are nominal values of these uncertain parameters respectively,

$$\bar{f}_2(x_1, x_2) = -\frac{\bar{K}x_1 + \bar{b}x_2}{m}, \quad \bar{g}_2 = \frac{A_p}{m}$$

$$\bar{f}_3(x_2, x_3) = -\frac{4\bar{\beta}_e A_p}{V_t}x_2 - \frac{4\bar{\beta}_e \bar{C}_{tl}}{V_t}x_3 \quad , \tag{3}$$

$$\bar{g}_3(x_3, u) = \frac{4\bar{\beta}_e \bar{C}_d w K_{sv}}{V_t \sqrt{\bar{\rho}}} \sqrt{p_s - \mathrm{sgn}(u)x_3}$$

and $\Delta_2(x_1, x_2) = f_2(x_1, x_2) - \bar{f}_2(x_1, x_2) - F_L(t)/m$, $\Delta_3(x_1, x_2, x_3) = f_3(x_2, x_3) - \bar{f}_3(x_2, x_3) + g_3(x_1, x_2, x_3) - \bar{g}_3(x_1, x_2, x_3)$ are the integrated elements of parametric uncertainties and the external load disturbance.

Due to limited boundaries of the parametric uncertainties and the external load mentioned in Remarks 1 and 2, the two uncertain nonlinearities $\Delta_2$, $\Delta_3$ are bounded by $|\Delta_2| < \Delta_{2\max}$, $|\Delta_3| < \Delta_{3\max}$, where $\Delta_{2\max}$, $\Delta_{3\max}$ are unknown bounded constants [28].

## 3. Prescribed Performance Constraint Control of EHS

### 3.1. Prescribed Performance Constraint

The prescribed performance constraint of position tracking error is regulated by a designed weighted performance function, which can guarantee not only the satisfactory dynamic performance but also the stable margin of EHS.

At first, the position tracking error is given by

$$e(t) = x_1(t) - y_d(t). \tag{4}$$

If the cylinder position $x_1$ is restricted in $x_{1\min} < x_1(t) < x_{1\max}$, where $x_{1\min}$ and $x_{1\max}$ are the maximal and minimal boundary of $x_1$, and the position demand $y_d$ has also two definite boundaries as $y_{d\min} \leq y_d \leq y_{d\max}$, then

$$e_{\min} < e(t) < e_{\max}, \tag{5}$$

where $e_{\min} = x_{1\min} - y_{d\max}$, $e_{\max} = x_{1\max} - y_{d\min}$.

**Definition 1.** *A continuous smooth function [32] $\rho(t) = (\rho(0) - \rho(\infty))e^{-\lambda t} + \rho(\infty)$ is called a weighted performance function if*
*(1) $\rho(t)$ is positive and monotonically decreasing;*
*(2) $\lim\limits_{x \to \infty} \rho(t) = \rho_\infty > 0$;*
*(3) $\rho(\infty) < \rho(0) < 1$.*

**Lemma 1.** *If a weighted performance function $\rho(t)$ is designed such that*

$$e_{\min} < e(t)/\rho(t) < e_{\max}, \tag{6}$$

*then $e(t)$ is restricted in $(e_{\min}, e_{\max})$ [31].*

Actually, if $e(t) \geq 0$, then $e(t) \leq e(t)/\rho_i(t) < e_{\max}$ due to $0 < \rho_i(t) < 1$. On the other hand, if $e(t) < 0$, then $e_{\min} < e(t)/\rho(t) < e(t)$. Thus, the position tracking error $e(t)$ is always restricted in the boundaries $(e_{\min}, e_{\max})$.

Secondly, according to *Lemma 1*, the PPC $\rho(t)e_{\min} < e(t) < \rho(t)e_{\max}$ can derive a new state errors as follow

$$z_1(t) = T^{-1}\left(\frac{e(t)}{\rho(t)}\right) = \ln\left(\frac{e_{\max}(e_{\min} - e(t)/\rho(t))}{e_{\min}(e_{\max} - e(t)/\rho(t))}\right), \tag{7}$$

where $T(\cdot)$ is a smooth function, $T^{-1}(\cdot)$ is its inverse function, $\ln(\cdot)$ is the natural logarithm function.

**Theorem 1.** *The smooth function $T(\cdot)$ is a monotonically increasing function [33], and holds the following properties*

$$
\begin{aligned}
e_{min} &< T(z_1) < e_{max} & T(0) &= 0 \\
\lim_{z_1 \to -\infty} T(z_1) &= e_{min} & \lim_{z_1 \to +\infty} T(z_1) &= e_{max}
\end{aligned}
\tag{8}
$$

**Proof.** From (33), the inverse function of $z_1$ is given by

$$
T(z_1) = \frac{e(t)}{\rho(t)} = \frac{e_{min}e_{max}(e^{z_1} - 1)}{e_{min}e^{z_1} - e_{max}}.
\tag{9}
$$

Since $e_{min} < 0$ and $e_{max} > 0$, the derivative of $T(z_1)$ yields

$$
\frac{dT}{dz_1} = \frac{e_{min}(e_{min} - e_{max})e^{z_1}}{(e_{min}e^{z_1} - e_{max})^2} > 0.
\tag{10}
$$

Hence, $T(z_1)$ is a monotonically increasing function. Furthermore, due to $\rho(t)e_{min} < e(t) < \rho(t)e_{max}$ with $0 < \rho(t) < 1$, then $e_{min} < T(z_1) < e_{max}$ is established. When $z_1 \to \pm\infty$, $T(z_1)$ is close to its up and down boundary $e_{max}$ and $e_{min}$ respectively. If $z_1 = 0$ is substituted into (7), then $T(0) = 0$. $\square$

*3.2. Controller Design Based on PPC*

Together with (7), the system state errors are defined as follows

$$
\begin{cases}
z_1 = \ln\left(\dfrac{e_{max}(e_{min} - e/\rho)}{e_{min}(e_{max} - e/\rho)}\right) \\
z_2 = x_2 - \alpha_1 \\
z_3 = x_3 - \alpha_2
\end{cases}
\tag{11}
$$

where $e$ is the position tracking error defined in (4), $\alpha_i(i = 1, 2)$ is the virtual control variable in controller design.

To avoid the explosion of complexity caused by the repeatedly calculated differentiations of $\dot{\alpha}_i(i = 1, 2)$ in the backstepping iteration, the dynamic surfaces of $z_{i+1}(i = 1, 2)$ are given as follows

$$
\tau_i\dot{\alpha}_i + \alpha_i = \beta_i, \quad \alpha_i(0) = \beta_i(0)
\tag{12}
$$

where $\beta_i(i = 1, 2)$ are the stabilizing functions to be designed, $\tau_i(i = 1, 2)$ are the time constants of the dynamic surfaces.

Thus, the output errors of two dynamic surfaces are defined as $S_i = \alpha_i - \beta_i(i = 1, 2)$. Substituting $S_i$ into (12), the virtual control derivatives $\dot{\alpha}_i = -S_i/\tau_i(i = 1, 2)$ are obtained.

Based on the system state errors (11) and the dynamic surface (12), the prescribed performance constraint controller $u$ is designed as follow

$$
\begin{cases}
\beta_1 = \ddot{y}_d + \dfrac{\dot{\rho}}{\rho}e - k_1\dfrac{z_1}{r} \\[2mm]
\beta_2 = -\dfrac{1}{\bar{g}_2}\left(k_2 z_2 + r z_1 + \bar{f}_2 + \dfrac{S_1}{\tau_1}\right) \\[2mm]
\alpha_i = -\displaystyle\int_0^t \dfrac{S_i}{\tau_i}dt, i = 1,2 \\[2mm]
S_i = \alpha_i - \beta_i, i = 1,2 \\[2mm]
u = -\dfrac{1}{\bar{g}_3}\left(k_3 z_3 + \bar{f}_3 + \bar{g}_2 z_2 + \dfrac{S_2}{\tau_2}\right)
\end{cases}
\tag{13}
$$

where the attenuated parameter $r$ is

$$
r = \frac{\partial T^{-1}}{\partial(e/\rho)}\frac{1}{\rho} = \frac{e_{\max} - e_{\min}}{(e_{\max} - e/\rho)(e/\rho - e_{\min})\rho} \geq \frac{e_{\max} - e_{\min}}{\left(\frac{e_{\max}-e_{\min}}{2}\right)^2} = \frac{4}{e_{\max} - e_{\min}} > 0.
\tag{14}
$$

**Theorem 2.** *Considering the stabilizing functions (13) together with their dynamic surfaces (12) for the EHS model (2) under Hypothesis 1 and Remarks 1 and 2, regardless of the system state errors $z_i(t)(i = 1,2,3)$ start from any initial values $-\infty < z_i(0) < \infty$, the generalized error $Z_g(t)$ including $z_i(i = 1,2,3)$ and $S_j(j = 1,2)$ is ultimate boundedness [42] and its convergence domain is an hypersphere $H_r$,*

$$
H_r \in \left\{ \sum_{i=1}^3 z_i^2 + \sum_{j=1}^2 S_j^2 = 2V(0)e^{-ct_f} + 2\delta/c \right\}
\tag{15}
$$

*where $\delta$ and $c$ are positive constants, $V(0)$ is the initial system state error, $\forall t > t_f$ ($t_f$ is a finite time).*

**Proof.** The candidate quadratic Lyapunov function of the EHS model (2) is given by

$$
V = \frac{1}{2}\sum_{i=1}^3 z_i^2 + \frac{1}{2}\sum_{j=1}^2 S_j^2.
\tag{16}
$$

For convenient proof, $V$ is rewritten into the cascade elements for the convenient controller design as follows

$$
\begin{cases}
V_1 = \dfrac{1}{2}z_1^2 + \dfrac{1}{2}S_1^2 \\[2mm]
V_2 = V_1 + \dfrac{1}{2}z_2^2 + \dfrac{1}{2}S_2^2 \\[2mm]
V_3 = V_2 + \dfrac{1}{2}z_3^2
\end{cases}
\tag{17}
$$

and the following inequalities are satisfied by Young's inequality

$$
|z_i S_i| \leq \frac{z_i^2 + S_i^2}{2} \qquad |S_i \dot{\beta}_i| \leq \frac{S_i^2 \left|\dot{\beta}_i\right|_{\max}^2}{2\sigma_i} + \frac{\sigma_i}{2} \qquad |z_{i+1}\Delta_{i+1}| \leq \frac{z_{i+1}^2 + \Delta_{i+1\,\max}^2}{2},
\tag{18}
$$

for $i = 1,2$, where $\sigma_i(i = 1,2)$ are positive constants, $\left|\dot{\beta}_i\right|_{\max}(i = 1,2)$ are the maximal boundaries of $\dot{\beta}_i(i = 1,2)$.

*Step 1*: Substituting (2), (11), (12) into the derivative of $V_1$, $\dot{V}_1$ yields

$$
\begin{aligned}
\dot{V}_1 &= z_1\dot{z}_1 + S_1\dot{S}_1 = z_1 r(x_2 - \dot{x}_{1d} - \frac{e(t)}{\rho(t)}\dot{\rho}(t)) + S_1(\dot{\alpha}_1 - \dot{\beta}_1) \\
&= z_1 r(z_2 + \beta_1 + S_1 - \dot{x}_{1d} - \frac{e(t)}{\rho(t)}\dot{\rho}(t)) + S_1(-\frac{S_1}{\tau_1} - \dot{\beta}_1)
\end{aligned}
\tag{19}
$$

If the stabilizing function $\beta_1$ in (13) is substituted into (19), and together with (18), then $\dot{V}_1$ is converted to

$$
\begin{aligned}
\dot{V}_1 &= \bar{\dot{V}}_1 + \frac{\sigma_1}{2} + r z_1 z_2 \\
\bar{\dot{V}}_1 &= -\Gamma_1 z_1^2 - \Omega_1 S_1^2
\end{aligned}
\tag{20}
$$

where

$$
\Gamma_1 = k_1 - \frac{r}{2}, \quad \Omega_1 = \frac{1}{\tau_1} - \frac{r}{2} - \frac{|\dot{\beta}_1|^2_{\max}}{2\sigma_1}.
\tag{21}
$$

If a constant gain $k_1$ and a time constant $\tau_1$ yield such that

$$
k_1 > \frac{r}{2}, \quad \frac{1}{\tau_1} > \frac{r}{2} + \frac{|\dot{\beta}_1|^2_{\max}}{2\sigma_1},
\tag{22}
$$

then $\bar{\dot{V}}_1 < 0$.

*Step 2*: The derivative of $V_2$ is given by

$$
\begin{aligned}
\dot{V}_2 &= \dot{V}_1 + z_2\dot{z}_2 + S_2\dot{S}_2 \\
&\leq \bar{\dot{V}}_1 + \frac{\sigma_1}{2} + r z_1 z_2 + z_2[\bar{f}_2 + \bar{g}_2(z_3 + \beta_2 + S_2) - \dot{\alpha}_1 + \Delta_2] + S_2(\dot{\alpha}_2 - \dot{\beta}_2) \\
&\leq \bar{\dot{V}}_1 + \frac{\sigma_1}{2} + z_2[r z_1 + \bar{f}_2 + \bar{g}_2(z_3 + \beta_2 + S_2) + \frac{S_1}{\tau_1} + \Delta_2] + S_2(-\frac{S_2}{\tau_2} - \dot{\beta}_2)
\end{aligned}
\tag{23}
$$

If the stabilizing function $\beta_2$ in (13) is substituted into (23), and together with (18), then $\dot{V}_2$ yields

$$
\begin{aligned}
\dot{V}_2 &\leq \bar{\dot{V}}_2 + \bar{g}_2 z_2 z_3 + \frac{\sigma_1}{2} + \frac{\sigma_2}{2} + \frac{\Delta^2_{2\,\max}}{2} \\
\bar{\dot{V}}_2 &= \bar{\dot{V}}_1 - \Gamma_2 z_2^2 - \Omega_2 S_2^2
\end{aligned}
\tag{24}
$$

where

$$
\Gamma_2 = k_2 - \frac{|\bar{g}_2|^2_{\max}}{2} - \frac{1}{2}, \quad \Omega_2 = \frac{1}{\tau_2} - \frac{|\bar{g}_2|^2_{\max}}{2} - \frac{|\dot{\beta}_2|^2_{\max}}{2\sigma_2}.
\tag{25}
$$

*Step 3*: Similarly, the derivative of $V_3$ is given by

$$
\begin{aligned}
\dot{V}_3 &= \dot{V}_2 + z_3\dot{z}_3 \\
&\leq \bar{\dot{V}}_2 + \bar{g}_2 z_2 z_3 + \frac{\sigma_1}{2} + \frac{\sigma_2}{2} + \frac{\Delta^2_{2\,\max}}{2} + z_3[\bar{f}_3 + \bar{g}_3 u - \dot{\alpha}_2 + \Delta_3] \\
&\leq \bar{\dot{V}}_2 + \frac{\sigma_1}{2} + \frac{\sigma_2}{2} + \frac{\Delta^2_{2\,\max}}{2} + z_3[\bar{g}_2 z_2 + \bar{f}_3 + \bar{g}_3 u + \frac{S_2}{\tau_2} + \Delta_3]
\end{aligned}
\tag{26}
$$

If the control variable $u$ is designed as the form in (13), then $\dot{V}_3$ yields

$$
\dot{V}_3 \leq \bar{\dot{V}}_2 - k_3 z_3^2 + \delta,
\tag{27}
$$

where $\delta = (\sigma_1 + \sigma_2 + \Delta^2_{2\,\max} + \Delta^2_{3\,\max})/2$ is a positive constant.

If a constant $c$ is defined as $c = \min\{2\Gamma_1, 2\Gamma_2, 2\Omega_1, 2\Omega_2, 2k_3\}$, from the definitions of $\bar{V}_1$ and $\bar{V}_2$, (27) is rewritten as

$$\dot{V}_3 \leq -cV_3 + \delta. \tag{28}$$

Integrating (28), $\dot{V}_3$ yields

$$V(t) \leq V(0)e^{-ct} + \int_0^t \delta e^{-c(t-\varepsilon)} d\varepsilon$$
$$\leq V(0)e^{-ct} + \delta(1 - e^{-ct})/c \tag{29}$$

According to (29), and letting $t \to t_f$, the error convergence domain $H_r$ in (15) is obtained. Furthermore, the size of the generalized error convergence domain $H_r$ mainly is decided by the element $\delta/c$. Thus, the increased control gains $k_i (i = 1, 2, 3)$ and the reduced constant $c$ can arbitrarily shrink the size of $H_r$ as $t \to \infty$. $\square$

Figure 2 shows the proposed prescribed performance constraint controller. The designed dynamic surface (12) is used to instead of the virtual control derivatives $\dot{\alpha}_i (i = 1, 2)$ and the virtual control variables $\alpha_i (i = 1, 2)$ are substituted by the stabilizing functions $\beta_i (i = 1, 2)$. The output-constraint (5) is converted to the time-varying performance constraint (6), which represents the position tracking error $e$ of EHS. Then this constraint is transformed into the new state error $z_1$ (7). The controller $u$ (13) is constructed to guarantee the dynamic performance of the EHS (2) under the hydraulic parametric uncertainties and the external load disturbance integrated in $\Delta_i (i = 2, 3)$.

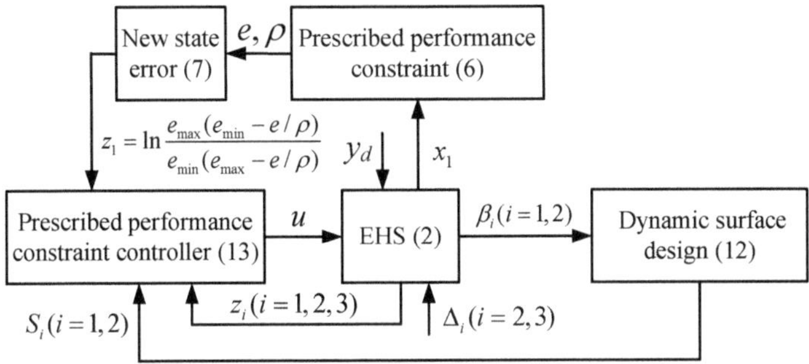


**Figure 2.** Block diagram of the prescribed performance constraint controller.

## 4. Comparison Results

To verify the proposed prescribed performance constraint control method, some known hydraulic parameters are $\bar{C}_d = 0.62$, $w = 0.024$ m, $x_{v\,max} = 7.9$ mm, $\bar{\beta}_e = 7000$ bar, $\bar{\rho} = 850$ kg/m$^3$, $\bar{K} = 1000$ N/m, $\bar{b} = 100$ Ns/m, $\bar{C}_{tl} = 2.5 \times 10^{-11}$ m$^3$/(s · Pa), $K_{sv} = 4.9 \times 10^{-6}$ m/V, $p_r = 2$ bar, $A_p = 2.01$ cm$^2$, $V_t = 1.74 \times 10^{-5}$ m$^3$, $m = 1.739$ kg, $\Delta C_d = 0.1\bar{C}_d$, $\Delta \beta_e = 0.5\bar{\beta}_e$, $\Delta K = 0.5\bar{K}$, $\Delta b = 0.5\bar{b}$, $\Delta \rho = -0.1\bar{\rho}$, $\Delta C_{tl} = 0.2\bar{C}_{tl}$, $F_{L\,max} = 500$ N. The time constants of two dynamic surfaces is $\tau_1 = \tau_2 = 10^{-3}$. Some control parameters are designed as $k_1 = 100$, $k_2 = 1200$ and $k_3 = 6000$, $x_{1\,min} = -50$ mm, $x_{1\,max} = 50$ mm, $y_{min} = -50$ mm, $y_{max} = 50$ mm, $\rho(0) = 0.95$, $\rho(\infty) = 0.03$, $\lambda = 0.3$.

In addition, to compare with the traditional control scheme, Proportional-Integral (PI) controller is also adopted in this EHS such that

$$u = k_p(y_d - x_1) + k_i \int (y_d - x_1) dt \tag{30}$$

where the control gains $k_p = 150$ and $k_i = 10$ have been well tuned to guarantee the fast response of the cylinder position.

### 4.1. Compared Results with Nominal Hydraulic Parameters

Firstly, the nominal hydraulic parameters is adopted in simulation with the uncertain nonlinearities $\Delta_2 = \Delta_3 = 0$. The cylinder position demands are selected as $y_d = 25(\sin(0.8\pi t) + \sin(0.4\pi t) + \sin(0.2\pi t))$ mm and $y_d = 25(\sin(1.6\pi t) + \sin(0.8\pi t) + \sin(0.4\pi t))$ mm. The initial states of two control schemes are $x_1(0) = 20$ mm, $x_2(0) = 0$ mm/s, $x_3(0) = 0$ bar. The proposed controller comparison with PI controller are shown in Figures 3–6. The controller based on prescribed performance constraint has the steady tracking errors $\Delta x_1 = 0.01$ mm in low frequency demand and $\Delta x_1 = 0.05$ mm in high frequency demand respectively, which is better than the PI controller $\Delta x_1 = 0.5$ mm and $\Delta x_1 = 2$ mm in corresponding frequency demand. Since the constraint holding technique is adopted in the proposed controller, the tracking error of the cylinder position is not always beyond the prescribed constraint $\rho(t)e_{\min} < e(t) < \rho(t)e_{\max}$. Thus, these comparison results indicate the advantage of this prescribed performance constraint technique.

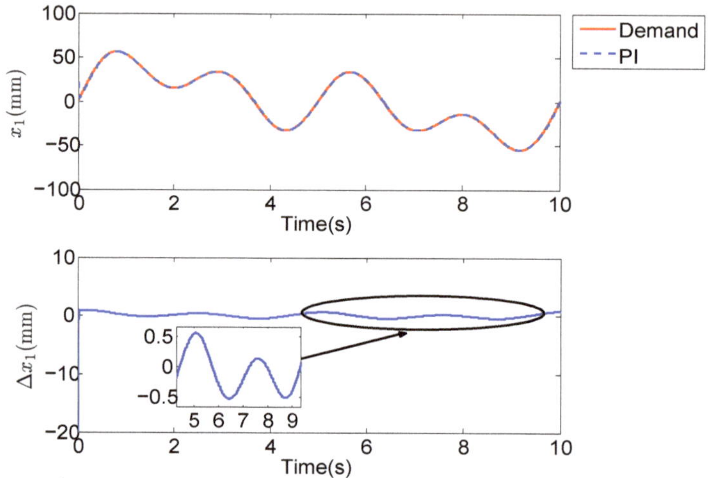


**Figure 3.** The cylinder position responses $x_1$ by PI controller with $\Delta_2 = \Delta_3 = 0$, the demand $y_d = 25(\sin(0.8\pi t) + \sin(0.4\pi t) + \sin(0.2\pi t))$ mm.

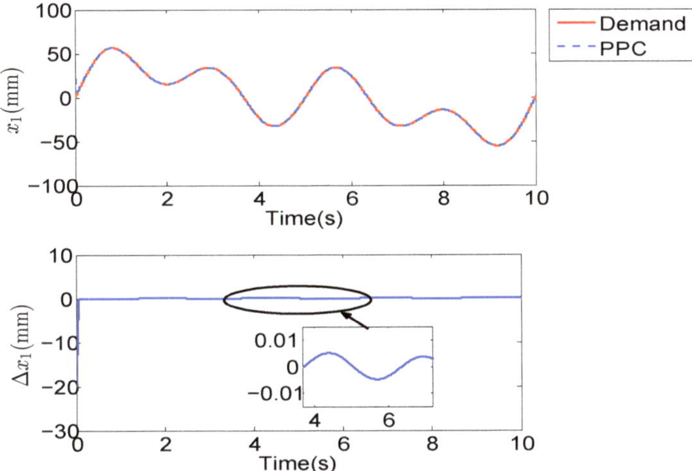

**Figure 4.** The cylinder position responses $x_1$ by prescribed performance constraint controller with $\Delta_2 = \Delta_3 = 0$, the demand $y_d = 25(\sin(0.8\pi t) + \sin(0.4\pi t) + \sin(0.2\pi t))$.

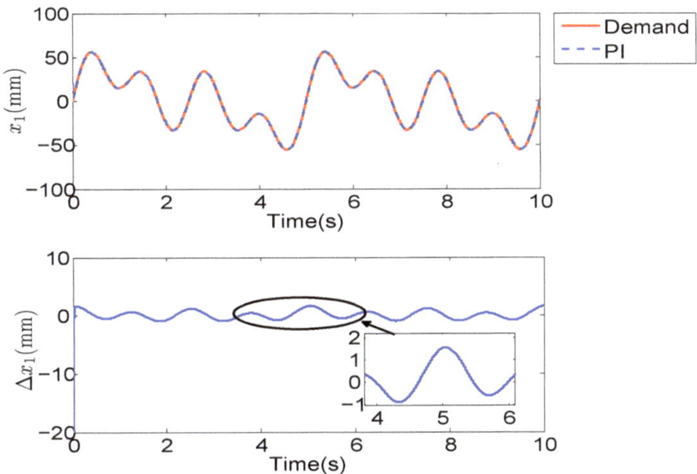

**Figure 5.** The cylinder position responses $x_1$ by PI controller with $\Delta_2 = \Delta_3 = 0$, the demand $y_d = 25(\sin(1.6\pi t) + \sin(0.8\pi t) + \sin(0.4\pi t))$.

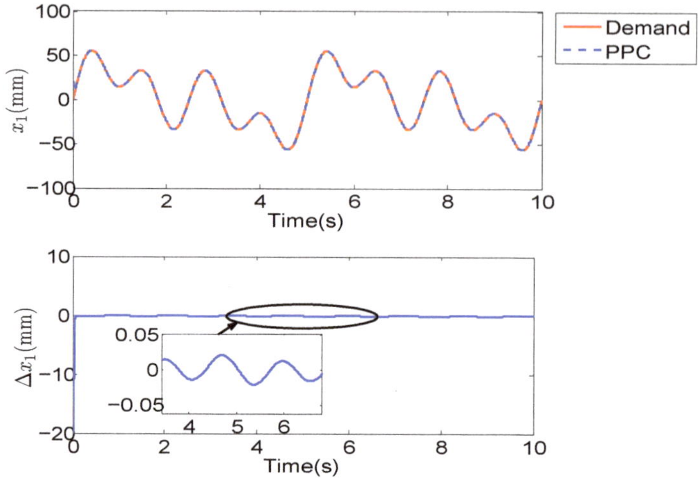



**Figure 6.** The cylinder position responses $x_1$ by prescribed performance constraint controller with $\Delta_2 = \Delta_3 = 0$, the demand $y_d = 25(\sin(1.6\pi t) + \sin(0.8\pi t) + \sin(0.4\pi t))$.

## 4.2. Compared Results with Uncertain Nonlinearities

To verify the dynamic response performance of the proposed prescribed performance constraint controller, the frequency of the cylinder position demands and the initial states are same to Section 4.1. The hydraulic parametric uncertainties $\Delta C_d$, $\Delta \beta_e$, $\Delta K$, $\Delta b$, $\Delta \rho$, $\Delta C_{tl}$ are all injected in the EHS model (2). Furthermore, the external load are assumed to be $F_L(t) = F_{L\max} \sin(2\pi t)$. The comparison results of two controllers are shown in Figures 7–10. When the total uncertain nonlinearities $\Delta_2$ and $\Delta_3$ are injected in EHS, the prescribed performance constraint controller has the steady tracking error $\Delta x_1 = 2$ mm both in low and high frequency demand. However, the position tracking error of PI controller is $\Delta x_1 = 5$ mm. These results indicate the prescribed performance constraint controller has higher dynamic response performance than the PI controller under the hydraulic parametric uncertainties and the external load disturbance.

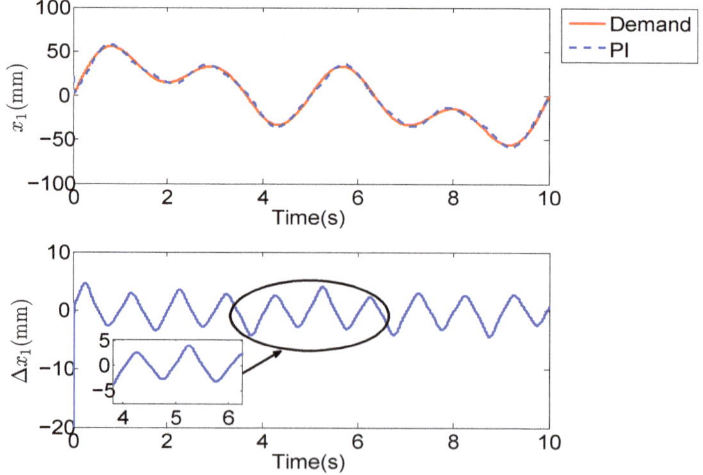



**Figure 7.** The cylinder position responses $x_1$ by PI controller with hydraulic parametric uncertainties and the external load, the demand $y_d = 25(\sin(0.8\pi t) + \sin(0.4\pi t) + \sin(0.2\pi t))$ mm.

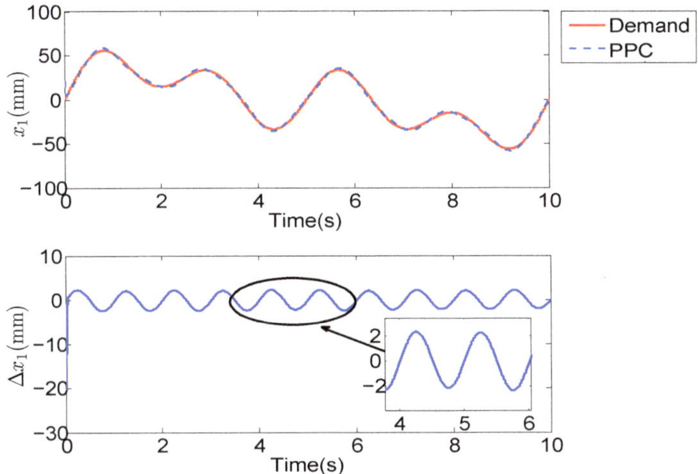

**Figure 8.** The cylinder position responses $x_1$ by prescribed performance constraint controller with hydraulic parametric uncertainties and the external load, the demand $y_d = 25(\sin(0.8\pi t) + \sin(0.4\pi t) + \sin(0.2\pi t))$ mm.

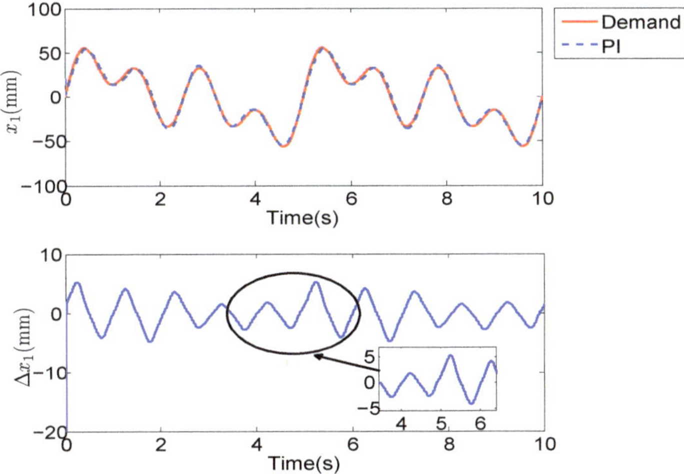

**Figure 9.** The cylinder position responses $x_1$ by PI controller with hydraulic parametric uncertainties and the external load, the demand $y_d = 25(\sin(1.6\pi t) + \sin(0.8\pi t) + \sin(0.4\pi t))$ mm.

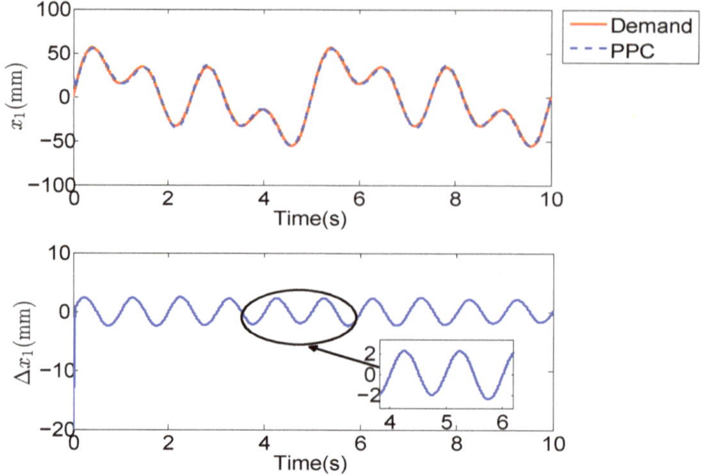

**Figure 10.** The cylinder position responses $x_1$ by prescribed performance constraint controller with hydraulic parametric uncertainties and the external load, the demand $y_d = 25(\sin(1.6\pi t) + \sin(0.8\pi t) + \sin(0.4\pi t))$ mm.

## 5. Conclusions

In this study, a prescribed performance constraint controller is proposed for electro-hydraulic system to improve the output position accuracy of EHS. Firstly, the EHS model is constructed as a state-space strict-feedback model with uncertain nonlinearities. Secondly, according to the required boundary of the tracking position error, the technique of prescribed performance constraint is used to design the time-varying error boundary to not only regulate the large initial error in satisfactory time but also consider the control capability of EHS. Furthermore, the dynamic surface is adopted to replace the repeatedly calculated differentiations of the virtual control variables in backstepping design. The comparative results with the PI controller verify that the proposed controller has better performance than PI controller.

**Acknowledgments:** The authors are grateful to the anonymous reviewers and the editor for their constructive comments. This research is supported by National Natural Science Foundation of China (51775089, 61305092, 51205045), the Fundamental Research Funds for the Central Universities, China (ZYGX2016J160), the Open Foundation of the State Key Laboratory of Fluid Power and Mechatronic Systems (GZKF-201515).

**Author Contributions:** Qing Guo and Dan Jiang conceived and designed the architecture of this paper; Yili Liu wrote the draft; Qiang Wang and Wenying Xiong performed the simulation; Jie Liu and Xiaochai Li checked the literature and the typos of this paper.

**Conflicts of Interest:** The authors declare no conflict of interest.

### Notation

| | |
|---|---|
| EHS | Electro-hydraulic system |
| BLF | Barrier Lyapunov function |
| PPC | Prescribed performance constraint |
| $K_{sv}$ (m/V) | Gain of the servo valve |
| $u$ (V) | control voltage of the servo valve |
| $C_d$ (–) | Discharge coefficient |
| $w$ (m) | Area gradient of the servo valve |
| $p_s, p_r$ (Pa) | Supply pressure and return pressure |
| $p_L$ (Pa) | Cylinder load pressure |

| | |
|---|---|
| $x_{v\max}$ (mm) | Maximal spool position of the servo valve |
| $\rho$ (kg/m$^3$) | Density of hydraulic oil |
| $C_{tl}$ (m$^3$/(s·Pa)) | Coefficient of the total leakage of the cylinder |
| $\beta_e$ (bar) | Effective bulk modulus |
| $A_p$ (m$^2$) | Annulus area of the cylinder chamber |
| $V_t$ (m$^3$) | Half-volume of the cylinder |
| $m$ (kg) | Load mass |
| $b$ (Ns/m) | Viscous damping coefficient of hydraulic oil |
| $K$ (N/m) | Load spring constant |
| $F_L$ (N) | External load of the electro-hydraulic system |
| $\dot{y}_d$ (m/s), $\dot{y}$ (m) | Desired and actual velocities of the cylinder |
| $\Delta C_d$ (–) | Parametric uncertainty of discharge coefficient |
| $\Delta\beta_e$ (bar) | Parametric uncertainty of effective bulk modulus |
| $\Delta\rho$ (kg/m$^3$) | Parametric uncertainty of density of hydraulic oil |
| $\Delta C_{tl}$ (m$^3$/(s·Pa)) | Parametric uncertainty of coefficient of the total leakage of the cylinder |
| $\Delta b$ (Ns/m) | Parametric uncertainty of viscous damping coefficient of hydraulic oil |
| $\Delta K$ (N/m) | Parametric uncertainty of load spring constant |

## References


1. Fales, R.; Kelkar, A. Robust control design for a wheel loader using $H_\infty$ and feedback linearization based methods. *ISA Trans.* **2009**, *48*, 313–320.
2. Zhao, J.; Wang, J.; Wang, S. Fractional order control to the electrohydraulic system in insulator fatigue test device. *Mechatronics* **2013**, *23*, 828–839.
3. Yao, J.; Jiao, Z.; Shang, Y.; Huang, C. Adaptive nonlinear optimal compensation control for electro-hydraulic load simulator. *China J. Aeronaut.* **2010**, *23*, 720–733.
4. Yang, Y.; Ma, L.; Huang, D. Development and repetitive learning control of lower limb exoskeleton driven by electro-hydraulic actuators. *IEEE Trans. Ind. Electron.* **2017**, *64*, 4169–4178.
5. Merritt, H. *Hydraulic Control Systems*; John Wiley & Sons: New York, NY, USA, 1967.
6. Shen, G.; Zhu, Z.; Zhao, J.; Zhu, W.; Tang, Y.; Li, X. Real-time tracking control of electro-hydraulic force servo systems using offline feedback control and adaptive control. *ISA Trans.* **2017**, *67*, 356–370.
7. Guo, Q.; Yin, J.; Yu, T.; Jiang, D. Saturated adaptive control of an electrohydraulic actuator with parametric uncertainty and load disturbance. *IEEE Trans. Ind. Electron.* **2017**, *64*, 7930–7941.
8. Shi, Y.; Wang, Y.; Cai, M.; Zhang, B.; Zhu, J. Study on the aviation oxygen supply system based on a mechanical ventilation model. *Chin. J. Aeronaut.* **2017**, doi:10.1016/j.cja.2017.10.008.
9. Shi, Y.; Zhang, B.; Cai, M.; Zhang, D. Numerical Simulation of volume-controlled mechanical ventilated respiratory system with two different lungs. *Int. J. Numer. Methods Biomed. Eng.* **2016**, *33*, 2852, doi:10.1002/cnm.2852.
10. Niu, J.; Shi, Y.; Cai, M.; Cao, Z.; Wang, D.; Zhang, Z.; Zhang, D.X. Detection of sputum by interpreting the time-frequency distribution of respiratory sound signal using image processing techniques. *Bioinformatics* **2017**, doi:10.1093/bioinformatics/btx652.
11. Ren, S.; Shi, Y.; Cai, M.; Xu, W. Influence of secretion on airflow dynamics of mechanical ventilated respiratory system. *IEEE/ACM Trans. Comput. Biol. Bioinform.* **2017**, doi:10.1109/TCBB.2017.2737621.
12. Niu, J.; Shi, Y.; Cao, Z.; Cai, M.; Wei, C.; Jian, Z.; Xu, W. Study on air flow dynamic characteristic of mechanical ventilation of a lung simulator. *Sci. China Technol. Sci.* **2017**, *45*, 1–8.
13. Ren, S.; Cai, M.; Shi, Y.; Xu, W.; Zhang, X.D. Influence of Bronchial Diameter Change on the airflow dynamics Based on a Pressure-controlled Ventilation System. *Int. J. Numer. Methods Biomed. Eng.* **2017**, doi:10.1002/cnm.2929.
14. Shi, Y.; Zhang, B.; Cai, M.; Xu, W. Coupling Effect of Double Lungs on a VCV Ventilator with Automatic Secretion Clearance Function. *IEEE/ACM Trans. Comput. Biol. Bioinform.* **2017**, doi:10.1109/TCBB.2017.2670079.

15. Shi, Y.; Wu, T.; Cai, M.; Wang, Y.; Xu, W. Energy conversion characteristics of a hydropneumatic transformer in a sustainable-energy vehicle. *Appl. Energy* **2016**, *171*, 77–85.
16. Shi, Y.; Wang, Y.; Liang, H.; Cai, M. Power characteristics of a new kind of air-powered vehicle. *Int. J. Energy Res.* **2016**, *40*, 1112–1121.
17. Shi, Y.; Niu, J.; Cao, Z.; Cai, M.; Zhu, J.; Xu, W. Online estimation methodfor respiratory parameters based on a pneumatic model. *IEEE/ACM Trans. Comput. Biol. Bioinform.* **2016**, *13*, 939–946.
18. Zhu, B.; Meng, C.; Hu, G. Robust consensus tracking of double-integrator dynamics by bounded distributed control. *Int. J. Robust Nonliner Control* **2016**, *26*, 1489–1511.
19. Zhu, B.; Liu, H.H.T.; Li, Z. Robust distributed attitude synchronization of multiple three-DOF experimental helicopters. *Control Eng. Pract.* **2015**, *36*, 87–99.
20. Zhang, Y.; Su, S.; Savkin, A.; Celler, B.; Nguyen, H. Multiloop integral controllability analysis for nonlinear multiple-input single-output Processes. *Ind. Eng. Chem. Res.* **2017**, *56*, 8054–8065.
21. Tee, K.P.; Ge, S.S.; Tay, E.H. Barrier lyapunov functions for the control of output-constrained nonlinear systems. *Automatica* **2009**, *45*, 918–927.
22. Tee, K.P.; Ren, B.; Ge, S.S. Control of nonlinear systems with time varying output constraints. *Automatica* **2011**, *47*, 2511–2516.
23. He, W.; Chen, Y.; Yin, Z. Adaptive neural network control of an uncertain robot with full-state constraints. *IEEE Trans. Cybern.* **2016**, *46*, 620–629.
24. He, W.; David, A.O.; Yin, Z.; Sun, C. Neural network control of a robotic manipulator with input deadzone and output constraint. *IEEE Trans. Syst. Man Cybern. A Syst.* **2016**, *46*, 759–770.
25. He, W.; Huang, H.; Ge, S.S. Adaptive neural network control of a robotic manipulator with time-varying output constraints. *IEEE Trans. Cybern.* **2017**, *47*, 3136–3147.
26. He, W.; Dong, Y. Adaptive fuzzy neural network control for a constrained robot using impedance learning. *IEEE Trans. Neural Netw. Learn. Syst.* **2017**, doi:10.1109/TNNLS.2017.2665.
27. Ren, B.; Ge, S.S.; Tee, K.P.; Lee, T.H. Adaptive neural control for output feedback nonlinear systems using a barrier lyapunov function. *IEEE Trans. Neural Netw.* **2010**, *21*, 1339–1345.
28. Won, D.; Kim, W.; Shin, D.; Chung, C.C. High-gain disturbance observer-based backstepping control with output tracking error constraint for electro-hydraulic systems. *IEEE Trans. Control Syst. Technol.* **2015**, *23*, 787–795.
29. Qiu, Y.; Liang, X.; Dai, Z. Backstepping dynamic surface control for an anti-skid braking system. *Control Eng. Pract.* **2015**, *42*, 140–152.
30. Guo, Q.; Zhang, Y.; Celler, B.G.; Su, S.W. State-constrained control of single-rod electrohydraulic actuator with parametric uncertainty and load disturbance. *IEEE Trans. Control Syst. Technol.* **2017**, doi:10.1109/TCST.2017.2753167.
31. Bechlioulis, C.P.; Rovithakis, G.A. Adaptive control with guaranteed transient and steady state tracking error bounds for strict feedback systems. *Automatica* **2015**, *45*, 532–538.
32. Zhang, Z.; Duan, G.; Hou, M. Robust adaptive dynamic surface control of uncertain non-linear systems with output constraints. *IET Control Theory Appl.* **2017**, *11*, 110–121.
33. Zhang, Z.; Duan, G.; Hou, M. Longitudinal attitude control of a hypersonic vehicle with angle of attack constraints. In Proceedings of the 10th Asian Control Conference, Kota Kinabalu, Malaysia, 31 May–3 June 2015.
34. Swaroop, D.; Hedrick, J.; Yip, P.; Gerdes, J. Dynamic surface control for a class of nonlinear systems. *IEEE Trans. Autom. Control* **2000**, *45*, 1893–1899.
35. Wang, D.; Huang, J. Neural network-based adaptive dynamic surface control for a class of uncertain nonlinear systems in strict-feedback form. *IEEE Trans. Neural Netw.* **2005**, *16*, 195–202.
36. Hou, M.; Duan, G. Robust adaptive dynamic surface control of uncertain nonlinear systems. *Int. J. Control Autom. Syst.* **2011**, *9*, 161–168.
37. Kim, W.; Shin, D.; Won, D.; Chung, C.C. Disturbance-observerbased position tracking controller in the presence of biased sinusoidal disturbance for electrohydraulic actuators. *IEEE Trans. Control Syst. Technol.* **2013**, *21*, 2290–2298.
38. Yao, J.; Jiao, Z.; Ma, D. High-accuracy tracking control of hydraulic rotary actuators with modeling uncertainties. *IEEE/ASME Trans. Mechatron.* **2014**, *19*, 633–641.
39. Guo, Q.; Zhang, Y.; Celler, B.G.; Su, S.W. Backstepping control of electro-hydraulic system based on extended-state-observer with plant dynamics largely unknown. *IEEE Trans. Ind. Electron.* **2016**, *63*, 6909–6920.

40. Yao, J.; Jiao, Z.; Ma, D. A practical nonlinear adaptive control of hydraulic servomechanisms with periodic-like disturbances. *IEEE/ASME Trans. Mechatron.* **2015**, *20*, 2752–2760.
41. Guo, Q.; Sun, P.; Yin, J.; Yu, T.; Jiang, D. Parametric adaptive estimation and backstepping control of electro-hydraulic actuator with decayed memory. *ISA Trans.* **2016**, *62*, 202–214.
42. Krstic, M.; Kanellakopoulos, I.; Kokotovic, P.V. *Nonlinear and Adaptive Control Design*; John Wiley & Sons: New York, NY, USA, 1995.

Article

# A Performance Test and Internal Flow Field Simulation of a Vortex Pump

Ping Tan [1], Yi Sha [1,*], Xiaobang Bai [2], Dongming Tu [1], Jien Ma [3,4,5,*], Wenjun Huang [3,4,5] and Youtong Fang [3,4,5]

[1] School of Automation and Electrical Engineering/ School of Mechanical & Automotive Engineering, Zhejiang University of Science and Technology, Hangzhou 310023, China; tankor@zju.edu.cn (P.T.); 221601852001@zust.edu.cn (D.T.)
[2] College of Energy and Power Engineering, Lanzhou University of Technology, Lanzhou 730050, China; baixiaobang@163.com
[3] College of Control Science and Engineering, Zhejiang University, Hangzhou 310027, China; huangwj@supcon.com (W.H.); youtong@zju.edu.cn (Y.F.)
[4] College of Electrical Engineering, Zhejiang University, Hangzhou 310027, China
[5] China Academy of West Region Development, Zhejiang University, Hangzhou 310027, China
* Correspondence: Shayi01@sina.com (Y.S.); jienma@126.com (J.M.)

Received: 17 September 2017; Accepted: 1 December 2017; Published: 7 December 2017

**Abstract:** Vortex pumps have good non-clogging performance and are widely used in the fluid transportation of food, sewage treatment, and mineral and coal slurry transportation. In order to design and manufacture a vortex pump with good performance and establish a method of optimum design, we must master the internal flow rules of the pump. Based on the self-design vortex pump (32WB8-12) experiment, the discharge-pump head ($q_v$-$H$), discharge-pump shaft power ($q_v$-$P$), discharge-pump efficiency ($q_v$-$\eta$), and discharge-critical net positive suction head ($q_v$-$NPSH_c$) curves are obtained, and the $q_v$-$NPSH_c$ curve shows an opposite tendency compared with the centrifugal pump. With the mathematical model selected with respect to the optimal condition, the three-dimensional internal flow within the vortex pump has been numerically simulated by a renormalization group $k$-$\varepsilon$ (RNG $k$-$\varepsilon$) turbulence model. The static pressure ($p_s$) and velocity distribution of the impeller and the middle section of the volute at 0°, 90°, 180°, and 270° are obtained, and the performance curves have been fitted using a CFX-calculated energy parameter. It was illustrated that the velocity field is relatively disordered and the flow in the impeller region is of a forced vortex character. The flow in the volute is similar to that of the combined vortex with backflow, which is a non-axisymmetric unsteady flow with quite high turbulence intensity. These factors are the main reasons for the relatively low efficiency of the vortex pump. The measurement of flow field in volute with a five-hole probe was conducted, and it is demonstrated that the numerical results are in good agreement with the flow field measurement data. An upward pressure gradient forms in the portal area of the impeller, and it is confirmed that the lowest pressure point is located in the upper position of the impeller hub. It is revealed that for the vortex pump to have advanced suction and anti-cavitation performance, the lowest pressure in the pump should be $-4 \times 10^4$ Pa and it should be located at the center of the vortex chamber cavity.

**Keywords:** vortex pump; computational fluid dynamics (CFD); five-hole spherical probe; *NPSH*

---

## 1. Introduction

The vortex pump evolved from a centrifugal pump with a semi-open impeller, and it is a type of non-clogging pump with an impeller located completely within the impeller chamber behind the vortex chamber, as shown in Figure 1. Its unique structure makes it suitable for conveying solid–liquid mixtures containing solid and long fibers [1–3], and in particular, it shows excellent performance in the hydraulic transportation of gases containing manure, biogas, digesters, waste liquid, and agricultural products such as potato, etc., whilst solid media were not damaged. At present, if we compare the vortex pump with the centrifugal pump, the biggest disadvantage is the low efficiency of the vortex pump [4,5]. In recent years, researchers have carried out studies on pump design and numerical simulation to improve the performance of the vortex pump. Gerlach et al. analyzed the how variations in geometric parameters influence vortex pump characteristics, and investigated many articles considering parameters in both single-parameter and multi-parameter experiments [6,7]. Moreover, Gerlach et al. performed a numerical study of the internal flow field of a vortex pump, and made a comparison with the Hamel–Oseen vortex model. The study suggests that a vortex similar to the Hamel–Oseen vortex is only present at the strong part load operation [8]. Through a numerical simulation and experimental analysis, Jiang and Zhu studied the influence of different types of hems and blades on the performance of the vortex pump [9,10]. Li carried out a numerical simulation and experiment analysis for the gas–liquid two-phase vortex pump [11]. Because of the change in structure from a centrifugal pump to a vortex pump, the internal flow of the vortex pump is complicated, and the flow through the outlet of the pump is the intersection of the flow through the vortex chamber and the impeller. The 32WB8-12 vortex pump was built based on continuous research findings from the 1980s. In order to design and manufacture a vortex pump with good performance and establish a method of optimum design, we must master the internal flow rules of the pump. So, this paper carries out research work on a performance test and internal flow field simulation of the vortex pump. At present, the internal flow velocity field of the vortex pump can be measured by the particle image velocimetry (PIV) technique, but the pressure field of the high-speed rotating impeller cannot be measured without contact. Therefore, the numerical calculation of the flow field is an effective way to obtain the internal flow rules. In order to determine whether the calculation results are correct and can reflect the actual objective, we need scientific experimental verification. The volute chamber of the vortex pump is relatively wide, so it is possible to measure the flow field by probe contact. As long as verification of the numerical calculation results of the vortex chamber flow field through comparative analysis reaches a certain level of precision, it can be extended to deduce that the numerical calculation of the internal flow field of the whole vortex pump gives realistic results. Our paper is based on this idea.

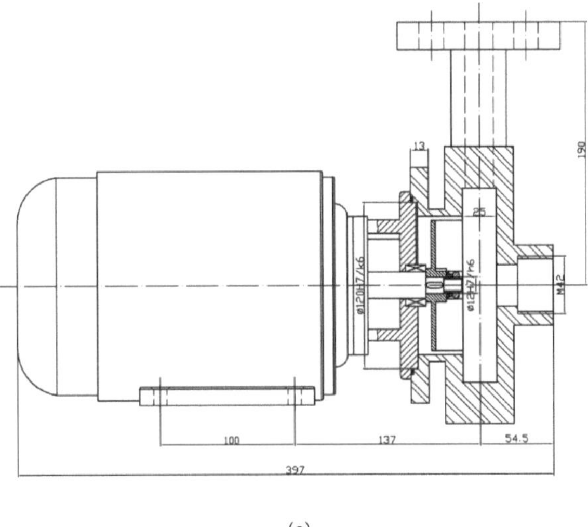

(**a**)

**Figure 1.** *Cont.*

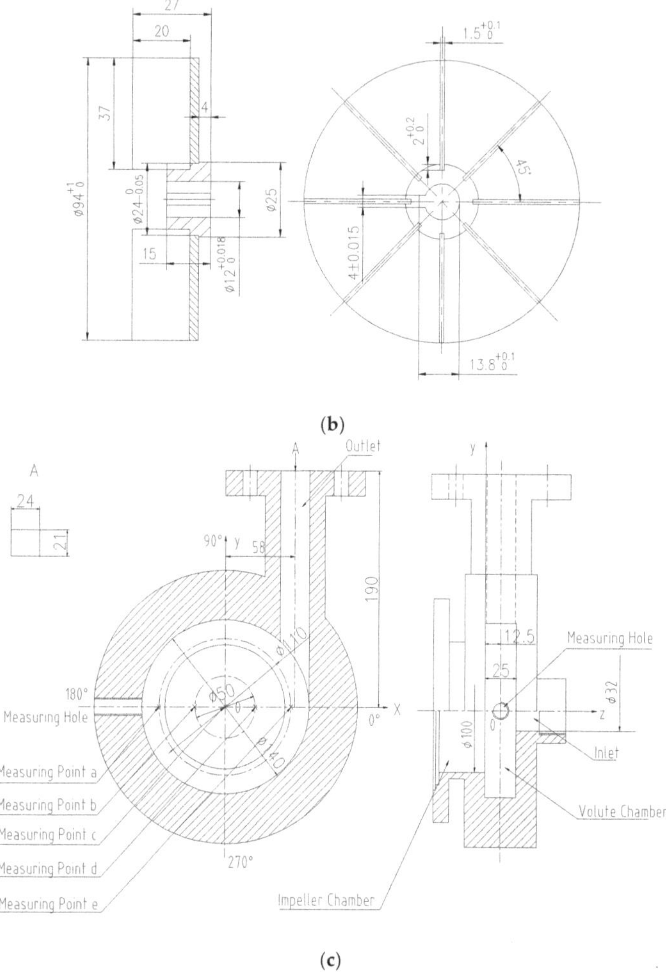



**Figure 1.** Structure and distribution of the hydraulic geometric sketch. (**a**) Structure diagram of the vortex pump; (**b**) Impeller component and geometric parameters; (**c**) Chamber component and geometric parameters.

The flow field inside the vortex pump was simulated by ANSYS® CFX software, and the performance curve was calculated based on the design, development, and performance test of the model pump. The flow field in the vortex chamber was measured by a five-hole probe, and the correctness of the numerical results was verified. After a comprehensive analysis, the internal flow rules of the vortex pump and the constraint relationship between the pump performance and the flow characteristics were clarified.

## 2. Design and Test Scheme of the Model Pump

For the study of the vortex pump, considering factors such as the scale of the test platform, this paper designed and developed the 32WB8-12 pump (Zhejiang University of Science and Technology, Hangzhou, Zhejiang, China) model. The hydraulic design and geometric parameters are shown in Table 1, and its structure and impeller, pressurized water chamber parts, coordinate system, and the position of the measuring hole are shown in Figure 1.

**Table 1.** Vortex pump design and hydraulic parameters.

| Design Parameters | | Hydraulic and Geometric Parameters | | | |
|---|---|---|---|---|---|
| Discharge/$q_v$ (m³/h) | 9 | Specific speed/$n_s$ | 80.67 | Diameters of impeller/$D_2$ (mm) | 94 |
| Head/$H$ (m) | 12 | Chamber diameter/$D_V$ (mm) | 140 | Chamber throat area/$F_{thr}$ (mm²) | 504 |
| Rotational speed/$n$ (r/min) | 2850 | Width of blade/$b_2$ (mm) | 20 | Width of chamber (mm) | 25 |
| Pump efficiency/$\eta$ (%) | 50 | Volute chamber lines | ring-shaped | Base circle diameter of chamber/$D_3$ (mm) | 100 |
| Shaft power/$P$ (kW) | 0.55 | Number of blades/$Z$ (piece) | 8 | Diameter of volute/$D_j$ (mm) | 32 |
| Net positive suction head/$NPSH_r$ (m) | 4 | Blade thickness/$\delta$ (mm) | 1.5 | Blade airfoil lines | radialized |

## 3. Model Pump Performance

The pump is directly connected to the motor, and the power of the pump shaft was measured by the electronic method. The no-load motor is in line with GB/T 12785-91 standards (Test Methods for Submersible Motor Pumps), the pump performance test was carried out according to GB/T 3216-2005 (Rotodynamic Pumps—Hydraulic Performance Acceptance Tests, Grades 1 and 2), and the frequency converter was used to adjust to the rated speed. Figure 2 is the scheme of the experimental test bench. The technical data for instrumentation are provided in Table 2.

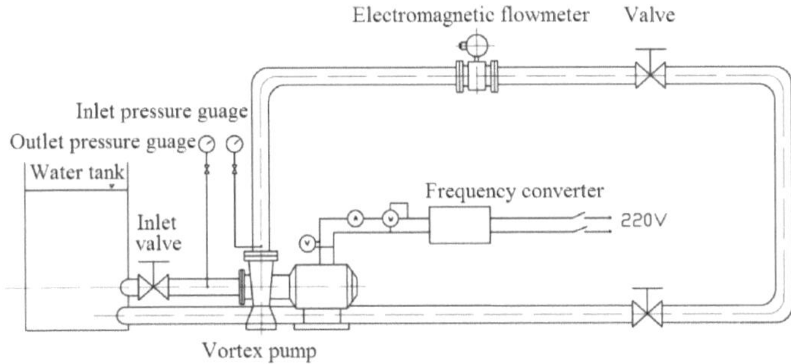


**Figure 2.** Scheme of the experimental test bench.

**Table 2.** The technical data for instrumentation.

| Parameter | Instrument | Precision | Parameter | Parameter | Precision |
|---|---|---|---|---|---|
| Pump head/$H$ | pressure gauge | 0.4 | Current/$A$ | ammeter | 0.5 |
| | vacuum gauge | 0.4 | Voltage/$V$ | voltmeter | 0.5 |
| Discharge/$q_v$ | electromagnetic flow meter | 0.5 | Shaft power/$P$ | wattmeter | 0.5 |
| Rotation rate/$n$ | digital tachometer | 0.5 | Resistance/$R$ | digital resistance meter | 0.1 |

Pump experimental performance curves are shown in Figure 3 ($\eta_{gr}$ is the unit efficiency). The optimal operating conditions of the pump are: discharge ($q_v$) is 9.00 m³/h, pump head ($H$) is 12.88 m, maximum efficiency ($\eta_{max}$) is 53.13%, and specific speed $n_s$ is 76.5. The pump $q_v$-$H$ curve is flat, higher than in the design condition. The $q_v$-$P$ curve rises faster than other curves. The discharge-critical net positive suction head ($q_v$-$NPSH_c$) curve is parabolic-diminishing and shows the opposite trend compared with the centrifugal pump's $q_v$-$NPSH_c$ curve. The critical $NPSH$ of small flow is larger, but the critical $NPSH_c$ decreases with the increased discharge.

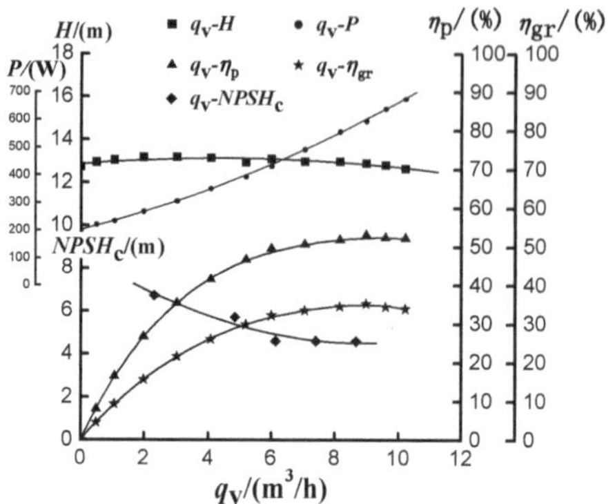

**Figure 3.** Experimental pump performance curves.

$\eta_p$: pump efficiency; $\eta_{gr}$: unit efficiency; $H$: pump head in m; $P$: shaft power in W; $NPSH_c$: critical net positive suction head in m; $q_v$: discharge in m³/h; $q_v$-$H$: discharge-pump head; $q_v$-$P$: discharge-pump shaft power in w; $q_v$-$\eta_p$: discharge-pump efficiency; $q_v$-$\eta_{gr}$: discharge–unit efficiency; $q_v$-$NPSH_c$: discharge-critical net positive suction head.

## 4. Numerical Calculation of Pump's Internal Flow Field

### 4.1. Governing Equation

The flow field was calculated based on selecting the best condition of pump performance test. The internal flow field of the vortex pump is very complex, which is three-dimensional turbulent flow dominated by strong swirling flow. It is affected by many factors, such as curvature, rotation, and so on. Choosing the turbulence model for the calculation of the internal flow field is very important. The previous calculation used the standard $k$-$\varepsilon$ two-equation model [12–14], which is a local equilibrium model that cannot accurately predict turbulence with drastic changes in mean flow, such as separation flow and swirling flow. Later, the researchers applied the modified $k$-$\varepsilon$ and the Reynolds stress equation model—which was derived from the anisotropic conditions—to the numerical calculation of the strong cyclonic flow field. The renormalization group $k$-$\varepsilon$ turbulence model is called the RNG $k$-$\varepsilon$ model. The coefficient in the RNG $k$-$\varepsilon$ turbulence model is derived by using the RNG model theory. There is an additional item R in the $k$-$\varepsilon$ equation, which represents the effect of the average strain rate on the turbulent dissipation rate $\varepsilon$. The RNG $k$-$\varepsilon$ turbulence model can simulate the flow separation and vortex well. Moreover, because of the complexity of the Reynolds stress model and the large amount of computational resources, we chose the RNG $k$-$\varepsilon$ model to numerically calculate the three-dimensional flow in the vortex pump [15].

There are three expressions of the governing equation:

(1)　The continuity equation

$$\frac{\partial \rho}{\partial t} + \frac{\partial(\rho \overline{u_i})}{\partial x_i} = 0 \tag{1}$$

(2)　The momentum equations

$$\frac{\partial(\rho \overline{u_i})}{\partial t} + \frac{\partial\left(\rho \overline{u_i u_j}\right)}{\partial x_j} = -\frac{\partial \overline{p}}{\partial x_i} + \frac{\partial}{\partial x_j}\left[\mu \frac{\partial \overline{u_i}}{\partial x_j} - \rho \overline{u_i' u_j'}\right] \tag{2}$$

$$-\rho \overline{u_i' u_j'} = \mu_t \left[\frac{\partial \overline{u_i}}{\partial x_j} + \frac{\delta \overline{u_j}}{\delta x_i}\right] - \frac{2}{3}\left[\rho k + \mu_t \frac{\delta \overline{u_j}}{\delta x_i}\right]\delta_{ij} \tag{3}$$

(3)  The $k$ and $\varepsilon$ equation

$$\rho\frac{Dk}{Dt} = \frac{\partial}{\partial x_j}\left[\left(\mu + \frac{\mu_t}{\sigma_k}\right)\frac{\partial k}{\partial x_j}\right] + 2\mu_t\overline{S}_{ij}\frac{\partial \overline{u}_i}{\partial x_j} - \rho\varepsilon \tag{4}$$

$$\rho\frac{D\varepsilon}{Dt} = \frac{\partial}{\partial x_j}\left[\left(\mu + \frac{\mu_t}{\sigma_\varepsilon}\right)\frac{\partial \varepsilon}{\partial x_j}\right] + 2C_{1\varepsilon}\frac{\varepsilon}{k}\mu_t\overline{S}_{ij}\frac{\partial \overline{u}_i}{\partial x_j} - C_{2\varepsilon}^*\rho\frac{\varepsilon^2}{k} \tag{5}$$

where

$$\mu_t = C_\mu\frac{k^2}{\varepsilon} \tag{7}$$

$$S_{ij} = \frac{1}{2}\left(\frac{\partial \overline{u}_i}{\partial x_j} + \frac{\delta \overline{u}_j}{\delta x_i}\right) \tag{8}$$

$$C_{2\varepsilon}^* = C_{2\varepsilon} + \frac{C_\mu\eta^3\left(\frac{1-\eta}{\eta_0}\right)}{1 + \beta\eta^3} \tag{9}$$

$$\eta = \sqrt{2\overline{S}_{ij}\overline{S}_{ij}}\frac{k}{\varepsilon} \tag{10}$$

where $\delta_{ij}$ is unit tensor, $\eta_0 = 4.38$, $C_\mu = 0.0845$, $\beta = 0.012$, $C_{1\varepsilon} = 1.42$, $C_{2\varepsilon} = 1.68$, $\alpha_k = 0.7194$, and $\alpha_\varepsilon = 0.7194$.

The environment temperature is 25 °C. The medium is water, and the density $\rho$ is 997 kg/m$^3$. The dynamic viscosity $\mu$ is 0.89 mPa·s. The quantities with over-bar (such as $\overline{u}_i$) are mean values.

### 4.2. 3D Modeling and Grid Division

The 3D modeling software Pro/ENGINEER (Version 5.0, PTC, Needham, MA, USA, 2015) [16–19] was used to perform solid modeling of the vortex chamber of the vortex pump and impeller region. The computational domain consists of the impeller region and volute region, in which the volute region is the stationary region. The impeller region is the rotating region and the transmission coupling flow parameter between the impeller and volute region is processed by the frozen rotor method. The calculation of the impeller flow field is carried out in the relative coordinate system, while the vortex chamber flow field calculation is carried out in the absolute coordinate system and the integral calculation of the dynamic and static components is realized through the coupling calculations between the two parts. Because the interface grid between the computational domains is not exactly the same, the general grid interfaces (GGIs) are used to connect the parts. The computational grids are divided by ANSYS® ICEM Tetra tools (Version 14.5, ANSYS, Canonsburg, PA, USA, 2012). The tetrahedral mesh is divided into the vortex chamber region and the impeller region, respectively, and the residual error is $10^{-4}$. The number of mesh units is 616,369, and the number of nodes is 117,300, as shown in Figure 4.

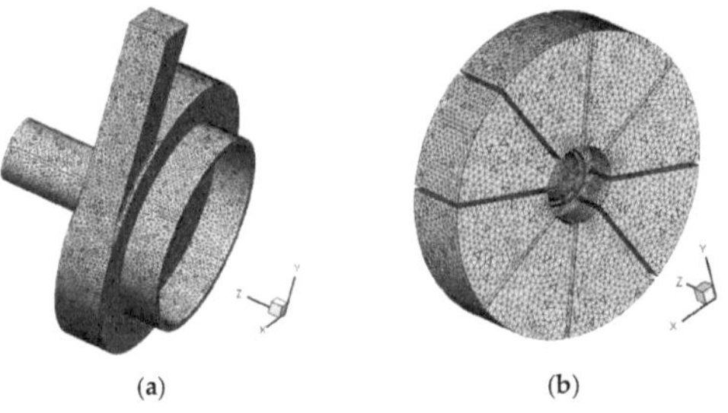

(a)          (b)

**Figure 4.** Simulation grids: (**a**) Volute; (**b**) Impeller.

### 4.3. Boundary Conditions

#### 4.3.1. Inlet Boundary Conditions

The inlet of the vortex pump calculation area is an axial inlet. The inlet axial velocity value is given at the inlet pipe according to the flow rate. Assuming that the inlet has no rotation, the tangential velocity and the radial velocity are 0, and the turbulence is fully developed and evenly distributed in the import section. A is the area of the inlet of the vortex pump. The inlet axial velocity $v_{in}$ is given at the inlet pipe according to the flow rate $q_v$.

$$v_{in} = \frac{q_v}{A} \tag{11}$$

The turbulent kinetic energy and turbulent energy dissipation rate at the inlet are set according to the flow and inlet cross-sectional area. The formula is as follows:

$$k_{in} = \frac{3}{2}(\overline{u}I)^2 \tag{12}$$

$\overline{u}$ is the mean flow velocity, and here $\overline{u} = v_{in}$.

$$\varepsilon_{in} = C_\mu^{\frac{3}{4}} \frac{k_{in}^{\frac{3}{2}}}{l} \tag{13}$$

where $I$ is turbulence intensity

$$I = u'/\overline{u} = 0.16\left(Re_{D_H}\right)^{-\frac{1}{8}} \tag{14}$$

From the above formula, $u'$ is the root-mean-square of the turbulent velocity fluctuations, $\overline{u}$ is the is the mean velocity, and $l$ is the inlet's length, $Re_{D_H}$ is the Reynolds number based on $D_H$, and $D_H$ represents the hydraulic diameter,

$$l = 0.07L \tag{15}$$

where $C_\mu$ is 0.09, and $L$ is the associated length. For the full development of the turbulence, desirable $L$ is equal to the hydraulzic diameter. Here, $L$ is equal to the inlet round pipe diameter.

Specific boundary conditions: velocity vector (0 m/s, 0 m/s, 3.11 m/s), turbulent kinetic energy $k_{in} = 0.021$ m²/s², and turbulent kinetic energy dissipation $\varepsilon_{in} = 0.22$ m²/s³.

#### 4.3.2. Outlet Boundary Conditions

Assuming that the outlet flow is fully developed for turbulence, the specific boundary conditions are as follows:

The outlet type is known as Outlet, and the average static pressure $p_s = 101{,}325$ Pa.

#### 4.3.3. Wall Boundary Conditions

In the non-slip boundary condition (No Slip) of the solid wall, the wall function method is used to calculate the turbulence in the vicinity of the wall. The influence of the wall roughness on the flow field is neglected in the calculation.

## 5. Numerical Results

### 5.1. The Flow Distribution of the Static Pressure Inside the Pump

Figures 5–8 are all based on the calculation result of CFX software (Version 14.5, ANSYS, Canonsburg, PA, USA, 2012). Based on the coordinate system in Figure 1, Figure 5 shows the static pressure distribution on the axial cross-section where Z is −10 mm. This axial section is also the middle cross-section of the impeller blade. According to the static pressure value and its position (radial direction and $r = \sqrt{x^2 + y^2}$), Figure 6 is generated. Figure 6 shows the static pressure $p_s$ variable curves along the 0°, 90°, 180°, and 270° radial directions. Figures 5 and 6 show that the center of the impeller is the lowest static pressure area in this axial cross-section, which is negative pressure (vacuum). The pressure in the center of the impeller is also lower than the pressure in the center of the chamber, so the run-through flow from chamber to impeller can be formed. When the fluid flow enters into the region near the blade, centrifugal force applies work on fluid flow and the pressure is increasing. So, in the region near the impeller edge, the pressure is higher than other regions.

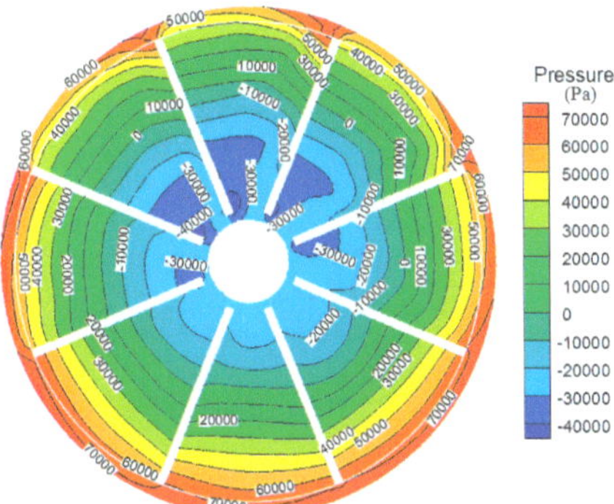

**Figure 5.** Static pressure distribution for the axial cross-section $Z = -10$ mm.

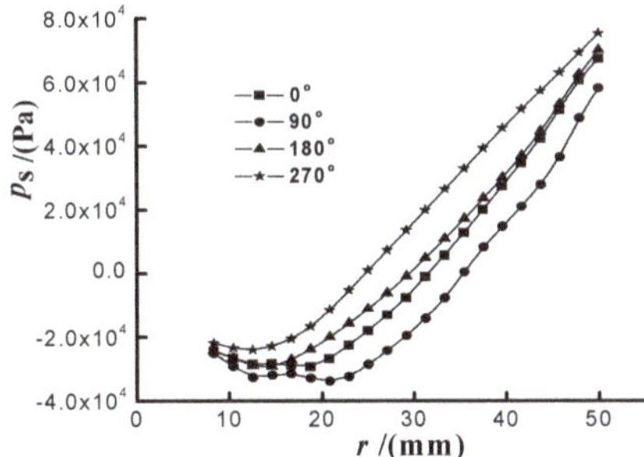

**Figure 6.** Static pressure distribution for the axial cross-section $Z = -10$ mm along the radial direction.

Figure 7 shows the static pressure distribution for the axial cross-section where $Z$ is 12.5 mm, which is also based on the coordinate system in Figure 1. This axial section is the middle axial cross-section of the chamber. Figure 8 is generated according to the static pressure value and its position (radial direction). Figure 8 shows the static pressure $p_s$ variable curves along the 0°, 90°, 180°, and 270° radial directions. Figures 7 and 8 show that the center of the chamber is the lowest pressure area in this axial cross-section, which is negative pressure (vacuum). So, the fluid flow from inlet to chamber can be formed.

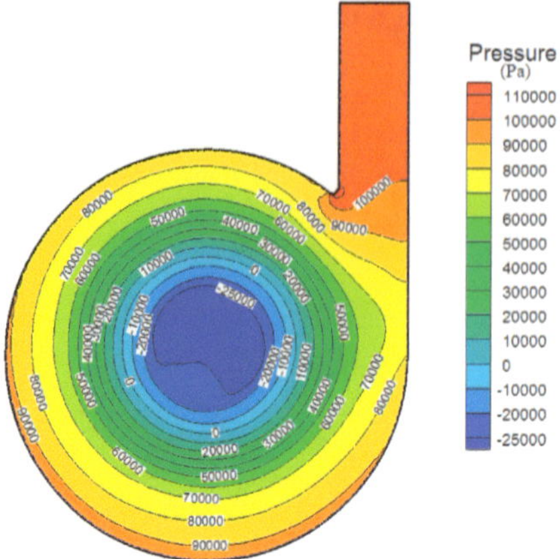

**Figure 7.** Static pressure distribution for the axial cross-section $Z$ = 12.5 mm.

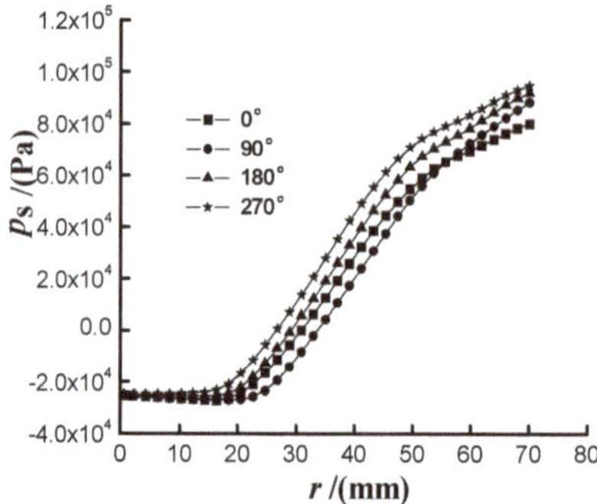

**Figure 8.** Static pressure distribution for the axial cross-section $Z$ = 12.5 mm along the radial direction.

*5.2. Flow Velocity Distribution in the Pump*

Figures 9 and 10 are also the calculation result of CFX software. Based on the coordinate system in Figures 1, 9 and 10 respectively show the absolute velocity $v$, the circumferential velocity $v_u$, the radial velocity $v_r$, and the axial velocity $v_z$ of two different axial cross-sections. Absolute velocity $v$ is the vector synthesis result of $v_u, v_r, v_z$.

Figure 9 shows the velocity variable on the axial cross-section, where $Z$ is −10 mm. This axial section is the middle cross-section of the impeller blade. Figure 9a–d show the $v, v_u, v_r, v_z$ variable curves along the 0°, 90°, 180°, and 270° radial directions on the axial cross-section where $Z$ is −10 mm.

Figure 10 shows the velocity variable on the axial cross-section, where $Z$ is 12.5 mm. This axial section is the middle axial cross-section of the chamber. Figure 10a–d show how the $v, v_u, v_r, v_z$ change along the 0°, 90°, 180°, and 270° radial directions on the axial cross-section, where $Z$ is 12.5 mm.

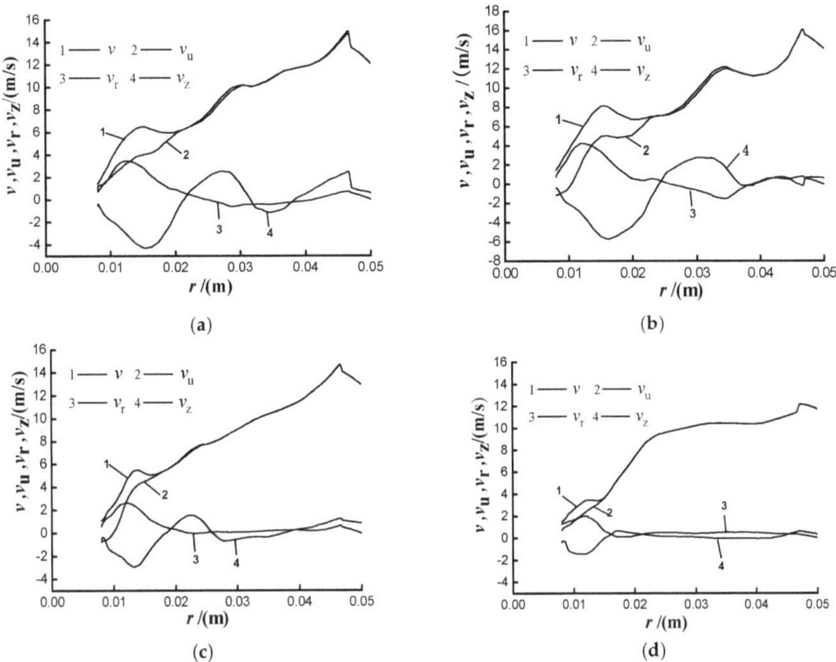

**Figure 9.** Velocity variable curves for the axial cross-section $Z = -10$ mm along the radial direction. (**a**) 0°; (**b**) 90°; (**c**) 180°; (**d**) 270°.

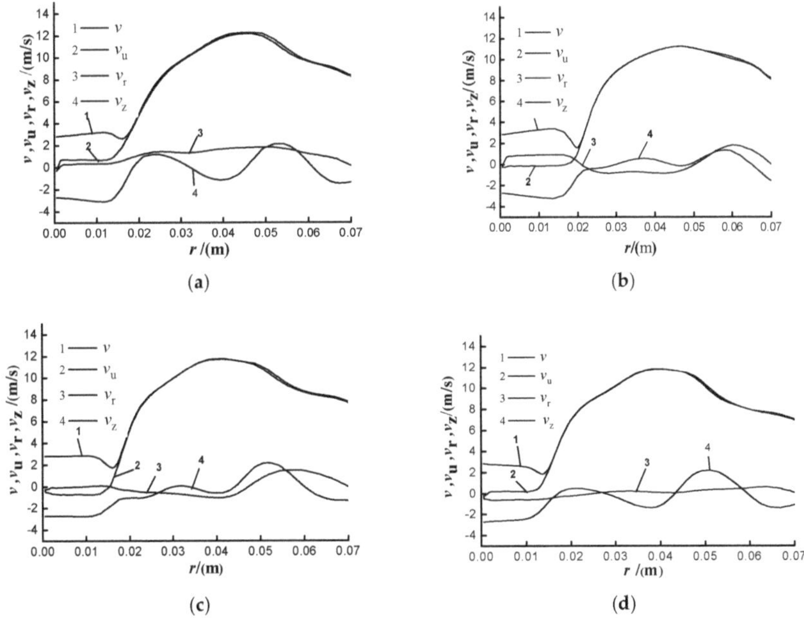

**Figure 10.** Velocity variable curves for the axial cross-section $Z = 12.5$ mm along the radial direction. (**a**) 0°; (**b**) 90°; (**c**) 180°; (**d**) 270°.

### 5.3. Numerical Simulation of the Pump Performance Curve

The discharge calculation software CFX was used to calculate the discharge at 2850 r/min. The calculated performance and measured curves are shown in Figure 11.

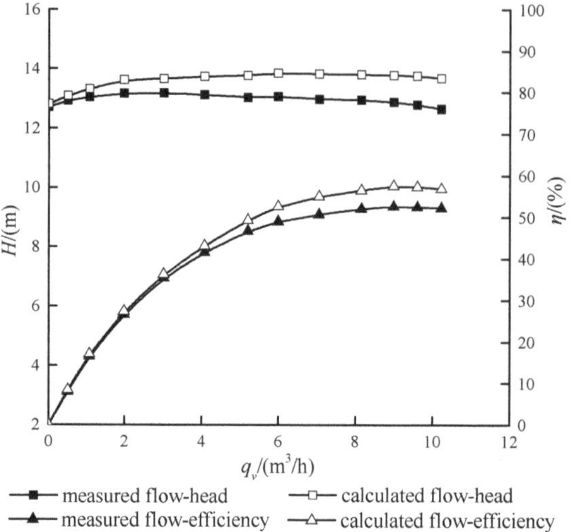


**Figure 11.** Performance comparison between experiments and simulation.

Disk frictional loss and volume loss are difficult to determine, and the computational fluid dynamics (CFD) numerical simulation pumper efficiency only considers hydraulic efficiency. The calculation model is simplified, for instance, by neglecting the clearance between the impeller back shroud, water pressure chamber, and surface roughness, etc.

## 6. Experimental Verification

In this paper, five spherical probes were used to extend into the vortex chamber, and five points were measured. The location of the measuring points is shown in Figure 12. The measuring system and principle of the probe are shown in Figure 11. The flow field calculation, measurement data, and the curve contrast are shown in Table 3 and Figure 13. Curves in Figure 13 are based on the coordinate system in Figure 1.

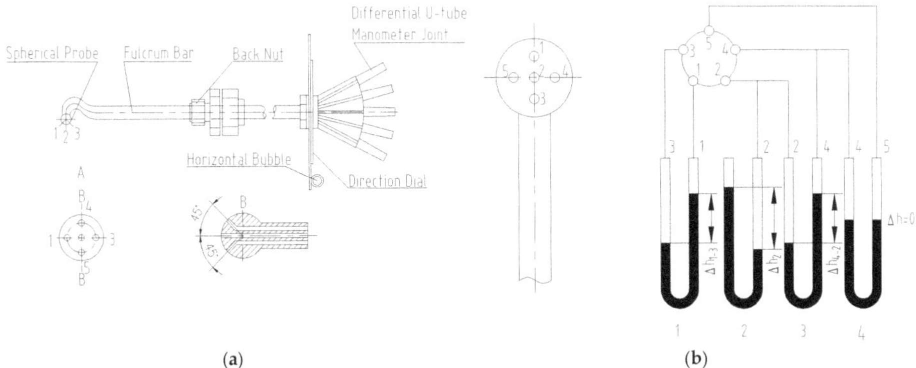



**Figure 12.** Probe measuring system. (**a**) Structure diagram of the five-hole spherical probe; (**b**) Connection system diagram of five-hole spherical probe and differential pressure meter.

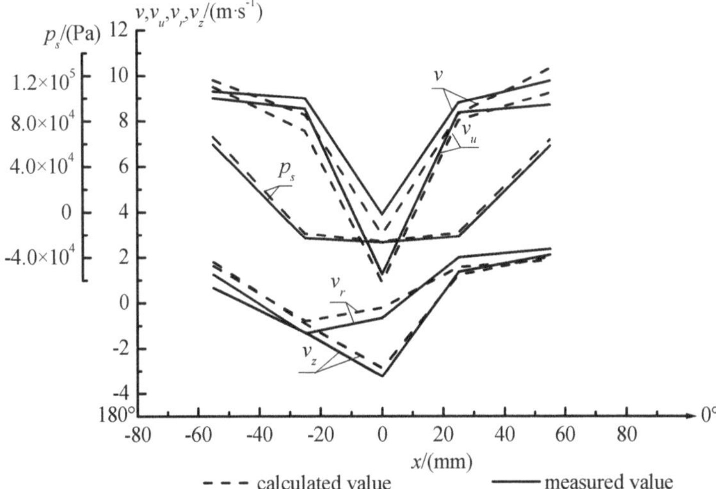



**Figure 13.** Calculated and measured curves of flow field curves on the axial-cross-section $Z$ = 12.5 mm.

It can be seen from the comparison and analysis that the numerical calculation is consistent with the experimental results, which can reflect the internal flow pattern of the vortex pump. The numerical simulation results are reliable, and can be applied to the qualitative and quantitative analysis of the internal flow field of the vortex pump.

**Table 3.** Calculated and measured values of flow field curves on the axial cross-section $Z = 12.5$ mm.

| Calculated and Measured Position | Discharge $q_v$/m$^3\cdot$h$^{-1}$ | Rotational Speed $n$/r$\cdot$min$^{-1}$ | Value Type | Absolute Velocity $v$/m$\cdot$s$^{-1}$ | Circumferential Velocity $v_u$/m$\cdot$s$^{-1}$ | Radial Velocity $v_r$/m$\cdot$s$^{-1}$ | Axial Velocity $v_z$/m$\cdot$s$^{-1}$ | Static Pressure $p_s$/Pa |
|---|---|---|---|---|---|---|---|---|
| $x = -55$ mm Measuring point a | 9.31 | 2851 | Calculated value | 9.806 | 9.512 | 1.612 | 1.782 | 66,126 |
| | | | Measured value | 9.312 | 9.002 | 1.237 | 0.657 | 59,051 |
| $x = -25$ mm Measuring point b | 9.30 | 2851 | Calculated value | 8.267 | 7.567 | −0.816 | −0.939 | −19,216 |
| | | | Measured value | 9.005 | 8.536 | −1.341 | −1.321 | −23,064 |
| $x = 0$ Measuring point c | 9.32 | 2850 | Calculated value | 3.006 | 0.905 | −0.223 | −2.872 | −26,138 |
| | | | Measured value | 3.902 | 1.264 | −0.663 | −3.242 | −26,842 |
| $x = 25$ mm Measuring point d | 9.31 | 2849 | Calculated value | 8.306 | 8.026 | 1.572 | 1.216 | −18,162 |
| | | | Measured value | 8.817 | 8.368 | 2.003 | 1.362 | −21,447 |
| $x = 55$ mm Measuring point e | 9.31 | 2850 | Calculated value | 10.312 | 9.248 | 1.865 | 2.016 | 63,636 |
| | | | Measured value | 9.769 | 8.705 | 2.355 | 2.100 | 58,100 |

## 7. Comprehensive Analysis of Research Results

Based on the CFD numerical calculation and experimental results, we can carry out a comprehensive analysis of the internal flow rules of the vortex pump

Due to the unique structure and working principle of the vortex pump (i.e., the use of a semi-open impeller), the impeller and vortex chambers are fully open and connected, and the vortex chamber and pump outlet are connected through a channel. Carrying on with the flow field analysis, we must consider the concept of potential strength; i.e., the flow field of gravity and centrifugal inertia forces. The unequal mass force of the impeller is an important reason for the asymmetric distribution of the impeller flow field and the degree of disorder.

In the center of the impeller, because the hub occupies the central space, in order to form a vacuum range in the 10 mm $< r <$ 20 mm annular area ($r$ represents the radius with the center of the impeller as the origin), the lowest point of static pressure is in the upper $y =$ 20 mm position, up to $-3.6 \times 10^4$ Pa, and it can be inferred that bubbles are first produced in this region, while the lower static pressure remained at $-2.7 \times 10^4$ Pa or so, forming clear dropped gradient upward pressure. The pressure in the lower half of the whole impeller chamber and the vortex chamber is greater than the pressure at the corresponding position of the upper part, and the static pressure reaches a maximum of about $1.1 \times 10^5$ Pa at the outlet of the pump body and becomes the main body of the pump head. The vacuum area is greater than the pump inlet area (pump inlet $d =$ 32 mm). The main reason is the gravity field mass force potential function $W = gz$; the potential energy at the lower position is greater than that at the top. In addition, the lower gravity is in accordance with the centrifugal inertial force, increasing the pressure energy in the region. The structure is mainly due to the influence of the pump's outlet.

In the range of 20 mm $< r <$ 48 mm, the static pressure of the impeller chamber and the vortex chamber grows linearly, which is influenced by the impeller and has a forced vortex property. In the range of 50 mm $< r <$ 70 mm, the upward trend of static pressure is slowed down, which indicates that the imposition of the impeller is reduced and the steady growth of the static pressure is due to the increase in the volume of the vortex chamber and the partial kinetic energy. The hydraulic design of the radius of the volute chamber and the width of the channel is calculated by this principle, and the cross-sectional area of the flow passage is found in the determination of the optimum velocity in the flow passage.

In the range of $r >$ 20 mm, the center speed of $v_u$ and the absolute speed $v$ of the impeller almost coincide with each other, and have greater values than the speed at the center of the vortex chamber. That is, the velocity of the volute chamber lags behind the rotation velocity of the impeller, and the vortex chamber $v_u$ and $v$ almost coincide. It can be seen that the width of the volute chamber should not be designed too wide under the premise of ensuring passing capacity; otherwise, the delay of the chamber velocity will increase.

Within the radius of the impeller, $v_u$ is approximately equal to the circumferential velocity of the impeller with the same radius. The $v_{umax}$ is approximately 14.3 m/s, which is equal to the circumferential velocity ($v_u =$ 14.326 m/s) near the outer edge of the impeller. It can be shown that numerical simulation achieves sufficient accuracy. In the range of 20 mm $< r <$ 48 mm, the center of $v_u$ and $v$ of the impeller rise almost linearly, and the center of the vortex chamber rises to approximately a parabola. In the range of $r >$ 48 mm, the center of the vortex chamber is of approximately parabolic descent, and the partial kinetic energy is converted to pressure energy.

Because of the pressure gradient, the $r <$ 20 mm $v_z$ range of axial velocity distribution is not uniform, and the maximum velocity is in the position of $90°$ section $y =$ 15 mm. At a maximum of about $v_{zmax} \approx$ 5.0 m/s for the impeller, the volute chamber value is $v_{zmax} \approx$ 3.0 m/s. In the $r <$ 20 mm range, the radial velocity $v_r$ varies irregularly, while in the $r >$ 20 mm range, the radial velocity $v_r$ fluctuates little.

The calculated pump head is slightly higher than the measured value. This is due to the neglect of mechanical loss and volumetric loss. The maximum hydraulic efficiency of the vortex pump is about 60%. Low hydraulic efficiency (i.e., flow loss) causes the main part of pump energy loss [20].

## 8. Conclusions

In a small flow state, the $q_v$-$NPSH_c$ curve of vortex pump is the opposite to the $q_v$-$NPSH_c$ of the centrifugal pump, and $NPSH_c$ value is large, that is, the vortex pump has poor anti-cavitation. At present, in ISO9906 (rotodynamic pumps—hydraulic performance acceptance tests—grades 1, 2 and 3), the evaluation methods of the centrifugal pumps are used for assessment of the rotodynamic pumps. Hence, the rationality of the evaluation methods is worthy of further study.

The experimental performance test of the prototype of the vortex pump (32WB8-12) indicates the 32WB8-12 pump has good performance and its optimized hydraulic model is worth popularizing. The numerical calculation is verified by the local experiment, and it is concluded that the numerical calculation of the internal flow field of the vortex pump gives realistic results. The pressure distribution in the vortex chamber presents a uniform increasing trend, and the velocity distribution appears to be in a turbulent state, which is characterized by a combined vortex. It is suggested that the flow rate be controlled to improve the efficiency of the vortex pump [21].

The stable pressure gradient is formed from the inlet of the vortex pump to the center of the impeller. The lowest pressure in the vortex chamber is $-2.5 \times 10^4$ Pa and it is located at the center of the cavity of vortex chamber. The area of low pressure is greater than the pump entrance area. Therefore, the 32WB8-12 vortex pump has advanced suction and anti-cavitation performance.

Probes were used for measuring the pressure and velocity field synchronously, which had to be installed in the flow and caused slight interference to the original flow field and generated some error. In the following research, non-contact laser PIV technology will be applied to measure velocity, as it can obtain the velocity field accurately without disturbing the flow field. Non-contact laser PIV can be combined with the probe to measure the accurate internal flow field of the vortex pump. Furthermore, we will establish a flow model based on studies of precise numerical simulation technology and carry out studies on the two-phase flow medium and cavitation fluid field of the vortex pump. Thus, we can reveal the internal mechanism of the excellent performance of the vortex pump.

**Author Contributions:** The following statements should be used P.T. and Y.S. conceived and designed the experiments; D.T. and X.B. performed the experiments; X.B. and W.H. analyzed the data; J.M. and Y.F. contributed reagents/materials/analysis tools; P.T. and Y.S. wrote the paper.

**Conflicts of Interest:** The authors declare no conflict of interest.

## References

1.  Ohba, H.; Nakashima, Y.; Shiramoto, K. A Study on Internal Flow and Performance of a Vortex Pump Part 1: Theoretical Analysis. *Bull. JSME* **2008**, *26*, 999–1006. [CrossRef]
2.  Schivley, G.P. An Analytical and Experimental Study of a Vortex Pump. *Trans. ASME Ser. D* **1970**, *92*, 889–900. [CrossRef]
3.  Wen, Q.Y.; Guo, Y.H. Design of vortex pump for conveying potato. *Pump Technol.* **1998**, *2*, 25–26.
4.  Chen, H.X. Measurement of rotating flow field within the impeller of vortex pump. *Trans. Chin. Soc. Agric. Mach.* **1996**, *27*, 49–54. (In Chinese)
5.  Sha, Y.; Liu, X.S. Numerical calculation on gas-liquid two-phase hydrotransport and flow field measurement in volute with probes of a vortex pump. *Trans. Chin. Soc. Agric. Eng.* **2014**, *30*, 93–100.
6.  Gerlach, A.; Thamsen, P.; Wulff, S.; Jacobsen, C. Design Parameters of Vortex Pumps: A Meta-Analysis of Experimental Studies. *Energies* **2017**, *10*, 58. [CrossRef]
7.  Gerlach, A.; Thamsen, P.U.; Lykholt-Ustrup, F. Experimental investigation on the performance of a vortex pump using winglets. In Proceedings of the 16th International Symposium on Transport Phenomena and Dynamics of Rotating Machinery, Honolulu, HI, USA, 10–15 April 2016.
8.  Gerlach, A.; Preuss, E.; Thamsen, P.U.; Lykholt-Ustrup, F. Numerical simulations of the internal flow pattern of a vortex pump compared to the Hamel-Oseen vortex. *J. Mech. Sci. Technol.* **2017**, *31*, 1711–1719. [CrossRef]
9.  Jiang, D.L.; Lv, J.X.; Dai, L.; Su, B.W. A Numerical Simulation of and Experimental Research on Optimum Efficiency of Vortex Pump. *China Rural Water Hydropower* **2012**, *4*, 92–94, 98. (In Chinese)

10. Zhu, R.S.; Chen, J.J.; Wang, X.L.; Su, B.W. Numerical Simulation and Experimental of Influence of Hem and High-low Blade on Performance of Vortex Pump. *Flue Mach.* **2012**, *40*, 1–5.

11. Li, Y.; Zhu, Z.C.; He, W.Q.; Wang, Y.P.; Cui, B.L. Numerical Simulation and Experiment Analyses for the Gas-liquid Two-phase Vortex Pump. *J. Therm. Sci.* **2010**, *19*, 47–50. [CrossRef]

12. Xia, P.H.; Liu, S.H.; Wu, Y.L. Numerical simulation of steady flow in vortex pumps. *J. Eng. Thermophys.* **2006**, *27*, 420–422.

13. Shi, W.D.; Wang, Y.Z.; Kong, F.Y.; Sha, Y.; Yuan, H.Y. Numerical simulation of internal flow field within the volute of vortex pump. *Trans. Chin. Soc. Agric. Eng.* **2005**, *21*, 72–75.

14. Steinmann, A.; Wurm, H.; Otto, A. Numerical and experimental investigations of the unsteady cavitating flow in a vortex pump. In Proceedings of the 9th International Conference on Hydrodynamics, Shanghai, China, 11–15 October 2010.

15. Wei, Y.J.; Tseng, C.C.; Wang, G.Y. Turbulence and cavitation models for time-dependent turbulent cavitating flows. *Acta Mech. Sin.* **2011**, *27*, 473–487. [CrossRef]

16. Sha, Y.; Yang, M.G.; Kang, C.; Wang, J.F.; Chen, H.L. Design method and characteristic analysis of vortex pump. *Trans. Chin. Soc. Agric. Eng.* **2004**, *36*, 124–127.

17. Sha, Y.; Shi, W.D.; Wang, Z.L.; Ji, H.S. Hydraulic design of non-clogging pump and experimental research on its characters. *Trans. Chin. Soc. Agric. Mach.* **2005**, *36*, 62–66.

18. Chen, H.X. Research on Turbulent Flow within the Vortex Pump. *J. Hydrodyn. Ser. B* **2004**, *16*, 701–707.

19. Wu, J.; Sha, Y.; Xu, X. Experimental investigation on speed performance and volute flow of vortex pump. *J. Zhejiang Univ. Eng. Sci.* **2010**, *44*, 1811–1817.

20. Sha, Y.; Liu, X.S. Performance test on solid-liquid two-phase flow hydrotransport of vortex pump. *Trans. Chin. Soc. Agric. Eng.* **2013**, *29*, 76–82.

21. Quan, H.; Fu, B.H.; Li, R.N.; Zhang, T.; Han, W.; Li, J. Research Stage and Development Tendency of vortex pump. *Fluid Mach.* **2016**, *44*, 36–40.

*Article*

# Modeling and Dynamic Analysis on the Direct Operating Solenoid Valve for Improving the Performance of the Shifting Control System

Xiangyang Xu [1,2], Xiao Han [1,2], Yanfang Liu [1,2,*], Yanjing Liu [1,2] and Yang Liu [3]

[1]  School of Transportation Science and Engineering, Beihang University, Beijing 100191, China; xxy@buaa.edu.cn (X.X.); hanxiaoook@163.com (X.H.); liuyanjing@buaa.edu.cn (Y.L.)

[2]  Beijing Key Laboratory for High-efficient Power Transmission and System Control of New Energy Resource Vehicle, Beihang University, 37 Xueyuan Road, Haidian District, Beijing 100191, China

[3]  Research Department of Vehicle Chasis, Beijing Institute of Space Launch Technology, Beijing 100076, China; kaka19881019@126.com

*  Correspondence: liuyf@buaa.edu.cn; Tel.: +86-139-1009-9728

Received: 7 November 2017; Accepted: 1 December 2017; Published: 5 December 2017

**Abstract:** The dynamic characteristics and energy loss in a shifting control system is important and necessary in the performance improvement of an automatic transmission. The direct operating solenoid valve has been considered as a potential component applying in the shifting control system in vehicle. The previous method can solve only a specific physical field or use the test results of the magnetic force as input curve. The paper presents a numerical approach for solving the multi-domain physical problem of the valve. A precise model of the direct acting solenoid valve considering different physical field is developed. An experimental study is also performed to evaluate and confirm the simulation. Based on the model, the influences on the dynamic characteristics of the valve are analyzed by calculating forces acting on the valve. The systematic analysis of forces and energy loss characteristics are performed for three different flow conditions varying clearance height from 10 μm to 30 μm. The results demonstrate that the pressure response time can be improved with smaller clearance between the spool and the sleeve. Moreover, the leakage of the shifting control system employing the direct acting solenoid valve can be reduced by 60% compared to the conventional two-stage pilot valve in our previous product.

**Keywords:** direct operating solenoid valve; energy loss; forces; response pressure; leakage flow

---

## 1. Introduction

To shift from one gear to another in an automatic transmission, one clutch needs to be released and another needs to be applied. The shifting control system of the automatic transmission controls the clutch pressure and the corresponding transmitted torque during shifting process. The dynamic characteristics of the shifting control system are directly related to the shifting quality. Figure 1 is the conventional structure of a shifting control system consisting of a relieve valve, a pilot solenoid valve and a clutch control valve. In this structure, the automatic transmission fluid passes through several valves and complicated oil lines from the main line to the clutch. The long passage causes the delayed pressure response, low control accuracy and big leakage.

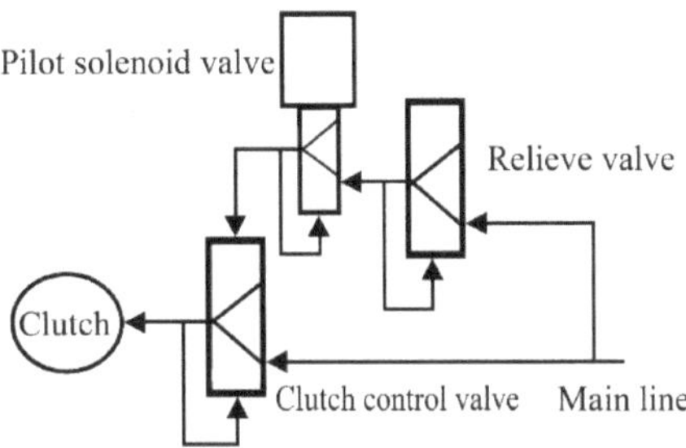


**Figure 1.** Conventional structure of the shifting control system.

With the development of the industry, a new pressure valve called direct operating solenoid valve could take place of the conventional two-stage structure for its small leakage, large flow, high reliability and high control accuracy. Figure 2 is the structure of the shifting control system applying the direct operating solenoid valve. It is obvious that simple structure and light weight of the system can be obtained by utilizing this kind of structure which has been more and more adopted in the latest automatic transmission. The direct operating valves are produced by many well-known worldwide companies such as Tosok, Bosch, Highlight and Narchi. Improving the characteristics of the valve can directly contribute to the performance of the shifting control system. Hence, a growing interest in direct operating valves emerges from scientific centers.

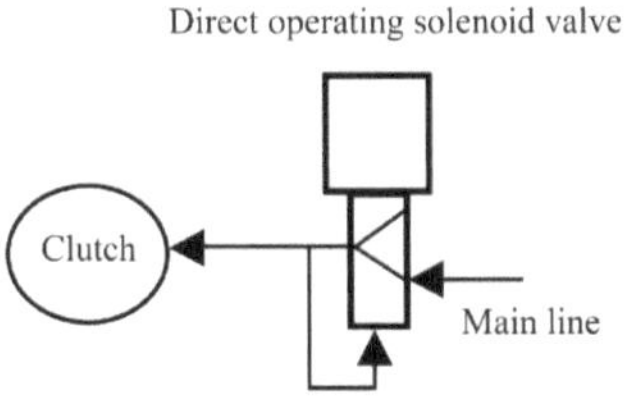

**Figure 2.** Structure of the shifting control system applying the direct operating solenoid valve.

Presently, many studies on the solenoid valve are focused on a specific physical field such as the magnetic field or the flow field. Shuai Wu et al. [1] proposed a new direct drive valve which has high frequency voice coil motor and advanced digital controller. A digital controller and a hybrid controller were developed to control the valve motion and were indicated to improve the control accuracy and robustness of the system through test. The static characteristics of the electromagnetic linear actuators were analyzed by A. Schultz [2]. 3D electromagnetic fields analysis of the electromagnet was carried out using ANSYS. Influence factors such as working area, coil turns, core number, armature depth and iron material that affects the magnetic force were discussed in detail. The influencing law to the magnetic force was obtained [3]. Liu Qianfeng et al. [4] reported that the electromagnetic force of the direct action solenoid valve was mostly influenced by the current. Optimization of the design parameters of the valve was achieved by complex method. The control accuracy of the magnetic force could be well improved based on the studies. Yujeong Shin et al. [5] adapted Ansys Maxwell electromagnetic analysis software to model the electromagnetic dynamics and selected the best optimization model using the verified approximation model. However, the dynamic characteristic of the valve could not be adequately enhanced if we only study the magnetic force because a sensitive pressure response is caused by the interaction of the resultant force.

Multi-physics coupling models were reported in many studies. The valves were usually designed as compact constructions, which contain a mechanical part, a hydraulic part and an electronic controller in a single module. Gee Soo Lee et al. [6] carried out a three-dimensional numerical simulation employing a moving mesh with dynamic layering meshes for varying boundary conditions to investigate the flow dynamic behavior and pressure characteristics of a variable force solenoid valve. Klaus Mutschler et al. [7] presented an alternative approach using network simulation methods to model a dispensing valve using a simulation software SABER. The model could correctly predict the dispensed liquid volume in dependence of the main parameters like pressure and opening time. Another multi-physics modeling method was introduced by Liu Yanfang et al. [8]. They successfully predicted the dynamic characteristics of a proportional solenoid valve and used it for valve design. Liu Z. et al. [9] optimized the structure of a large flow solenoid valve by multi-physics modeling method. Yi Xiong et al. [10] extended the dynamic model of the high-response dual proportional solenoid valve discontinuous projection-based adaptive robust control to synthesize controller to deal with the parametric uncertainties and uncertain nonlinearities. Both the simulation and experimental results showed the proposed controller was effective. The nonlinear characteristic, the orifice area and the dynamic model of the solenoids were also analyzed using Matlab by Yaguang Zhu and Bo Jin [11]. Meanwhile, Liu Lei et al. [12] analyzed the influences of throttle nozzles' diameter and other parameters on the dynamic and static characteristics of a low-pressure large-flow pilot-operated solenoid valve via AMESim simulation software. And optimization was conducted in the model. Chang-Dae Park et al. [13] provided a convenient method of design verification of solenoid operated valve. Lan Wang et al. [14] found that firing current, holding current, spring pre-tightening force and spring stiffness had great effects on the dynamic response characteristics of solenoid valve. An electromagnetic mathematical model of a high-speed solenoid valve was developed in Fortran language. Jianhui Zhao et al. [15] discovered that the electromagnetic energy conversion characteristics of the HSV were affected by the drive current and the total reluctance, consisting of the gap reluctance and the reluctance of the iron core and armature soft magnetic materials. Paul D. Walker et al. [16] established mathematical models of the integrated electrohydraulic solenoid valve and wet clutch piston assembly in the Simulink environment of Matlab. The dependency of the system to system variables on input pressure and the influence of air content on dynamic response of the valve were investigated.

Though an amount of work has been done on the direct operating solenoid valve, there still exist several shortcomings in previous studies. The magnetic force obtained from the tests is input directly into the model rather than real-time computing which would result in the capability of analyzing the magnetic force. Forces, including the magnetic force, play a very important role in a direct operating valve because they have strong relation to both the pressure response and the leakage. Therefore, it is essential to introduce a multi-physics method which could compute all forces in real time and predict the dynamic characteristics of the direct operating solenoid valve.

In this paper, a multidisciplinary approach was conducted to obtain the dynamic characteristics of the direct operating solenoid valve. According to the approach, the simulation model of the valve was developed using the minimum element method. Secondly, experiments have been conducted on the test rig in order to verify the numerical results. The experimental results showed good agreement with the simulated ones, which confirm the accuracy of the simulation model. Thirdly, based on the model, the influencing factors including forces, critical size and leakage flow on the performance of the valve were detailed and analyzed. Finally, suggestions are proposed to improve the performance of the direct operating solenoid valve, which could be effective for the valve design.

## 2. Valve Structure and Principle

The direct operating solenoid valve is composed of mechanical subsystem, electro-magnetic subsystem and hydraulic subsystem as shown in Figure 1. The electro-magnetic subsystem is composed of coil, armature and tube. The mechanical subsystem consists of a sleeve in which a spool and a spring are located inside. The position of the spool is mainly determined by an electromagnetic force, a spring force and a feedback force arising from the pressure difference of the orifice.

The solenoid valve under study is a direct operating solenoid valve manufactured by Tosok (Kanagawa Prefecture, Japan). The valve is kept at its closed position by the return spring if there is no input current (Figure 3a). The control port is connected to the exhaust port.

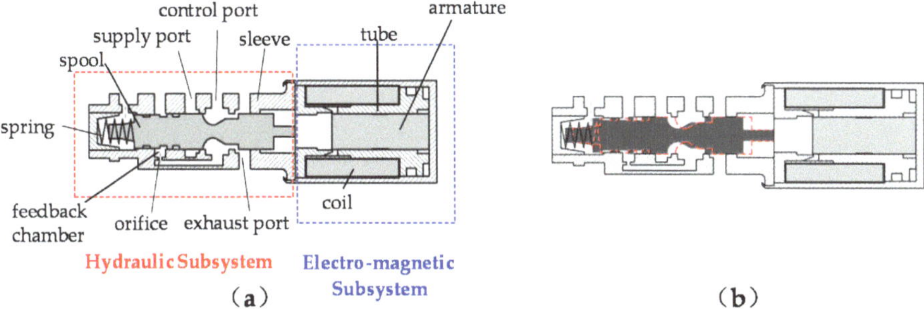


**Figure 3.** (**a**) Structure of the direct operating solenoid valve (**b**) Working principle of the valve.

When the solenoid is energized, the magnetic force actuates the spool to move against the return spring. Then the supply port opens where the transmission fluid flows through to the control port. Meanwhile, the exhaust port is closed, and the fluid enters the control port to reach precise control pressure and finally flows to the clutch as shown in Figure 3b. Likewise, the fluid discharging from the clutch reenters the valve through the control port and flows to the tank as the electric current disappears and the spool returns to its initial position.

## 3. Modeling

A direct operating solenoid valve is a multi-physics system with characteristic of strong nonlinearity. Different subsystems of the valve are investigated and expressed in nonlinear state equations.

In the model, the outflow direction is assumed to be positive. The variables in mathematical model are scalars during calculation. However, the flow direction will be prejudged before solving the system equation.

### 3.1. Electro-Magnetic Subsystem

The magnetic force is the driving force of the direct operating solenoid valve which is generated by the magnetic circuit when electric current is input. The magnetic circuit is formed of a fixed core surrounded by the coil turns and an armature connected to the spool which moves with the spool under the effect of the exerted magnetic force.

According to Kirchhoff's Voltage Law, the following equation must be held for each loop in the magnetic circuit:

$$\sum V_{m,n} = \sum \Phi_n \cdot R_{m,n} = 0 \tag{1}$$

where $V_{m,n}$ is the magnetic voltage, $\Phi_n$ is the magnetic flux, $R_{m,n}$ is the magnetic resistance.

With the definition of the magnetic flux, the following statement is applied to any connection between these components of the magnetic circuit:

$$\sum \Phi_i = 0 \tag{2}$$

From the first equation of Maxwell, the electric field could be expressed as follows:

$$\oint H \cdot ds = \theta \tag{3}$$

where $H$ is the magnetic field strength.

The relationship between magnetic field and electric field is

$$V_m = \theta = w \cdot i \tag{4}$$

where $w$ is the winding number, $i$ is the electric current.

In the electromagnetic field, the electric voltage difference $V_{Ei}$ consists of the voltage drop and the induced voltage.

$$V_{Ei} = R_{Ei} \cdot i + w \cdot \frac{d\Phi}{dt} \tag{5}$$

where $R_{Ei}$ is the electric resistance.

According to the following fundamental relations:

The magnetic field intensity $B$ is

$$B = \frac{\Phi}{A_a} \tag{6}$$

where $A_a$ is the area of air gap.

The magnetic field strength is

$$H = \frac{V_m}{l} \tag{7}$$

where $l$ is the air gap length.

The magnetic resistance $R_m$ can be stated as

$$R_m = \frac{l}{\mu_0 \cdot A_c} \tag{8}$$

where $\mu_0$ is the permeability of vacuum.

The cross-sectional area $A_c$ is

$$A_c = \frac{\pi}{4} \cdot d^2 \tag{9}$$

where $d$ is the diameter of air gap.

The magnetic voltage is

$$V_m = R_m \cdot \Phi \tag{10}$$

$$H = \frac{B}{\mu_0} \tag{11}$$

Hence, the static magnetic force is

$$F_{ms} = \frac{1}{2} \cdot \Phi_{Air}^2 \cdot R_m = \frac{1}{2} \cdot \Phi_{Air}^2 \cdot \frac{l}{\mu_0 \cdot A} \tag{12}$$

where $\Phi_{Air}$ is magnetic flux of airgap.

However, the air gap length will keep changing when the spool starts to move with the magnetic force. Thus, the dynamic magnetic force $F_{md}$ is calculated as

$$F_{md} = \frac{1}{2} \cdot \Phi_{Air}^2 \cdot \frac{dR_m}{dl} \tag{13}$$

### 3.2. Hydraulic Subsystem

The electromagnetic and mechanical parts of the valve are used to control the flow through the valve orifice by controlling the spool position. The flow balance of the supply line can be described as

$$\dot{p}_{out} = \frac{\beta}{V_{out}} \left( Q_{sup} - Q_{con} - Q_{fb} \right) \tag{14}$$

where $\dot{p}_{out}$ is the output pressure, $\beta$ is the buck modulus of the hydraulic fluid, $V_{out}$ is the volume of the output port and $Q_{sup}$, $Q_{con}$, $Q_{fb}$ is the flow through the supply port, control port and the feedback orifice, respectively.

The flow through the supply port is as follows [17–20]

$$Q_{sup} = C_d A_{sup}(x_v) \sqrt{\frac{2}{\rho}|p_{sup} - p_{con}|} \tag{15}$$

where $C_d$ is the unitless discharge coefficient, $A_{sup}$ is the area of the supply port, $x_v$ is the spool position, $\rho$ is the fluid density, $p_{sup}$ is the supply pressure and $p_{con}$ is the control pressure.

The flow through the control port is given as

$$Q_{con} = C_d A_{con}(x_v) \sqrt{\frac{2}{\rho}p_{con}} \tag{16}$$

where $A_{con}$ is the area of the control port.

As the control port feeds to the clutch piston cylinder, the initial pressure $P_{clu}$ should be 0. Thus, the differential pressure in Equation (16) should be $|P_{con}\text{-}P_{clu}|$, simplified as $P_{con}$.

There exists an orifice which connects with the control port to regulate the output pressure because of the choking effect. The feedback force from the orifice is large enough to be one of the most important influence factors of the spool kinstate. Equation (17) is used to calculate the orifice flow.

$$Q_{fb} = C_d \pi r_{fb}^2 \sqrt{\frac{2}{\rho}|p_{con} - p_{fb}|} \tag{17}$$

where $r_{fb}$ is the radius of the orifice, $p_{fb}$ is the feedback pressure.

The dynamic behavior of pressure in the orifice can be illustrated as

$$\dot{p}_{fb} = \frac{\beta}{V_{fb}}\left(Q_{fb} + A_{fb} \cdot \dot{x}_v\right) \tag{18}$$

where $V_{fb}$ is the volume of the chamber from the control port to the pressure feedback area.

### 3.3. Dynamics Motion

The spool of the direct operating solenoid valve is subjected to the magnetic force, feedback force, friction force and compression spring force. Deduced from Newton's second law, the dynamic equation of the direct operating valve can be expressed as the following second order differential equation [21–25].

$$M_v \cdot \ddot{x}_v = \left[F_m - F_{fb} - F_{vs} - D_v\dot{x}_v - F_{sp}\right] \tag{19}$$

where, $M_v$ is the spool mass, $F_m$ is the magnetic force, $F_{fb}$ is the feedback force, $F_{vs}$ is the viscous force, $D_v$ is the damping coefficient, $F_{sp}$ is the spring force.

Figure 4 is the force diagram of the direct operating solenoid valve. The magnetic force is the force which is generated by the electric coil acting on the armature and is transferred on the spool. The feedback force arises when the feedback pressure in the feedback chamber acts on the end face of the spool. The viscous force appears when the spool has a tendency to move. The spring force originates from the compression of the spring.

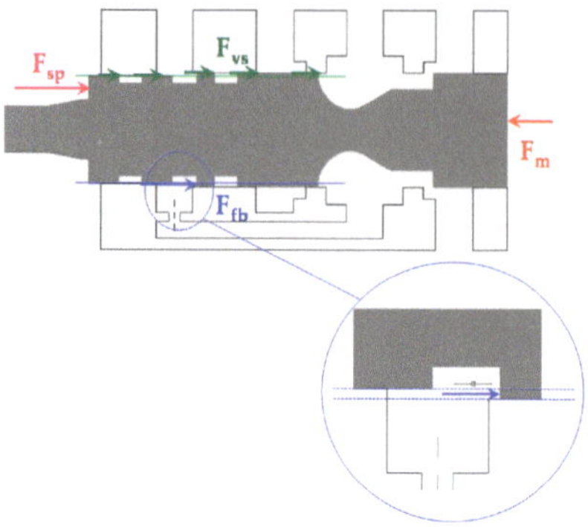


**Figure 4.** Force diagram of the direct operating solenoid valve.

$$F_{vs} = \tau_{\omega} \cdot A_{\tau} \tag{20}$$

where, $\tau_{\omega}$ is the wall shear stress, $A_{\tau}$ is the contact area for wall shear.

$$A_{\tau} = \omega \cdot l_c \tag{21}$$

$$\tau_{\omega} = (h/2l_c) \cdot p \tag{22}$$

where, $\omega$ is the clearance width, $l_c$ is the clearance length, $h$ is the clearance height.

The compression spring is fixed at the end of the spool, and the spring force is

$$F_{sp} = k_v(x_v + x_{v0}) \tag{23}$$

where $k_v$ is the spring coefficient.

The feedback force from orifice can be derived as

$$F_{fb} = A_{fb}p_{fb} \tag{24}$$

Simulation of the direct operating solenoid valve is developed based on the above mathematical models. The model parameters are defined according to the real geometric structure.

### 3.4. The Simulation Model

The dynamic characteristics of the direct operating solenoid valve can be accurately solved by Simulation X based on the above mathematical method. The simulation model of the valve was developed adopting the minimum element method in which model parameters were defined according to the real geometric structure. And Figure 5 is a general view of the created simulation model in which the solenoid valve is detailed and modeled.

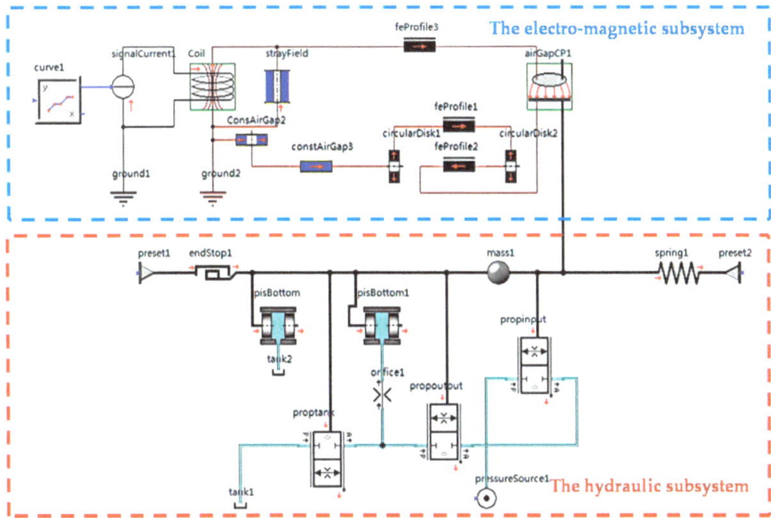


**Figure 5.** Modeling of the direct operating solenoid valve.

## 3.5. Validation of Model

To evaluate the performance of the simulation model of the direct operating solenoid valve, the laboratory testing is conducted to make the comparison with simulation results. Figure 6 provides the schematic diagram of the test rig 12. It is comprised of a heat exchanger 1 manufactured by Siemens(Berlin, Germany), an oil filter 2, a motor (also by Siemens), an oil pump 3 whose rated power is 13.6 kW, a reservoir, a TCU 5, an external computer 4 and a fixture 11 where the direct operating solenoid valve inserted in. The pressure of the control port is measured by a pressure sensor 9 (Sensata, Attleboro, MA, USA) which has a measuring range of 0–2.5 MPa. And the flow rate is obtained by a flow meter 6 (Hydrotechnik, Limburg, Germany) ranging from 7.5 L/min to 30 L/min. The automatic transmission fluid used in the test is Dexron IV from local supplier.

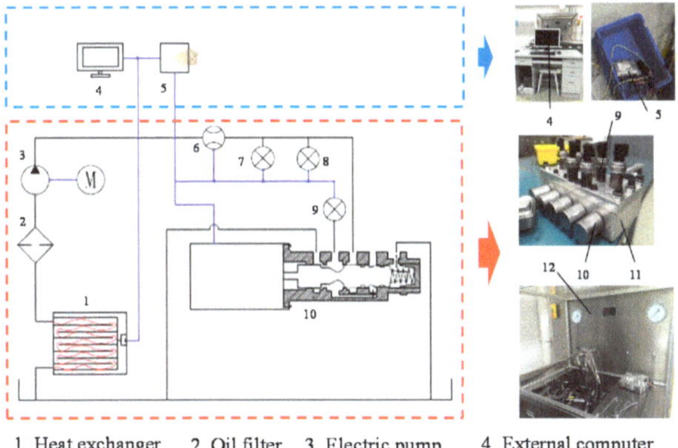


1 Heat exchanger    2 Oil filter    3 Electric pump    4 External computer
5 TCU   6 Flow meter  7.8.9 Pressure sensors   10 Direct operating valve
11 Fixture   12 Test bench

**Figure 6.** Schematic diagram and photos of experimental test rig.

The principle of the hydraulic circuit can be simply described as follows. High pressure hydraulic fluid was supplied from the mechanic pump driven by the motor to the direct operating solenoid valve 10 installed in a tailor made fixture 11. The control pressure of the solenoid valve spool was controlled by the input signal sent by the computer and transferred by the transmission control unit (TCU) to the electromagnetic unit. The control pressure, which is a most important feature to verify the simulation model as well as analyze the valve dynamic characteristics, was displayed in the computer collected by the pressure sensor and converted by TCU.

Comparison was made between the measured and simulated pressure at the control port. The input current which is extracted from the actual control signal is shown in Figure 7. The current begins with a peak up to 600 mA to activate the spool within 0.01 s. After 0.14 s, the electric current returns to 300 mA and holds for 1 s. From 1.16 s to 11.16 s, the electric current increases linearly to 1000 mA and maintains for another 1 s. Finally, the electric current declines to 0 from 12.16 s to 28.29 s.

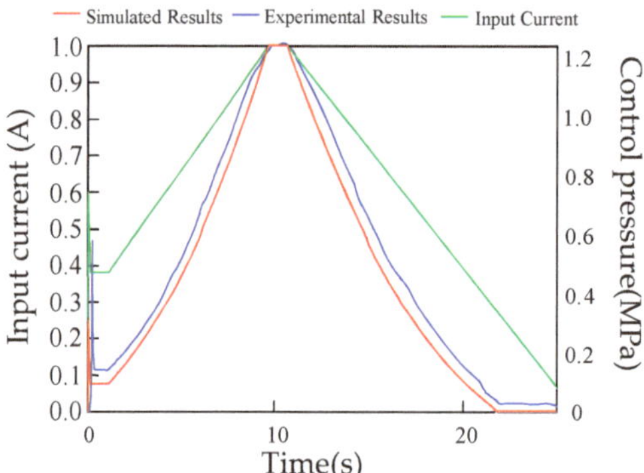


**Figure 7.** Comparison of experimental and numerical control pressure in the direct operating solenoid valve.

Both the simulated and measured control pressure at the control port is illustrated in Figure 7. It is obvious that the maximum pressure in the simulation result is the same with that in the experimental one. And there is a good agreement between the two curves with about only 4% error of the slope. Moreover, the pressure response time and the time when the pressure disappears are completely consistent. Therefore, the experimental result confirms the correctness of the model. The pressure difference is within 0.1 MPa, which is probably caused by the equipment error.

## 4. Results and Discussion

The response time and the leakage flow are two of the most important characteristic indicators of the solenoid valve. Analysis has been conducted for the two characteristics of the direct operating solenoid valve.

### 4.1. Analysis of Pressure Response

Figure 8 shows the curves of the control pressure and the spool displacement. In the initial opening process of the direct operating solenoid valve from 0–0.38 s, the pressure greatly rises up because the spool suddenly moves fast. And the movement of the spool keeps the opening area of the control port almost the same so that the pressure increases smoothly during the next 0.5 s. Finally, the pressure goes down with the closing command and the spool returns back.

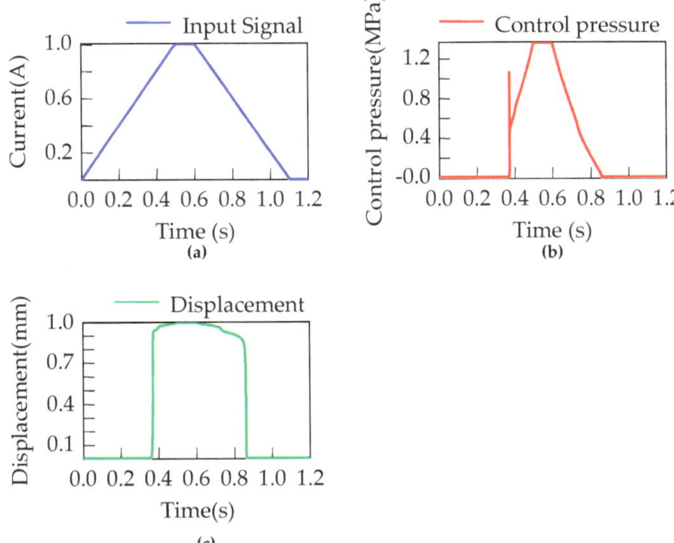

Figure 8. Control pressure and displacement of the valve. (a) Input signal (b) Control pressure (c) Spool displacement.

Since the motion of the valve is determined by the resultant force on the valve, all forces acting on the spool as well as the spool displacement are depicted in Figure 9. The spool remains still in the range of 0 to 0.4 s because the increasing magnetic force is smaller than the composition of the spring force and the viscous force. And there is no feedback force due to the closure of the input port. Once the magnetic force is big enough, the spool moves immediately after there appears the feedback force.

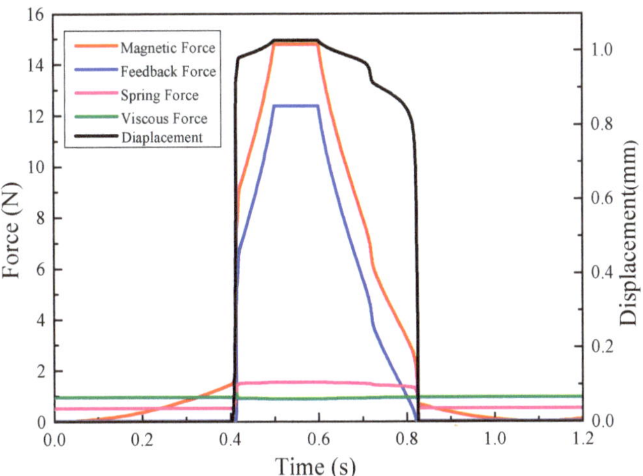

Figure 9. Forces on the spool and valve displacement.

Further analysis on the forces is conducted combining Figures 9 and 10. There are spring force called preload force before the spool starts to move, and viscous force acting on the spool when without current. The preload force comes from the spring compression. When leakage flows through the sleeve and spool, the viscous force is generated by the leakage from the supply port to the control port. When the electric current is input, the magnetic force grows up and becomes larger than the preload force at 0.32 s.

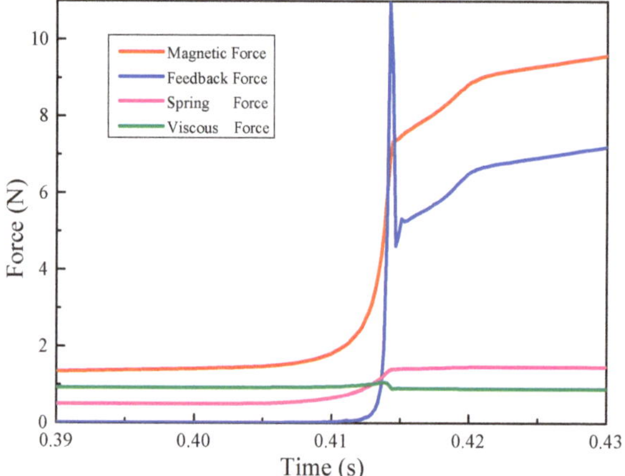


**Figure 10.** Forces acting on the spool during 0.4–0.43 s.

As the magnetic force grows large enough to overcome the preload force combined with the viscous force, the force balance is thrown off and the spool starts to move against the spring and the spring is compressed, which causes the spring force to rise. When the supply port is connected to the control port, the fluid flows through to the clutch. Meanwhile, the feedback force appears at 0.41 s to hinder the opening of the control port and changes with the control pressure. At the opening phase of the control port, the feedback force heavily fluctuates due to the sharp increase of the feedback pressure. Since different force changes at different speed, the acceleration is not constant and the displacement of the spool varies. It is worth noting that the magnetic force will vary with the input current during the spool movement and remain the similar variation trend with the control pressure.

It can be observed that the active force on the spool is the magnetic force. The passive forces on the spool are the spring force, the viscous force and the feedback force. The motion state of the valve depends on the resultant force acting on the spool and can be divided into three phases: (1) Stationary state. There is only spring force acting on the spool. (2) Moving tendency. The magnetic force, spring force and the viscous force apply on the spool. However, the magnetic force is in the opposite direction to the spring and viscous forces. The control port is still closed. (3) Motion state. The feedback force arises in the feedback chamber since the control port is connected to the supply port. The spool begins to move under the action of the magnetic force, the spring force, the viscous force and the feedback force.

The pressure and flow response characteristics are important indicators of the performance of the direct operating solenoid valves. A faster response characteristic is not the better one, because the pressure overshoot will be amplified if the response is too fast. However, the response delay will lead to the reduction of the control accuracy. The response characteristics are mainly dependent on the spool motion state which mostly relies on the resultant force on the valve.

In order to improve the pressure response time, the spool motion should be started earlier in order to reach force balance. This target can be reached by increasing the active force or decreasing the passive force. The spool will move earlier if the preload force and the viscous force diminished. Moreover, the control port could be more quickly opened if the spring stiffness decreases. Nevertheless, the spring was not analyzed in this paper as it is an optional product. More research will be done on the magnetic force and the viscous force in the further.

### 4.2. Analysis of Forces

The magnetic force is the main driving force acting on the solenoid valve which would significantly influence the pressure response time. Meanwhile the response time will change with the variation of the magnetic force.

The critical structure of the magnetic part that impacts on the magnetic force was analyzed using the model proposed in this paper. The air gap in the magnetic part transfers forces generated by the magnetic part to the mechanical part. Different diameters of the air gap are set in the model and the results are shown in Figure 11.

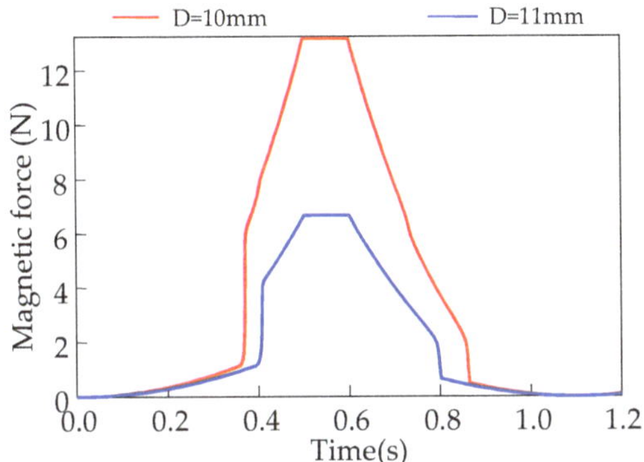


**Figure 11.** Magnetic force in different air gap.

It can be obviously observed that smaller air gap leads to bigger peak value and faster change of the magnetic force. The magnetic forces are almost accordant at the beginning and the ending of the time. However, the force in 10 mm rises up earlier than that in 11 mm. Higher maximum value in 10 mm gives rise to larger slope of the magnetic force curve.

Figure 12 presents the static characteristics of the magnetic force in different current. It indicates that the magnetic force is almost unchanged with the displacement at the same current. Slight skewing is confined within 0.5 N. Because the magnetic force is generated by the electric current and has no connection with the displacement, slight skewing increases with the growing current. However, the motion of the armature will have a small influence on the magnetic field which could result in the slight change of the magnetic force during the spool movement.

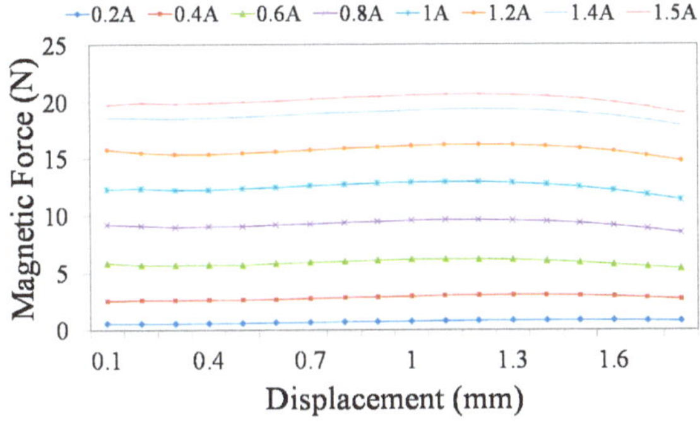


**Figure 12.** The static characteristics of the magnetic force as a function displacement for different electric current.

The static viscous force appears when the spool has the tendency of motion. And the dynamic force derives from the relative movement between the spool and the sleeve when the fluid flows through. Although small, the viscous force cannot be neglected because the force balance which determines the pressure response characteristics will change with it. Figure 13 illustrates the effect of the viscous force on the control pressure.

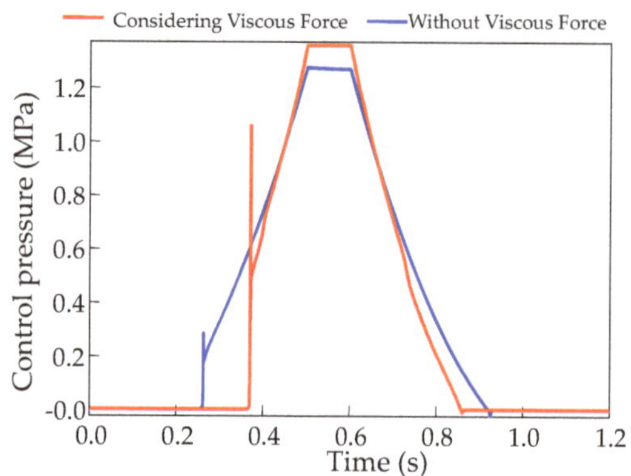

**Figure 13.** Control pressure in the valve.

The control pressure almost equals to 0 before 0.37 s when considering the viscous force, which means that the viscous force hinders the rise of the control pressure.

Different clearance height is considered in the model and the results are shown in Figures 14–16. The static viscous force grows with the increase of the clearance height. The static viscous force in 10 µm is 0.32 N while the viscous force in 30 µm is 0.94 N which indicates the viscous force will increase when the size of the clearance is increased. Meanwhile the dynamic viscous force will rapidly increase or decline when there is sudden change in the spool velocity. However, the greater the spool velocity changes, the smaller the dynamic viscous force varies because fast spool velocity thins the oil film.

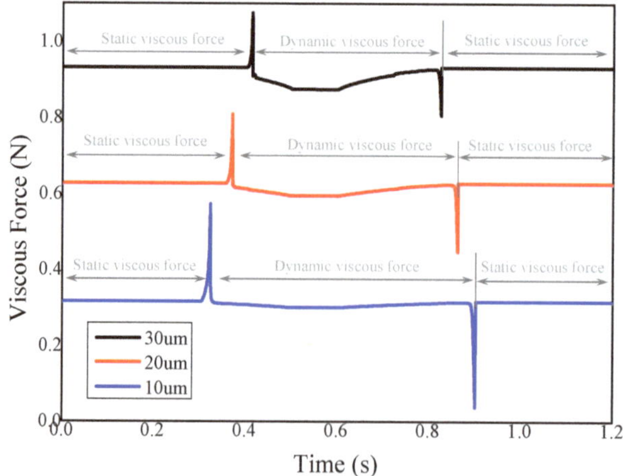

**Figure 14.** Viscous force for different clearance.

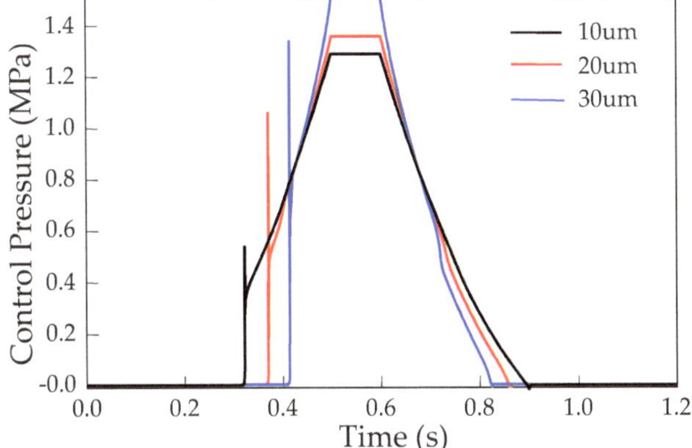

**Figure 15.** Control pressure for different clearance.

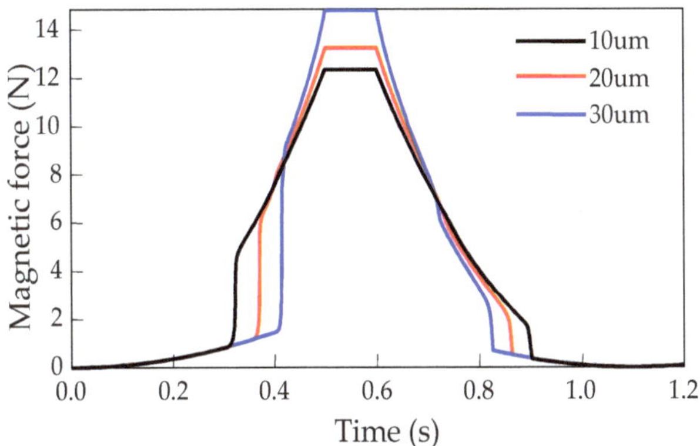

**Figure 16.** Magnetic force for different clearance.

One noticeable observation is that the magnetic force response becomes slower with the increase of the clearance. On the contrary, the height of the clearance contributes to the maximum value of the magnetic force. The same is true for the pressure response and the maximum pressure. Comparing the magnetic force with the pressure response in the same clearance, it can be found that both of them have the same variation tendency. Hence, the magnetic force has a direct influence on the control pressure.

Figures 15 and 17 show the relation between the feedback force and the control pressure. The feedback force and its growth rate rise with the growing clearance height which indicates that the initial velocity and opening area of the valve are augmented due to the variation of the equilibrium point of the resultant force. What is more, the spool displacement rises which can also lead to larger valve opening area. Therefore, the clearance height is a key factor that influences on the control pressure, because the control pressure is in direct proportion to the valve opening area.

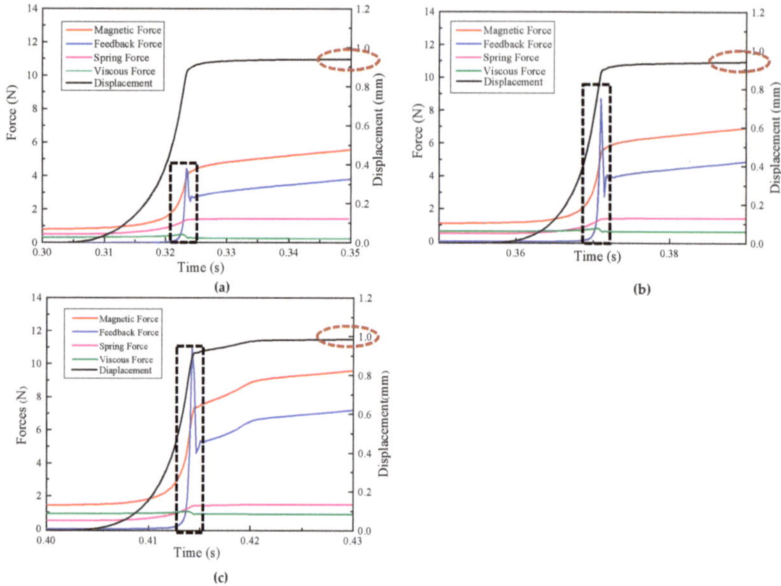

**Figure 17.** (**a**) Forces on the valve and the spool displacement in 10 μm (**b**) Forces on the valve and the spool displacement in 20 μm (**c**) Forces on the valve and the spool displacement in 30 μm.

Figure 17 shows that the growth rate of the magnetic force increases with the clearance height while that of the viscous force and that of the spring force remain unchanged before the spool starts to move. As the magnetic force is much larger than the other two forces, the positive acceleration of the valve with higher clearance is larger than that with smaller one (Figure 18), which would induce longer displacement of the spool (Figure 19). During spool motion, the control pressure grows with the valve opening area, leading to the increasing pressure in the feedback chamber connected with the control port combining Figures 15 and 19. Hence, the feedback force and its growth rate both rise with the growing clearance height (Figure 17). Since the feedback force hinders the spool movement, the negative acceleration goes up with the increasing clearance height. According to Newton's second law, the value of the spool acceleration could be quite high because the spool is only 7.39 g.

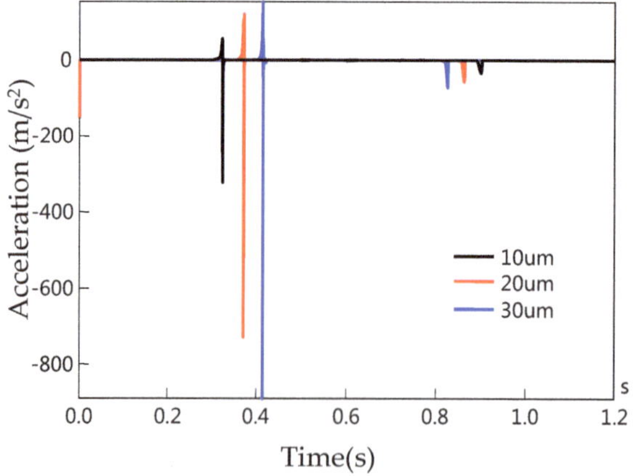

**Figure 18.** Spool acceleration for different clearance.

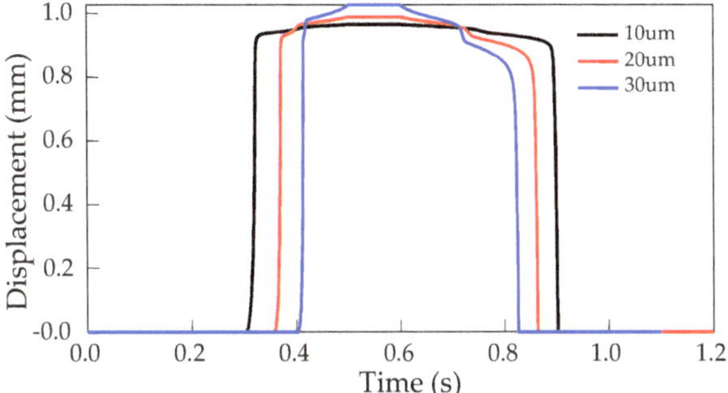

**Figure 19.** Spool displacement for different clearance.

The force of inertia mainly influences the stability and the responsiveness of the valve during spool movement. If the force of inertia is too small, the spool will be very sensitive to the change of the resultant force, resulting in frequent changes of the spool motion state. The force of inertia helps the spool maintain a relatively stable movement trend and prevent vibration when the dynamic forces acting on the spool are constantly changing. In the hydraulic control circuit of an automatic transmission, the vibration of the spool can easily give rise of the pressure fluctuation which greatly reduces the shifting quality. In addition, it is hard for the hydraulic control circuit to converge because of the superposition of spool vibration.

On the contrast, if the force of inertia is too large, the motion response of the valve will be delayed because the spool is insensitive to the change of the resultant force. The delayed motion response will result in both delayed control response and the increase of the overshoot in the hydraulic circuit. Therefore, the force of inertia is of great importance because it directly affects the control quality on the spool motion state.

Thus, the maximum control pressure is enhanced when the displacement increased. The maximum control pressure will increase from 1.2 MPa to 1.5 MPa if the displacement grows from 0.9 mm to 1.1 mm. It could be easily deduced from the analysis that the magnetic force plays a main role in all the forces which could influence the control pressure of the valve.

The starting time of the control pressure delays 0.1 s in the 30 μm comparing to that in the 10 μm as shown in Figure 15. As the movement of the valve starts only if the force balance is reached, the valve with small clearance height starts earlier than that with bigger one. The time that all forces reach to a balance is when the control pressure appears because the control port has been opened due to the displacement of the spool. Therefore, both the leakage and the pressure response should be taken into consideration when designing the clearance.

### 4.3. Analysis of Leakage Flow

Leakage is one of the inevitable problems for the automatic transmission which would give rise to pressure loss, waste of ATF, fluid insufficient, etc. The leakage flow has an important influence on the direct operating solenoid valve. Firstly, the leakage flow will directly affect the response characteristics of the solenoid valves, including pressure response and output flow. If the leakage flow is too large, the actual output flow will be less than the control flow which can't meet the requirement of both the flow and the pressure. At this point, the oil filling time of the clutch will be extended which leads to the delay of the pressure response time. As a result, the shifting time will be prolonged and the shifting quality will be reduced. Secondly, large leakage in the solenoid valve will reduce the efficiency of the automatic transmission. Thereby, it affects the fuel consumption of the vehicle.

Although a small leakage has been achieved in the direct operating solenoid valve, it cannot be completely avoided. The leakage in the direct operating solenoid valve is mainly caused by the clearance between the spool and the sleeve. The clearance height which is the most important parameter affecting the leakage flow in the solenoid design is usually between 20 μm–40 μm. A good clearance design is actually accurate design of the machining tolerance and the assembly tolerance of both the valve spool and the sleeve. Small clearance height will raise the requirements of the processing technique and the assembly technique, so it will significantly increase the cost. Meanwhile, it is averse to the quality control because high processing requirements will increase the product reject ratio which brings down the production efficiency.

It is hard for small clearance to keep the oil slick between the spool and the sleeve. Once the oil slick is broken during relative motion, the surface of the components is easily scratched because of dry friction. In addition, small clearance would frequently result in clamping stagnation. Thus, the solenoid valve falls into control failure. Although big clearance could not only decrease the cost but cut down the risk of clamping stagnation, the unavoidable large leakage will greatly reduce energy efficiency and bring about poor pressure response.

Since the clearance height between the spool and the sleeve has a direct effect on the performance of the direct operating solenoid valve. It is necessary to balance the energy efficiency, manufacturing technique and quality control. In order to analyze the effect of the clearance, the leakage from different height of clearance is depicted in Figure 20.

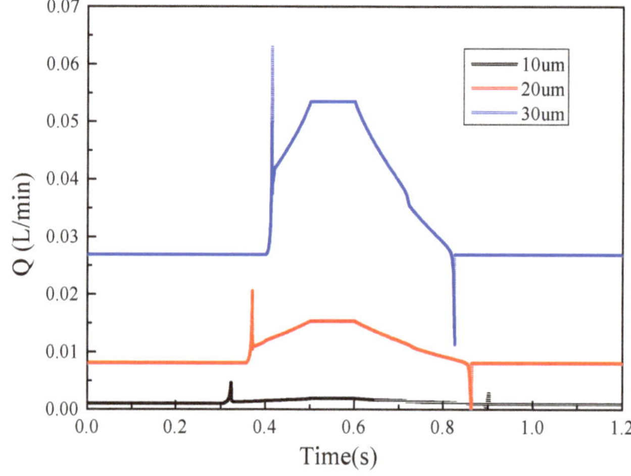


**Figure 20.** Leakage flow characteristics in the direct operating solenoid valve for different clearance.

Figure 20 describes that the maximum leakage in 30 μm is 0.06 L/min and the leakage reduces rapidly with the decrease of the clearance. It is observed that the solenoid valve with smaller clearance height has not only less leakage but faster pressure response when considering both Figures 15 and 20. The leakage flow is of same variation trend with the control pressure at the same clearance height. To make a further study, the clearance between the spool and the sleeve could be considered as an orifice when controlling the other variables constant. The pressure difference between the control port and the exhaust port will increase with the growing control pressure, and vice versa. According to Equation (18),

$$Q_{orifice} = C_d \pi r_{orifice}^2 \sqrt{\frac{2}{\rho}|\Delta p|} \tag{25}$$

where $Q_{orifice}$ is the flow through the orifice, $r_{orifice}$ is the radius of the orifice, $\Delta p$ is the pressure difference at both ends of the orifice.

Hence, the flow is in proportion to the pressure difference at both ends of the orifice. As a result, the leakage flow changes with the control pressure.

In our initial 8AT product, the leakage in the conventional clutch control unit consisting of two solenoid valves and one mechanical valve is usually 0.15 L/min which can be reduced by 60% if the control unit is replaced by a direct operating solenoid valve.

## 5. Conclusions

This paper proposed a new numerical approach solving the multi-domain physical problem of the direct operating solenoid valve. A precise model of the valve was developed to investigate the performance of the direct operating solenoid valve. The accuracy of the simulation results was verified by the experimental data in the test rig. Effect of various forces, including the magnetic force, the spring force, the viscous force, the feedback force on the pressure response of the direct operating solenoid valve were investigated. Furthermore, the influences on the dynamic characteristics of the valve were analyzed based on the model. The leakage flow in different clearance has been compared. As a consequence, the conclusions can be drawn:

(1) The simulation results of this study agreed with the experimental results. Thus, the mathematical model developed in this study was effective and accurate.

(2) Both the magnetic force and the viscous force had significantly influence on the pressure response. The pressure response time would be shortened if the magnetic force responded faster or the viscous force was reduced.

(3) To improve the response time of the solenoid valve, the clearance height should be reduced. The resultant force on the spool in 10 $\mu$m would reach the equilibrium point 0.1 s earlier than that in 30 $\mu$m.

(4) The clearance height was proved to have great influence on the leakage of the solenoid valve. The leakage increases with the growing clearance height, which showed the leakage in 30 $\mu$m was triple the amount of that in 20 $\mu$m.

(5) The leakage of the shifting control system employing the direct acting solenoid valve can be reduced by 60% compared to the conventional two-stage pilot valve in our previous product.

This paper could benefit the researchers dealing with the direct operating solenoid valve design for pressure smooth control and energy conservation in the shifting control system of an automatic transmission.

**Acknowledgments:** This work was supported by the National Natural Science Foundation of China (grant number 51405010).

**Author Contributions:** Xiao Han and Yanfang Liu designed and conducted the experiments and wrote the paper. Xiangyang Xu is the research advisor, who proposed the structure of the paper. Yanjing Liu and Yang Liu were responsible in improving of the quality of this article.

**Conflicts of Interest:** The authors declare no conflict of interest.

## References

1. Wu, S.; Jiao, Z.; Yan, L.; Zhang, R.; Yu, J.; Chen, C.Y. Development of a direct-drive servo valve with high-frequency voice coil motor and advanced digital controller. *IEEE/ASME Trans. Mechatron.* **2014**, *19*, 932–942. [CrossRef]

2. Schultz, A. Transient Simulation to Improve the Solenoid Dynamic in Pneumatic Valves. Ph.D. Thesis, Institute for Fluid Power Drives and Controls of RWTH, Aachen, Germany, 2004.

3. Wang, Z. Optimum Research on the Great Flow and Fast Response Electromagnetic Valve Used in the Medium Pressure Common Rail System of Diesel Engine. Master's Thesis, Wuhan University of Technology, Wuhan, China, 2006.

4. Liu, Q.; Bo, H.; Qin, B. Design and analysis of Direct Action Solenoid Valve. *At. Energy Sci. Technol.* **2010**, *44*, 706–711.

5. Shin, Y.; Lee, S.; Choi, C.; Kim, J. Shape optimization to minimize the response time of direct-acting solenoid valve. *J. Magn.* **2015**, *20*, 193–200. [CrossRef]

6.  Lee, G.S.; Sung, H.J.; Kim, H.C.; Lee, H.W. Flow force analysis of a variable force solenoid valve for automatic transmissions. *J. Fluids Eng.* **2010**, *132*, 031103. [CrossRef]

7.  Mutschler, K.; Dwivedi, S.; Kartmann, S.; Bammesberger, S.; Koltay, P.; Zengerle, R.; Tanguy, L. Multi physics network simulation of a solenoid dispensing valve. *Mechatronics* **2014**, *24*, 209–221. [CrossRef]

8.  Liu, Y.; Dai, Z.; Xu, X.; Tian, L. Multi-domainmodelingandsimulationofproportionalsolenoidvalve. *J. Cent. South Univ.* **2011**, *18*, 1589–1594. [CrossRef]

9.  Liu, Z.; Han, X.; Liu, Y. Dynamic Simulation of Large Flow Solenoid Valve. In Proceedings of the ASME 2016 International Mechanical Engineering Congress and Exposition, Phoenix, AZ, USA, 11–17 November 2016; ASME: New York, NY, USA, 2016; p. V009T012A015.

10. Xiong, Y.; Wei, J.; Feng, R. Adaptive robust control of a high-response dual proportional solenoid valve with flow force compensation. *Proc. Inst. Mech. Eng. Part I J. Syst. Control Eng.* **2015**, *299*, 3–26. [CrossRef]

11. Zhu, Y.; Jin, B. Analysis and modeling of a proportional directional valve with nonlinear solenoid. *J. Braz. Soc. Mech. Sci. Eng.* **2015**, *2*, 1–8. [CrossRef]

12. Liu, L.; Zhang, D.; Zhao, J. Design and research for the water medium low pressure large-flow pilot operated solenoid valve. *Strojniski Vestnik* **2014**, *60*, 10.

13. Park, C.D.; Lim, B.J.; Chung, K.Y. Design verification methodology for a solenoid valve for industrial applications. *J. Mech. Sci. Technol.* **2015**, *29*, 677–686. [CrossRef]

14. Wang, L.; Li, G.X.; Xu, C.L.; Xi, X.; Wu, X.J.; Sun, S.P. Effect of characteristic parameters on the magnetic properties of solenoid valve for high-pressure common rail diesel engine. *Energy Convers. Manag.* **2016**, *127*, 656–666. [CrossRef]

15. Zhao, J.; Fan, L.; Liu, P.; Grekhov, L.; Ma, X.; Song, E. Investigation on electromagnetic models of high-speed solenoid valve for common rail injector. *Math. Probl. Eng.* **2017**, *2017*, 1–10. [CrossRef]

16. Walker, P.D.; Zhu, B.; Zhang, N. Nonlinear modeling and analysis of direct acting solenoid valves for clutch control. *J. Dyn. Syst. Meas. Control* **2014**, *136*, 562–576. [CrossRef]

17. Shi, Y.; Wu, T.; Cai, M.; Wang, Y.; Xu, W. Energy conversion characteristics of a hydropneumatic transformer in a sustainable-energy vehicle. *Appl. Energy* **2016**, *171*, 77–85. [CrossRef]

18. Shi, Y.; Wang, Y.; Liang, H.; Cai, M. Power characteristics of a new kind of air-powered vehicle. *Int. J. Energy Res.* **2016**, *40*, 1112–1121. [CrossRef]

19. Shi, Y.; Wu, T.; Cai, M.; Liu, C. Modelling and study on the output flow characteristics of expansion energy used hydropneumatic transformer. *J. Mech. Sci. Technol.* **2016**, *30*, 1163–1170. [CrossRef]

20. Cai, M.; Wang, Y.; Shi, Y.; Liang, H. Output dynamic control of a late model sustainable energy automobile system with nonlinearity. *Adv. Mech. Eng.* **2016**, *8*, 1–11. [CrossRef]

21. Xu, Q.; Cai, M.; Shi, Y. Dynamic heat transfer model for temperature drop analysis and heat exchange system design of the air-powered engine system. *Energy* **2014**, *68*, 877–885. [CrossRef]

22. Xu, Q.; Shi, Y.; Yu, Q.; Cai, M. Virtual prototype modeling and performance analysis of the air-powered engine. *ARCHIVE Proc. Inst. Mech. Eng. Part C J. Mech. Eng. Sci.* **2014**, *228*, 2642–2651.

23. Yu, Q.; Cai, M.; Shi, Y. Working characteristics of two types of compressed air engine. *J. Renew. Sustain. Energy* **2016**, *8*, 397–411. [CrossRef]

24. Shi, Y.; Cai, M. Dimensionless study on output flow characteristics of expansion energy used pneumatic pressure booster. *J. Dyn. Syst. Meas. Control* **2013**, *135*, 021007. [CrossRef]

25. Yu, Q.; Cai, M.; Shi, Y.; Yuan, C. Dimensionless study on efficiency and speed characteristics of a compressed air engine. *J. Energy Resour. Technol.* **2015**, *137*, 044501–044509. [CrossRef]

 *applied sciences*

*Article*

# Theoretical Analysis for the Flow Ripple of a Tandem Crescent Pump with Index Angles

**Hua Zhou, Ruilong Du, Anhuan Xie * and Huayong Yang**

State Key Laboratory of Fluid Power and Mechatronic Systems, Zhejiang University, Hangzhou 310027, China; hzhou@sfp.zju.edu.cn (H.Z.); duruilong@zju.edu.cn (R.D.); yhy@sfp.zju.edu.cn (H.Y.)
* Correspondence: xieanhuan@zju.edu.cn; Tel.: +86-571-8795-1646

Received: 16 October 2017; Accepted: 3 November 2017; Published: 8 November 2017

**Featured Application: This paper presents a theoretical approach for lowering the outlet flow ripple of a crescent pump by applying a tandem crescent pump consisting of two gear pairs with an index angle between them.**

**Abstract:** This paper presents a theoretical approach for lowering the outlet flow ripple of a crescent pump by applying a tandem crescent pump consisting of two gear pairs with an index angle between them. The outlet flow of the tandem pump is obtained by summing the flow produced by the two gear pairs, and the flow ripple of the tandem pump can be attenuated by properly selecting the design parameters in terms of the index angle and the displacement ratio between the two gear pairs. A lumped parameter model is presented for evaluating the crescent pump's flow ripples, and experiments were performed on a single crescent pump to validate the model from the aspects of the steady-state flow-pressure characteristics and the outlet pressure ripples. In this way, the main causes of the flow ripple could be identified by comparing the kinematic flow with the actual flow evaluated by the model. Additionally, simulation results suggested that a tandem pump with an index angle of 13.85° and displacement ratio of 0.5 could lead to a more than 45% decrease in the outlet flow ripple than a single pump with the same displacement in a wide range of operating conditions.

**Keywords:** crescent pumps; tandem crescent pumps; flow ripple; index angle

## 1. Introduction

Crescent pumps are widely used in many fluid power applications such as injection molding machines, automotive applications, and robotic systems due to their advantages in terms of compactness, low flow ripple, and low noise level [1–3]. Today, with the increase in demand for injection quality and control accuracy, it is hoped that the flow ripple of the crescent pump can be reduced [4]. This paper focused on a theoretical approach for lowering the flow ripple by applying a tandem crescent pump comprised of two sets of gear pairs with an index angle between them.

Figure 1 depicts a schematic of the tandem crescent pump with two sets of indexed gear pairs. As suggested by the name, the tandem pump is composed of two gear pairs, namely the front gear pair and the back gear pair, and each gear pair can fulfill the function of fluid delivery via meshing as typical crescent pumps. Figure 1 also illustrates the basic working principle of the crescent pump (displacement pump): sealing chambers are formed by the floating plates, the crescent fillers, and the gear pair (the gear shaft and the ring gear), and by meshing the gears, fluid is sucked into the suction chamber due to the increase of the suction chamber's volume, delivered to the discharge chamber, and then discharged from the discharge chamber due to the decrease of the discharge chamber's volume.

As shown in Figure 1, the back gear shaft is connected to the front gear shaft via a spline coupling, which enables the same angular velocity and an index angle between the two gear shafts (view A-A and B-B). It should be noted that the flowrate by the two gear pairs can be varied by setting different widths of the gear pairs (parameter '*b*'), and a design parameter, namely, the displacement ratio is introduced, which is defined as the ratio of the displacement by the back gear pair to that by the front gear pair. By sharing the same inlet and outlet, the outlet flow of the tandem pump is obtained by superposing the flow produced by the two gear pairs, and it is expected that the outlet flow ripple of the tandem pump can be attenuated by properly selecting the index angle and the displacement ratio.

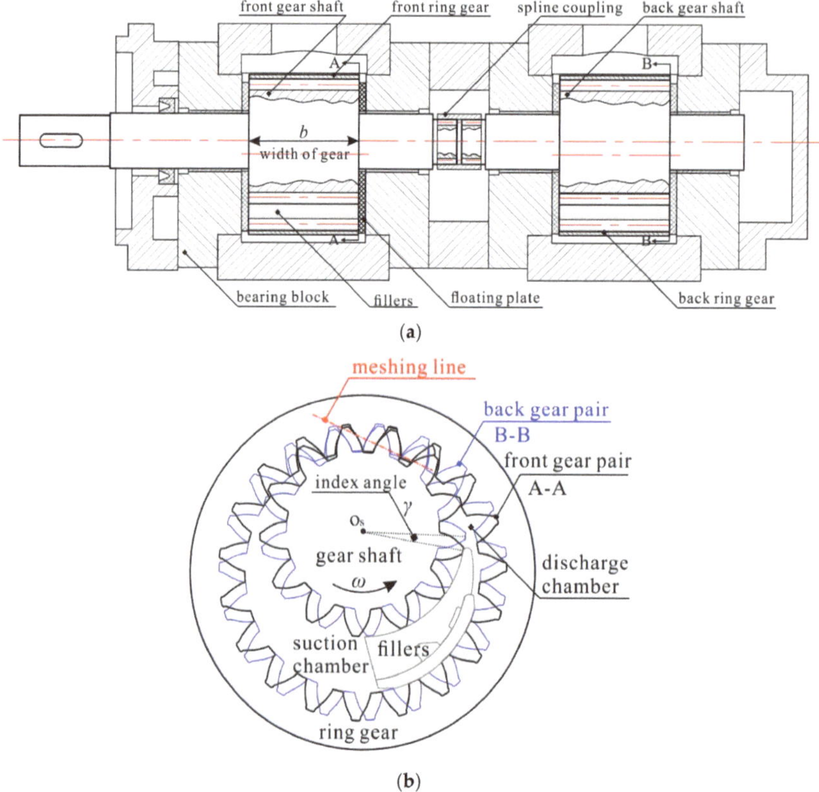



**Figure 1.** (**a**) A schematic of the tandem crescent pump design; and (**b**) a schematic of the two sets of gear pairs with an index angle (view A-A and B-B).

Usually, the internal gear pumps (IGPs) are classified into two types based on the existence of the crescent fillers: the gerotor pumps without fillers and the crescent pumps with fillers, and the flow characteristics has represented a main focus in previous studies. After a thorough review of the literature, it was found that most publications concerning IGPs were related to gerotor pumps, with only a small number focused on crescent pumps. With respect to the gerotor pumps, Mimmi et al. [5,6] addressed the theoretical flow characteristics from a kinematic aspect (based on the theory of gearing), focusing on the influence of the gears' geometric parameters. Gamez-Montero et al. [7,8] investigated the flow ripple characteristics via a mathematical model by means of the bond graph technique, and experimentally validated the model by measuring the instantaneous flow using the 'secondary source' method. Additionally, the same research group in Reference [9] further studied the flow ripple via a 3D computational fluid dynamics (CFD) model by means of ANSYS Fluent, allowing the analysis of the influence of the interteeth clearance and teeth contact for a better estimation of the instantaneous flow. Hsieh [10,11] built a CFD model for the gerotor pump using Pumplinks, presenting a novel geometrical design that enabled variable clearance between the inner and outer rotors for the purpose of lowering the flow ripples. Manco et al. [12] and Schweiger et al. [13] studied the flow ripple and the pressure ripple via a lumped parameter model by means of LMS Amesim, experimentally validating the model by measuring the steady-state flowrate and the outlet pressure ripples. In the work of Manco et al. [12], they suggested that a bi-rotor pump with indexed rotors had the potential to attenuate the flow ripple. Pellegri et al. [14,15] conducted a comparison between the lumped parameter and the CFD approach for a unique insight into the pump's performance (volumetric efficiency, flow and pressure ripples, etc.), with a specific focus on the rotors' radial micro-motions.

With respect to the crescent pumps, Ichikawa [16] addressed the mathematical expressions for the ideal delivery from a kinematic aspect, focusing on the variations of the trapped volume. Zhou et al. [17] studied the theoretical flow characteristics based on the theory of gearing, and presented a set of conjugated involute gears that enabled better fluid delivery capacity. Additionally, the same research group in Reference [18] extended the study to the trapped volume performances via a discretization approach, with a specific focus on the gears' geometric parameters. Rundo [19] investigated the theoretical flowrate of the crescent pumps with respect to different combinations of the gear pair's tooth numbers (the gear shaft and the ring gear), suggesting that the increase of the gear shaft's tooth number helped attenuate the flow ripple. Moreover, the same research group in Reference [20] extended their research to cover the internal leakage and the outlet pressure ripple via a lumped parameter model built in LMS Amesim, validating the model through a comparison to experimental results on the steady-state flowrate and the outlet pressure ripples. Hence, it can be seen that few attempts have been made to study the flow ripple with respect to the tandem crescent pump design.

Concerning the tandem pump design, it has been previously applied in other hydraulic pumps, particularly in axial piston pumps and in external gear pumps. Manring et al. [21,22] addressed the theoretical torque ripple and flow ripple of a tandem axial piston pump with two identical rotating groups under different index angles from a kinematic aspect, suggesting that an index angle of $10°$ resulted in the greatest reduction of the flow ripple (roughly 75%) with respect to a nine-piston rotating group. Xu et al. [23] studied the tandem piston pump's flow ripple via a lumped parameter model using LMS Amesim, suggesting that an index angle of $20°$ resulted in the greatest reduction of the flow ripple. In Xu et al.'s work, they suggest that the reason for the difference in index angle from that in Manring et al.'s work was due to the flow through the triangular grooves on the valve plate, which has been shown to have a great influence on the flow ripple (which was not considered in the work of Manring et al.). Battarra et al. [24] presented a tandem external gear pump comprised of a set of spur gears and a set of helical gears, sharing the same driving and driven shafts. However, the presented tandem external gear pump did not share the same outlet, thus the flow produced by the two gear pairs was delivered to two different hydraulic systems, which is different to the aforementioned tandem axial piston pump and the tandem crescent pump presented in this work.

Hence, few published works can be found on tandem crescent pump design. This paper focused on the flow ripple of the tandem crescent pump via a lumped parameter model built in Matlab. It was expected that the tandem crescent pump could lead to a decrease in the flow ripple through the proper selection of the design parameters in terms of the index angle and the displacement ratio. The rest of the paper is organized as follows: the simulation model is proposed in Section 2 focusing on the evaluations of the flow and pressure in the sealing chambers around the gear circumference; the validation of the model is conducted in Section 3 from the aspects of the steady-state flowrate and the outlet pressure ripples; the numerical results are presented in Section 4 regarding the pump's outlet flow and the related flow ripples; and discussions and conclusions are made in Section 5 based on the numerical results.

## 2. Simulation Model

Figure 2 depicts the sealing chambers around the gear circumference divided by the meshing gears and the crescent fillers: the suction chamber, the transitional chamber, the discharge chamber and the trapped chamber (if it exists), where the suction chamber is formed by the gear profiles between the meshing point $C_2$ and the intersections $A_1$ and $A_2$; the transitional chamber denotes the tooth space (TS) located in the transitional stage from the suction chamber to the discharge chamber; the discharge chamber is formed by the gear profiles between the meshing point $C_1$ and the intersections $B_1$ and $B_2$; and the trapped chamber is formed by the gear profiles between the two meshing points $C_1$ and $C_2$. Flow exchange was observed between the adjacent sealing chambers via the clearances and the grooves machined on the floating plate, namely the triangular grooves and the relief grooves, due to the pressure difference, as shown in Figure 3. To account for the flow characteristics, an evaluation of the pressure evolution around the gear circumference, where the pressure in suction chamber can be treated as the inlet pressure (atmospheric pressure) was needed; the pressure with respect to time in the other three chambers can be evaluated by applying the mass conservation equation.

$$\frac{\mathrm{d}p}{\mathrm{d}t} = \frac{\beta}{V}\left(\sum q_{\mathrm{in}} - \sum q_{\mathrm{out}} - \frac{\mathrm{d}V}{\mathrm{d}t}\right) \tag{1}$$

In Equation (1), $\beta$ denotes the bulk modulus of the fluid. It should be noted that in this work, the oil temperature was set as a constant value (40 °C); hence the oil bulk modulus $\beta$ was dependent on the pressure, as addressed by Xu et al. [25]. Regarding the other terms, $V$ denotes the chamber's volume; $q_{\mathrm{in}}$ and $q_{\mathrm{out}}$ denote the flow into and out of the chamber, respectively; and $\mathrm{d}V/\mathrm{d}t$ denotes the time derivative of the chamber's volume.

By applying Equation (1) to each chamber, the pressure can be estimated after evaluations of the terms in the right hand side of Equation (1), and the related flow characteristics can be obtained. Details are described in the following sub-sections on the evaluations of the related terms and their implementation in Matlab.

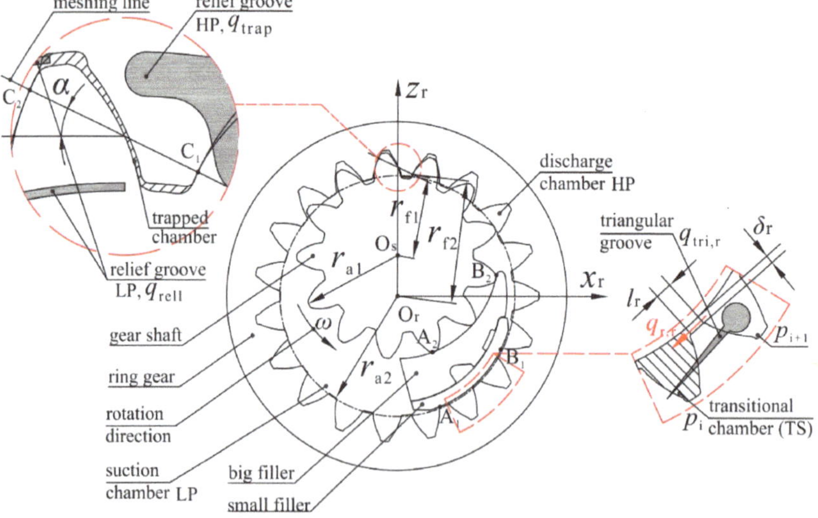


**Figure 2.** Definition of the sealing chambers around the gear circumference.

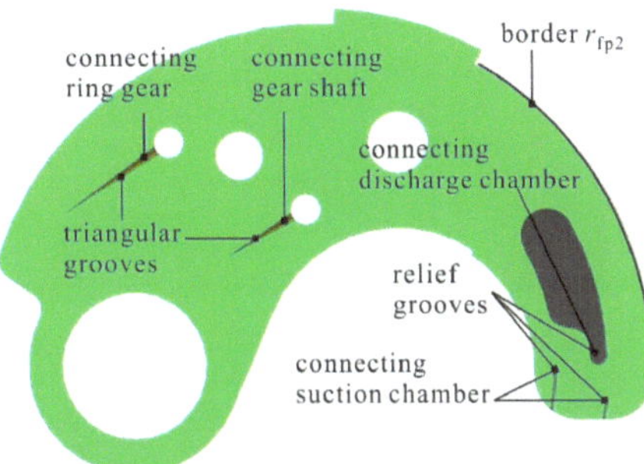



**Figure 3.** The triangular grooves and the relief grooves machined on the floating plate.

### 2.1. The Discharge Chamber

Figure 4 depicts a simple hydraulic system for the estimation of the crescent pump's flow-pressure characteristics, which is comprised of a throttle valve utilized as the pump's load, and a delivery line with a constant diameter utilized as the connection between the pump's outlet and the throttle valve. As observed, the pressure in the discharge chamber corresponds to the pump's outlet pressure. Regarding Equation (1), the term $V$ denotes the volume of the discharge chamber and the delivery line, the term $q_{in}$ denotes the flow into the discharge chamber from the trapped chamber ($q_{trap}$ in Figure 2) via the relief groove on the floating plate that connects the discharge chamber and the trapped chamber (Figure 3). In this scenario, the term $q_{in}$ yields

$$q_{in} = q_{trap} = C_d A_{rel} \sqrt{2|p_{trap} - p|/\rho} \cdot \text{sign}(p_{trap} - p) \tag{2}$$

where $q_{trap}$ is the trapped flow; $A_{rel}$ is the flow area of the relief groove; $p_{trap}$ is the trapped pressure; $\rho$ is the oil density dependent on the pressure (the oil temperature was set as a constant and not taken into account in this work), as addressed by Ivantysyn et al. [26]; $C_d$ is the discharge coefficient dependent on the flow velocity and is difficult to be determined. In this work, the value of $C_d$ was set as 0.7 according to the study by Ma et al. [27].

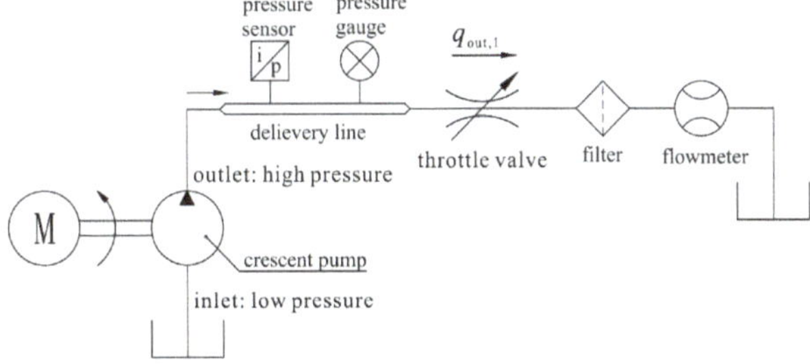



**Figure 4.** A hydraulic system for estimating the crescent pump's flow-pressure characteristics.

Regarding the term $q_{out}$, it denotes the flow out of the discharge chamber, which consists of three terms: the valve flow through the throttle valve in the hydraulic system ($q_{out,1}$ in Figure 4), the triangular flow into the transitional chamber through the triangular grooves on the floating plate (Figure 3), and the internal leakage of the pump. The first part can be given as

$$q_{out,1} = q_{val} = C_d A_{val} \sqrt{2p/\rho} \tag{3}$$

where $q_{val}$ is the flow through the throttle valve; and $A_{val}$ is the flow area of the throttle valve.

As Figure 3 shows, two triangular grooves are machined on the floating plate, therefore, the triangular flow is further split into two terms accordingly: the flow ($q_{\text{tri,s}}$) into the gear shaft's tooth space (TS) and the flow ($q_{\text{tri,r}}$) into the ring gear's TS. The term $q_{\text{tri,r}}$ is depicted in Figure 2, and can be given as

$$q_{\text{tri,r}} = C_{\text{d}} A_{\text{tri}} \sqrt{2(p_{i+1} - p_i)/\rho} \tag{4}$$

where $A_{\text{tri}}$ is the flow area of the triangular groove; and $p_{i+1}$ and $p_i$ are the fluid pressures in the adjacent tooth spaces.

Noting that the term $q_{\text{tri,s}}$ can be evaluated in the same way as Equation (4), the triangular flow yields

$$q_{\text{out,2}} = q_{\text{tri}} = q_{\text{tri,s}} + q_{\text{tri,r}} \tag{5}$$

With respect to the internal leakage of the pump, it can be split into three terms according to different types of clearances inside the pump: the lateral leakage through the lateral clearances between the gears' lateral sides and the floating plates, the radial leakage through the radial clearances between the gears' tooth tips and the fillers, and the ring-gear/case leakage through the clearance between the ring gear and the case.

Figure 5 depicts the lateral leakage that goes through the gear shaft's lateral side ($q_{\text{l,s}}$) and the ring gear's lateral side ($q_{\text{l,r}}$), thus the lateral leakage is further split into two terms accordingly. As observed, due to the complexity of the gears' profiles, the lateral leakage is evaluated utilizing the annular areas bounded by the gears' pitch circles and the floating plate's borders. Using these quantities, the lateral leakage through the ring gear's side ($q_{\text{l,r}}$) yields

$$q_{\text{l,r}} = \frac{\psi_{\text{d,r}} \cdot \delta_1^3}{12\mu} \cdot \frac{(p - p_{\text{in}})}{\ln\left(r_{\text{fp2}}/r_{\text{p2}}\right)} \tag{6}$$

where $\psi_{\text{d,r}}$ is the central angle of the discharge chamber; $\delta_1$ is the lateral clearance between the gears' lateral sides and the floating plates; $p_{\text{in}}$ is the inlet pressure; and $\mu$ denotes the dynamic viscosity of oil dependent on the pressure (the oil temperature was set as a constant and not taken into account in this work), as addressed by Ivantysyn et al. [26].

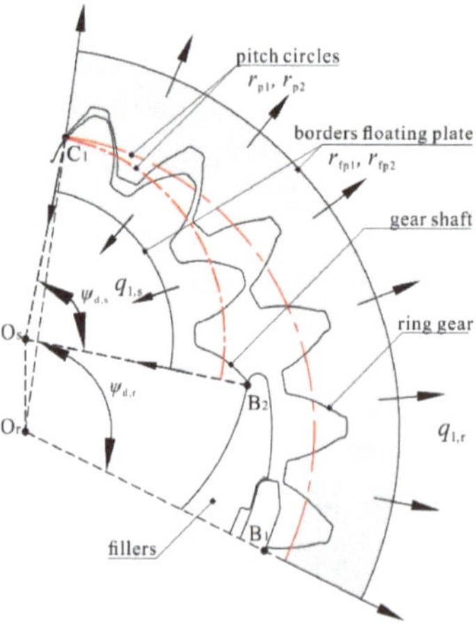


**Figure 5.** Lateral leakage that goes through the gear shaft's side and the ring gear's side.

Noting that the term $q_{l,s}$ can be evaluated in the same way as Equation (6) and there exist two lateral sides of the gears, the lateral leakage yields

$$q_l = 2 \cdot (q_{l,s} + q_{l,r}) \tag{7}$$

Concerning the radial leakage, it can also be split into two terms like the lateral leakage: the flow through the gear shaft's tooth tip ($q_{r,s}$), and the flow through the ring gear's tooth tip ($q_{r,r}$). The term $q_{r,r}$ is depicted in Figure 2, and can be given as

$$q_{r,r} = \frac{b\delta_r^3 (p_{i+1} - p_i)}{12\mu \cdot l_r} - \frac{b\delta_r \omega_r r_{a2}}{2} \tag{8}$$

where $b$ is the width of the gear; $\delta_r$ is the radial clearance between the gears' tooth tips and the fillers; $l_r$ is the length of the tooth tip; $\omega_r$ is the angular velocity of the ring gear; and $r_{a2}$ is the addendum radius of the ring gear.

Noting that the term $q_{r,s}$ can be evaluated in the same way as Equation (8), the radial leakage yields

$$q_r = q_{r,s} + q_{r,r} \tag{9}$$

Figure 6a depicts the cross section of the case for the analysis of the ring-gear/case leakage. It can be seen that rectangular sealing areas formed in the ring-gear/case interface surrounding the high pressure outlets. Moreover, fluid film formed in the sealing areas for the purpose of sealing, bearing and lubricating, and there was leakage flow in the film due to the pressure difference between the outlets and the borders of the sealing areas. Noting the small ratio between the film height (micrometer level) and the other two dimensions (millimeter level), the sealing area was unwrapped on a plane, as shown in Figure 6b, and was treated as a rectangular hydrostatic bearing. In this scenario, the ring-gear/case leakage ($q_{rc}$) through the rectangular sealing area can be evaluated with Equation (10) according to the study by Hamrock et al. [28].

$$q_{rc} = 2 \cdot \left( \frac{L}{6\mu b_B} + \frac{B}{12\mu b_{L1}} + \frac{B}{12\mu b_{L2}} \right) \cdot \delta^3 \cdot p_{out} \tag{10}$$

where $b_B$, $b_{L1}$, $b_{L2}$, $B$ and $L$ are the geometric parameters of the sealing area depicted in Figure 6b; and $\delta$ is the radial clearance between the ring gear and the case.

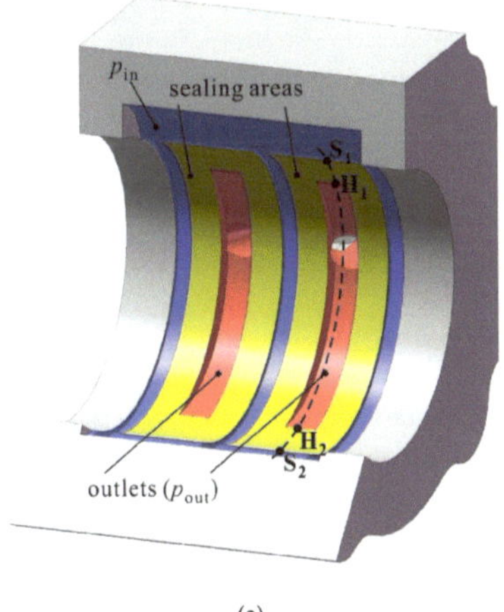


(a)

**Figure 6.** *Cont.*

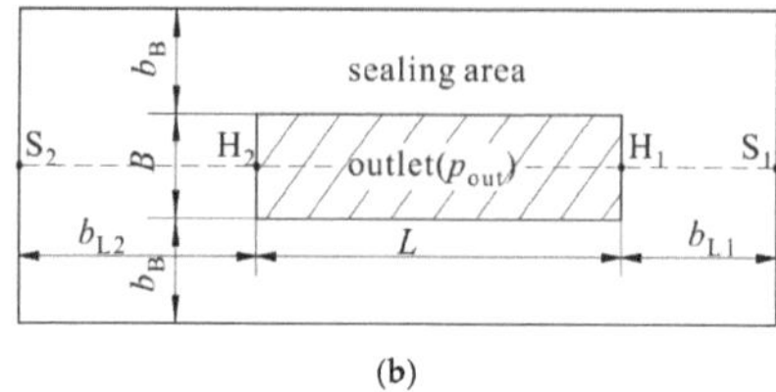


(b)

**Figure 6.** (**a**) Cross section of the case depicting the rectangular sealing areas surrounding the outlets; and (**b**) the rectangular sealing area unwrapped on a plane.

Hence, the internal leakage of the pump yields by summing all the leakages is

$$q_{out,3} = q_{leak} = q_l + q_r + q_{rc} \tag{11}$$

and the term $q_{out}$ in Equation (1) yields

$$q_{out} = q_{out,1} + q_{out,2} + q_{out,3} \tag{12}$$

Regarding the term $dV/dt$, it denotes the time derivative of the discharge chamber's volume, which can be interpreted as the kinematic flow of the crescent pump derived from the kinematic relations according to the work by Zhou et al. [17].

$$-\frac{dV}{dt} = \frac{dV_{kin}}{dt} = Q_{kin} = \frac{b\omega}{2}\left[\left(r_{a1}^2 - r_{f1}^2\right) - \frac{z_1}{z_2}\left(r_{a2}^2 - r_{f2}^2\right)\right] \tag{13}$$

where $V_{kin}$ is the volume of the discharged fluid when the pump operates for a period of time $t$ from the kinematic aspect; $Q_{kin}$ is the kinematic outlet flow; $\omega$ is the angular velocity of the gear shaft; $r_{a1}$ and $r_{a2}$ are the addendum radii of the gear shaft and the ring gear, respectively; $r_{f1}$ and $r_{f2}$ are the distances between the contact point and the centers of the gear shaft and the ring gear, respectively; and $z_1$ and $z_2$ are the tooth numbers of the gear shaft and the ring gear, respectively.

### 2.2. The Trapped Chamber

According to the theory of gearing, one or two meshing points are formed during the meshing process. Under the circumstance of two meshing points, a trapped chamber is formed by the gears' profiles between the two meshing points, as shown in Figure 2. Regarding the terms in Equation (1), the term $V$ denotes the trapped chamber's volume, which is characterized first by a decrease, then followed by an increase; the term $dV/dt$ denotes the variations of the trapped chamber's volume, which can be evaluated as

$$\frac{dV_{trap}}{dt} = \left(\frac{dV_{kin}}{dt}\right)_{C_2} - \left(\frac{dV_{kin}}{dt}\right)_{C_1} \tag{14}$$

As shown in Figures 2 and 3, relief grooves are machined on the floating plate to connect the trapped chamber to the discharge chamber during the volume decreasing stage, and to the suction chamber during the volume increasing stage. Hence, the terms $q_{in}$ and $q_{out}$ are contributed by two types of flow: the flow through the relief grooves, which can be evaluated by Equation (2) ($p = p_{out}$ when connected to the discharge chamber, $p = p_{in}$ when connected to the suction chamber); and the lateral leakage, which can be evaluated by Equations (6) and (7).

### 2.3. The Transitional Chamber

As stated above, the transitional chamber denotes the gears' TS located in the transitional stage, which can be divided into two categories: the gear shaft's TS and the ring gear's TS. As shown in

Figures 2 and 3, triangular grooves are machined on the floating plate to connect the transitional chamber to the discharge chamber, which enables flow into the transitional chamber due to the pressure difference between the discharge chamber and the transitional chamber for the purpose of increasing the fluid pressure in the transitional chamber from the inlet pressure to the outlet pressure smoothly. Hence, regarding the terms in Equation (1), the term $V$ denotes the volume of the TS, the term $dV/dt$ yields zero since the volume of the TS does not vary during the transitional stage, and the terms $q_{in}$ and $q_{out}$ are contributed by three types of flow: the flow through the triangular grooves, which can be evaluated by Equation (4); the lateral leakage, which can be evaluated by Equation (6); and the radial leakage, which can be evaluated by Equation (8).

## 2.4. Simulation Procedure

Figure 7 depicts the solution algorithm for the analysis of the tandem crescent pump's flow characteristics implemented in Matlab. It starts with the input parameters of the pump's geometric parameters and the initial values including the initial pressures in the sealing chambers and the initial oil properties, as shown in box 1. Concerning the tandem pump design, it was comprised of two sets of gear pairs characterized by two design parameters, namely the index angle ($\gamma$) and the displacement ratio ($\zeta$), as shown in boxes 2 and 3 (in the blue dashed box).

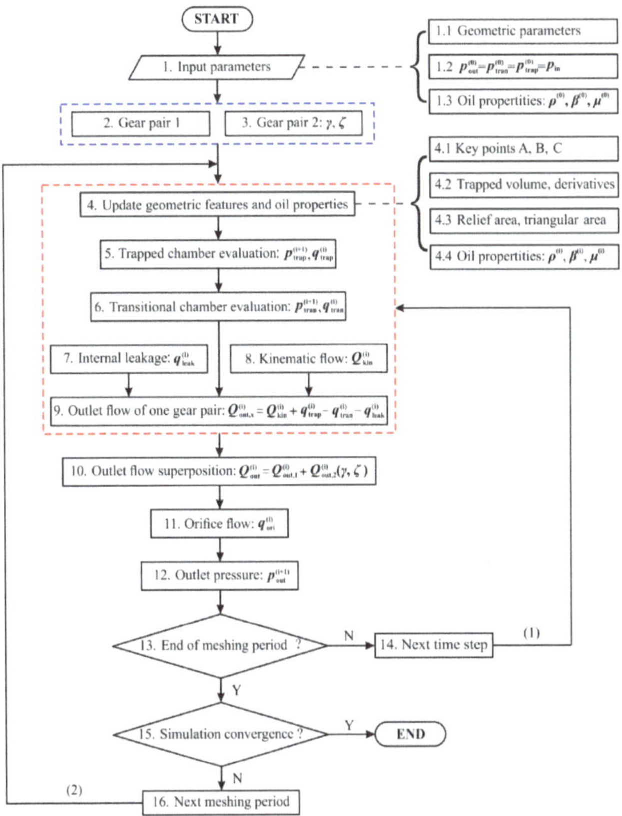

**Figure 7.** Solution algorithm for analyzing the tandem crescent pump's flow characteristics.

The key part of the algorithm is represented by the evaluation of the outlet flow of one gear pair depicted in the red dashed box. With respect to a certain time step (the *i*th time step), the geometric features and the oil properties that vary over time need to be updated first, as shown in box 4: the key points (points A, B, C in Figure 2) for dividing the sealing chambers, the trapped chamber's volume and its time derivative for the evaluation of the trapped pressure, the flow areas of the relief grooves and the triangular grooves on the floating plate for the evaluations of the trapped flow and the triangular flow, respectively, and the oil properties ($\rho$, $\beta$, $\mu$), which are dependent on the pressure (boxes 4.1 to 4.4).

Under the circumstance of two meshing points ($C_1$ and $C_2$ in Figure 2, the trapped chamber exists), noting that the lateral leakage of the trapped chamber can be evaluated by Equations (6) and (7) (with different central angles), the trapped pressure in the next time step (the $(i + 1)$th time step) can be evaluated by applying the mass conservation equation written as Equation (15), as shown in box 5.

$$p_{\text{trap}}^{(i+1)} = \frac{\beta_{\text{trap}}^{(i)}}{V_{\text{trap}}^{(i)}} \left( -\sum q_{\text{trap}}^{(i)} - \sum q_{\text{leak,trap}}^{(i)} - \left( \frac{dV_{\text{trap}}}{dt} \right)^{(i)} \right) \cdot \Delta t + p_{\text{trap}}^{(i)} \tag{15}$$

where $\Delta t$ is the time interval between time steps.

The pressure in the transitional chamber in the $(i + 1)$th time step can be evaluated by Equation (16), the same way as that in the trapped chamber, noting that the term $dV/dt$ yields zero and the leakage of the transitional chamber consists of two parts, namely the lateral leakage and the radial leakage, as shown in box 6.

$$p_{\text{tran}}^{(i+1)} = \frac{\beta_{\text{tran}}^{(i)}}{V_{\text{TS}}} \left( \sum q_{\text{tri}}^{(i)} - \sum q_{\text{leak,tran}}^{(i)} \right) \cdot \Delta t + p_{\text{tran}}^{(i)} \tag{16}$$

where $V_{\text{TS}}$ is the volume of the tooth space; and $p_{\text{tran}}$ is the transitional pressure.

As stated above, the internal leakage of the pump consists of three parts: the lateral leakage and the radial leakage of the discharge chamber, and the ring-gear/case leakage, which can be evaluated by Equations (6) to (10) by leveraging the updated points B and C and the $i$th outlet pressure as shown in box 7. Noting that the time derivative of the discharge chamber's volume is interpreted as the kinematic flow in Equation (13) (as shown in box 8), the outlet flow of one gear pair yields (box 9)

$$Q_{\text{out,x}}^{(i)} = Q_{\text{kin}}^{(i)} + q_{\text{trap}}^{(i)} - q_{\text{tri}}^{(i)} - q_{\text{leak}}^{(i)} \tag{17}$$

Hence, the outlet flow of the tandem pump yields by summing the flow produced by the two gear pairs (box 10).

$$Q_{\text{out}}^{(i)} = Q_{\text{out,1}}^{(i)} + Q_{\text{out,2}}^{(i)}(\gamma, \zeta) \tag{18}$$

Noting that the valve flow can be evaluated by Equation (3), the outlet pressure (pressure in the discharge chamber) in the $(i + 1)$th time step can be evaluated by Equation (19) as shown in box 12.

$$p_{\text{out}}^{(i+1)} = \frac{\beta_{\text{d}}^{(i)}}{V_{\text{d}}^{(i)}} \left( Q_{\text{out}}^{(i)} - Q_{\text{val}}^{(i)} \right) \cdot \Delta t + p_{\text{out}}^{(i)} \tag{19}$$

where $V_{\text{d}}$ is the volume of the discharge chamber and the delivery line.

Hence, it can be seen that in the $i$th time step, the $i$th flow characteristics and the $(i + 1)$th pressures in the sealing chambers are evaluated; then in the $(i + 1)$th time step, the $(i + 1)$th pressures are used to evaluate the $(i + 1)$th flow characteristics and the $(i + 2)$th pressures until the end of a meshing period (boxes 13 and 14, loop1). At the end of the meshing period, the flow characteristics, in terms of the outlet flow, the triangular flow, the trapped flow, and the internal leakage, serve as the criteria to judge the convergence of the simulation, which is characterized by the fact that the flow characteristics do not differ from one meshing period to another defined by Equation (20); the simulation stops after the convergence of the simulation (boxes 15 and 16, loop2).

$$\sum |q_{\text{new}} - q_{\text{old}}| \, / \sum |q_{\text{new}}| \leq q_{\text{err}} \tag{20}$$

In Equation (20), $q_{\text{new}}$ denotes the flow in the present meshing period; $q_{\text{old}}$ denotes the flow in the previous meshing period; and $q_{\text{err}}$ denotes the error between the flow in the two successive meshing periods (converged to $10^{-8}$).

It is worthwhile to note that the time interval between time steps ($\Delta t$) was to the order of $10^{-8}$ s and the consuming time of the simulation process was roughly seven minutes with an Intel® Xeon® CPU E3-1230 v3 and 16.0 GB RAM.

## 3. Experimental Validation

Figure 8 depicts the test rig for the experimental campaign on a 40 cc/rev single crescent pump. It is worth noting that the layout of the test rig was in accordance with the hydraulic circuit displayed in Figure 4. As observed, the crescent pump was driven by a servo motor (0–2000 rpm), and its outlet pressure was built up by a throttle valve. The working medium of the pump was L-HM 46 mineral oil, and the oil temperature was maintained within (40 ± 3) °C, consistent with the oil temperature in the simulation model. The test rig enabled measurements of the steady-state outlet flowrate via a gear type flowmeter and the outlet pressure ripples via a high frequency pressure sensor. Table 1 provides the main features of the flowmeter and the pressure sensor which have been calibrated before experiments, and Table 2 provides the main geometric parameters of the crescent pump.

**Figure 8.** Test rig for measuring the single crescent pump's steady-state flowrate and outlet pressure ripples.

**Table 1.** Main features of the flowmeter and the pressure sensor.

| Sensor | Type | Main Feature |
|---|---|---|
| Flowmeter | Kracht®, Germany, VC5F1PV | scale: 1–250 L/min, 0.3% accuracy (from measured value) |
| Pressure sensor | Shuangqiao®, China, CYG1401F | scale: 0–35 MPa, 0.5% FS accuracy, 0.5% nonlinearity, 100 KHz natural frequency |

**Table 2.** Main geometric parameters of the crescent pump.

| Parameter | Notation | Value | Unit |
|---|---|---|---|
| Module of the gear | $m$ | 3 | mm |
| Tooth numbers of the gear shaft and the ring gear | $z_1, z_2$ | 13, 19 | – |
| Pressure angle of the gear | $\alpha_0$ | 22 | ° |
| Operating pressure angle of the gear pair | $\alpha$ | 24.87 | ° |
| Width of the gear | $b$ | 55 | mm |
| Addendum and pitch radii of the gear shaft | $r_{a1}, r_{p1}$ | 23.43, 19.5 | mm |
| Addendum and pitch radii of the ring gear | $r_{a2}, r_{p2}$ | 26.95, 28.5 | mm |
| Radii of inner and outer borders of the floating plate | $r_{fp1}, r_{fp2}$ | 15.5, 38.25 | mm |

Figure 9 depicts the comparison between the simulated and experimental results concerning the steady-state flow characteristics with respect to different operating conditions (500–2000 rpm, 5–20 MPa). As observed, there existed a linear relationship between the flowrate and the operating speed, and the outlet flowrate yielded roughly 40 L/min at 1000 rpm, in accordance with the pump's displacement (40 cc/rev). Furthermore, it was also observed that the outlet flowrate and the volumetric efficiency yielded a decrease as the outlet pressure increased (at a certain speed), which was expected since higher outlet pressure leads to greater internal leakage. A clear example can be given by the case of working at 500 rpm, where the volumetric efficiency drops from 0.98 at 5 MPa to 0.92 at 20 MPa. Moreover, it can be seen that a good agreement was found between the simulated and experimental results regarding the flowrate and the volumetric efficiency, noting that the accuracy of the flowmeter was within 0.3% (from measured value).

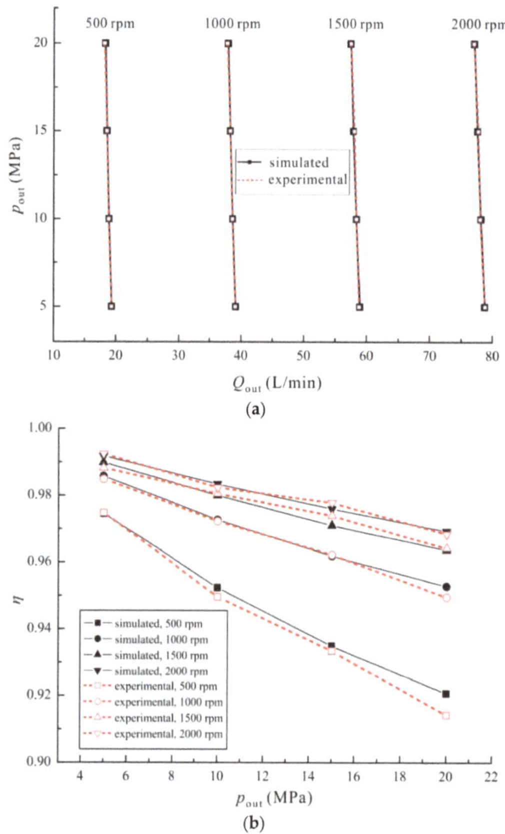

**Figure 9.** Steady-state flow characteristics regarding different operating speeds (500–2000 rpm) and outlet pressures (5–20 MPa): (**a**) outlet flowrate; and (**b**) volumetric efficiency.

Figure 10 depicts the comparison between the simulated and experimental results on the outlet pressure ripples for one shaft revolution with respect to different operating conditions (500 rpm, 5 MPa; 500 rpm, 20 MPa; 2000 rpm, 20 MPa). As observed, 13 outlet pressure ripples existed in a shaft revolution, which is believed to be due to the tooth number of the gear shaft (driving gear) which was 13. Apart from that, it was visible that the pressure ripples yielded greater amplitudes under the circumstance of low operating speed and high outlet pressure (500 rpm, 20 MPa). Moreover, it can be seen that that a good agreement was found between the simulated outlet pressure ripples and the experimental results regarding different operating conditions, noting that the accuracy of the pressure sensor was within 0.5% (full scale) and the nonlinearity of the pressure sensor was within 0.5%.

Figure 11 depicts the comparison of the frequency spectra of the simulated and experimental outlet pressure ripples at 500 rpm, 20 MPa. As observed, the primary frequency concerning the simulated results yielded 108.33 Hz with an amplitude of 0.253 MPa, and the primary frequency concerning the experimental results yielded 108.33 Hz with an amplitude of 0.250 MPa, suggesting a good match of the frequency spectra. Aside from that, the primary frequency of the simulated and experimental results was in accordance with the analytical value given by Equation (21).

$$f = \frac{z_1 \cdot n}{60} = \frac{13 \times 500}{60} = 108.33 \, \text{Hz} \tag{21}$$

Judging from the analysis above, a good match was found between the simulated and experimental results regarding the steady-state outlet flowrate and the outlet pressure ripples, therefore justifying the capability of the proposed model regarding the analysis of the pump's outlet flow characteristics.

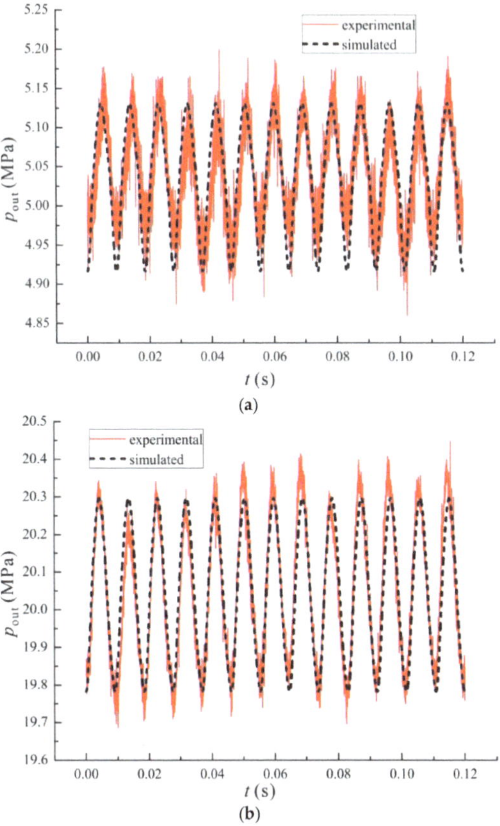


**Figure 10.** *Cont.*

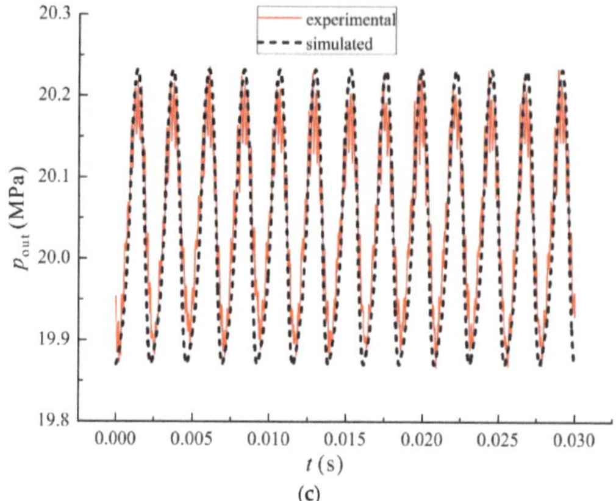

(c)

**Figure 10.** Comparison of outlet pressure ripples regarding different operating conditions: (**a**) 500 rpm, 5 MPa; (**b**) 500 rpm, 20 MPa; and (**c**) 2000 rpm, 20 MPa.

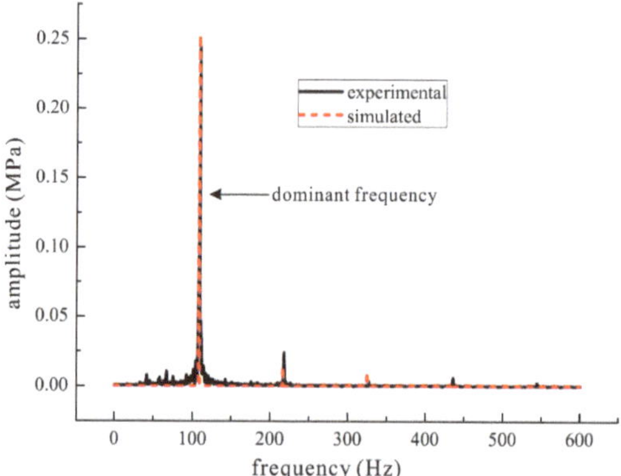

**Figure 11.** Frequency spectra of simulated and experimental outlet pressure ripples at 500 rpm, 20 MPa.

## 4. Numerical Results

In this section, the numerical results from the simulation model will be presented in terms of the outlet flow of the tandem crescent pump, and the related flow ripples with respect to different design parameters and operating conditions.

### 4.1. Outlet Flow of the Crescent Pump

Figure 12 depicts the three types of outlet flow from one gear pair under 2000 rpm and 20 MPa: the kinematic outlet flow ($Q_{kin}$) defined by Equation (13), the with-trapped outlet flow ($Q_{wt}$) by taking the trapped flow into consideration defined by Equation (22), and the actual outlet flow ($Q_{out}$) evaluated by the proposed model defined by Equation (23).

$$Q_{wt} = Q_{kin} + q_{trap} \tag{22}$$

$$Q_{out} = Q_{kin} + q_{trap} - q_{leak} - q_{tri} \tag{23}$$

Referring to Figure 12, it can be seen that the three types of flow were subject to ripples with the period of approximately 27.7°, which was expected since the tooth number of the gear shaft (driving gear) was 13, as shown in Equation (24). The kinematic outlet flow ($Q_{kin}$) yielded between 74.88 and 80.14 L/min; the with-trapped outlet flow ($Q_{wt}$) yielded between 78 and 80.14 L/min; and the actual outlet flow ($Q_{out}$) yielded between 75.25 and 78.33 L/min.

$$\lambda = 360°/z_1 = 27.7° \tag{24}$$

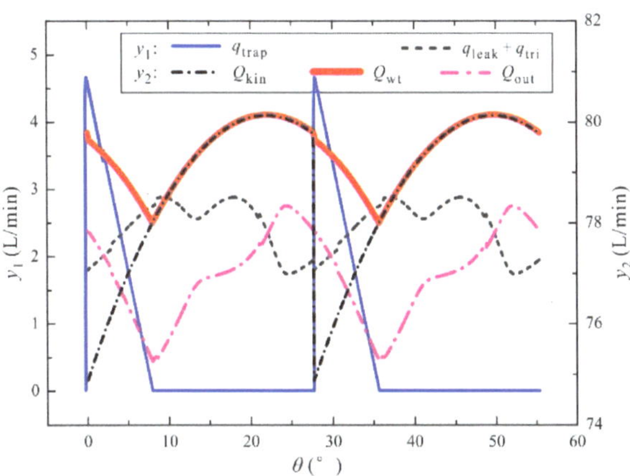

**Figure 12.** Three types of outlet flow from one gear pair under 2000 rpm and 20 MPa.

Figure 13 depicts the outlet flow from two gear pairs with an index angle ($\gamma$) under 2000 rpm and 20 MPa ($\zeta = 1$). It can be seen that within a period, the flow from one gear pair was characterized by an ascending stage followed by a descending stage, and one could observe a peak value and a valley value. The flow from the two gear pairs were in the same shape since the displacement was $\zeta = 1$. A phase difference was also observed between the flow from the two gear pairs due to the index angle.

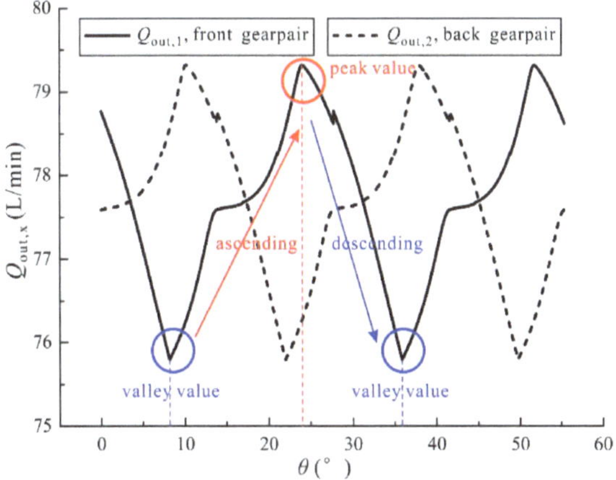


**Figure 13.** Outlet flow from two indexed gear pairs under 2000 rpm and 20 MPa ($\zeta = 1$).

Figure 14 depicts the outlet flow from the tandem pump ($\zeta = 1$) under 2000 rpm and 20 MPa with respect to different index angles (0, $\lambda/4$, $\lambda/2$), noting that the index angle $\gamma$ was bounded between 0 and $\lambda$. It can be seen that the outlet flow of the tandem pump exhibited a periodic behavior with the same period of 27.7° and roughly the same mean outlet flowrate though the index angle varied. However, the outlet flow exhibited different shapes and different ripples as the index angle varied. The outlet flow yielded between 151.59 and 158.64 L/min when $\gamma = 0$, between 153.25 and 156.96 L/min when $\gamma = \lambda/4$, and between 154.01 and 156.34 L/min when $\gamma = \lambda/2$.

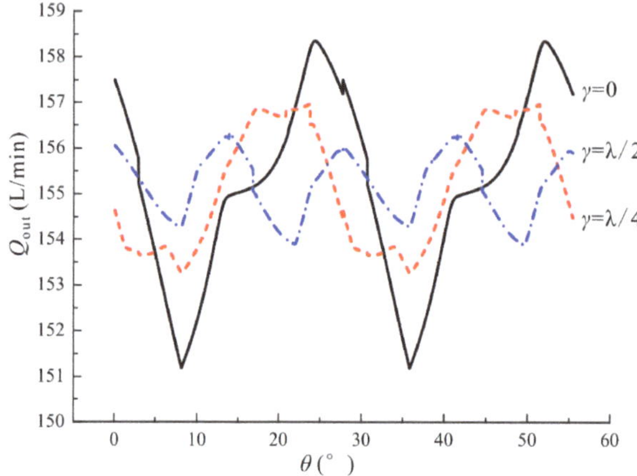

**Figure 14.** Outlet flow from the tandem pump ($\zeta = 1$) under 2000 rpm and 20 MPa regarding different index angles.

Figure 15 depicts the outlet flow from the tandem pump ($\gamma = \lambda/2$) under 2000 rpm and 20 MPa with respect to different displacement ratios (0, 0.5, 1), noting that the displacement ratio $\zeta$ was bounded between 0 and 1. Expectedly, the outlet flow exhibits a periodic behavior with the period of 27.7°. It could also be seen that under the same operating conditions, the pump yielded different mean outlet flowrate and different flow ripples as the displacement ratio varied. The outlet flow yielded between 75.25 and 78.33 L/min when $\zeta = 0$, between 115.02 and 116.88 L/min when $\zeta = 0.5$, and between 153.91 and 156.34 L/min when $\zeta = 1$.

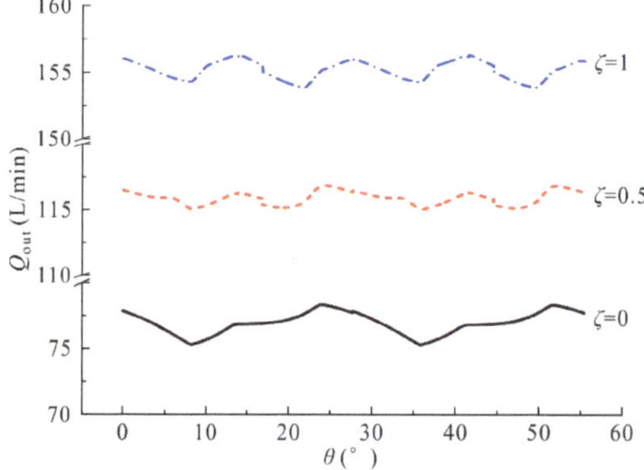

**Figure 15.** Outlet flow from the tandem pump ($\gamma = \lambda/2$) under 2000 rpm and 20 MPa regarding different displacement ratios.

### 4.2. Flow Ripple under Different Design Parameters

Figure 16 depicts the flow ripple ($\delta_q$) of the tandem pump ($\zeta = 1$) under 2000 rpm and 20 MPa regarding different index angles (from 0 to $\lambda$), noting that the flow ripple was defined by Equation (25), following the work addressed by Ivantysyn et al. [26].

$$\delta_q = \frac{Q_{max} - Q_{min}}{0.5 \cdot (Q_{max} + Q_{min})} \tag{25}$$

where $Q_{max}$ is the maximum flowrate; and $Q_{min}$ is the minimum flowrate.

Referring to Figure 16, it was clear and consistent that with respect to a certain index angle, the tandem pump's kinematic flow ($Q_{kin}$) yielded a greater ripple while the with-trapped flow ($Q_{wt}$) yielded a lower ripple than the actual flow ($Q_{out}$). Additionally, as the index angle increased (from 0 to $\lambda$), one observed a decrease followed by an increase in the flow ripples of the three types of flow, and the flow obtained the minimum flow ripple when $\gamma = \lambda/2$ (in red rectangles). Concerning the actual flow ($Q_{out}$), the maximum flow ripple was roughly 4.62% when $\gamma = 0$, and the minimum flow ripple was roughly 1.57% when $\gamma = \lambda/2$.

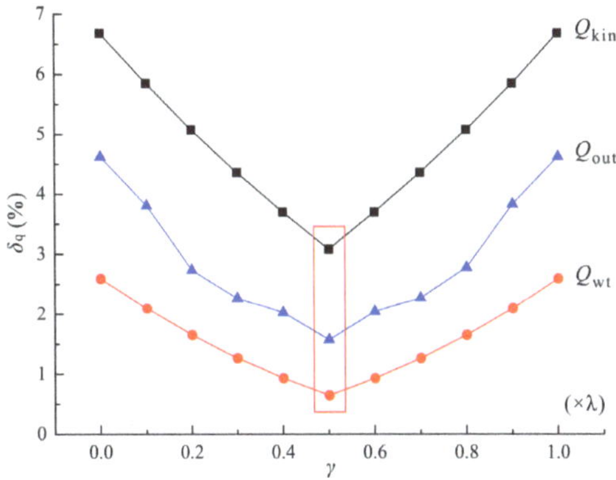


**Figure 16.** Flow ripple of the tandem pump ($\zeta = 1$) under 2000 rpm and 20 MPa regarding different index angles.

Figure 17 depicts the flow ripple of the tandem pump at 20 MPa regarding different displacement ratios (from 0.1 to 1) under the same mean outlet flowrate. The operating speed for the tandem pump with the displacement ratio $\zeta = 1$ was 2000 rpm, and the operating speeds for other displacement ratios were set as Equation (26) for the purpose of maintaining the same mean outlet flowrate.

$$n_{\zeta = \zeta_0} = n_{\zeta = 1} \cdot \frac{1 + 1}{1 + \zeta_0} \tag{26}$$

Referring to Figure 17, it can be seen that with respect to a certain index angle ($\gamma$), the flow ripple first exhibited a decrease, then followed by an increase as the displacement ratio ($\zeta$) increased, and a minimum value of the flow ripple could be achieved by properly selecting the displacement ratio. With respect to a certain displacement ratio ($\zeta$), the condition of $\gamma = \lambda/2$ yielded the minimum flow ripple when $\zeta$ was greater than 0.4. Hence, as observed, the condition of $\gamma = \lambda/2$ and $\zeta = 0.5$ yielded the minimum flow ripple under the same outlet pressure and mean outlet flowrate (in red squares), and the related flow ripple yielded 1.41%.

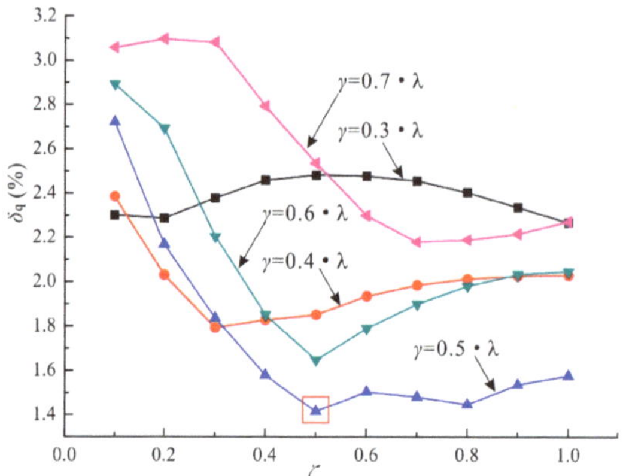

**Figure 17.** Flow ripple of the tandem pump at 20 MPa regarding different displacement ratios under the same mean outlet flowrate ($Q_{out}$).

### 4.3. Flow Ripple under Different Operating Conditions

Figure 18 depicts the outlet flow ripples of the tandem pump ($\gamma = \lambda/2$ and $\zeta = 0.5$) and the single pump under different operating conditions (500–3000 rpm, 0–30 MPa). It should be noted that the displacement of the tandem pump was 60 mL/rev given that the displacements of the two gear pairs were 40 mL/rev and 20 mL/rev, respectively; and the displacement of the single pump was 60 mL/rev for the purpose of maintaining the same mean outlet flowrate as the tandem pump when working under the same operating condition.

For these two types of pumps (tandem, single), it can be seen that with respect to a certain operating speed, the flow ripple increased as the outlet pressure increased. Furthermore, there was a clear and consistent trend that with respect to a certain outlet pressure, the flow ripple decreased as the operating speed increased. Moreover, the decrease of the flow ripple was noticed by applying the tandem pump with $\gamma = \lambda/2$ and $\zeta = 0.5$. For instance, when working under 500 rpm and 30 MPa, the flow ripple of the single pump was 8.52% and the tandem pump was 3.72%; when working under 3000 rpm and 30 MPa, the flow ripple of the single pump was 4.42% and the tandem pump was 1.89%.

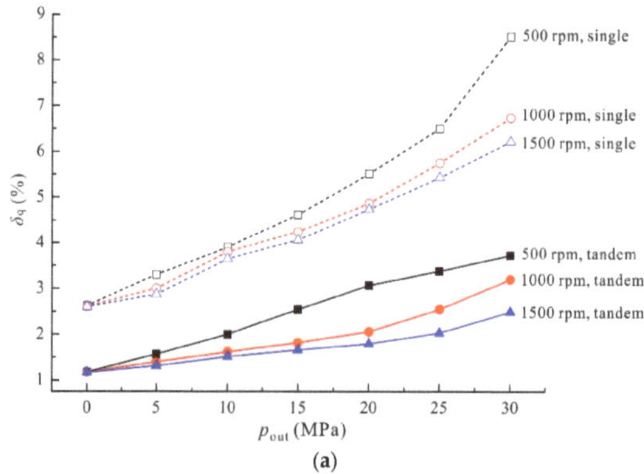

(a)

**Figure 18.** *Cont.*

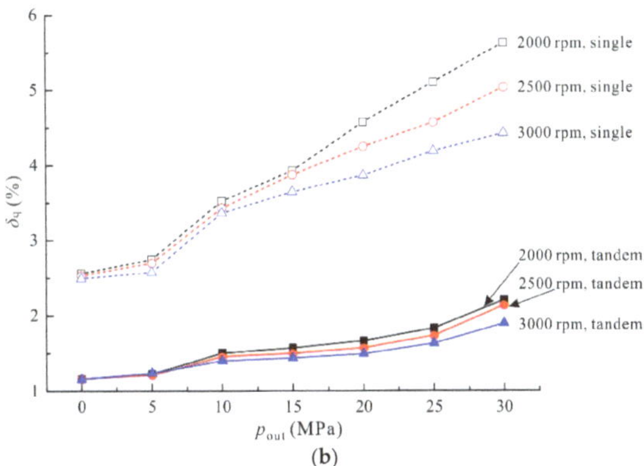



**Figure 18.** Flow ripples of the tandem pump and the single pump under different operating conditions: (**a**) 500–1500 rpm; and (**b**) 2000–3000 rpm.

## 5. Discussion and Conclusions

In this section, we discuss the outlet flow of the pump to identify the main causes of the flow ripple. Additionally, the influence of the design parameters is discussed for the purpose of selecting the proper design parameters to enable a decrease of the crescent pump's flow ripple.

### 5.1. Main Causes of the Flow Ripple

As observed in Figure 12, by applying Equation (25), the flow ripple of the kinematic flow ($Q_{kin}$) yielded 6.8%, which is believed to be caused by the kinematic relations between the meshing gears. As stated above, the trapped flow ($q_{trap}$) was forced into the discharge chamber through the relief groove on the floating plate, thus leading to an increase in the outlet flow. Consequently, the with-trapped outlet flowrate ($Q_{wt}$) yielded a greater mean outlet flowrate than that of $Q_{kin}$. The flow ripple of $Q_{wt}$ in Figure 12 yielded 2.7%, an approximately 60% decrease than that of $Q_{kin}$, indicating that the trapped flow plays an important role in decreasing the flow ripples. Regarding the internal leakage ($q_{leak}$) and the triangular flow ($q_{tri}$), it was clear that they led to a decrease in the outlet flow since they are the flow out of the discharge chamber. The flow ripple of $Q_{out}$ in Figure 12 yielded 4%, an approximate decrease of 41% than that of $Q_{kin}$, but a 60% increase than that of $Q_{wt}$, indicating that the internal leakage ($q_{leak}$) and the triangular flow ($q_{tri}$) led to an increase in the flow ripples.

Hence, it can be seen that the trapped flow led to a decrease in the flow ripple, while the triangular flow and the internal leakage of the pump led to an increase in the flow ripple; which was also consistent with the cases depicted in Figure 16.

### 5.2. Influence of Design Parameters

#### 5.2.1. Influence of the Index Angle

As observed in Figure 14, the index angle ($\gamma$) had little influence on the mean outlet flowrate; however, it had a great influence on the outlet flow ripples. Referring to Figure 13, it can be seen that the index angle enabled a phase difference between the flow from the two gear pairs. Noting that the flow of the tandem pump was obtained by summing the flow from the two gear pairs, it was workable to attenuate the flow ripple of the tandem pump by properly selecting the index angle, therefore allowing the peak value of the flowrate from one gear pair to coincide with the valley value of the flowrate from the other gear pair. This is believed to be the reason why the index angle of 13.85° ($\lambda/2$) led to the greatest decrease in the flow ripple. As shown in Figure 16, the flow ripple was reduced by 66% from the maximum value ($\gamma = 0$) by properly selecting the index angle ($\gamma = 13.85°$).

### 5.2.2. Influence of the Displacement Ratio

The displacement ratio ($\zeta$) led to different displacements of the pump, thus resulting in a different mean outlet flowrate in Figure 15 when working under the same operating condition. Under the circumstance of the same mean outlet flowrate, there was a value of the displacement ratio which enabled the greatest decrease in the flow ripple, and the value varied with respect to different index angles, as shown in Figure 17.

Referring to Figure 17, it can be seen that the index angle of $13.85°$ ($\lambda/2$) and the displacement ratio of 0.5 led to the greatest decrease in the outlet flow ripple, which was 10% lower than that of $\gamma = 13.85°$ and $\zeta = 1$ (flow ripple: 1.57%), and 70% lower than that of $\gamma = 0$ and $\zeta = 1$ (flow ripple: 4.62%).

### 5.3. Influence of Operating Conditions

With respect to a certain operating speed, the internal leakage of the pump increased as the outlet pressure increased, and this is believed to be the reason why the flow ripple increased (the internal leakage leads to an increase in the flow ripple as discussed above). With respect to a certain outlet pressure, the mean outlet flowrate increased as the operating speed increased, and this is believed to be the reason why the flow ripple decreased (the denominator in Equation (25) became greater while the numerator remained roughly the same).

Referring to Figure 18, the tandem pump led to a more than 45% decrease in the flow ripple than the single pump with respect to the depicted operating conditions (500–3000 rpm, 0–30 MPa). Hence, it can be seen that a tandem crescent pump with proper design parameters ($\gamma = 13.85°$ and $\zeta = 0.5$) can result in a significant decrease in the outlet flow ripple (more than 45%) than a single pump with the same displacement across a wide range of operating conditions (500–3000 rpm, 0–30 MPa).

**Acknowledgments:** The authors are grateful to the anonymous reviewers and the editor for their constructive comments. The authors would also like to thank Xiaoping Ouyang from Zhejiang University for the improvement of the paper. Additionally, the authors would like to acknowledge the support of the National Natural Science Foundation of China (Grant No. 51521064) and Science and Technology Innovation Team of Independent Design Projects of Zhejiang Province (Grant No. 2013TD01).

**Author Contributions:** Hua Zhou and Ruilong Du proposed the idea of the tandem crescent pump for the purpose of decreasing the outlet flow ripple; Ruilong Du and Anhuan Xie conceived and designed the experiments; Huayong Yang contributed to the test rig; Ruilong Du and Anhuan Xie performed the experiments and analyzed the data; Ruilong Du wrote the Matlab code, conducted the numerical simulations and wrote the paper.



### Notation

| | |
|---|---|
| $A_{\text{val, rel, tri}}$ | flow area of the throttle valve, the relief groove, and the triangular groove (m$^2$) |
| $b$ | width of the gear (m) |
| $b_{\text{B, L1, L2}}$ | parameters of the rectangular sealing area around the outlets (m) |
| $C_d$ | discharge coefficient (-) |
| $f$ | frequency (Hz) |
| $l$ | length of the sealing area in the ring-gear/case interface(m) |
| $l_r$ | length of the tooth tip (m) |
| $m$ | module of the gear (m) |
| $n$ | operating speed (rpm) |
| $O_{s,r}$ | centers of the gear shaft and the ring gear (-) |
| $p$ | pressure (Pa) |
| $p_{\text{in, out}}$ | inlet or outlet pressure (Pa) |
| $p_{\text{tran}}$ | transitional pressure (Pa) |
| $p_{\text{trap}}$ | trapped pressure (Pa) |
| $q_{\text{in, out}}$ | flowrate into and out of a chamber (m$^3$/s) |
| $q_l$ | lateral leakage of the lateral interface of the gears' lateral sides and the floating plates (m$^3$/s) |

| | |
|---|---|
| $q_{\text{leak}}$ | total internal leakage (m$^3$/s) |
| $q_{\text{val, rel, tri}}$ | flowrate through the throttle valve, the relief groove and the triangular groove (m$^3$/s) |
| $q_{\text{tri,s; tri,r}}$ | triangular flow for the gear shaft and the ring gear (m$^3$/s) |
| $q_{\text{r}}$ | radial leakage of the radial interface of the tooth tips and the crescent fillers (m$^3$/s) |
| $q_{\text{rc}}$ | leakage of the ring-gear/case interface (m$^3$/s) |
| $q_{\text{trap}}$ | trapped flow, flow out of the trapped chamber through the relief groove (m$^3$/s) |
| $Q_{\text{max, min}}$ | maximum and minimum flowrate (m$^3$/s) |
| $Q_{\text{kin}}$ | kinematic outlet flow (m$^3$/s) |
| $Q_{\text{wt}}$ | with-trapped flow by adding the kinematic flow and the trapped flow (m$^3$/s) |
| $Q_{\text{out}}$ | actual outlet flow by adding the kinematic flow, the trapped flow, the leakage and the triangular flow (m$^3$/s) |
| $r_{\text{a1, a2}}$ | addendum radii of the gear shaft and the ring gear (m) |
| $r_{\text{f1, f2}}$ | distances between the contact point and the centers of the gear shaft and ring gear (m) |
| $r_{\text{fp1, fp2}}$ | radii of the inner and the outer border of the floating plate (m) |
| $r_{\text{p1, p2}}$ | pitch radii of the gear shaft and the ring gear (m) |
| $t$ | time (s) |
| $V$ | volume (m$^3$) |
| $V_{\text{d}}$ | volume of the discharge chamber and the delivery line (m$^3$) |
| $V_{\text{kin}}$ | volume of the discharged fluid when the pump operates for a period of time $t$ from the kinematic aspect (m$^3$) |
| $V_{\text{trap}}$ | volume of the trapped chamber (m$^3$) |
| $V_{\text{TS}}$ | volume of the tooth space (m$^3$) |
| $z_{1,2}$ | tooth number of the gear shaft and the ring gear (-) |

*Greek symbols*

| | |
|---|---|
| $\alpha_0$ | pressure angle of the gear (°) |
| $\alpha$ | operating pressure angle of the gear pair (°) |
| $\beta$ | fluid bulk modulus (Pa) |
| $\gamma$ | index angle (°) |
| $\delta$ | radial clearance between the ring gear and the case (m) |
| $\delta_l$ | lateral clearance between the gears' lateral sides and the floating plates (m) |
| $\delta_q$ | flow ripple (-) |
| $\delta_r$ | radial clearance between the gears' tooth tips and the fillers (m) |
| $\Delta t$ | time interval (s) |
| $\zeta$ | displacement ratio (-) |
| $\eta$ | volumetric efficiency (-) |
| $\theta$ | rotation angle of the gear shaft (°) |
| $\lambda$ | circumferential angle of one tooth in the gear shaft ($\lambda = 360°/z_1$) |
| $\mu$ | fluid viscosity (Pa·s) |
| $\rho$ | fluid density (kg/m$^3$) |
| $\varphi$ | angle circumference the sealing area in the ring-gear/case interface (rad) |
| $\psi$ | sector angle of the annular sector for estimation of the lateral leakage (rad) |
| $\omega, \omega_r$ | angular velocity of the gear shaft and the ring gear (rad/s) |

## References


1. Zhou, H.; Song, W. Optimization of floating plate of water hydraulic internal gear pumps. In Proceedings of the 8th JFPS International Symposium on Fluid Power, Okinawa, Japan, 25–28 October 2011.
2. Shi, Y.; Wu, T.; Cai, M.; Yixuan, W.; Weiqing, X. Energy conversion characteristics of a hydropneumatic transformer in a sustainable-energy vehicle. *Appl. Energy* **2016**, *171*, 77–85. [CrossRef]
3. Shi, Y.; Wang, Y.; Cai, M.; Zhang, B.; Zhu, J. Study on the aviation oxygen supply system based on a mechanical ventilation model. *Chin. J. Aeronaut.* **2017**. [CrossRef]
4. Yang, H.; Pan, M. Engineering research in fluid power: A review. *J. Zhejiang Univ. Sci. A* **2015**, *16*, 427–442. [CrossRef]
5. Mimmi, G.; Pennacchi, P. Involute gear pumps versus lobe pumps: A comparison. *J. Mech. Des.* **1997**, *119*, 458–465. [CrossRef]

6.  Bonandrini, G.; Mimmi, G.; Rottenbacher, C. Theoretical analysis of internal epitrochoidal and hypotrochoidal machines. *Proc. Inst. Mech. Eng. C J. Mech. Eng. Sci.* **2009**, *223*, 1469–1480. [CrossRef]

7.  Gamez-Montero, P.J.; Codina, E. Flow characteristics of a trochoidal-gear pump using bond graphs and experimental measurement. Part 1. *Proc. Inst. Mech. Eng. I J. Syst. Control Eng.* **2007**, *221*, 331–346. [CrossRef]

8.  Gamez-Montero, P.J.; Codina, E. Flow characteristics of a trochoidal-gear pump using bond graphs and experimental measurement. Part 2. *Proc. Inst. Mech. Eng. I J. Syst. Control Eng.* **2007**, *221*, 347–363. [CrossRef]

9.  Gamez-Montero, P.J.; Castilla, R.; del Campo, D.; Ertürk, N.; Raush, G.; Codina, E. Influence of the interteeth clearances on the flow ripple in a gerotor pump for engine lubrication. *Proc. Inst. Mech. Eng. D J. Automob. Eng.* **2012**, *226*, 930–942. [CrossRef]

10. Hsieh, C.-F. Influence of gerotor performance in varied geometrical design parameters. *J. Mech. Des.* **2009**, *131*, 121008. [CrossRef]

11. Hsieh, C.F. Flow characteristics of gerotor pumps with novel variable clearance designs. *J. Fluids Eng.* **2015**, *137*, 041107. [CrossRef]

12. Manco, S.; Nervegna, N.; Rundo, M. A contribution to the design of hydraulic lube pumps. *Int. J. Fluid Power* **2002**, *3*, 21–32. [CrossRef]

13. Schweiger, W.; Schoefmann, W.; Vacca, A. Gerotor pumps for automotive drivetrain applications: A multi domain simulation approach. *SAE Int. J. Passeng. Cars-Mech. Syst.* **2011**, *4*, 1358–1376. [CrossRef]

14. Pellegri, M.; Vacca, A.; Frosina, E.; Buono, D.; Senatore, A. Numerical analysis and experimental validation of gerotor pumps: A comparison between a lumped parameter and a computational fluid dynamics-based approach. *Proc. Inst. Mech. Eng. C J. Mech. Eng. Sci.* **2016**. [CrossRef]

15. Pellegri, M.; Vacca, A. Numerical simulation of gerotor pumps considering rotor micro-motions. *Meccanica* **2017**, *52*, 1851–1870. [CrossRef]

16. Ichikawa, T. Characteristics of internal gear pump. *Bull. JSME* **1959**, *2*, 35–39. [CrossRef]

17. Zhou, H.; Song, W. Theoretical flowrate characteristics of the conjugated involute internal gear pump. *Proc. Inst. Mech. Eng. C J. Mech. Eng. Sci.* **2013**, *227*, 730–743. [CrossRef]

18. Song, W.; Chen, Y.; Zhou, H. Investigation of fluid delivery and trapped volume performances of Truninger gear pump by a discretization approach. *Adv. Mech. Eng.* **2016**, *8*, 1–15. [CrossRef]

19. Rundo, M. Theoretical flow rate in crescent pumps. *Simul. Model. Pract. Theory* **2017**, *71*, 1–14. [CrossRef]

20. Rundo, M.; Corvaglia, A. Lumped parameters model of a crescent pump. *Energies* **2016**, *9*, 876. [CrossRef]

21. Manring, N.D.; Mehta, V.S.; Raab, F.J.; Graf, K.J. The shaft torque of a tandem axial-piston pump. *J. Dyn. Syst. Meas. Control* **2007**, *129*, 367–371. [CrossRef]

22. Mehta, V.S. Torque Ripple Attenuation for an Axial Piston Swash Plate Type Hydrostatic Pump: Noise Considerations. Ph.D. Thesis, University of Missouri, Columbia, SC, USA, 2006.

23. Xu, B.; Ye, S.; Zhang, J. Effects of index angle on flow ripple of a tandem axial piston pump. *J. Zhejiang Univ. Sci. A* **2015**, *16*, 404–417. [CrossRef]

24. Battarra, M.; Mucchi, E.; Dalpiaz, G. A model for the estimation of pressure ripple in tandem gear pumps. In Proceedings of the ASME 2015 International Design Engineering Technical Conference and Computers and Information in Engineering Conference, Boston, MA, USA, 2–5 August 2015.

25. Xu, B.; Zhang, J.; Yang, H. Simulation research on distribution method of axial piston pump utilizing pressure equalization mechanism. *Proc. Inst. Mech. Eng. C J. Mech. Eng. Sci.* **2013**, *227*, 459–469. [CrossRef]

26. Ivantysyn, J.; Ivantysynova, M. *Hydrostatic Pumps and Motors*, 1st ed.; Academia Books International: New Delhi, India, 2001; ISBN 81-85522-16-2.

27. Ma, J.; Fang, Y.; Xu, B.; Yang, H. Optimization of cross angle based on the pumping dynamics model. *J. Zhejiang Univ. Sci. A* **2010**, *11*, 181–190. [CrossRef]

28. Hamrock, B.J.; Schmid, S.R.; Jacobson, B.O. *Fundamentals of Fluid Film Lubrication*, 2nd ed.; Marcel Dekker, Inc.: New York, NY, USA, 2004; ISBN 0-8247-5371-2.

*Article*

# A New Method for Analysing the Pressure Response Delay in a Pneumatic Brake System Caused by the Influence of Transmission Pipes

Fan Yang [1], Gangyan Li [1,*], Jian Hua [1], Xingli Li [1] and Toshiharu Kagawa [2]

[1]  School of Mechanical and Electronic Engineering, Wuhan University of Technology, Wuhan 430070, China; yang_fan@whut.edu.cn (F.Y.); huajian@whut.edu.cn (J.H.); xinglili@whut.edu.cn (X.L.)
[2]  Laboratory for FIRST, Tokyo Institute of Technology, Yokohama 226-8503, Japan; kagawa@pi.titech.ac.jp
*  Correspondence: gangyanli@whut.edu.cn

Received: 29 July 2017; Accepted: 9 September 2017; Published: 13 September 2017

**Abstract:** This study aims to propose an analysis method for resolving the pressure response of a pneumatic brake circuit considering the effect of a transmission pipe. Pneumatic brake systems (PBS) are widely used in commercial vehicles. The pressure response characteristic of the PBS is the key factor affecting braking performance. By using the thermodynamics of a variable-quality system, the pressure response model of the brake chamber is established, which includes the dynamic model of the pipe considering the unsteady friction and heat transfer. The partial-differential control equations of pipe are solved by introducing the constrained interpolation profile (CIP) method, and a virtual chamber model is proposed to set the boundary condition so as to solve the pressure response in the brake chamber simultaneously. Thus, the regularity of the brake pressure response is obtained by considering the influence of the pipe. Lastly, the model is verified experimentally. The present study indicates that the main factors that affect the pressure response delay are the pipe length and the combination forms of the sonic conductances of the orifices inlet and outlet. Furthermore, it helps to verify that the CIP method is an effective way of solving the pressure response of a brake circuit because of its high accuracy. The present study serves as a foundation for the design and analysis of a PBS.

**Keywords:** pneumatic brake; pressure response; pipeline; brake chamber; CIP method

## 1. Introduction

The brake system is the key component guaranteeing the safety of vehicles, and its performance affects driving safety, braking stability and passenger comfort. At present, pneumatic brake systems (PBSs) are widely used in buses and heavy trucks for their cleanliness, simple structure and high reliability. However, many nonlinear factors, including the compressibility of air and the unsteady characteristics of braking, make it difficult to calculate the characteristics of PBSs accurately. Hence, their design is based mostly on experience or experiments [1], which harms their efficiency and increases their maintenance cost. Therefore, researchers are always looking for ways to enhance the performance of PBSs [2,3]. An electro-pneumatic proportional valve was proposed to replace the antilock brake system so as to control the brake pressure more accurately [4], but it has been rarely used in vehicles because of its high cost. Europe and America have proposed the mandatory installation of emergency brake systems on vehicles for active safety [5], but their spread in Asian countries has been slow because of cost and problems with the technology. Consequently, if the performance of PBSs could be enhanced by optimising their control algorithm or improving the accuracy of their analysis model, this would broaden their application prospects and avoid spiralling hardware costs.

When the driver treads on the brake pedal to decelerate or stop a vehicle, there is time consumption that includes the response time of the driver, the transmission time of force and signals, and the action time of the actuator. All of these factors cause brake hysteresis more often when compared to the brake expectation, and the hysteresis is the response delay of the brake system. The response time delay consists of the delay of pedal valve, pneumatic brake circuit, actuator, and control networks. Wang et al. [6] had proved that a pipe in a pneumatic brake system account for 30% of the overall time delay by experiment but without theoretical analysis. Therefore, the time delay caused by a pipe is an important part of the delay of a pneumatic brake circuit. In this paper, we mainly focused on the time delay caused by the influence of a pipe but did not study the other factors. Because of the time delay of the pneumatic brake system, the time delay will cause the longitudinal or lateral braking distance deviation [7]. Especially, for the emergency braking condition, a tiny deviation of braking distance may lead to vehicle crash and the death of a human. Table 1 gives the braking distance with different time delays referred to in [8]. It is obvious that the brake distance increases considerably when the time delay increases. Thus, ISO7346 and GB/T7258 clearly define the range of time delay to guarantee safe driving. Therefore, decreasing the response time delay or identifying the time delay and controlling it with proper control methods are effective ways to decrease the brake distance and increase driving safety.

**Table 1.** Influence degree of time delay on braking distance.

| Time Delay (s) | Increasing of Brake Distance (m) | Influence Degree (%) | Remarks |
|---|---|---|---|
| 0.9 | 15.0 | 40.9 | |
| 0.7 | 11.7 | 31.8 | The whole brake distance is assumed to satisfy the force law, and it is equal to 36.7 m |
| 0.5 | 8.3 | 22.7 | |
| 0.3 | 5.0 | 13.6 | |

At present, the low cost and high efficiency of computer simulation means that it is the main way of studying the performance of brake systems. For example, He et al. [9–11] used AMESim, SIMULINK and MWorks to establish a model of a brake system and study its pressure response characteristics. However, the algorithms of these commercial software packages are closed, which makes it difficult to use them to develop brake systems with different requirements. Meanwhile, simulations using such software are based on time sampling, which makes it difficult to tackle problems that are both temporally and spatially dependent. Thus, the distributed-parameter components, such as the pipe, are equivalent by restriction or capacity-restriction (it sacrifices the accuracy artificially). Furthermore, adiabatic and isothermal models [9,10] are mostly used for modelling the brake chamber, which simplifies the analysis but also reduces its accuracy. The brake chamber is a particular air cylinder, and Tokashiki et al. [12] have shown that the pipe affects the dynamic characteristics of this cylinder via the temperature changes caused by the pipe. Qin [7] demonstrated experimentally the existence of a pipe led time delay in a PBS, and proposed compensating for it in the design of the control system. However, that approach is not a universal method because it lacks a theoretical basis. Therefore, it is important to consider the influence of the pipe so as to calculate the brake force response accurately.

To calculate the pressure response in the pipe, it is necessary to solve the partial-differential equations (PDEs) represented by the continuity, kinematic and energy equations. Zielke [13] first solved these equations by the method of characteristics considering the friction. However, he neglected advection and temperature change, so his approach is suitable only for low-pressure, low-speed problems. In addition, Zielke's method has to solve a convolution integral repeatedly, which is inefficient. Kitagawa et al. [14] then proposed an optimised method in which the convolution integral is replaced by an exponential function, which improves the efficiency markedly. Hashimoto et al. [15] proposed a grid differential scheme to calculate the large pressure surge in the pipe, but that approach is only first-order accurate and the time step must be sufficiently small to avoid divergence.

Wei et al. [16,17] also used the method of characteristics to solve the pressure response in the brake circuit of a train. However, it is complex to solve the characteristics curve and the time step must be strictly equal to the spatial step divided by the speed, which are restrictions that make such an approach difficult to apply universally. The key difficulty in calculating the pressure response in the pipe is to solve the PDE control equations accurately. Yabe et al. [18] and Takewaki et al. [19] proposed the constrained interpolation profile (CIP) method with third-order accuracy, which is good for solving hyperbolic PDEs. The CIP method uses a cubic interpolation to calculate the values between grid nodes, and has obtained good results in many applications. For example, it has been used for weather forecasting and for simulating dam failures and hull shock waves [20,21], all cases in which natural phenomena were predicted accurately.

Through the analysis above, the deficiencies in previous studies of PBSs can be summarised as follows. Either the influence of the pipe was neglected or the pipe was replaced by a simple restriction, both of which approaches reduce the accuracy. Also, the conventional methods used for solving the PDEs were either of low accuracy or were inefficient. Furthermore, heat exchange between the system and its surroundings was neglected, which also reduces the accuracy. In contrast, in the present study, the accurate CIP method is introduced and the heat exchange is considered comprehensively. We couple the models of the pipe and the brake chamber and solve them simultaneously, bringing regularity to the influence of the pipe parameters on the response of the brake pressure. The present study provides a good reference for the design and analysis of a PBS.

## 2. Modelling the Brake Pressure Response

A typical pneumatic brake circuit is shown in Figure 1. It consists of a compressor, an air filter, a compressed-air reservoir, a pedal valve, pipes and four brake chambers, and so on. This pneumatic brake circuit consists of four subcircuits, which are front brake circuit, rear brake control circuit, rear brake circuit, and parking brake circuit. When the driver treads on the pedal, the pedal valve opens and the compressed air flows into the front and rear brake chambers through the brake control valve and pipe. The rear brake chamber is relatively farther from the air tank, and then a relay valve is used to decrease the time delay. In these four subcircuits, the similarity is that all of the subcircuits use long pipes to transmit the air. Moreover, the time delay of a pipe is the key point that should be first solved. The operation pressure is generally 0.6–0.8 MPa for a pneumatic brake system, and the diameter of a transmission pipe is 6–16 mm. When analysing the brake circuit, we can study the subcircuit one by one. In this study, we mainly focused on the calculation method of the time delay caused by a transmission pipe; thus, we will simplify the system into one circuit to describe the modelling method clearly.

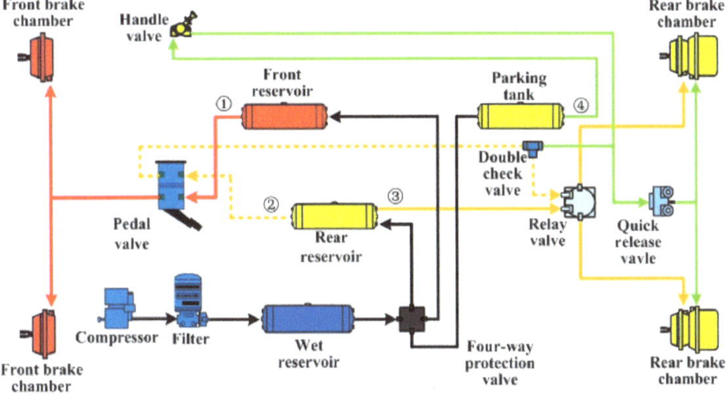


**Figure 1.** Schematic of the pneumatic brake circuit of a bus.

Although the brake control valve has a complex internal structure, the air flows through it quickly and its volume is small, so it can be thought of as lumped parameter components. The main effects of the brake control valve on the braking process are its flow-rate characteristics and the time delay of the mechanism. Therefore, when modelling the brake system, the brake control valve can be dealt with as an orifice [22]. For a pneumatic transmission circuit, the flow-rate characteristics of the circuit depend mainly on the smallest component and its upstream components; the downstream components have little effect [23]. In this paper, we mainly focused on the modelling and solving method of the pipe. Conversely, compared with the pipe, the sonic conductance of the control valve is smaller. Therefore, the valve can be considered as an orifice with a constant effective area.

Thus, the simple circuit shown in Figure 2 is adopted as the research object in the present paper, which is composed of an air source, two orifices, a pipe, and a brake chamber. In the actual application, orifice 1 depends on the control valve and orifice 2 is the inlet orifice of the brake chamber. According to ISO6358 [24], the flow-rate characteristics of the components can be measured by the isothermal chamber (ITC) discharge method, and the time delay of the brake control valve can be obtained according to ISO19973 [25]. The resultant flow-rate characteristics of the components in a circuit can be calculated by referring to [23,26].

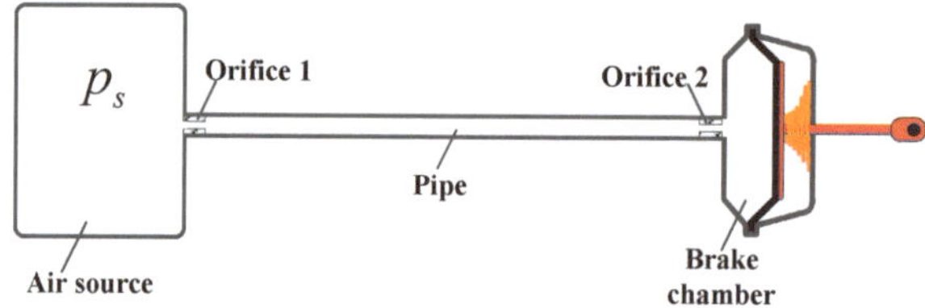


**Figure 2.** Simplified pneumatic brake circuit.

The following assumptions are used to facilitate the analysis model when modelling the PBS [27,28]:

(a)  The working fluid (air) of the system follows all ideal gas laws;
(b)  There is no leakage from the fittings or the brake chamber;
(c)  The airflow in the pipe is a one-dimensional flow, and the parameters in each section are uniform.

### 2.1. Flow-Rate Characteristics of the Orifice

According to ISO6358, the sonic conductance $C$ and the critical pressure ratio $b$ are used as the main parameters to describe the flow-rate characteristics [24]. Thus, if we know the flow-rate characteristics of an orifice, the flow-rate output from the tank or the input to the brake chamber can be obtained according to the definition of the flow-rate characteristics. However, we will use numerical computation in the following discussion, so we are introducing a corrected formula proposed by Kassa [29] so as to avoid numerical divergence if the step size becomes too large. Hence, the mass flow rate of air through an orifice can be given as

$$G = \begin{cases} k_1 p_1 \left(1 - \frac{p_2}{p_1}\right)\sqrt{\frac{\theta_0}{\theta_1}} & 0.995 \le \frac{p_2}{p_1} \le 1 \\ C p \rho_0 \sqrt{\frac{\theta_0}{\theta_1}} \left(1 - \left(\frac{\frac{p_2}{p_1} - b}{1 - b}\right)^2\right)^{0.5} & b < \frac{p_2}{p_1} < 0.995 \\ C p_1 \rho_0 \sqrt{\frac{\theta_0}{\theta_1}} & \frac{p_2}{p_1} \le b \end{cases} \tag{1}$$

where $k_1$ is the linear gain obtained by

$$k_1 = \frac{1}{1 - 0.995} \cdot C \cdot \rho_0 \sqrt{1 - \left(\frac{0.995 - b}{1 - b}\right)^2} \tag{2}$$

*2.2. Modelling the Pipe*

In a PBS, the pipe is the medium for air transmission and is the key component that causes time delays in the system. The length of the vehicle body could be up to 20 m for a large bus or a truck, so it is necessary to consider the influence of the pipe when it transmits pressure. The pipe is sufficiently long for both the pressure and temperature to be spatially and temporally dependent, so we deal with the pipe as a distributed-parameter component to enhance the analysis accuracy. Figure 3 shows the analysis model used for the pipe, in which each element in the dashed line is the control volume, and it is also the grid we will use in the next section.

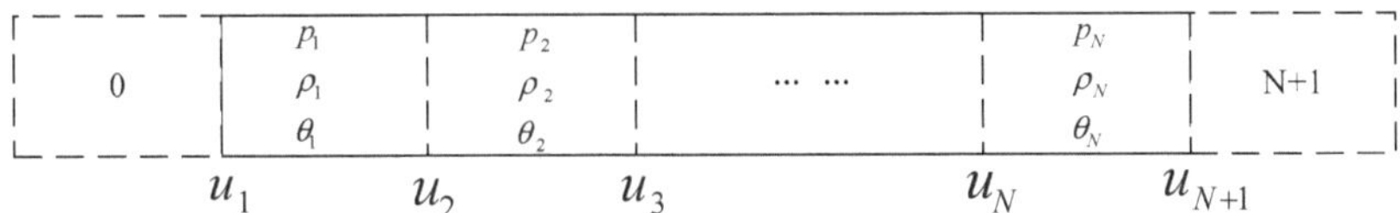

**Figure 3.** Analysis model of pipe.

The diameters of the pipes commonly used in pneumatic brake circuits are in the range 6–16 mm, which is very small compared with the pipe length. Hence, the flow in the pipe can be thought of as being a one-dimensional flow. According to gas dynamics, the air in the control volume satisfies the conservation of mass, momentum, and energy [30]. So, the control volume can be described by the control equations as follows:

$$\frac{\partial \rho}{\partial t} + \frac{\partial(\rho u)}{\partial x} = 0 \tag{3}$$

$$\frac{\partial u}{\partial t} + u\frac{\partial u}{\partial x} = -\frac{1}{\rho}\frac{\partial p}{\partial x} - \frac{\lambda}{2D}u^2 \tag{4}$$

$$\frac{\partial}{\partial t}\left[\rho A\left(e + \frac{u^2}{2}\right)\right] + \frac{\partial}{\partial x}\left[\rho u A\left(e + \frac{u^2}{2} + \frac{p}{\rho}\right)\right] = \rho A q \tag{5}$$

To enhance the model accuracy, in Equations (3)–(5), we consider the friction loss and heat exchange, respectively, in terms of a friction loss coefficient $\lambda$ and the heat convection $q$ between the control volume and its surroundings. Coefficient $\lambda$ is a function of the Reynolds number according to the flow regime and is expressed as follows:

$$\lambda = \begin{cases} 64/Re & Re < 2.5 \times 10^3 \\ 0.3164Re^{-0.25} & 2.5 \times 10^3 \leq Re < 1.0 \times 10^5 \\ 0.0032 + 0.221Re^{-0.237} & Re \geq 1.0 \times 10^5 \end{cases} \tag{6}$$

$Re$ is the Reynolds number given by $Re = uD/v$, where $u$ is the gas velocity, $D$ is the diameter of pipe, and $v$ is kinematic viscosity.

And then the heat exchange $q$ is given by

$$q = \frac{4h(\theta_a - \theta)}{D\rho} \tag{7}$$

where $h = 0.046(\kappa/D)Re^{0.8}Pr^{0.4}$ and $Pr$ is the Prandtl number, which is equal to 0.72 for air.

By using Equations (3)–(7), we comprehensively consider the factors related to nonlinear effects, whereupon the pressure, density, and temperature in each control volume can be solved accurately. Furthermore, we can obtain the regularity for these physical quantities that change temporally and spatially. However, it is difficult to solve these hyperbolic PDEs analytically because they include many nonlinear factors, such as heat exchange and friction. The methods used most commonly to solve PDEs are up-wind difference schemes, central difference schemes, and the Lax–Wendroff method [31]. However, these all deliver only either first-order or second-order accuracy, and their stability is very poor for a large pressure surge. Thus, to enhance the accuracy and stability, we introduce the CIP method with its third-order accuracy; the details will be explained later.

### 2.3. Modelling the Brake Chamber

Figure 4a shows the structure of the brake chamber, which consists of a diaphragm, a push plate, a push rod, and a spring, thereby forming two chambers. When braking is applied, the rod-less chamber is charged, the air expands to push the piston and push rod, and then the rod chamber is compressed. However, there are four holes in the rod chamber with diameters of 6–8 mm so that the air can be discharged to the atmosphere as quickly as possible, and this means that there is no back pressure in the rod chamber. Thus, we need only consider the rod-less chamber when modelling the brake chamber.

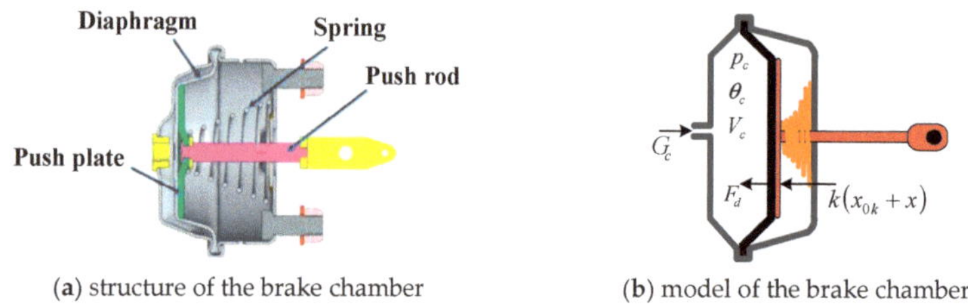


(**a**) structure of the brake chamber          (**b**) model of the brake chamber

**Figure 4.** Structural diagram of brake chamber.

The main parameters are shown in Figure 4b. By differentiating the gas state equation $pV = mR\theta$ with respect to time, the pressure response in the rod-less chamber is obtained as

$$\frac{dp_c}{dt} = \frac{1}{V_c}\left[G_cR\theta_c - p_c\frac{dV_c}{dt} + \frac{p_cV_c}{\theta_c}\frac{d\theta_c}{dt}\right] \tag{8}$$

To improve the accuracy of the model, we consider here the heat exchange in the brake chamber. Thus, according to the first law of thermodynamics, the energy flow in or out of the chamber always satisfies

$$dU = dH + dW + Q \tag{9}$$

where $U$ is the internal energy, $H$ is the enthalpy that flows in the chamber, $W$ is the work done by the rod, and $Q$ is the heat exchanged between the inner chamber and the surroundings through the wall, which can be expressed as follows according to Newton's law of cooling:

$$Q = hS_h(\theta_a - \theta_c) \tag{10}$$

The heat-exchange coefficient $h$ can be measured using the stop method [32]. However, it changes during the charging process, so we adopt an average value here of $h = 20$. In addition, the model is adiabatic for $h = 0$, tending to isothermal has $h \to \infty$.

By differentiating Equation (9) with respect to time and combining it with the equations $U = c_v\theta$, $H = c_p\theta$ and the Mayer equation $c_p - c_v = R$, we obtain the temperature response as

$$\frac{d\theta_c}{dt} = \frac{1}{m_cc_v}\left[c_vG_c(\theta_a - \theta_c) + R\theta_aG_c - p_c\frac{dV_c}{dt} + hS_{hc}(\theta_a - \theta_c)\right] \tag{11}$$

The volume $V_c$ of the diaphragm changes as that of a cone rather than a cylinder, and is given approximately by

$$V_c = A\left(x_0 + \frac{\pi}{3}x\right) \tag{12}$$

The force on the piston is shown diagrammatically in Figure 4b, from which the kinematic equation of the push rod can be obtained as follows according to Newton's second law of motion:

$$\begin{cases} a = (p_c - p_a)A_p - k(x_{0k} + x) - F_d \\ \frac{du}{dt} = \frac{1}{M}\left[(p_c - p_a)A_p - k(x_{0k} + x) - F_d\right] \\ (a > 0 \cap x = s) \cup (0 < x < s) \\ \frac{du}{dt} = 0 \quad (a < 0 \cap x = 0) \cup x = s \end{cases} \tag{13}$$

where $A_p$ is the effective loaded area, which changes during the stroke of the piston [33] and is expressed as

$$A_p = \begin{cases} A & 0 \le x \le \frac{2}{3}s \\ \frac{7}{5}A - \frac{3A}{5s}x & \frac{2}{3}s < x \le s \end{cases} \tag{14}$$

In Equation (13), $F_d$ is the deforming force on the diaphragm, which also changes during the stroke and is expressed as

$$F_d = \begin{cases} F_{d0} + \frac{4F_{d0}}{s}x & 0 \le x \le \frac{1}{4}s \\ 2F_{d0} & \frac{1}{4}s \le x \le \frac{3}{4}s \\ 0 & \frac{3}{4}s \le x \le s \end{cases} \tag{15}$$

Through the three models established above, we obtain an analytical model for the brake pressure response. The method for obtaining solutions to this model is discussed in detail in Section 3.

## 3. Simulations and Experiments on Brake Pressure Response

### 3.1. Solving the Pipe Control Equations

#### 3.1.1. CIP Method

First proposed by Yabe [18,19] in the 1980s, the CIP method is an effective way to solve hyperbolic PDEs. It uses the function value and its derivative on a spatial grid, and then a cubic interpolation is constructed so that the values between nodes can be solved inversely. The CIP method is an explicit scheme of numerical computation with third-order accuracy [20]. Equation (16) is a typical hyperbolic PDE:

$$\frac{\partial f}{\partial t} + u\frac{\partial f}{\partial x} = 0 \tag{16}$$

If the speed $u$ is positive and constant, the analytical solution is $f(t, x) = f(t - \Delta t, x - u\Delta t)$. This means that the value of a node at the next time step can be obtained from the current one; however, the location is not on the node but is shifted by an amount $u\Delta t$ from it. As we know, $(x - u\Delta t)$ is located between two nodes, so we need to solve the value at $(x - u\Delta t)$ by interpolation. The highly accurate interpolation expression proposed by Yabe is given as

$$f(x) = a_i(x - x_{i-1})^3 + b_i(x - x_{i-1})^2 + f'_{i-1}(x - x_{i-1}) + f_{i-1} \tag{17}$$

The coefficients in Equation (17) can be obtained using Equations (18)–(21):

$$a_i = \frac{\left(f'^*_i + f'^*_{iup}\right)}{\Delta^2} + \frac{2\left(f^*_i - f^*_{iup}\right)}{\Delta^3} \tag{18}$$

$$b_i = \frac{3\left(f^*_{iup} - f^*_i\right)}{\Delta^2} - \frac{\left(2f'^*_i + f'^*_{iup}\right)}{\Delta} \tag{19}$$

when $u_i < 0$, $\Delta = -\Delta x$, $iup = i - 1$; and if $u_i > 0$, $\Delta = \Delta x$, $iup = i + 1$.

$$f_i^{n+1} = a_i \xi^3 + b_i \xi^2 + f_i'^* \xi + f_i^* \tag{20}$$

$$f_i'^{n+1} = 3a_i \xi^2 + 2b_i \xi + f_i'^* \tag{21}$$

where $\xi = -u_i \Delta t$.

To verify the accuracy and suitablity of the CIP method, a rectangular wave, as shown in Figure 5a, was used to assess the advantages of the CIP method. The control equation is expressed as Equation (16). Moreover, the velocity of propagation is $u = 1$, the spacial step is $\Delta x = 1$, the time step is $\Delta t = 0.1$, and the results after 400 steps are shown in Figure 5b–d that includes the CIP method, the up-wind difference method, and the Lax–Wendroff method. Clearly, the CIP method is best suitable for a rectangular wave with large pressure difference. The up-wind method with first-order accuracy is the worst, and the Lax–Wendroff method with second-order accuracy can relativly express the wave-type well, but it has oscillation and divergence in the process. The pressure wave in the pneumatic brake system is large when the brake control valve opens or closes. Thus, it is better to introduce the CIP method for its stability and high accuracy for solving the pressure response with large pressure fluctuation.

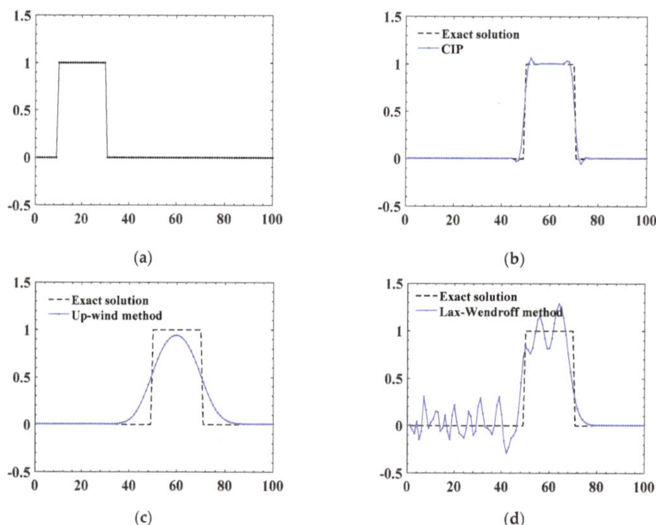

**Figure 5.** Comparison of different difference schemes. (**a**) Initial value; (**b**) CIP method; (**c**) Up-wind method; (**d**) Lax–Wendroff method

### 3.1.2. Arrangement of the Control Equations

We rearranged Equations (3)–(5) with the CIP method, we rearrange them into advection and non-advection terms. Substituting the gas state equation $e = p/[\rho(\kappa - 1)]$ into Equation (5), the new control equations can be written as

$$\frac{\partial \rho}{\partial t} + u \frac{\partial \rho}{\partial x} = -\rho \frac{\partial u}{\partial x} \tag{22}$$

$$\frac{\partial u}{\partial t} + u \frac{\partial u}{\partial x} = -\frac{1}{\rho} \frac{\partial p}{\partial x} - \frac{\lambda}{2D} u^2 \tag{23}$$

$$\frac{\partial p}{\partial t} + u \frac{\partial p}{\partial x} = -\kappa p \frac{\partial u}{\partial x} + \rho(\kappa - 1)\left(q + u \frac{\lambda}{2D} u^2\right) \tag{24}$$

where the left-hand sides hold the advection terms and the right-hand sides hold the non-advection terms. We then rewrite Equations (22)–(24) in vector form as

$$\frac{\partial \vec{f}}{\partial t} + u\frac{\partial \vec{f}}{\partial x} = \vec{G} \tag{25}$$

where

$$\vec{f} = \begin{pmatrix} \rho \\ u \\ P \end{pmatrix}, \vec{G} = \begin{pmatrix} -\rho\frac{\partial u}{\partial x} \\ -\frac{1}{\rho}\frac{\partial P}{\partial x} - \frac{\lambda}{2D}u^2 \\ -\kappa P\frac{\partial u}{\partial x} + \rho(\kappa-1)\left(q + u\frac{\lambda}{2D}u^2\right) \end{pmatrix}$$

### 3.1.3. Solving the Control Equations

According to the principle of the CIP method, we differentiate Equation (25) with respect to the spatial variable $x$, whereupon the derivative form of the equation is obtained as

$$\frac{\partial \vec{f'}}{\partial t} + u\frac{\partial \vec{f'}}{\partial x} = \vec{G'} - \vec{f'}\frac{\partial u}{\partial x} \tag{26}$$

The CIP method is used to solve hyperbolic PDEs of the form shown in Equation (4). Hence, we should separate the equation into advection and non-advection terms.

Figure 6 describes the detailed solving principle, in which $t(n)$ indicates a parameter at the current time and $t(n+1)$ indicates the same parameter but at the next time step. The real process involves a physical quantity being transmitted from $t(n)$ to $t(n+1)$ in one time step; however, we can deal with it in two steps. The first step is non-advection transmission, which transmits from $t(n)$ to $t^*$; the second step is advection transmission, which transmits from $t^*$ to $t(n+1)$. The value at $x_i - u_i\Delta t$ can be obtained by using Equation (17).

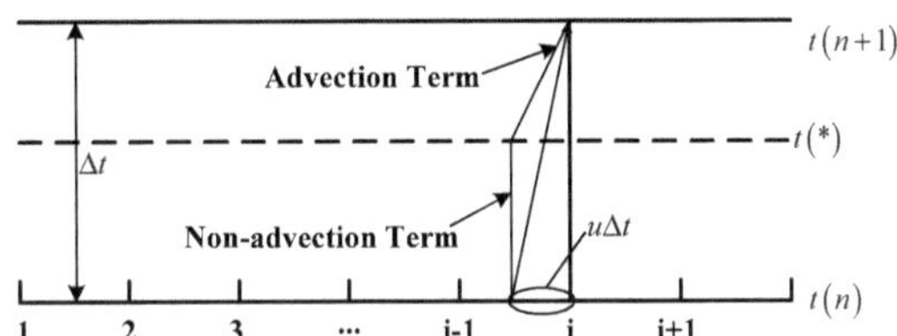


**Figure 6.** Principle of CIP method.

In Figure 6, the non-advection term is

$$\frac{\partial \vec{f}}{\partial t} = \vec{G} \tag{27}$$

$$\frac{\partial \vec{f'}}{\partial t} = \vec{G'} - \vec{f'}\frac{\partial u}{\partial x} \tag{28}$$

The values of $f_i^*$ and $f'^*$ can be solved directly with an up-wind method as

$$f_i^* = f_i^n + \Delta t G_i^n \tag{29}$$

$$f_i'^* = f_i'^n + \frac{\left(f_{i+1}^* - f_{i+1}^n\right) - \left(f_{i-1}^* - f_{i-1}^n\right)}{2\Delta x} - f_i'^n\frac{u_{i+1}^n - u_{i-1}^n}{2\Delta x}\Delta t \tag{30}$$

where

$$G_i^n = \begin{cases} -\rho_i^n \frac{u_i^n - u_{i-1}^n}{\Delta x} \\ -\frac{1}{\rho_i^n} \frac{p_{i+1}^n - p_i^n}{\Delta x} - \frac{\lambda}{2D} u_i^{n2} \\ -\kappa p_i^n \frac{u_i^n - u_{i-1}^n}{\Delta x} + \rho_i^n (\kappa - 1)\left(q_i^n + u_i^n \frac{\lambda}{2D} u_i^{n2}\right) \end{cases} \tag{31}$$

The advection term is

$$\frac{\partial \vec{f}}{\partial t} + u \frac{\partial \vec{f}}{\partial x} = 0 \tag{32}$$

$$\frac{\partial \vec{f'}}{\partial t} + u \frac{\partial \vec{f'}}{\partial x} = 0 \tag{33}$$

The quantities $f_i^*$ and $f_i'^*$ are obtained from Equations (29) and (30). Substituting the values into Equations (18)–(21) allows Equations (32) and (33) to be solved using the CIP method.

### 3.2. Outlet Boundary Settings

According to the principle of the CIP method and Figure 6, the values on the inlet and outlet boundaries cannot be derived but should instead be set directly. Therefore, as shown in Figure 3, we add one more grid at the inlet and outlet separately. Grid zero stands for the air source, and grid $N + 1$ is the actuator linked to the pipe, which is the brake chamber in this paper. The parameters of grid zero are always the same as those of the air source in the tank, and we focus instead on the settings for grids $N$ and $N + 1$.

To solve the pressure response in grid $N$, we deal with it here as a virtual small chamber whose volume is $A\Delta x$; the other parameters are shown in Figure 7. The volume is so small that there is no time to exchange heat with the surroundings, so we handle it as an adiabatic chamber. The air flow into the chamber is $G_{in}$ and the air consumption downstream is $G_c$, so the pressure response in the chamber can be obtained as

$$V \frac{dp}{dt} = k(G_{in} - G_c)R\theta \tag{34}$$

Rewriting Equation (34) in difference form, the pressure is

$$p_N^{n+1} = \kappa \frac{(G_{in} - G_c)R\theta_a}{A\Delta x} \Delta t + p_N^n \tag{35}$$

where

$$G_{in} = \rho_{N-1}^n u_{N-1}^n A \tag{36}$$

$$G_c = p_N^n C \rho_0 \sqrt{\frac{\theta_0}{\theta_a}} \varphi(p_c^n / p_N^n) \tag{37}$$

$$u_N^{n+1} = \frac{G_c}{\rho_N^n A} \tag{38}$$

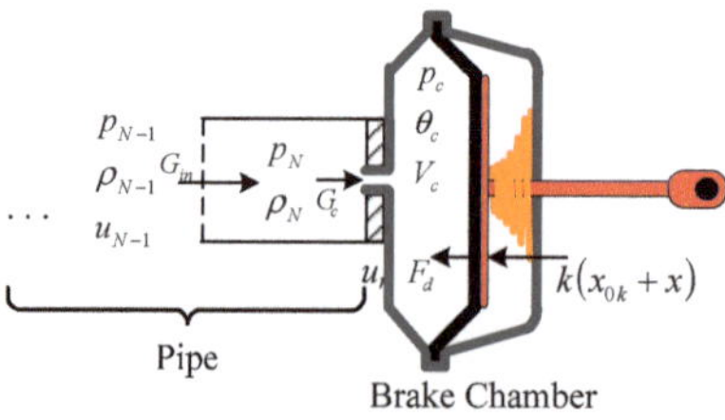


**Figure 7.** Model of the outlet boundary.

### 3.3. Discretisation of the Control Equations of the Brake Chamber

To solve the pressure response of the brake chamber simultaneously with that of the pipe, the brake chamber is handled as grid $N + 1$ and Equations (8), (11) and (13) are rewritten as difference schemes:

$$\theta_c^{n+1} = \frac{1}{m_c c_v}\left[c_v G_c(\theta_a - \theta_c^n) + R\theta_a G_c - \frac{\pi}{3}p_c^n A u_p^n + hS_{hc}(\theta_a - \theta_c^n)\right]\Delta t + \theta_c^n \tag{39}$$

$$p_c^{n+1} = \frac{1}{V_c}\left[G_c R\theta_c^n - \frac{\pi}{3}p_c^n u_p^n A + \frac{p_c^n V_c}{\theta_c^j}\frac{\theta_c^{n+1} - \theta_c^n}{\Delta t}\right]\Delta t + p_c^n \tag{40}$$

$$u_p^{n+1} = \frac{\Delta t}{M}\left[(p_c^n - p_a)A_p - k\left(x_{0k} + x^{n+1}\right) - F_d\right] + u_p^n \tag{41}$$

where $V_c = \frac{\pi}{3}Ax^{n+1}$ and $m_c = \frac{p_c^{n+1}V_c}{RT_c^{n+1}}$.

The control equations of the brake circuit were discrete in Sections 3.1–3.3 , including solving the pipe with the CIP method and the difference scheme used for the brake chamber. Lastly, the solution process is summarized in Figure 8 that is realized by a MATLAB m-file, which leads to obtaining the pressure response in the pipe and brake chamber simultaneously. The main function CIPVec called in Figure 8 can be found in Appendix A.

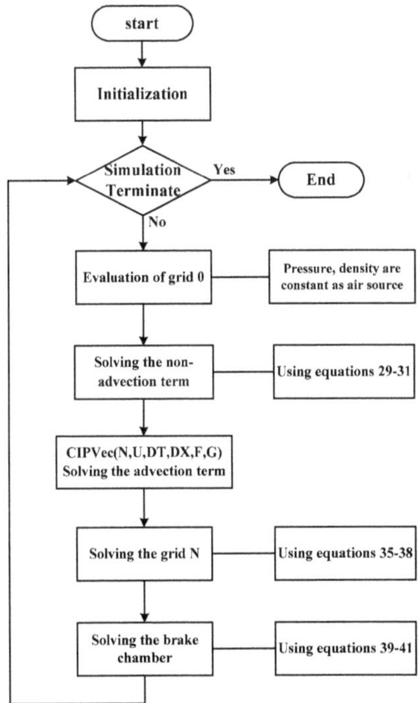


**Figure 8.** Flow-chart of brake circuit simulation.

### 3.4. The Experiment on the Brake Pressure Response

In this experiment, we focused on verifying the mathematical model and the influence of the pipe, so we used a solenoid valve instead of the brake control valve to simplify the measurements. The orifices in Figure 2 are replaced by the solenoid valve and fittings. A brake-circuit test rig was built, as shown in Figure 9. It consists of an air source, a precision regulator (IR3020, SMC), a buffer tank (27 L), a solenoid valve (SY5100, SMC), two pressure sensors (PSE540, SMC), a pipe (T1075, SMC), a brake chamber (C3519VS05D, DONGFENG), and a data acquisition card (NI6009). The accuracy of the pressure sensor is ±0.5% F.S.

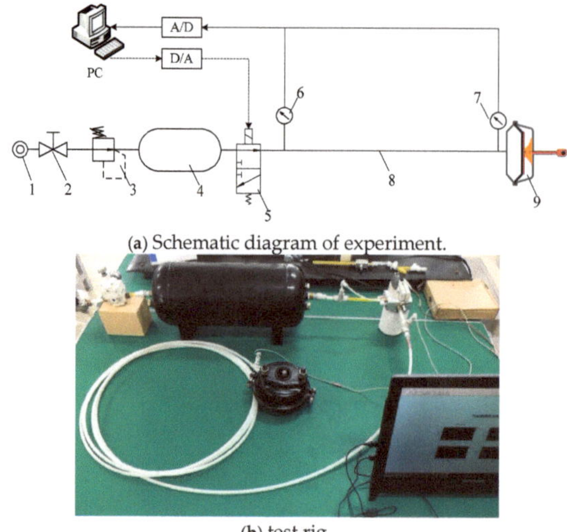

(a) Schematic diagram of experiment.

(b) test rig.

**Figure 9.** Schematic diagram of pneumatic brake circuit. (**a**) Schematic diagram of experiment. 1—Air source; 2—shut-off valve; 3—precision regulator; 4—buffer tank; 5—solenoid valve; 6,7—pressure sensor; 8—pipe; 9-brake—chamber; and (**b**) test rig.

## 4. Results and Discussion

### 4.1. Experimental Results and Analysis

Figure 10 shows the comparison of the experiment and simulation results, and the main parameters settings in simulation are given in Table 2.

**Table 2.** Simulation parameters of brake pressure response.

| Components | Brake Chamber | | | | Pipe | | Solenoid Valve | |
|---|---|---|---|---|---|---|---|---|
| Parameters | $s$ [mm] | $k$ [N/m] | $x_0$ [mm] | $d$ [cm] | $L$ [m] | $D$ [mm] | $C$ [m³/(s·Pa)] | $b$ |
| Values | 80 | 3000 | 2.7 | 16 | 10 | 7.5 | $2.27 \times 10^{-8}$ | 0.38 |

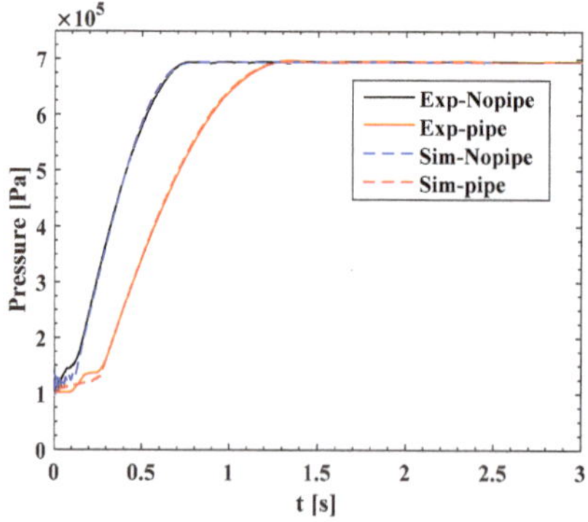


**Figure 10.** Pressure response curves in brake chamber.

Figure 10 indicates that the charge time in the brake chamber is 0.7 s without a pipe and 1.2 s with one. Thus, the time delay is increased by 0.5 s by the pipe restricting the air transmission. For a vehicle being driven at high speed, 0.5 s could cause a serious traffic accident if the brakes cannot be applied promptly. Therefore, the pipe effect must be considered carefully when designing a brake system, and the time delay should be reasonably compensated. In addition, the simulation results are consistent with the experimental ones, which also helps to validate the mathematical models developed in this paper. However, if we look at the results carefully, there is a little deviation at the beginning of the charge. This is because the parameters of the rubber piston are complex and difficult to measure correctly, and the deforming force in the extension process is complex too. Thus, we used an approximate model to describe the piston that causes the deviation.

If we combine Figures 10 and 11, it is obvious that the location of the deviation is exactly coincident with the process of the rod extending. Once the rod has extended fully, the rubber deformation has no effect, and the simulation coincides well with the experiment. Furthermore, the pressure in the initial part is very small and it does not begin to brake, so the deviation has little effect on the predicted braking force. Thus, all of this verifies that the CIP method is a precision method for predicting the pressure in the pipe. It also demonstrates that the model and method proposed in this paper are sufficiently accurate to calculate the pressure response of the brake system.

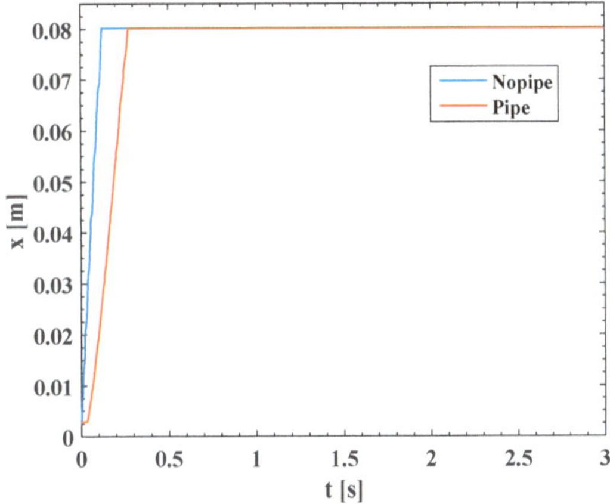


**Figure 11.** Displacement curves of piston in brake chamber.

Now that the model has been verified experimentally, we turn our attention to modifying the values of various structural parameters in the simulation to study their effects.

*4.2. Influence of Pipe Length*

The layout of the brake circuit varies according to the type of vehicle, which in turn could affect the brake pressure response or time delay. Figure 12 shows pressure response curves for different pipe lengths for an inner diameter of 7.5 mm. As the length is increased from 6 m to 14 m, the time delay increases by 0.15 s for every 4 m of pipe. This indicates that the longer the pipe, the more the time delay. Therefore, when we design a brake circuit the pipe should be as short as possible, we should reduce the number of bends, and we should arrange the circuit reasonably, all of which can help to minimise the time delay.

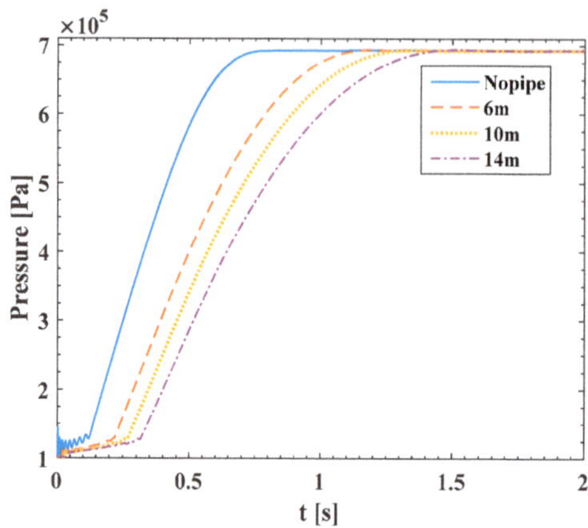

**Figure 12.** Pressure curves for different pipe lengths.

### 4.3. The Influence of the Pipe Diameter

The most common used pipe diameters for a PBS are in the range 6–16 mm. Figure 13 shows pressure response curves when the diameter is increased from 4 mm to 14 mm. In the simulation, the sonic conductance of the orifice was kept constant to focus on the effect of diameter. As shown, when the diameter varies from 4 to 6 mm, the times delay decreases significantly. However, when the diameter increases from 7.5 to 12 mm, the time delay slightly decreases. However, the time delay increases but does not decrease as the diameter increases from 14 to 16 mm. Clearly, the pipe is similar as a capacitance or resistance according to different pipe length. When the diameter is small, the resistance is with a dominant status compared to capacitance. However, as the diameter increases, the volume increases. Moreover, the condensance gets to be the main influencing factors. Therefore, the time delay decreases first and then increases. However, in actual applications, pipes with a diameter in the range of 4–6 mm are mostly used as control pipes and those in the range of 7.5–12 mm are used as transmission pipes. Therefore, the influence of diameter is relatively small compared to that of length, and so the diameter effect can be neglected. However, for the restriction of the brake control valves, the pipe diameter should be chosen to match the fittings. We can use a smaller diameter if that satisfies the requirements, but we should not use too narrow a pipe with too many transfer joints, because that would increase the local pressure loss.

### 4.4. Influence of Inlet/Outlet Sonic Conductance

Figure 13 indicates that the influence of the pipe diameter is minimal. However, for the pipe restriction, the diameter of the fittings is limited, so we should consider the influence of the inlet or outlet sonic conductance. Figure 14 shows pressure response curves for different orifice sizes. Here, $O_1$ is orifice 1 in Figure 2 and $O_2$ is orifice 2. By comparing curves 1 and 2, we see that the time delay increases significantly as the orifice shrinks. If we compare curves 3 and 4, we see that the pressure response depends on the orifice arrangement: when the larger orifice is in the front, the pressure response is quicker. The main reason for this phenomenon is that the orifice size restricts the flow rate; there will be a large pressure drop if the section area changes abruptly, whereupon the flow rate and charging speed will decrease simultaneously. Thus, when we design a brake circuit, the arrangement of the components should reduce gradually to decrease the local pressure drop.

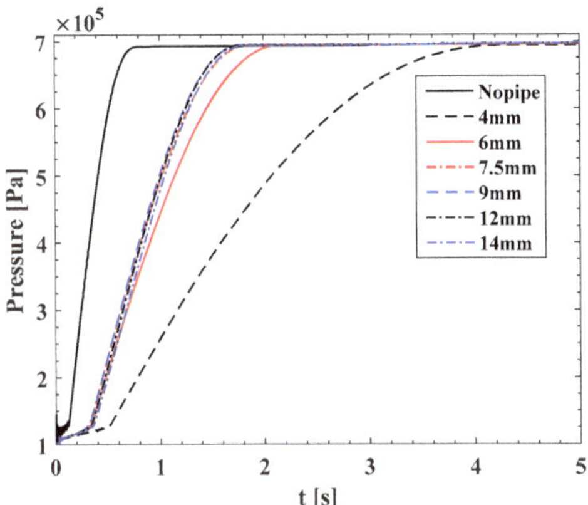

**Figure 13.** Pressure curves for different pipe diameters.

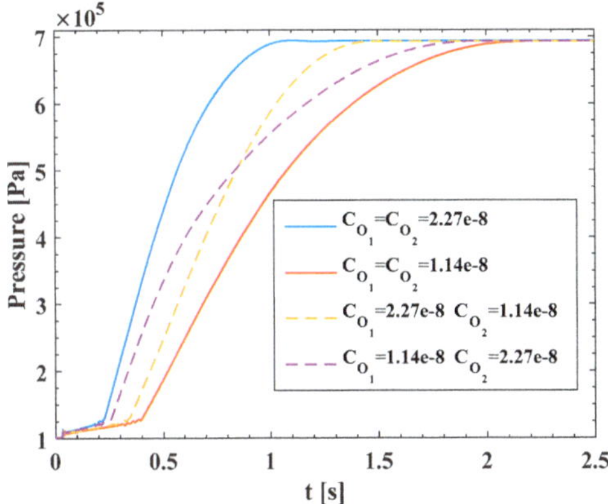

**Figure 14.** Pressure curves for different sonic conductances.

### 4.5. Influence of the Thermodynamic Model

In the former, we have identified the influence of the pipe when the brake chamber is charged. However, if the brake chamber is modelled using different thermodynamic assumptions, the pressure response is different. The most-used model is either the adiabatic model or the isothermal model. To understand the influence of the thermodynamic model, we compare different ones here. As shown in Figure 15, when heat exchange is taken into consideration, the model is more like the adiabatic one. Figure 15 also shows the temperature change to be as high as 70 K, in which case the isothermal model produces large deviations and cannot reflect the real pressure response in the brake chamber.

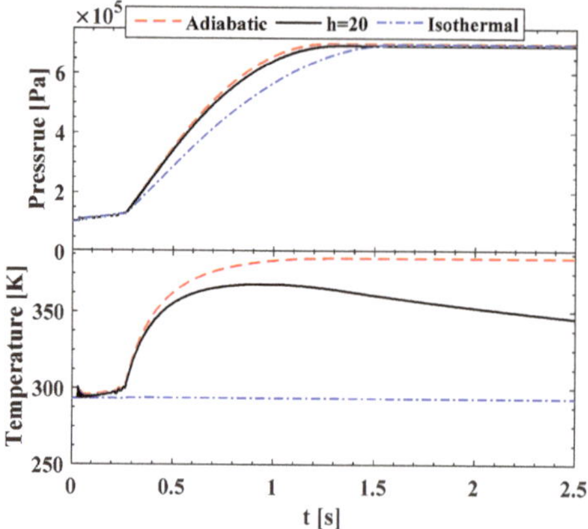

**Figure 15.** Pressure and temperature curves for different thermodynamics models.

### 4.6. Comphensive Analysis of Structure Parameters

As shown in Figure 12, the pressure response time gets larger when the pipe length increases. However, here we focus more on the relationship between time delay and pipe length. In other words, we want to find out time delay that occurs due to different parameter configurations. Though the range of actuation pressure and pipe length is limited, the combination forms of these structure parameters are various. Then, we introduced the response surface method [34] to predict the regularity between different simulation conditions. Here a simulation experiment with three factors (Length, $C_{O_1}$, and $C_{O_2}$) and three levels (as shown in Table 3) was designed by Design-Expert [35]. In these experiment groups, the diameter was constant at 7.5 mm, and the heat exchange model was used. The experiment was run 17 times, as shown in Table 3, and the response times were obtained, as shown in Figure 16.

Figure 16 shows that the response time by contour and the white labels in the graph are the response times. Clearly, the response time varies significantly when the configuration changes. For example, when the pipe length is 10 m, the response time varies from 0.94 s to 2.66 s. It is also seen that increasing the sonic conductance of an orifice decreases the response time. However, if the orifice 1 is constant and relatively smaller, the response time will vary slightly by increasing the orifice 2. This graph also presents a reference for design and analysis.

**Table 3.** Arrangement of the simulation experiment.

| No. | Length [m] | $C_{O1}$ [$\times 10^{-8}$ m$^3$/(s·Pa)] | $C_{O2}$ [$\times 10^{-8}$ m$^3$/(s·Pa)] | Response Time [s] |
|---|---|---|---|---|
| 1 | 14.0 | 2.28 | 1.14 | 1.95 |
| 2 | 10.0 | 3.42 | 1.14 | 1.65 |
| 3 | 14.0 | 1.14 | 2.28 | 2.85 |
| 4 | 14.0 | 2.28 | 3.42 | 1.52 |
| 5 | 14.0 | 3.42 | 2.28 | 1.18 |
| 6 | 10.0 | 2.28 | 2.28 | 1.36 |
| 7 | 10.0 | 1.14 | 1.14 | 2.66 |
| 8 | 6.0 | 1.14 | 2.28 | 1.98 |
| 9 | 10.0 | 2.28 | 2.28 | 1.37 |
| 10 | 10.0 | 2.28 | 2.28 | 1.37 |
| 11 | 6.0 | 2.28 | 3.42 | 1.07 |
| 12 | 6.0 | 3.42 | 2.28 | 0.96 |
| 13 | 10.0 | 3.42 | 3.42 | 0.94 |
| 14 | 10.0 | 2.28 | 2.28 | 1.37 |
| 15 | 10.0 | 1.14 | 3.42 | 2.35 |
| 16 | 6.0 | 2.28 | 1.14 | 1.71 |
| 17 | 10.0 | 2.28 | 2.28 | 1.37 |

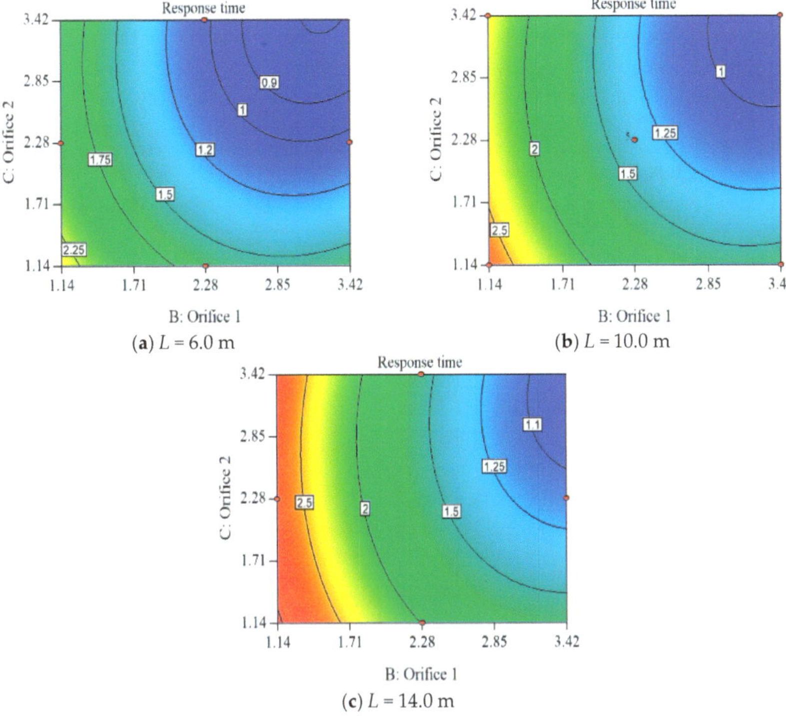



**Figure 16.** Pressure response time for different parameters configuration.

## 5. Conclusions

In this study, a pressure response model of a brake circuit was developed by considering the influence of the pipe. The pipe model took into consideration heat transfer and unsteady friction, and was solved by introducing the CIP method with its third-order accuracy. Meanwhile, the brake-chamber model considered heat transfer, thereby further improving the accuracy. Lastly, the model was verified experimentally and was shown to be highly accurate and suitable. The conclusions drawn from this study can be summarised as follows:

(1) Because there is a time delay associated with air flow in the pipe, the brake pressure response obviously lags. In this study, the time delay varies from 0.9 s to 2.85 s with a different parameters configuration; therefore, the time delay should be carefully considered when designing the brake circuit.

(2) The brake pressure response time increased significantly with pipe length, but the diameter of a pipe has less effect when the diameter is in the range of 7.5–12 mm. Moreover, the effect of pipe volume gets larger compared with resistance when the diameter further increases.

(3) The inlet and outlet sonic conductances affected the pressure response to different degrees. The time delay increases when the sonic conductance decreases, and they should be matched carefully.

(4) The brake-chamber model was the most accurate when considering the heat transfer. Compared with the isothermal model, the adiabatic model was more accurate for temperature changes up to 70 K.

(5) The CIP method could be used to calculate the pressure response in the pipe and was verified to be highly accurate and stable.

Here, we proposed the method and algorithm to solve the pressure response time of a pneumatic brake circuit. It is useful to simplify the process of designing brake systems. In the future, we intend to further develop this method and make the algorithms programmed. This will make it better suited to engineering applications, and it will also possess high efficiency and accuracy.

**Acknowledgments:** The authors acknowledge the SMC Corporation for the financial support and technical advice.

**Author Contributions:** Gangyan Li and Toshiharu Kagawa conceived this study and proposed the analysis method; Fan Yang designed the experiment; Jian Hua and Xingli Li performed the experiments and analyzed the data; Fan Yang wrote the paper and Gangyan Li modified the final paper.

**Conflicts of Interest:** The authors declare no conflict of interest.

## Nomenclature

| | |
|---|---|
| $A$ | area of piston [m$^2$] |
| $b$ | critical pressure ratio |
| $c_v$ | specific heat at constant volume [J/(kg·K)] |
| $c_p$ | specific heat at constant pressure [J/(kg·K)] |
| $C$ | sonic conductance [m$^3$/(s·Pa)] |
| $d$ | diameter of brake chamber [m] |
| $D$ | inner diameter of pipe [m] |
| $e$ | inner energy [J] |
| $F_d$ | deforming force [N] |
| $G$ | mass flow rate [kg/s] |
| $h$ | coefficient of heat convection [W/(m$^2$·K)] |
| $H$ | enthalpy [J] |
| $k$ | spring constant [N/m] |
| $L$ | length of pipe [m] |
| $M$ | mass of piston [kg] |
| $p$ | pressure [Pa] |
| $R$ | gas state constant = 287 [J/(kg·K)] |
| $s$ | stroke of push rod [m] |
| $S_h$ | heat exchange area [m$^2$] |
| $t$ | time [s] |
| $u$ | velocity [m/s] |
| $V_c$ | volume [m$^3$] |
| $x$ | displacement [m] |
| $\kappa$ | specific heat index = 1.4 |
| $\theta$ | temperature [K] |
| $\rho$ | density [kg/m$^3$] |
| Subscripts | |
| 0 | stand reference state (20 °C, 100 kPa) |
| 1 | upstream |
| 2 | downstream |
| $c$ | brake chamber |

## Appendix A

The main function of CIPVec called in Figure 8 is given as follows:

```
function [FN,GN] = CIPVec(NX,YU,DT,DX,F,G)
% NX-the numbers of grids
% YU-the velocity of advection
% DT-the length of time step
% DX-the length of spatial step
```

```
% F/G-the parameters under solving and its differential
% FN/GN-the values of F/G in next time step
for i = 2:NX − 1
    if YU(i) < 0
        ISG = 1;
    else
        ISG = −1;
    end
    IUP = i + ISG;
        XX = -YU(i)*DT;
        FDIF = (F(IUP) − F(i))/DX*ISG;
        XAM1 = (G(i) + G(IUP) − 2.0*FDIF)/(DX*DX);
        XBM1 = (3.0*FDIF − 2*G(i) − G(IUP))/DX*ISG;
    FN(i) = ((XAM1*XX + XBM1)*XX + G(i))*XX + F(i);
    GN(i) = (3.0*XAM1*XX + 2.0*XBM1)*XX + G(i);
end
```

## References


1. Fujino, K.; Taniguchi, K.; Yamamoto, N.; Youn, C.; Kagawa, T. Transient pressure and flow rate measurement of pneumatic power supply line in Shinkansen: Examination of unsteady characteristics in pneumatic supply system by experiment. In Proceedings of the 2010 SICE Annual Conference, Taipei, Taiwan, 18–21 August 2010; pp. 1664–1669.
2. Selvaraj, M.; Gaikwad, S.; Suresh, A.K. Modeling and Simulation of Dynamic Behavior of Pneumatic Brake System at Vehicle Level. *SAE Int. J. Commer. Veh.* **2014**, *11*, 1–9. [CrossRef]
3. Lopez, A.; Sherony, R.; Chien, S.; Li, L.; Qiang, Y.; Chen, Y. Analysis of the braking behaviour in pedestrian automatic emergency braking. In Proceedings of the 2015 IEEE 18th International Conference on Intelligent Transportation Systems (ITSC), Las Palmas, Spain, 15–18 September 2015; pp. 1117–1122.
4. Karthikeyan, P.; Chaitanya, C.S.; Raju, N.J.; Subramanian, S.C. Modelling an electropneumatic brake system for commercial vehicles. *IET Electr. Syst. Transp.* **2011**, *1*, 41–48. [CrossRef]
5. Fleming, B. Advances in automotive electronics [automotive electronics]. *IEEE Veh. Technol. Mag.* **2015**, *10*, 4–11. [CrossRef]
6. Wang, Z.; Zhou, X.; Yang, C.; Chen, Z.; Wu, X. An Experimental Study on Hysteresis Characteristics of a Pneumatic Braking System for a Multi-Axle Heavy Vehicle in Emergency Braking Situations. *Appl. Sci.* **2017**, *7*, 799. [CrossRef]
7. Qin, T. Research on Delay Time Analysis and Its Control Techniques of Bus Pneumatic Brake System. Ph.D. Thesis, Wuhan University of Technology, Wuhan, China, 2012.
8. Long, X.; Hu, X. Optimization on the response time of pneumatic brake systems of commercial vehicle. *Enterp. Sci. Technol. Dev.* **2013**, *365*, 44–46. (In Chinese)
9. Mithun, S.; Gayakwad, S. Modeling and simulation of pneumatic brake system used in heavy commercial vehicle. *IOSR J. Mech. Civ. Eng.* **2014**, *11*, 1–9. [CrossRef]
10. Modeling and Simulation Vehicle Air Brake System. Available online: https://www.modelica.org/events/modelica2011/Proceedings/pages/papers/17_3_ID_144_a_fv.pdf (accessed on 13 September 2017).
11. Li, S.; Bao, W. Analysis of transient hydraulic pressure pulsation in pipelines using MATLAB Simulink. *Eng. Mech.* **2006**, *23*, 184–188.
12. Tokashiki, L.; Fujita, T.; Kagawa, T. Simulation on pneumatic cylinder including pipes. *Hydraul. Pneum.* **1997**, *28*, 766–771. (In Japanese) [CrossRef]
13. Zielke, W. *Frequency Dependent Friction in Transient Pipe Flow*; University of Michigan: Ann Arbor, MI, USA, 1966.
14. Kitagawa, A.; Kagawa, T.; Takenaka, T. High speed and accurate computing method for transient response of pneumatic transmission line using characteristics method. *Trans. Soc. Instrum. Control Eng.* **1984**, *20*, 648–653. [CrossRef]
15. Hashimoto, K.; Imaeda, M.; Kikuchi, K. The analyses of transient responses of fluid lines by characteristics grid method. *Hydraul. Pneum.* **1985**, *16*, 140–146. (In Japanese) [CrossRef]

16. Wei, W.; Du, N. Influence of braking pipe on braking performance for heavy haul train. *J. Traffic Transp. Eng.* **2011**, *5*, 49–54.
17. Wei, W.; Wu, X. Influence of Brake Characteristics on Longitudinal Impulse of Train. *J. Dalian Jiaotong Univ.* **2012**, *2*, 1–5.
18. Yabe, T.; Aoki, T.; Sakaguchi, G.; Wang, P.Y.; Ishikawa, T. The compact CIP (Cubic-Interpolated Pseudo-particle) method as a general hyperbolic solver. *Comput. Fluids* **1991**, *19*, 421–431. [CrossRef]
19. Takewaki, H.; Nishiguchi, A.; Yabe, T. Cubic interpolated pseudo-particle method (CIP) for solving hyperbolic-type equations. *J. Comput. Phys.* **1985**, *61*, 261–268. [CrossRef]
20. Zhao, X.; Liu, B.; Liang, X. Constrained interpolation profile (CIP) method and its application. *J. Ship Mech.* **2016**, *20*, 393–402.
21. Zhao, X.; Fu, Y.; Zhang, D. Numerical simulation of flow past a cylinder using a CIP-based model. *J. Harbin Eng. Univ.* **2016**, *37*, 297–305.
22. Pang, P.T.; Agnew, D. *Brake System Component Characterization for System Response Performance: A System Level Test Method and Associated Theoretical Correlation*; SAE Technical Paper; SAE International: Detroit, MI, USA, 2004.
23. Yang, F.; Li, G.; Hu, J.; Liu, W.; Qi, W. Method for resultant and calculating the flow-rate characteristics of pneumatic circuit. In Proceedings of the 2015 International Conference on Fluid Power and Mechatronics (FPM), Harbin, China, 5–7 August 2015; pp. 379–384.
24. *Pneumatic Fluid Power—Determination of Flow-Rate Characteristic of Component Using Compressible Fluids—Part 1: General Rules and Test Methods for Steady-State Flow*; ISO 6358-1; International Organization for Standardization (ISO): Geneva, Switzerland, 2013.
25. *Pneumatic Fluid Power—Assessment of Component Reliability by Testing—Part 2: Directional Control Valves*; ISO 19973-2; International Organization for Standardization (ISO): Geneva, Switzerland, 2015.
26. Wang, T.; Peng, G.; Toshiharu, K. Measurement and Resultant Methods of Flow-rate Characteristics of Small Pneumatic Valves. *J. Mech. Eng.* **2009**, *45*, 290–297. (In Chinese) [CrossRef]
27. Niu, J.L.; Shi, Y.; Cao, Z.X.; Cai, M.L.; Chen, W.; Zhu, J.; Xu, W.Q. Study on air flow dynamic characteristic of mechanical ventilation of a lung simulator. *Sci. China Technol. Sci.* **2017**, *60*, 243–250. [CrossRef]
28. Shi, Y.; Jia, G.W.; Cai, M.; Xu, W.Q. Study on the dynamics of local pressure boosting pneumatic system. *Math. Probl. Eng.* **2015**, *2015*, 849047. [CrossRef]
29. Kaasa, G.-O.; Chapple, P.J.; Lie, B. Modeling of an electro-pneumatic cylinder actuator for nonlinear and adaptive control, with application to clutch actuation in heavy-duty trucks. In Proceedings of the 3rd FPNI PhD International Symposium on Fluid Power, Terrassa, Spain, 30 June–2 July 2004.
30. Matsuo, K. *Compressible Fluid Flow—Theory and Analysis of Internal Flow*; Rikogakusha Publishing: Tokyo, Japan, 1999; pp. 145–147.
31. Shiraishi, K.; Matsuoka, T. The simulation of acoustic wave propagation by using characteristic curves with CIP method. *Butsuri Tansa* **2006**, *59*, 261–274. (In Japanese) [CrossRef]
32. Guo, Z.; Li, X.; Kagawa, T. Heat transfer effects on dynamic pressure response of pneumatic vacuum circuit. In Proceedings of the ASME/JSME 2011 8th Thermal Engineering Joint Conference, Honolulu, HI, USA, 13–17 March 2011.
33. Qi, W.; Li, G.; Lu, Q.; Long, X. Influence of Bus Brake Chamber Inlet Diameter on Pressure Characteristic. *Hydraul. Pneum. Seals* **2016**, *36*, 21–25. (In Chinese)
34. Nouby, M.; Mathivanan, D.; Srinivasan, K. A combined approach of complex eigenvalue analysis and design of experiments (DOE) to study disc brake squeal. *Int. J. Eng. Sci. Technol.* **2009**, *1*, 254–271. [CrossRef]
35. Li, L.; Sai, Z.; Qiang, H. Application of response surface methodology in experiment design and optimization. *Res. Explor. Lab.* **2015**, *34*, 41–45. (In Chinese)

*Article*

# Analysis of the Energy Efficiency of a Pneumatic Booster Regulator with Energy Recovery

**Fan Yang [1,2,*], Kotaro Tadano [2], Gangyan Li [1,*] and Toshiharu Kagawa [2]**

1   School of Mechanical and Electronic Engineering, Wuhan University of Technology, Wuhan 430070, China
2   Future Interdisciplinary Research of Science and Technology, Tokyo Institute of Technology,
    Yokohama 226-8503, Japan; tadano@pi.titech.ac.jp (K.T.); kagawa@pi.titech.ac.jp (T.K.)
*   Correspondence: yang_fan@whut.edu.cn (F.Y.); gangyanli@whut.edu.cn (G.L.); Tel.: +86-152-7177-8967 (F.Y.)

Received: 27 June 2017; Accepted: 7 August 2017; Published: 9 August 2017

**Abstract:** Pneumatic booster regulators (PBR) are in great demand in modern pneumatic systems for their energy-saving abilities. A new booster regulator with energy recovery (VBA-R) was proposed, and its energy efficiency was investigated by introducing the concept of air power. On the basis of quality-alterable gas thermodynamics, an energy efficiency assessment and pressure response model for VBA-R was proposed. First, a model was solved using MATLAB/Simulink software, and an alternative experiment was designed to verify the mathematical model and performance improvement. The results showed that the simulation was consistent with the experiment. We also can conclude that, first of all, the energy efficiency decreases with the increasing of supply pressure and flow-rate consumption; a VBA-R has the highest efficiency when its diameter ratio is closest to 1.3. Finally, a recovery chamber helped to improve the performance of the VBA-R, which included a boost ratio improvement of 15–25% and an efficiency improvement of 5–10% compared with a conventional VBA booster regulator. This research lays the foundation for optimism regarding the proposed booster regulator.

**Keywords:** pneumatic booster regulator; energy recovery; air power; energy efficiency

## 1. Introduction

With the increasing awareness of the need for environmental protection, low-carbon and energy-saving technologies have become the current theme of development. Pneumatic systems, which account for industrial energy consumption of 10 to 20 percent [1,2], are inevitably discussed as an energy-saving topic. Low energy efficiency is the main factor restricting the further development of pneumatic systems. A study shows that energy consumption can be reduced by about eight percent for each 0.1 MPa reduction in the supply pressure of the compressor [3]. Therefore, more and more Japanese factories are beginning to reduce their air supply pressure to save energy. For example, the supply pressure of the Toyota factory was gradually decreased from 0.6 to 0.3 MPa. However, in practical applications, local high-pressure air is in demand due to the existence of heavy load or high-pressure equipment. Pneumatic booster regulator systems are widely used in factories as a main solution for satisfying local high-pressure requirements. At present, almost all commercial pneumatic booster regulators (PBR) have a structure in which compressed air exhausts directly to the atmosphere. It is obvious that energy is wasted, but it is hard to evaluate the real energy efficiency for lack of an appropriate method. Therefore, a way to evaluate the efficiency of PBR systems and improve their performance has become one of the problems of enterprise that should be solved urgently.

A PBR is also known as a pressure amplifier; it is driven by low-pressure air, and its output pressure is high. Pneumatic booster regulators can be categorised as either symmetric or asymmetric, with respect to their structure [4–6]. The maximum boost ratio of the symmetric type is two-fold, according to Pascal's law, whereas the boost ratio of the asymmetric type can be up to three or four

*Appl. Sci.* **2017**, *7*, 816; doi:10.3390/app7080816                                                      www.mdpi.com/journal/applsci

times, because the diameter ratio of the drive and boost chambers can be changed for different demands. However, it sacrifices the air flow rate, and more compressed air is exhausted to the atmosphere with each stroke. To improve the energy efficiency, Shi [7–9] proposes an expansion energy booster in which the air supply is cut off before the piston moves to the end of its stroke, and then the residual stroke is driven by air expansion. Similarly, many other means are also used to improve the energy efficiency in the driven cylinder, such as a differential drive [10], dual pressure supply, utilising expansion energy [11], recovering energy with a rubber bladder and storing the strain energy [12], or reusing exhaust air for power generation [13]. However, all of these make the circuit more complex, and when the recovery energy was directly input to the drive chamber of the cylinder, it led to velocity fluctuations, which are harmful to the life of the cylinder [14]. On the other hand, there is an urgent need to assess the energy efficiency of a PBR in order to evaluate the performance of a booster regulator. In recent decades, the consumption of air [$dm^3$/min(ANR)] or specific energy [kW·h/$m^3$(ANR)] have been commonly used to assess the energy of compressible air [15]. However, the consumption of air is a volume unit, so it does not reflect energy consumption. The specific energy is the ratio of the energy output of the air compressor to the flow-rate consumed by the component. This assessment method cannot reflect the energy consumption of the components independently because it considers the power consumed by the air compressor, and it is hard to analyse the losses of the intermediate links. In addition, both of these two methods do not reflect the effect of pressure. Other methods used for the energy assessment of compressed air include enthalpy and exergy [16]. However, enthalpy is dependent on mass and temperature and cannot reflect the effects of pressure, whereas exergy reflects the system's ability to conduct work from a different energy form. To describe the ability of compressed air to do work, a new concept of air power, similar to exergy but simpler, was proposed by Cai and Kagawa [17]. In their concept, the energy of air is related to the flow-rate, pressure and temperature. We will introduce this as a method for evaluating the energy efficiency of a booster regulator later in this paper.

The above analysis shows that there are some problems with pneumatic booster regulators, such as the exhaustion of pressurised air directly to the atmosphere, which leads to energy waste; thus, the factories are eager to propose a new method to enhance the efficiency. However, an effective way to evaluate the energy efficiency of a PBR is lacking. To overcome these problems, a new booster regulator with energy recovery was proposed in which the air tank was considered, the energy efficiency and dynamic models were established, and the regularity of change in energy efficiency was found, which provides a good reference for the design and operation of a booster regulator.

## 2. Working Principle of the Booster Regulator

Based on Pascal's law, low pressure can be transferred to high pressure for the area of difference. A typical commercial booster regulator (VBA series, SMC, Tokyo, Japan) is shown in Figure 1. It is composed of two drive chambers, two boost chambers and a movable piston. As shown in Figure 1, the piston moves from left to right. The pressurised air flows into drive chamber A and boost chamber B, and the air in drive chamber B discharges directly to the atmosphere. The force on the left side of the piston can be twice as great as that on the right side, so the pressure in boost chamber A can become twice as high as the supply pressure. When the piston moves to the end of its stroke, the switching valve changes its position, which causes the piston to change direction. Thus, the booster regulator can continuously exhaust high-pressure air.

However, the compressed air in the drive chamber is exhausted directly to the atmosphere with each stroke, so considerable energy is wasted over a long run-time. In order to reuse this energy to enhance the energy efficiency, and with reference to the expansion energy used (EEU) booster proposed in the reference [7–9], a new booster regulator was proposed, as shown in Figure 2 (it is called the VBA-R, corresponding to the old type of VBA). The drive and boost chambers are the same as a traditional booster regulator. However, two more chambers (called recovery chambers) were added to the middle to recover the exhausted energy, and a seven-way two-position (7/2) solenoid valve was introduced to control the air flow. There are also other elements, such as check valves, regulators and magnetic rings, and so on.

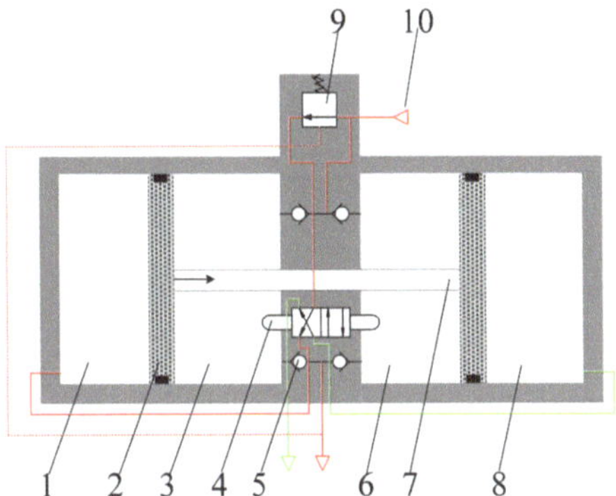

**Figure 1.** Schematic of booster regulator (VBA). 1-Drive chamber A, 2-piston, 3-booster chamber A, 4-switching valve, 5-check valve, 6-booster chamber B, 7-piston rod, 8-drive chamber B, 9-regulator, 10-air source. The pressure is higher in red lines than in green lines.

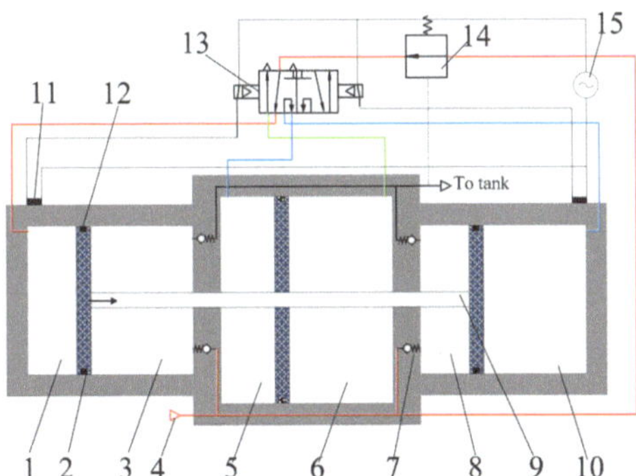


**Figure 2.** Structure schematic diagram of booster regulator with energy recovery (VBA-R). 1: drive chamber A, 2: piston, 3: booster chamber A, 4: air source, 5: recovery chamber A, 6: recovery chamber B, 7: check valve, 8: boost chamber B, 9: piston rod, 10: drive chamber B, 11: magnetic switch, 12: magnetic ring, 13: 7/2 solenoid valve, 14: regulator, 15: power source.

When the piston is located at the left end, the magnetic signal was detected and the position of the 7/2 solenoid valve was set as shown in Figure 2. The compressed air flows into drive chamber A and boost chamber B. The piston then moves from left to right, and the air in boost chamber A is further compressed so that the pressure becomes higher than the supply pressure. The compressed air in drive chamber B, instead of exhausting to the atmosphere, flows into recovery chamber A through the 7/2 solenoid valve, and the recovered air is reused to move the piston. It also helps to increase the area of difference so that the boost ratio can be further enhanced. After the piston moves to the right end, an impulse signal is detected once again, and the 7/2 solenoid valve changes its position so that the piston can move in a reciprocating manner and continuously discharge high-pressure air. When the booster regulator works with a different load, the boost ratio can be set by adjusting regulator 14 to output a different air pressure.

### 3. Energy Efficiency Evaluation Method for Booster Regulator with Energy Recovery (VBA-R)

The compressed air is recovered by recovery chambers, rather than being exhausted to the atmosphere, through the 7/2 solenoid valve; thus, it helps to raise the boost ratio. On the other hand, the energy is reused in order to improve the energy efficiency of the booster regulator. However, the evaluation methods of energy efficiency for PBRs should be proposed first, in order to assess the performance improvements. Here, we introduced the concept of air power to calculate the energy input or output of a booster regulator system.

*3.1. Concept of Air Power*

According to Cai and Kagawa [17], air power is defined as the work-producing potential of compressed air. In this definition, the atmosphere is set as the reference state. In the other words, the pneumatic system is supposed to work in the atmosphere. The advantage of air power is that it is independent of the air source and only depends on the current state of air. It is expressed as follows:

$$P = p_a Q_a \left[ \ln \frac{p}{p_a} + \frac{\kappa}{\kappa - 1} \left( \frac{\theta}{\theta_a} - 1 - \ln \frac{\theta}{\theta_a} \right) \right] \tag{1}$$

Equation (1) indicates that the air power depends on the pressure, flow rate and temperature. However, Cai [17] verified that, when the range of change of air temperature is less than 50 K, the change rate of air power is less than 3%. In modern pneumatic systems, an air dryer and air cooler are essential equipment, so the temperature is almost the same as the atmosphere, meaning the effects of temperature can be neglected. Thus, Equation (1) can be simplified as follows:

$$P = p_a Q_a \ln \frac{p}{p_a} \tag{2}$$

Then, the air power depends only on the pressure and flow rate and is zero when the pressure is the same as the atmosphere. Air power is different from hydraulic power due to the fact that air can be compressed. Therefore, in addition to transmission energy, it also includes expansion energy. However, we will not distinguish between these two kinds of air power because both of them are input to the booster regulator system.

*3.2. Energy Efficiency of Booster Regulator*

In practical applications, a tank is always used with a booster regulator to decrease pressure fluctuations. Therefore, the real output is the pressure in the tank, which is lower than the pressure in the chamber of the booster regulator (this will be proved later). To assess the energy efficiency of a booster regulator, considering the application and the definition of efficiency, the energy efficiency of a booster regulator is defined as the ratio of the energy output (air flow out of the tank) to the energy input (air flow in the boost and drive chambers). It is defined as:

$$\eta = \frac{E_t}{E_{bc} + E_{dc}} \tag{3}$$

Combining the air power proposed in Equation (2), the energy flow in the booster regulator in each cycle is:

$$E_{bc} + E_{dc} = \int_0^T p_a (Q_{bc} + Q_{dc}) \ln \frac{p_s}{p_a} dt = p_a (V_{bc} + V_{dc}) \ln \frac{p_s}{p_a} \tag{4}$$

It is obvious that the energy input depends only on the supply pressure, because the volume of each chamber is constant. When the supply pressure increases, more energy is consumed by the pneumatic booster regulator system.

Energy output, like the energy input, is the integral of air power flowing out the tank in each cycle, and it can be written as:

$$E_t = \int_0^T p_a Q_a \ln(p_t / p_a) dt \tag{5}$$

Pressure fluctuations in the tank lead to simultaneous fluctuations of the flow rate, so it is hard to calculate the energy output directly. However, if the tank is large enough, we could suppose that the fluctuation is small and can be neglected. When the system is running steadily, the flows in and out of the chambers are equal to the volume of the boost chamber. Therefore, we substitute Equations (4) and (5) into Equation (3), and the energy efficiency of the booster regulator system can be obtained as:

$$\eta = \frac{E_t}{E_{bc} + E_{dc}} = \frac{\ln(p_t / p_a)}{(1 + V_{dc}/V_{bc}) \ln(p_s / p_a)} \tag{6}$$

where $p_t$ is the pressure in the tank and the ratio of $p_t$ to $p_s$ is defined as the boost ratio, which reflects the boost capability of the booster regulator. Therefore, Equation (6) intuitively indicates that the energy efficiency depends only on the supply pressure and boost ratio because the volume of each chamber is constant. The greater the boost ratio or the lower the supply pressure is, the higher the energy efficiency is. To understand the energy efficiency, it is necessary to identify the boost ratio of the system. Therefore, we will discuss the pressure response characteristics of the VBA-R.

### 3.3. Pressure Response of VBA-R

To facilitate the model, the assumptions are given first as follows:

(1) Air is an ideal gas and it satisfies the ideal gas state equation;
(2) The initial temperature of air in each chamber is the same as the atmosphere, and so is the air source;
(3) The dead volume of each chamber is very small and can be set as the same;
(4) The air tank can exchange heat sufficiently with the surroundings, and it can be considered as an isothermal tank.

### 3.3.1. Flow Rate Characteristics Equation

The air flows in and out of the chambers through the check valve or throttle valve, and all of the mass flow rate of air can be calculated based on the equations given in ISO6358 [18]. However, when we used these equations in the numerical simulation, it was unstable for an accurate simulation and we had to set the steps to be very small in order to avoid divergence. Because of this numerical inaccuracy, the pressure ratio may surpass 1 when it is close to 1, which then leads to a negative value under the radical sign. This might lead to unknown errors or uncertain results, which would stop the simulation. To avoid these problems, we introduced the laminar flow model proposed by Kaasa [19] for use when the pressure ratio is approximately 1. In that model, subsonic flow is taken as laminar flow when the pressure ratio is greater than $\beta$ ($\beta$ ranges 0.995 to 0.999). Then, a redefined flow rate characteristic equation can be given as:

$$G = \begin{cases} k_1 p_1 \left(1 - \frac{p_2}{p_1}\right) \sqrt{\frac{\theta_0}{\theta_1}} & \frac{p_2}{p_1} \geq \beta \\ C p p_0 \sqrt{\frac{\theta_0}{\theta_1}} \left(1 - \left(\frac{\frac{p_2}{p_1} - b}{1 - b}\right)^2\right)^{0.5} & b < \frac{p_2}{p_1} < \beta \prime \\ C p_1 p_0 \sqrt{\frac{\theta_0}{\theta_1}} & \frac{p_2}{p_1} \leq b \end{cases} \tag{7}$$

where $k_1$ is the linear gain:

$$k_1 = \frac{1}{1 - \beta} \cdot C \cdot p_0 \sqrt{1 - \left(\frac{\beta - b}{1 - b}\right)^2} \tag{8}$$

### 3.3.2. Gas State Equation

The pressure, temperature and volume all change with the movement of the piston. However, these state quantities of air always satisfy the gas state equation according to assumption (1). By differentiating the state equation with time, the pressure response in the charge and discharge chambers can be obtained as:

$$V_c \frac{dp_c}{dt} = G_c R\theta_c - p_c S_c u + \frac{p_c V_c}{\theta_c}\frac{d\theta_c}{dt} \tag{9}$$

$$V_d \frac{dp_d}{dt} = -G_d R\theta_d + p_d S_d u + \frac{p_d V_d}{\theta_d}\frac{d\theta_d}{dt} \tag{10}$$

### 3.3.3. Energy Conservation Equation

Air flows in or out of the chambers whenever the booster regulator is working, and we have assumed that there is no leakage in the system. Therefore, the system can be considered as a variable-mass system that satisfies the energy conservation equation. Then the temperature response [20] in the chambers can be obtained as shown in Equations (11) and (12).

$$m_c c_v \frac{d\theta_c}{dt} = c_v G_c(\theta_a - \theta_c) + R\theta_a G_c - p_c S_c u + h S_{hc}(\theta_a - \theta_c) \tag{11}$$

$$m_d c_v \frac{d\theta_d}{dt} = -R\theta_d G_d + p_d S_d u + h S_{hd}(\theta_a - \theta_d) \tag{12}$$

### 3.3.4. Kinematic Equation of Piston

In the Figure 2, the pressure acts on the six surfaces of the piston. Therefore, the kinematic equation of the piston can be derived from Newton's second law:

$$M\frac{du}{dt} = (p_{da} - p_{db})S_{da} + (p_{ra} - p_{rb})S_{ra} + (p_{bb} - p_{ba})S_{ba} - F_f \tag{13}$$

The friction force of the piston in the pneumatic cylinder is so complex that many models have been used to describe it, such as the Coulomb sliding model, Viscous model and Striberk model [21]. Of these, the Striberk model has the best accuracy, but there are so many parameters that it is hard to confirm them because the booster regulator is a closed system. Therefore, we adopted a mixed friction model based on the Coulomb and Viscous models. This model has a high accuracy and has been verified in the motion control of a cylinder [22]. The friction force is:

$$F_f = cu + F_c \tag{14}$$

### 3.3.5. Pressure Response in Tank

A booster regulator is always used with a tank to restrain the pressure fluctuations. The flow rate output of a booster system depends on the pressure in the tank, which is also affected by the boost ratio of the booster regulator. Whenever the system is working, air flows out the tank to drive the load, but the tank is charged by the booster regulator at the same time. According to Equation (7), the output flow rate rises with increasing pressure in the tank, then the pressure in the tank decreases. Meanwhile, the output flow rate from the booster regulator dynamically decreases or increases with the changing pressure in the tank. Thus, the equilibrium pressure develops in the tank. This is a simple self-balance system, and if the air consumption grows with the downstream load, the pressure in the tank will dynamically adjust and will tend to balance after fluctuating for a period of time. The charge and discharge model of the tank is simplified, as shown in Figure 3. The inlet is a check valve; its opening depends on the pressure difference between the booster regulator and tank, and the flow rate changes dynamically as it opens. The outlet is a throttle valve, and we simulated different loads by adjusting its opening.

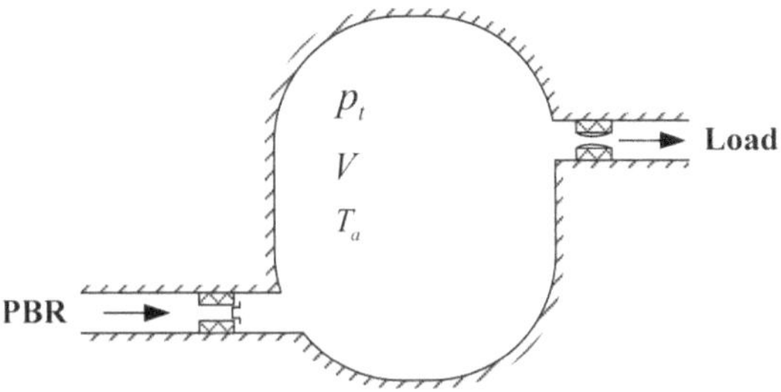


**Figure 3.** Pressure response model of tank.

The tank was regarded as an isothermal chamber (ITC) because the wall area was large enough to exchange heat sufficiently with the atmosphere. Then, the gas state equation could be simplified, as the change of temperature could be neglected. Thus, the pressure response in the tank could be obtained as Equation (15) by differentiating the gas state equation:

$$\frac{dp_t}{dt} = \frac{1}{V_t} G' R \theta_a \tag{15}$$

where $G'$ is the net flow rate of the tank, which is affected by the booster regulator, tank and downstream load. Equation (15) also indicates that the pressure differential is proportional to the net flow rate for an ITC.

The check valve (AKH10-00, SMC, Tokyo, Japan) was used in the experiment, and the air flowing into the tank depended on its flow-rate characteristics. According to the supplier catalogue, the cracking pressure of the valve is 5 kPa, its sonic conductance is $4.8 \times 10^{-8}$ (m$^3$/(s Pa)), and its critical pressure ratio is 0.5. We assume that the check valve will fully open once the pressure reaches the cracking pressure. The flow rate characteristics changes with the pressure difference and can then be expressed as:

$$C_{ch} = \begin{cases} 0 & (p_{br} - p_t < 5\text{kPa}) \\ 4.8 \times 10^{-8} & (p_{br} - p_t \geq 5\text{kPa}) \end{cases}$$

Although the critical pressure ratio is also affected by the pressure difference [23], it is very small, so we set it as constant in the simulation. By substituting these parameters into Equation (7), we can obtain the flow rate into or out of the tank. Therefore, the net flow rate can be calculated by:

$$G' = G_{ch} - G_{th} \tag{16}$$

When we calculated $G_{th}$, the downstream pressure was set as constant (atmosphere) because we used a throttle valve to simulate the load. However, in practical application, it should be set by considering the pressure of the load. By analysing Equations (15) and (16), we recognised that the pressure in the tank would fluctuate with variations of the net flow rate.

## 4. Simulation and Experiment on VBA-R

### 4.1. Simulation Model

The mathematical model was established in the previous section; this was solved with MATLAB/Simulink software (MathWorks, Natick, MA, USA). There are six chambers in the VBA-R, and each chamber has three control equations. Considering the air power, piston motion and pressure response in the tank, more than twenty equations need to be solved. It is difficult work to model all of the chambers one by one. However, when the piston moves in a left or right stroke, three chambers are charged while the others are discharged, and all of them interchange during each stroke. We also found that, whether the chambers' volumes changed or not, they could be described by a lumped parameter model with three control equations; the only differences were the structure parameters of the chambers. Thus, we could utilise this peculiarity and only two kinds of chambers needed to be modelled. The tank could be thought of as a particular chamber whose piston was fixed.

Therefore, the discharge chambers can be described by Equations (7), (10) and (12), and the charge chambers can be described by Equations (7), (9) and (11). For the fixed chambers, $u$ can be set to zero; for isothermal chambers, $d\theta/dt$ can be set to zero. We needed to model only five standard subsystems and redefine the structure parameters during the simulation, which helped to improve the efficiency of the modelling.

A flow chart of the simulation programme was created, as shown in Figure 4, and the simulation model was established using the standard subsystems. For the simulation, it was necessary to automatically judge the stroke end. Then, the variable step size was selected to improve the calculation efficiency, and the solver was set as ode45 (Dormand–Prince).

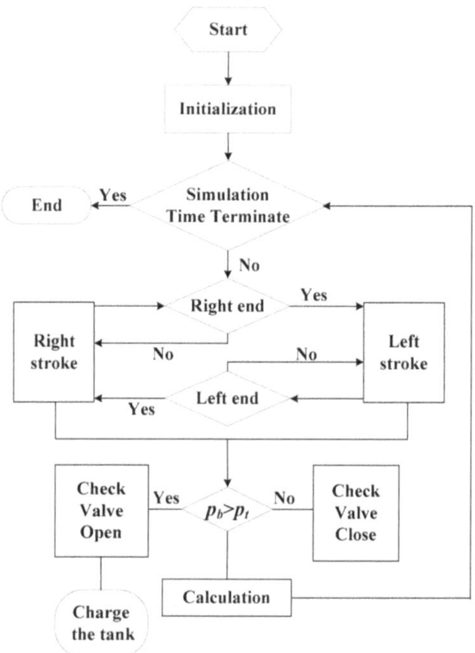


**Figure 4.** Flow chart of simulation programme.

*4.2. Energy Efficiency Test*

To assess the energy efficiency of the booster regulator, an air power metre (APM) [24] was used to measure the energy input and output. The core element of an APM is its quick laminar flow rate sensor, which can convert the pressure and flow rate into air power based on Equation (2). The real-time energy consumption can also be monitored by integrating the air power.

The experiment circuit was configured as shown in Figure 5. It consisted of a precision regulator (IR3020-02A, SMC, Tokyo, Japan), a booster regulator (VBA20A-03, SMC, Tokyo, Japan; VBA-R), two APMs (APM-L-800s, TOKYO METER, Tokyo, Japan), an air tank (VBAT10A1-15, SMC, Tokyo, Japan), a pressure sensor (AP-13S, KEYENCE, Osaka, Japan), a throttle valve (AS300, SMC, Tokyo, Japan), and a data acquisition card (TNS6810, INTERFACE, Hiroshima, Japan). The accuracy of the APM and pressure sensor were ±2% F.S. and ±0.5% F.S., respectively.

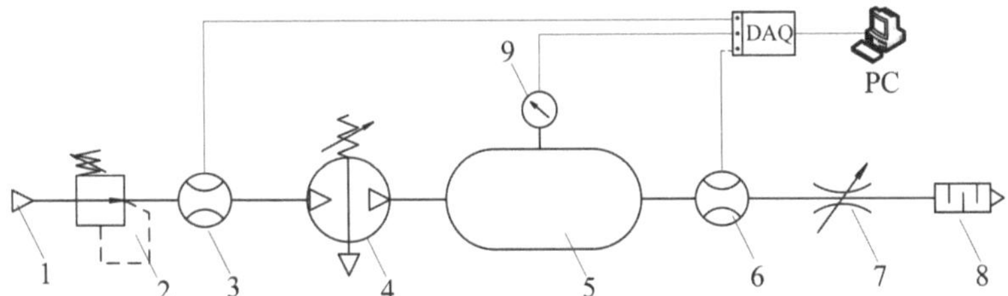


**Figure 5.** Schematic diagram of booster regulator test. 1: air source, 2: regulator, 3: air power metre (APM), 4: booster regulator, 5: air tank, 6: APM, 7: throttle valve, 8: silencer, 9: pressure sensor.

To verify the energy recovery performance considering the cost and period, we used an alternative method rather than manufacture a new VBA-R. The experimental apparatuses used to verify the energy recovery of the VBA-R are shown in Figure 6. According to the structure shown in Figure 2, a double-acting, double-rod cylinder (CG1WLN, SMC, Tokyo, Japan) and two double-acting, single-rod cylinders (CDG1YL, SMC, Tokyo, Japan) were assembled together. The 7/2 solenoid valve was replaced by two 5/2 solenoid valves (SV3200R, SMC, Tokyo, Japan), which were connected in parallel and controlled by a programmable logic controller (PLC).

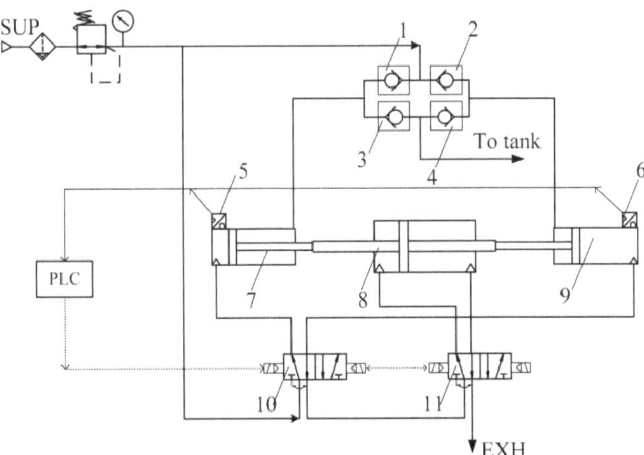


**Figure 6.** Schematic diagram of VBA-R. 1,2,3,4: check valve, 5,6: auto switch, 7,9: single-rod cylinder, 8: double-rod cylinder, 10,11: 5/2 solenoid valve.

The details of experiment apparatuses are listed in Table 1.

**Table 1.** Main equipment and their parameters.

| Components | Model | Parameters |
| --- | --- | --- |
| Cylinder | CDG1YL63 | $L$:100 mm; $D$:63 mm |
| | CG1WLN80 | $L$:100 mm; $D$:80 mm |
| Solenoid valve | SV3200R | C:4.5 $\times$ 10$^{-8}$ m$^3$/(s·Pa); $b$:0.3 |
| Check valve | AKH10-00 | C:4.8 $\times$ 10$^{-8}$ m$^3$/(s·Pa); $b$:0.3 |
| Throttle valve | AS300 | C:2.7 $\times$ 10$^{-8}$ m$^3$/(s·Pa); $b$:0.2 |
| PLC | FX1s-20MR | - |
| F.R.L Units | AC40D | 800 dm$^3$/min 0.05–0.7 MPa |
| Air power meter | APM-L-800s | 8–800 dm$^3$/min 0–1.2 kW |
| Pressure sensor | ISE80H | 0–1.0 MPa |
| Booster regulator | VBA20A-03 | 1 m$^3$/min 0.2–1.0 MPa |
| Tank | VBAT10A1-15 | 10 dm$^3$ |
| DAQ card | TNS6810 | - |

With this equipment, the configuration of the experiments corresponding to Figures 5 and 6 are shown in Figure 7.

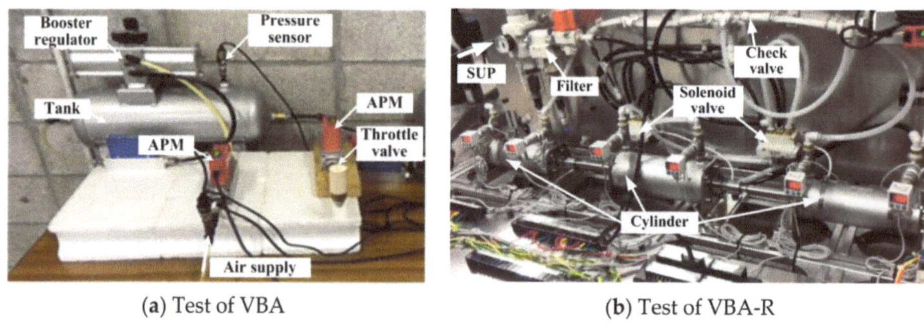


(a) Test of VBA    (b) Test of VBA-R

**Figure 7.** Test rig of energy efficiency test, (**a**) test of VBA and (**b**) test of VBA-R.

In the test circuit, the APM was mounted at the outlet of tank instead of the booster regulator. As we know, the cheque valve opens only when the pressure in the boost chamber is larger than that in the tank. In that situation, the cheque valve opened intermittently, and air was exhausted in an unsteady flow. The flow rate during unsteady flow is difficult to measure accurately. However, the flow rate output from the tank was almost steady for the constant load, which was easy to measure.

## 5. Results and Discussion

### 5.1. Experiment Results

With the experiment system, the main characteristics of the VBA-R were measured as shown in Figures 8 and 9. Figure 8 shows the pressure response in the tank: the pressure gradually increases and then becomes steady. There is a slightly greater time-delay in the response of the experiment than in the simulation at the beginning of air charging, which is due to the isothermal assumption in Equation (15). However, the deviation is within the uncertainty of the system (mainly sensor errors) and can be neglected. Therefore, we believe the experiment results are in good agreement with the simulation. Figure 9 shows the air power output from the tank, and the regularity of change is similar to the pressure.

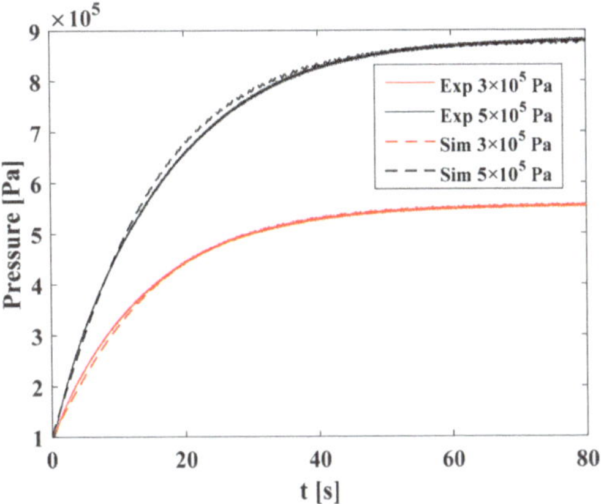

**Figure 8.** Pressure response in the tank.

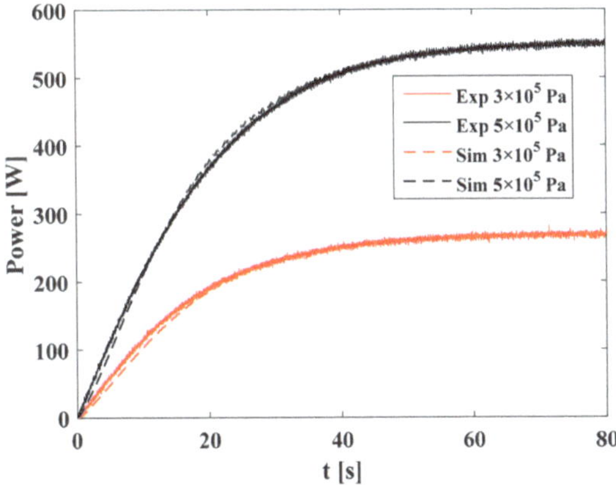

**Figure 9.** Air power output from the system.

The mathematical model was verified by experiments, and the simulation model was used to study the other characteristics and the influences of different parameters.

## 5.2. Characteristics of VBA-R

The simulation model was established in Section 4, and its main parameters are shown in Table 2. Then the characteristics of VBA-R were obtained and shown in Figures 10–13, which included the dynamic characteristics of the piston and the pressure response in the boost chamber and tank.

**Table 2.** Parameters and values used in the simulation.

| Parameters | Piston Diameter (mm) | Piston Rod Diameter (mm) | Stroke (mm) | Supply Pressure (kPa(abs)) | Atmosphere Temperature (K) | Sonic Conductance of Throttle Valve (m³/(s·Pa)) | Tank Volume (L) |
|---|---|---|---|---|---|---|---|
| VBA-R | $D_{ba} = 63,$ $D_{ra} = 80$ | 10 | 100 | 300 | 293 | $2.79 \times 10^{-9}$ | 5.0 |

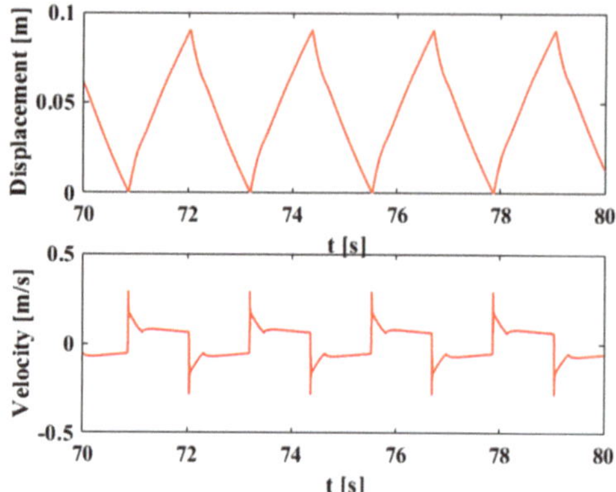

**Figure 10.** Curves of piston velocity and displacement.

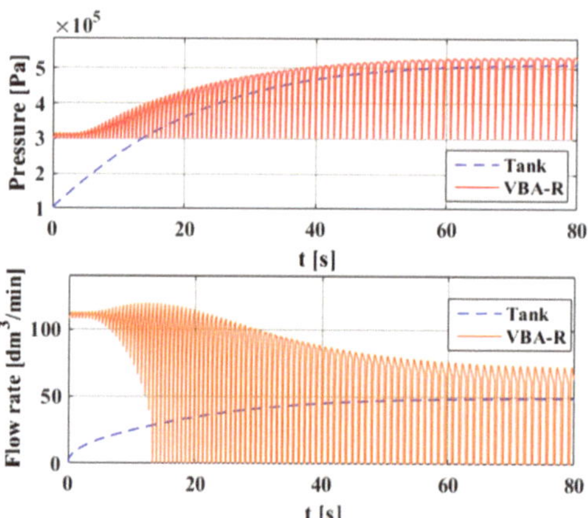

**Figure 11.** Pressure/flow rate response of VBA-R and tank.

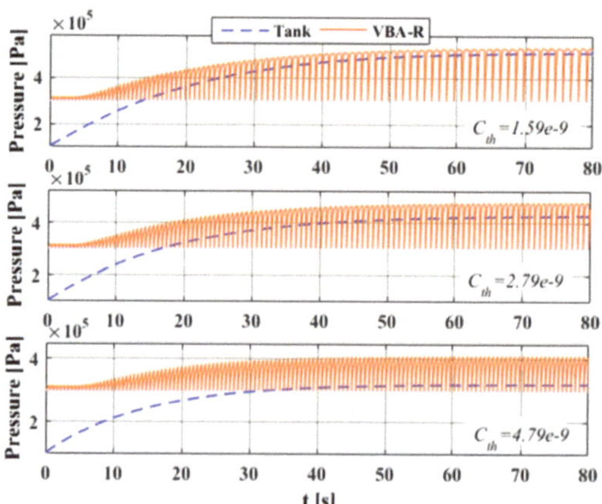



**Figure 12.** Pressure response with different loads.

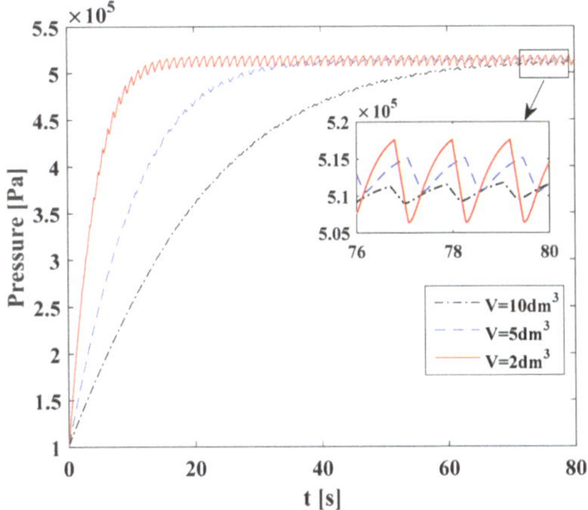

**Figure 13.** Pressure fluctuations in the tank.

Figure 10 gives the motion characteristics of the piston. When the piston changes its direction, a slight fluctuation occurs because we forced the piston to change its direction at the end of its stroke in the simulation. This would not happen in reality because of the existence of the air buffer.

Figures 11 and 12 reflect the outputs of the VBA-R and tank. In the Figure 11, it is clear that the outputs of the tank are smaller than those of the booster regulator for both pressure and flow rate. It also suggests that the outputs of the booster regulator are unsteady and discontinuous. Figure 12 further proves that the output pressure of the tank is less than that of the booster regulator, and with increasing load, the boost ratio decreases sharply. This must be seriously considered when choosing a booster regulator.

Figure 13 indicates that the pressure in the tank fluctuates, but it is much steadier than the booster regulator when compared with the pressure response in Figures 11 and 12. We also found that the fluctuations get smaller as the tank volume increases. This is a good reference for when we match a tank with a booster regulator.

### 5.3. Energy Efficiency of VBA-R

Equation (6) suggests that the energy efficiency was mainly affected by the supply pressure and boost ratio. However, when the supply pressure was constant, the boost ratio changed with the flow rate consumed by the load. Beyond this, the structure was considered to be proposed as a design reference. The parameters were changed to identify the influence of supply pressure and boost ratio.

### 5.3.1. Effect of Supply Pressure and Flow Rate

Based on the concept of air power, the pressure is inversely proportional to the flow rate based on energy conservation. Then, considering the flow rate characteristics, the energy efficiency was obtained as shown in Figure 14. As the supply pressure increases, the energy efficiency curves move down when the flow rate is small and they move up as the flow rate increases. To understand this phenomenon, we have to consider the results shown in Section 5.4.1 and Equation (6). According to Equation (6), the energy efficiency gets lower with increasing supply pressure or decreasing boost ratio. From Section 5.4.1, it can be seen that the boost ratio increases with increasing supply pressure, and the losses always increase with increasing supply pressure. Therefore, based on an overall consideration of these factors, the energy efficiency gets smaller with increasing supply pressure, but it is relatively stable for an improving boost ratio. In other words, the boost ratio can compensate for the energy efficiency loss caused by that increasing supply pressure.

Figure 14 also suggests that, when the pressure is constant, the more the flow rate is, the less the energy efficiency is. Therefore, it better to use a booster regulator with a small flow rate. If the flow consumption is large, a high supply pressure has a much better efficiency.

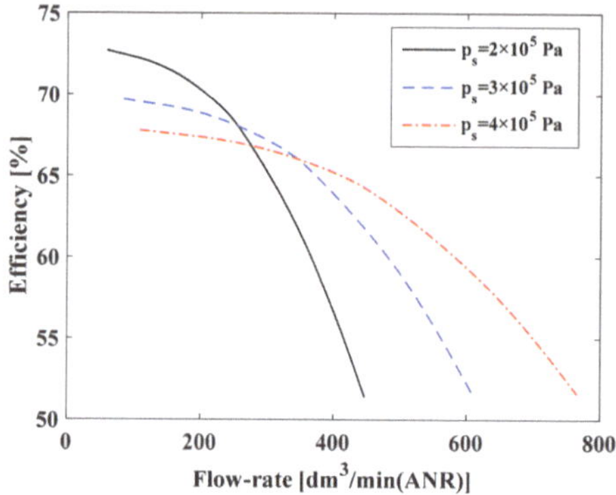


**Figure 14.** Energy efficiency of VBA-R.

## 5.3.2. Effect of Diameter Ratio

Figure 2 gives the structure of the VBA-R, which we have claimed helps energy recovery. However, when we considered this again, we found that the pressure in the drive chamber had simultaneously done negative work. With an increase of the recovery chamber diameter, the volume increased, which lead the recovery pressure to decrease. Therefore, it should be confirmed carefully whether the force on the piston in the recovery chamber is increasing or decreasing. This also affected the energy efficiency of the VBA-R, so we changed the diameters to study the influence of the diameter ratio (the ratio of the recovery chamber diameter to the drive chamber diameter, $D_2/D_1$) in order to properly select the design parameters.

The change of boost ratio with $D_2/D_1$ is shown in Figure 15, when all other parameters are constant. It is clear that the boost ratio increases first and then decreases, and the boost ratio reaches its greatest value when the diameter ratio is nearly 1.3. The variation of energy efficiency is depicted in Figure 16, in which the energy efficiency decreases with increasing supply pressure. The energy efficiency reaches its largest value for different supply pressures when the diameter ratio is closest to 1.3.

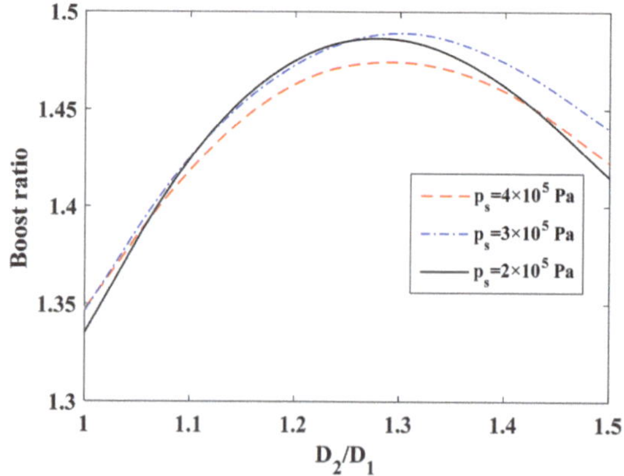


**Figure 15.** Relationship between boost ratio and diameter ratio.

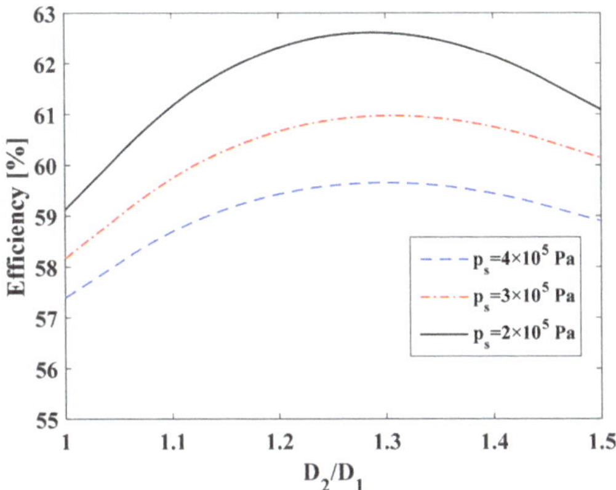

**Figure 16.** Relationship between energy efficiency and diameter ratio.

### 5.3.3. Effect of Operate Conditions

There are different demands in practical applications, and, according to the different operating conditions, we hope to achieve higher energy efficiency. We made a comparison between two kinds of operation conditions, as shown in Figure 17. In these two conditions, the energy efficiency difference was up to 20%. We analysed previously that, as the supply pressure or flow rate increased, the energy efficiency decreased. Thus, in a practical application, it is better to use a booster regulator with a high boost ratio.

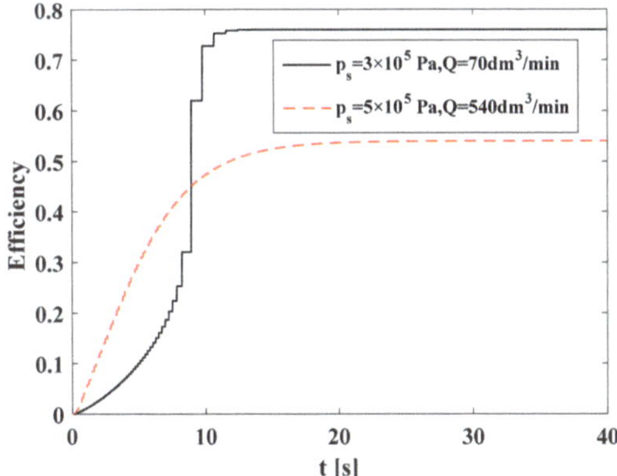

**Figure 17.** Efficiency under different operating conditions.

In Figure 18, the influence of boost ratio was further verified, and the result was consistent with the analysis according to Equation (6), which indicated that the energy efficiency gets higher with decreasing supply pressure or increasing boost ratio. However, the boost ratio gets smaller with decreasing supply pressure, so we should comprehensively consider these factors in order to obtain higher efficiency.

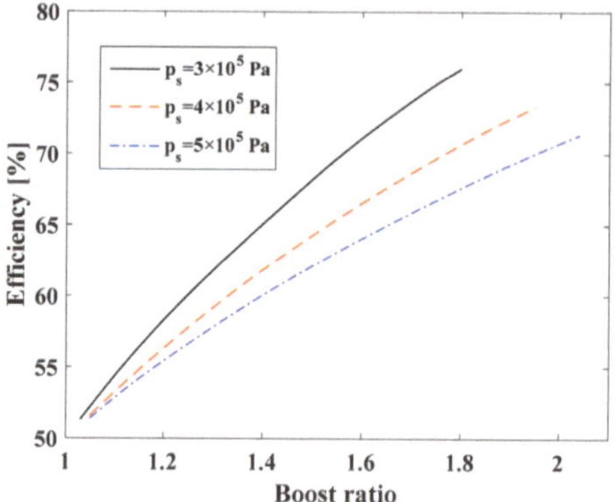

**Figure 18.** Efficiency under different boost ratios.

## 5.4. Comparison of VBA and VBA-R

The VBA-R was designed as an updated version of the VBA, so a performance comparison and assessment are essential. Flow-rate characteristics are the basic performance indicator that should be considered first for pneumatic components. We added energy efficiency as another index for the booster regulator: it also can be referred to in the design of other pneumatic components by introducing the concept of air power.

### 5.4.1. Comparison of Flow-Rate Characteristics

The flow rate characteristics of the VBA and VBA-R are shown in Figure 19. It is clear that, with the same supply pressure, the output pressure of the VBA-R is much higher. The largest boost ratio of the VBA is two-times as much as the supply pressure, but that of the VBA-R is even greater. This directly indicates that the recovery energy is used to push the piston, and then the boost ratio enhances it. However, as Figure 15 shows, we must select the diameter ratio properly to obtain the best performance.

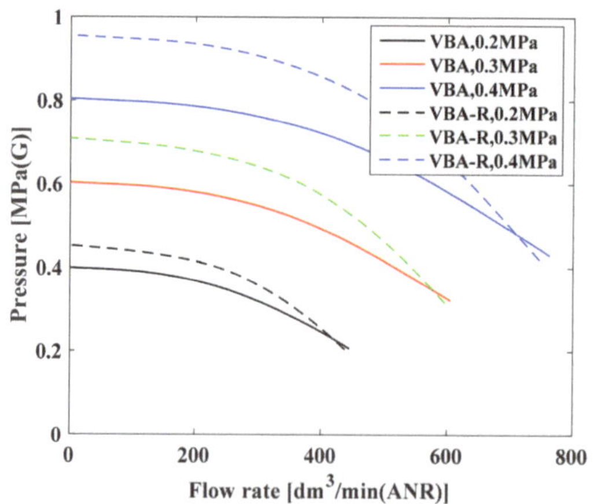

**Figure 19.** Flow rate characteristics of booster regulator system.

### 5.4.2. Comparison of Energy Efficiency

To verify that the recovery energy helps improve the energy efficiency, the VBA and VBA-R were compared as shown in Figure 20. It is clear that, as the supply pressure increases, the energy efficiency decreases. However, with the help of the recovery chamber, the efficiency of the VBA-R is always 5–10% higher than that of the VBA. Once again, this proves the performance improvement of the VBA-R.

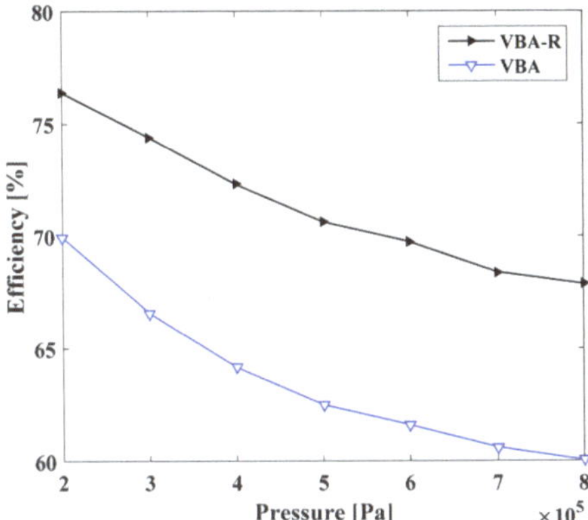


**Figure 20.** Comparison of VBA and VBA-R.

## 6. Conclusions

A new pneumatic booster regulator with energy recovery was proposed and its energy efficiency characteristics were modelled by considering the tank. It was consistent with the application of a booster regulator system. With this model, the energy efficiency was confirmed by using the concept of air power. Then, the effects of various factors on energy efficiency were comprehensively analysed, and the performance improvement of the VBA-R was verified by comparing it with the VBA. The conclusions can be summarised as follows:

(1) The system outputs, especially the pressure, were less than those of the booster regulator. When the air consumption increased, the boost ability decreased, and the tank output became much smaller;

(2) The energy efficiency was directly affected by the supply pressure, and it decreased with increasing supply pressure or decreasing boost ratio;

(3) When the supply pressure was constant, the energy efficiency decreased with the flow rate consumed by an increasing load;

(4) Both the boost ratio and energy efficiency increased first and then decreased with the increasing recovery chamber diameter. When the diameter ratio was close to 1.3, the VBA-R has the largest boost ratio and energy efficiency;

(5) The recovery chamber helped improve the boost ratio by 15–25% and energy efficiency by 5–10% under different operating conditions. Thus, it will save energy over long running times.

**Acknowledgments:** The research work presented in this paper is financially supported by SMC Corporation. And thanks for the support by Wuhan International Scientific and Technological Cooperation Project, No. 2016030409020224.

**Author Contributions:** Toshiharu Kagawa and Kotaro Tadano conceived and designed the experiments; Fan Yang performed the experiments and analyzed the data; Fan Yang wrote the paper and Gangyan Li modified the manuscript.

## Nomenclature

| | |
|---|---|
| $b$ | critical pressure ratio |
| $c$ | viscous friction coefficient |
| $C$ | sonic conductance [m$^3$/(s Pa)] |
| $c_v$ | volumetric specific heat [J/(kg K)] |
| $c_p$ | pressure specific heat [J/(kg K)] |
| $D$ | piston diameter [m] |
| $E$ | energy of air [J] |
| $F_f$ | frictional force [N] |
| $G$ | mass flow rate [kg/s] |
| $h$ | heat transfer coefficient [W/(m$^2$ K)] |
| $H$ | enthalpy [J] |
| $\kappa$ | specific heat index |
| $M$ | mass of piston [kg] |
| $\eta$ | energy efficiency |
| $R$ | gas state constant, 287 [J/(kg K)] |
| $\rho$ | density [kg/m$^3$] |
| $p$ | pressure [Pa] |
| $P$ | air power [W] |
| $q$ | heat exchange [J] |
| $Q$ | volume flow rate [m$^3$/s] |
| $S$ | section area of piston [m$^2$] |
| $t$ | time [s] |
| $T$ | cycle of booster regulator [s] |
| $\theta$ | temperature [K] |
| $u$ | velocity of piston [m/s] |
| $U$ | internal energy [J] |
| $V$ | volume of chamber [m$^3$] |
| Subscript | |
| $0$ | standard state of atmosphere [20 °C,100 kPa] |
| $1$ | upstream |
| $2$ | downstream |
| $ba/bb$ | boost chamber A/B |
| $da/db$ | drive chamber A/B |
| $ra/rb$ | recovery chamber A/B |
| $bc/dc$ | boost/drive chamber |
| $a$ | state of atmosphere |
| $c$ | charge chamber |
| $d$ | discharge chamber |
| $t$ | air tank |

## References


1. Cai, M.L.; Kagawa, T. Energy consumption assessment and energy loss analysis in pneumatic system. *Chin. J. Mech. Eng.* **2007**, *43*, 69–74. [CrossRef]
2. Li, T.C.; Wu, H.W.; Kuo, M.J. A study of gas economizing pneumatic cylinder. *J. Phys. Conf. Ser.* **2006**, *48*, 1227. [CrossRef]
3. Cai, M.L.; Kagawa, T. Simulation for energy savings in pneumatic system. In *Systems Modeling and Simulation*; Springer: Tokyo, Japan, 2007; pp. 258–261.
4. Wang, H.T.; Xiong, W.; Wang, X. Research on the static characteristics of air driven gas booster. *Proc. JFPS Int. Symp. Fluid Power* **2008**, *2008*, 715–718. [CrossRef]
5. Shi, Y.; Cai, M.L. Working characteristics of two kinds of air-driven boosters. *Energy Convers. Manag.* **2011**, *52*, 3399–3407. [CrossRef]

6.  Li, Z.Y.; Zhao, Y.Y.; Li, L.S.; Shu, P.C. Mathematical modeling of compression processes in air-driven boosters. *Appl. Therm. Eng.* **2007**, *27*, 1516–1521. [CrossRef]

7.  Shi, Y.; Jia, G.W.; Cai, M.L.; Xu, W.Q. Study on the dynamics of local pressure boosting pneumatic system. *Math. Probl. Eng.* **2015**, *2015*, 1–11. [CrossRef]

8.  Shi, Y.; Cai, M.L. Dimensionless study on output flow characteristics of expansion energy used pneumatic pressure booster. *J. Dyn. Syst. Meas. Control* **2013**, *135*, 1–8. [CrossRef]

9.  Shi, Y.; Cai, M.L. Dimensionless study on outlet flow characteristics of an air-driven booster. *J. Zhejiang Univ. Sci. A* **2012**, *13*, 481–490. [CrossRef]

10. Merkelbach, S.; Murrenhoff, H. Exergy based analysis of pneumatic air saving measures. In *ASME/BATH 2015 Symposium on Fluid Power and Motion Control*; American Society of Mechanical Engineers: Chicago, IL, USA, 2015; p. V001T01A007.

11. Harris, P.; Nolan, S.; O'Donnell, G.E. Energy optimisation of pneumatic actuator systems in manufacturing. *J. Clean. Prod.* **2014**, *72*, 35–45. [CrossRef]

12. Cummins, J.J.; Barth, E.J.; Adams, D.E. Modeling of a pneumatic strain energy accumulator for variable system configurations with quantified projections of energy efficiency increases. In *ASME/BATH 2015 Symposium on Fluid Power and Motion Control*; American Society of Mechanical Engineers: Chicago, IL, USA, 2015; p. V001T01A055.

13. Luo, X.; Wang, J.H.; Sun, H.; Derby, J.W.; Mangan, S.J. Study of a new strategy for pneumatic actuator system energy efficiency improvement via the scroll expander technology. *IEEE/ASME Trans. Mechatron.* **2013**, *18*, 1508–1518. [CrossRef]

14. Luo, X.; Sun, H.; Wang, J.H. An energy efficient pneumatic-electrical system and control strategy development. In Proceedings of the 2011 American Control Conference, San Francisco, CA, USA, 29 June–1 July 2011.

15. Shi, Y.; Wu, T.C.; Cai, M.L.; Wang, Y.X.; Xu, W.Q. Energy conversion characteristics of a hydropneumatic transformer in a sustainable-energy vehicle. *Appl. Energy* **2016**, *171*, 77–85. [CrossRef]

16. Trujillo, J.A. Energy Efficiency of High Pressure Pneumatic Systems. Ph.D. Thesis, Polytechnic University of Catalonia, Barcelona, Spain, 2015.

17. Cai, M.; Kawashima, K.; Kagawa, T. Power assessment of flowing compressed air. *J. Fluids Eng.* **2006**, *128*, 402–405. [CrossRef]

18. ISO6358-1. *Pneumatic Fluid Power—Determination of Flow-Rate Characteristics of Components Using Compressible Fluids—Part 1: General Rules and Test Methods for Steady-State Flow*; International Organization for Standardization (ISO): Geneva, Switzerland, 2013.

19. Kaasa, G.O.; Chapple, P.J.; Lie, B. Modeling of an electro-pneumatic cylinder actuator for nonlinear and adaptive control, with application to clutch actuation in heavy-duty trucks. In Proceedings of the 3rd International Ph.D. Symposium on Fluid Power, Terrassa, Spain, 30 June–2 July 2004.

20. Okawa, Y.; Youn, C.; Kawashima, K.; Kagawa, T. Flow rate measurement via isothermal discharge method for hydrogen. *Int. J. Hydrogen Energy* **2012**, *37*, 18882–18887. [CrossRef]

21. Harris, P.G.; O'Donnell, G.E.; Whelan, T. Modelling and identification of industrial pneumatic drive system. *Int. J. Adv. Manuf. Technol.* **2012**, *58*, 1075–1086. [CrossRef]

22. Li, J.; Kawashima, K.; Fujita, T.; Kagawa, T. Control design of a pneumatic cylinder with distributed model of pipelines. *Precis. Eng.* **2013**, *37*, 880–887. [CrossRef]

23. Baojun, H.; Fujita, T.; Kawashima, K.; Kagawa, T. Influence of pressure condition change on the flow rate characteristics of pneumatic valve. *J. Jpn. Fluid Power Syst. Soc.* **2001**, *32*, 143–149.

24. Cai, M.L.; Kagawa, T. Design and application of air power meter in compressed air systems. In Proceedings of the Eco Design 2001: Second International Symposium; Environmentally Conscious Design and Inverse Manufacturing, Tokyo, Japan, 11–15 December 2001.

*Article*

# An Experimental Study on Hysteresis Characteristics of a Pneumatic Braking System for a Multi-Axle Heavy Vehicle in Emergency Braking Situations

**Zhe Wang [1,2,*], Xiaojun Zhou [1,2], Chenlong Yang [1,2], Zhaomeng Chen [1,2] and Xuelei Wu [3]**

[1]   State Key Laboratory of Fluid Power and Mechatronic Systems, Zhejiang University, Hangzhou 310027, China; cmeesky@163.com (X.Z.); yclzju@163.com (C.Y.); chernzm@zju.edu.cn (Z.C.)
[2]   Key Laboratory of Advanced Manufacturing Technology of Zhejiang Province, Zhejiang University, Hangzhou 310027, China
[3]   Beijing Institute of Space Launch Technology, Beijing 100076, China; xlwuworks@vip.sohu.com
*   Correspondence: wzhe@zju.edu.cn; Tel.: +86-0571-8795-2516

Received: 27 June 2017; Accepted: 2 August 2017; Published: 6 August 2017

**Featured Application: The obtained data and model from the research can be useful for the design of a pneumatic braking system and the development of advanced brake control strategy with respect to multi-axle heavy vehicles in the future.**

**Abstract:** This study aims to investigate the hysteresis characteristics of a pneumatic braking system for multi-axle heavy vehicles (MHVs). Hysteresis affects emergency braking performance severely. The fact that MHVs have a large size and complex structure leads to more nonlinear coupling property of the pneumatic braking system compared to normal two-axle vehicles. Thus, theoretical analysis and simulation are not enough when studying hysteresis. In this article, the hysteresis of a pneumatic brake system for an eight-axle vehicle in an emergency braking situation is studied based on a novel test bench. A servo drive device is applied to simulate the driver's braking intensions normally expressed by opening or moving speed of the brake pedal. With a reasonable arrangement of sensors and the NI LabVIEW platform, both the delay time of eight loops and the response time of each subassembly in a single loop are detected in real time. The outcomes of the experiment show that the delay time of each loop gets longer with the increase of pedal opening, and a quadratic relationship exists between them. Based on this, the pressure transient in the system is fitted to a first-order plus time delay model. Besides, the response time of treadle valve and controlling pipeline accounts for more than 80% of the loop's total delay time, indicating that these two subassemblies are the main contributors to the hysteresis effect.

**Keywords:** hysteresis effect; pneumatic braking system; multi-axle heavy vehicle; emergency braking

---

## 1. Introduction

Multi-axle heavy vehicles (MHVs) are usually used as land transport platforms for large-scale equipment. Compared with ordinary vehicles, MHVs have large inertia and a long wheelbase, and they usually work in complicated and changeable driving conditions. Hence, good braking performance is an important guarantee for fast maneuvering and safe driving. At present, the pneumatic braking system (PBS) is widely employed in MHVs with the advantages of large braking force, good reliability and easy maintenance [1].

Due to compressibility of gas, PBS will inevitably perform a hysteresis effect, which may easily lead to serious traffic accident in an emergency braking condition [2]. For example, when a vehicle runs at a normal speed of 20 m/s (approximately 70 km/h), if the braking time is delayed 0.5 s, the braking distance will increase by nearly 10 m. This delay could lead to heavy longitudinal loads along the vehicle instantaneously [3]. The consequences will be more serious for the MHV owing to its large inertia.

Currently, electronic brake systems (EBSs), electronic brake force distribution (EBD) and other advanced control technologies have been successfully applied to the majority of passenger cars, while the applications on MHV are still relatively narrow and immature [4]. In this case, scholars have undertaken a large amount of related research recently, and various technologies have been proposed and successfully implemented for better pneumatic braking performance of MHVs, including sliding mode controllers [5–7], nonlinear model predictive controllers [8], indirect adaptive robust controllers (IARC) [9], and novel braking actuators [10]. Nevertheless, the braking performance of MHV is still much poorer than that of the passenger car [7]. The reason may be the following. Most of these studies have paid much attention only to the optimization and improvement of braking control strategies [11]. Actually, the response time of valves as well as the long pipeline length tremendously limit the performance of the pneumatic brake system, making it harder to control [9,12]. Therefore, the properties of PBS itself is the basis for studying braking control strategies.

Hysteresis is an inherent property of the PBS, which with no doubt has a significant impact on braking performance. Research on hysteresis characteristics can be divided into the system level and component level. Most literature has focused on the latter, such as treadle valves [13], brake pipelines [14], relay valves [15], proportional valves [16], and other subassemblies. However, subassemblies in a loop are connected to each other and affect each other; analysis of individual subassembly seems to be insufficient. In terms of the system level, Selvaraj [17] and He [18] established simulation models of the PBS on two-axle vehicles using AMEsim (Advanced Modeling Environment for performing Simulation of engineering systems) and MWorks, respectively. Pugi and his team [3,19] have conducted in-depth research on the pneumatic braking plant of a railway vehicle and established a complete parametric simulation library using Matlab-Simulink, which makes customizable design possible. As for MHVs, existing literature concerned with the hysteresis effect is rare, and the few accessible studies mainly concentrate on road tests [20]. However, road tests imply high risk, have high costs, and involve a limited range of working conditions. In addition, sensors are not easily placed or replaced on subassemblies after the vehicle is completely assembled. Hence, a test bench for studying hysteresis characteristics of PBS for MHV is necessary.

A PBS on an MHV is usually composed of a service brake circuit, a parking brake circuit and an auxiliary braking circuit. When a driver encounters an emergency braking condition, pushing the brake pedal is probably the most common behavior. Therefore, the opening and moving speed of the pedal directly reflect the driver's braking intentions, and they have the most direct influence on the hysteresis effect of the PBS since the brake pedal is the power source of the service brake circuit [21,22]. Hence, this paper pays more attention to the service brake circuit.

In this paper, a novel test bench of PBS for an 8-axle vehicle is described in detail. The influence of opening and moving speed of the brake pedal, as well as the response time of subassemblies on hysteresis effect in service braking circuit is analyzed. This research may provide a reference for further experimental study of hysteresis effect of the PBS, and the obtained data and laws may be useful for system design or the development of advanced control strategy on MHVs in the future.

## 2. System Principle

A PBS in an MHV is briefly shown in Figure 1. The system is composed of service brake circuit and parking brake circuit. Usually, double circuits are employed in the service brake circuit, just for avoiding the braking failure in case one of the two circuits has stochastic fault. Each circuit in the system mainly consists of gas reservoirs, control valves, brake chambers and brakes. The gas source of the whole vehicle is provided by an air compressor and stored in a main reservoir with large capacity as well as several normal ones with relatively small capacity. Control valves mainly comprise a hand brake valve and a treadle valve located in the cab, several relay valves, and delay relay valves arranged in the rear axle of the chassis. Drum brakes, which can offer more powerful braking force compared to disc brakes, are widely used in MHVs [23].

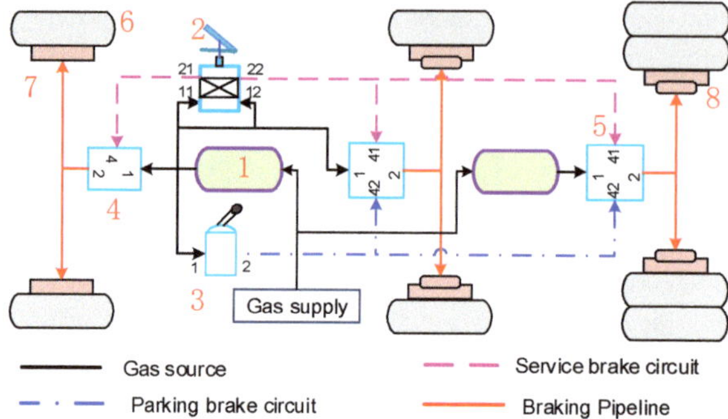



**Figure 1.** Schematic diagram of a pneumatic braking system (PBS) on a multi-axle heavy vehicle (MHV) (1—gas reservoir; 2—Treadle valve; 3—Hand brake valve; 4—Relay valve; 5—Delay relay valve; 6—Wheel; 7—Brake chamber; 8—Spring brake chamber).

The structure of the service brake circuit of a PBS on MHV can be simplified as shown in Figure 2. This circuit is made up of three parts: the gas supply part, pneumatic transmission part and mechanical actuator part. In this paper, we focus on the pneumatic transmission part because hysteresis is mainly caused by the long-distance pressure transmission in this part [24]. A complete working procedure of the service brake can be divided into the brake process, brake keeping process and brake release process. Taking the S-cam drum brake as an example, the working procedure can be described as follows. When encountering an emergency braking situation, the driver steps on the brake pedal quickly and the treadle valve is opened soon; this behavior meters out the compressed gas from the supply ports (port 11 and 12) of the treadle valve to its delivery ports (port 21 and 22). Then, the compressed gas travels to port 4 of the relay value through controlling pipeline, making the relay value open. Next, high-pressure gas moves into the brake chamber through braking pipeline and presses the diaphragm, pushing the plunger out. The ejected plunger then drives the brake arm, rotating the S-cam and expanding the brake shoes. As a result, the pads produce a friction torque, compelling the rotating hub to stop quickly. When the brake pedal is released by the driver, the gas is rapidly exhausted from the brake chamber, resulting in the opposite rotating direction of the S-cam. Subsequently, the contact between the pad and the hub is broken and the brake is finally released.

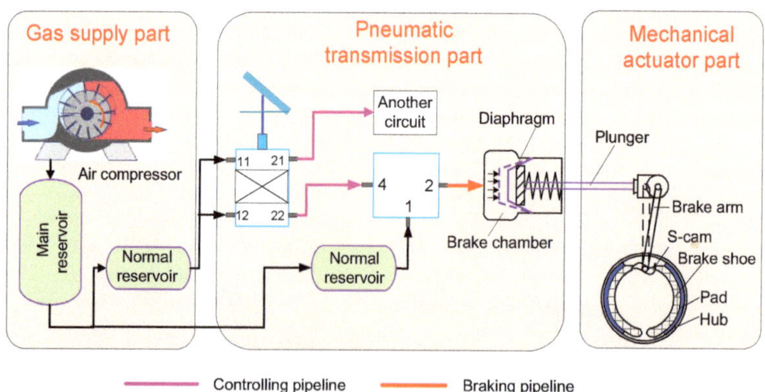



**Figure 2.** Structure of the service brake circuit of a PBS.

## 3. Test System Design

### 3.1. General Setup

Referring to a real-world 8-axle vehicle, the test bench of PBS for an eight-axle vehicle is designed as shown in Figure 3. The test bench is composed of two parts, the operation desk and control desk.

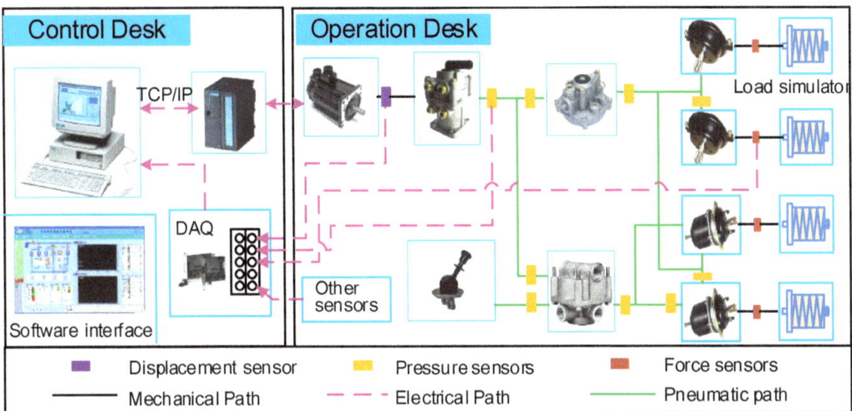


**Figure 3.** Schematic diagram of the test bench (all the sensors are connected to the data acquisition cards). TCP: transmission control protocol; IP: internet protocol; DAQ: data acquisition.

(1) The operation desk consists of gas supply devices, servo drive device, valves, pipelines, and load simulators; the test bench is shown in Figure 4. The gas supply device consists of several reservoirs and a pressure-regulating valve used for setting the working pressure of the system, as displayed in Figure 4b. A low-friction and high-precision linear servo drive device is used to drive the treadle valve, so as to accurately simulate the driver's intensions expressed by pedal opening and moving speed. It is composed of a servo motor and screw transmission, as shown in Figure 5. Sixteen load simulators (manual control) made up of disc springs are applied to simulate different load torques of 16 wheels. The force sensor arranged between the brake chamber and load simulator is used for detecting the brake force. In order to avoid function failure or error accumulation of subassemblies, these used key pneumatic components are calibrated beforehand on a specific test bench, as shown in Figure 4c.

(2) The control desk is composed of hardware and software. The hardware includes an Advantech 610H industrial personal computer (IPC) (Advantech, Taipei, China), a Panasonic programmable logic controller (PLC) ( Panasonic, Osaka, Japan), NI PCI-6229 data acquisition (DAQ) cards (National Instruments, Austin, TX, USA) and fast-response sensors. Communication between the PLC and the host computer is implemented through the object linking and embedding for process control (OPC) protocol using the transmission control protocol (TCP)/internet protocol (IP) interface. The test software is developed on the NI LabVIEW platform (version 14.0, National Instruments, Austin, TX, USA, 2014). The software has rich functions such as convenient data acquisition, processing, preservation and real-time display.

### 3.2. Arrangement of Sensors

The layout of gas transmission path of PBS on the test bench is shown in Figure 6. There are double service brake circuits, the front circuit including four loops (dark red) and the rear circuit including the other four loops (magenta); each loop corresponds to an axle. The parking brake circuit (blue) is also described in the figure. Due to the symmetrical arrangement of two brake chambers on one bridge, sensors are placed only in one side of the bridge.

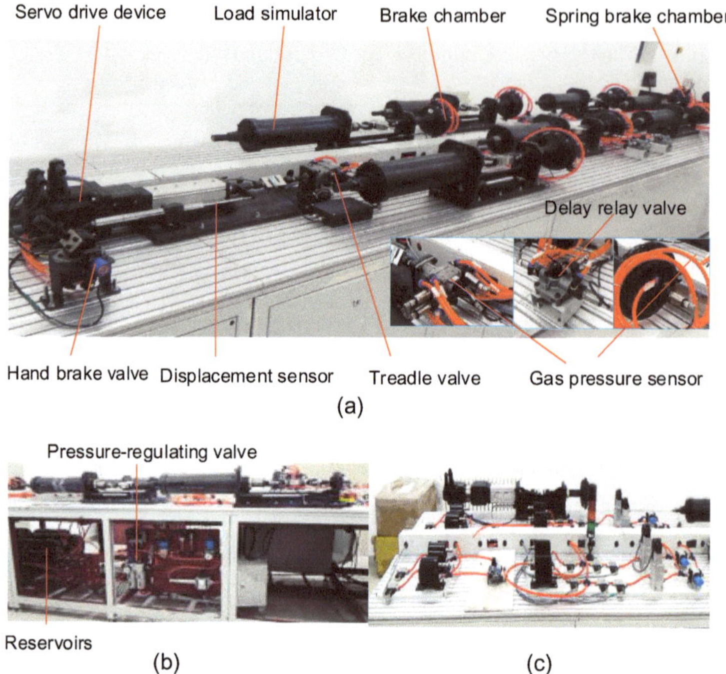



**Figure 4.** Operation desk of the test bench (**a**) The main operation zone (**b**) Gas supply devices (**c**) A specific test bench for calibrating the pneumatic subassemblies.

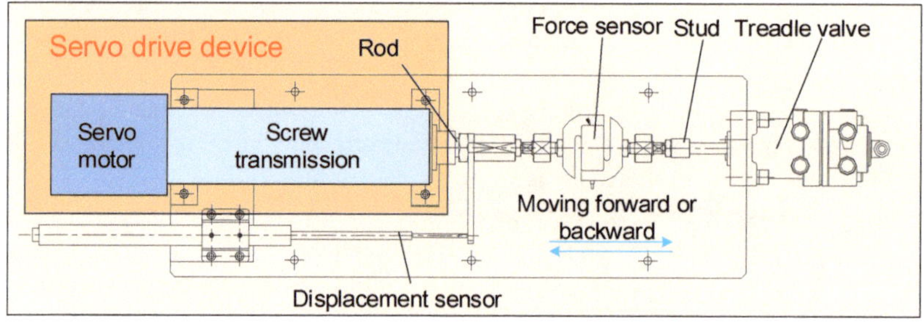



**Figure 5.** A servo drive device used for simulating driver's braking intensions.

Hysteresis characteristics of PBS on an MHV are mainly displayed in two aspects. On one hand, the delay time of each loop is different because of different gas transmission path. On the other hand, the delay time of a certain loop is the accumulation of response time of each subassembly. Both the times need to be detected on this test bench. In order to measure the delay time of each loop in service brake circuit, pressure sensors marked $f_A, f_B, f_C, f_D$ and $r_E, r_F, r_G, r_H$, are respectively placed on the entrance of brake chamber in the front four loops and rear four loops,as shown in Figure 6. To detect response time of each subassembly, pressure sensors marked $f_i, r_i (i = 1, 2, 3, 4)$ are respectively laid on the C-loop in the front circuit and E-loop in the rear circuit (Note: $f_4, r_4$ are the same as $f_C, r_E$. A displacement sensor marked TD (Treadle displacement sensor) is used for measuring pedal propulsion displacement, as shown in Figure 7.

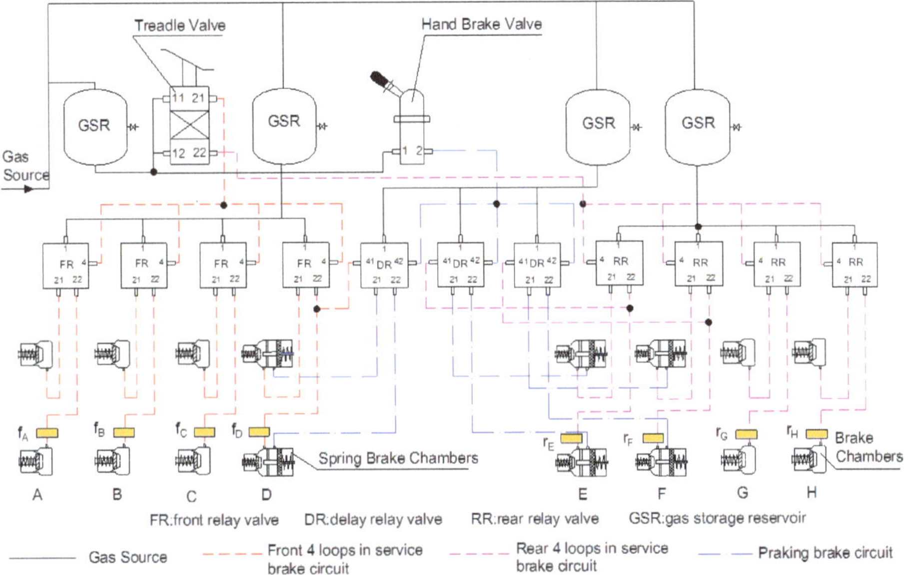


**Figure 6.** Layout of gas transmission path of the PBS on the test bench. GSR: gas storage reservoir; FR: front relay valve; DR: delay relay valve; RR: rear relay valve. A, B, C, D and E, F, G, H are front four loops and rear four loops in the service brake circuit,respectively.

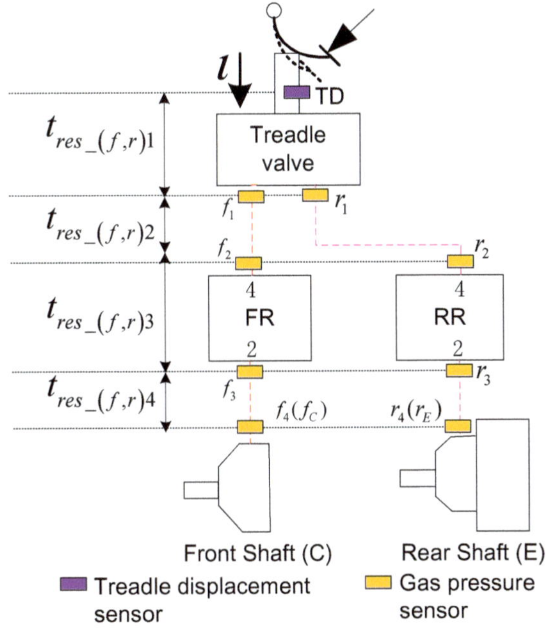


**Figure 7.** Arrangement of sensors. TD: treadle displacement sensor.

### 3.3. Key Parameters of the Test Bench

Control valves in this experiment include the relay valve, treadle valve, hand brake valve and delay relay valve; they are all produced by Westinghouse Air Brake Company (WABCO) and have been offered by a vehicle research insitute. Brake chambers of the QF series are produced by the Sanjiang Mechanical Company (Yibin, China). In order to obtain meaningful experimental results, these key subassemblies are the same as the real ones equiped on a certain eight-axle heavy vehicle. Gas supply pressure of the circuit is set at 800 kpa by the regulate valve shown in Figure 4b,which is the rated working pressure according to the National Standard GB12676-1999. Considering the size on a real-world eight-axle vehicle, the length of controlling pipeline is chosen as 4 m for the front circuit and 10 m for the rear one on the test bench. Meawhile, the lengths of braking pipeline in loop A, D, E and H are chosen as 1 m ,and in loop B, C, F and G as 2 m. Other key parameters are shown in Table 1.

**Table 1.** Key parameters of test bench. (FS: full scale).

| Name | Type | Range | Precision |
|---|---|---|---|
| Servo driver | TJE075-S100 | 0–40 mm | 0.5% FS (measurement) 1% FS (control) |
| Pressure sensor | MEAS | 0–1.6 Mpa | 0.3% FS |
| Force sensor | PST2T | 0–20000 N | 0.5% FS |
| Displacement sensor | DA35 | 0–35 mm | 0.5% FS |
| pipeline | Φ10PA | 0–2 Mpa | - |
| Control valves | WABCO | | - |

### 3.4. Testing Conditions

In order to study the influence of opening and moving speed of the brake pedal on hysteresis characteristics in service circuit, different propulsion displacements and action time of the servo drive device were set in different testing conditions. As for the brake process and brake release process in a certain condition, the two parameters were set the same except for the opposite moving direction of the servo motor. The detailed testing groups are shown in Table 2. Firstly, in the first to the fifth groups, the propulsion speeds ($v_T = l/t$) were approximately equal while the displacements were different, just for studying the relationship between hysteresis and opening. Secondly, in the fifth to the sixth group the propulsion displacements were the same while speeds were different, just for studying the relationship between hysteresis and moving speed. Given the fact that the rated stroke of pushrod in the tested treadle valve is 15 mm, maximum propulsion displacement of the servo drive device was set to 14 mm for protecting the valve. Each experiment group was repeated five times for less random error, and their mean values were used for subsequent calculations.

**Table 2.** Testing conditions.

| Group | Propulsion Displacement ($l$/mm) (Brake and Brake Release) | Action Time ($t$/s) (Brake and Brake Release) |
|---|---|---|
| 1 | 6 | 0.20 |
| 2 | 8 | 0.27 |
| 3 | 10 | 0.33 |
| 4 | 12 | 0.40 |
| 5 | 14 | 0.46 |
| 6 | 14 | 2.50 |

## 4. Results and Discussion

According to GB12676-1999(Road vehicle-braking systems-structure, performance and test methods. China) and GB7258-2012(Safety specifications for power-driven vehicles operating on roads. China), for a PBS, the delay time of a certain loop can be defined as follows,

$$
\begin{cases}
\Delta_{del\_up} = t_{\gamma_p\_up} - t_{start\_up} \\
\Delta_{del\_down} = t_{\gamma_p\_down} - t_{start\_down}
\end{cases},
\tag{1}
$$

where $\Delta_{del\_up}$ and $\Delta_{del\_down}$ are respectively the delay time in brake process and brake release process. $t_{start\_up}$ is the time when the brake pedal starts to move to begin the brake process, while $t_{start\_down}$ is the time when the brake pedal starts to move to begin the brake release process. Both of these are measured by a TD sensor. $t_{\gamma_p\_up}$ and $t_{\gamma_p\_up}$ are the times when the pressure in the brake chamber reach 75% and 15% of the steady value, and their corresponding pressures are marked as $\gamma_{p\_up}$ and $\gamma_{p\_down}$, respectively. For a clear description, the working procedure of a PBS is displayed in Figure 8.

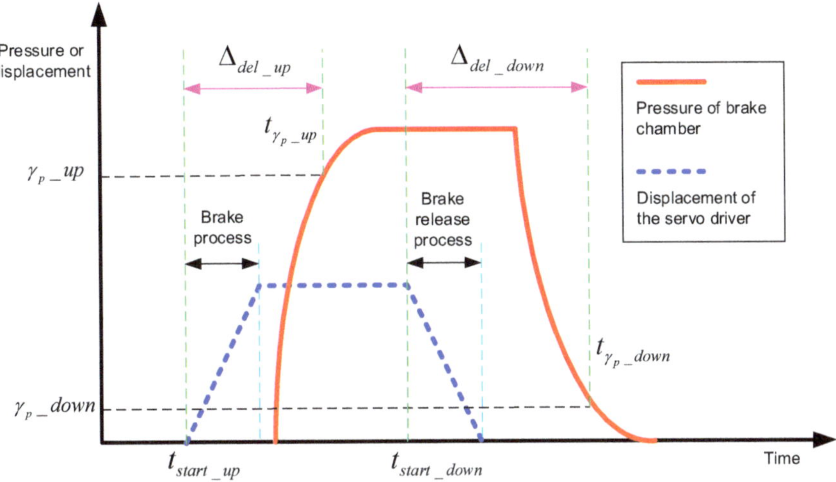


**Figure 8.** The working procedure of the PBS.

### 4.1. Hysteresis Effect of 8 Axle

The experimental results obtained under the testing condition of group 1 are shown in Figure 9, including the original data curve and the calculated delay time of eight axles based on Formula (1). The standard deviation is used to weigh the random error. Moreover, the delay times of eight axles under the testing conditions of all the testing groups are calculated as shown in Table 3. From this table, we know that the delay time of each loop is different from each other in both the brake process and brake release process. The asynchronous response of the multi-axle vehicle leads to different braking torques on different axes at the same braking moment, which may seriously affect the braking stability. The hysteresis effect in brake release process is relatively more obvious. This effect may lead to an unexpected once-more brake action due to a slow gas exhaust of the loop, which will result in difficult restarting of the vehicle. Therefore, when developing the hysteresis compensation control strategy for multi-axle vehicle, different delay times of every loop both in the brake process and brake release process should be taken into consideration.

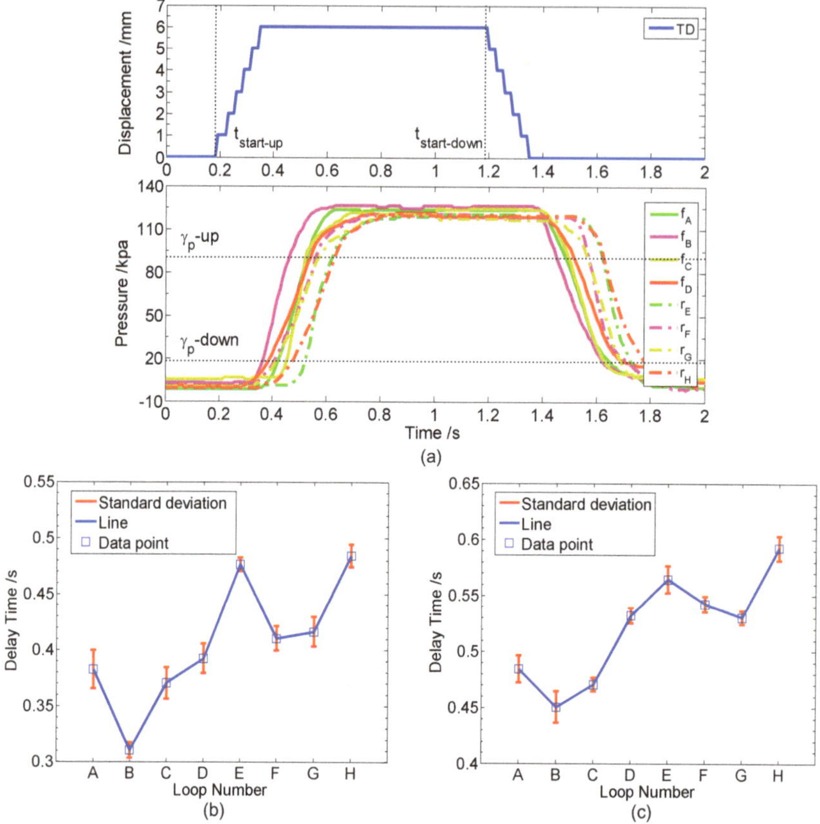

**Figure 9.** Hysteresis characteristic of PBS under the testing condition of group 1. (**a**) Original data curves; (**b**) Delay time of eight loops in the brake process; (**c**) Delay time of eight loops in the brake release process.

**Table 3.** Delay time of eight loops under the testing condition of all the groups.

| | | \multicolumn{12}{c}{Testing Groups} | | | | | | | | | | |
| Loop | | 1 | | 2 | | 3 | | 4 | | 5 | | 6 | |
| | | Up | Dw | Up | Dw | Up | Dw | Up | Dw | Up | Dw | Up | Dw |
|---|---|---|---|---|---|---|---|---|---|---|---|---|---|
| | A | 0.38 | 0.48 | 0.42 | 0.62 | 0.48 | 0.72 | 0.58 | 0.88 | 0.67 | 1.02 | 2.49 | 2.66 |
| | B | 0.31 | 0.45 | 0.34 | 0.51 | 0.42 | 0.73 | 0.50 | 0.84 | 0.62 | 0.93 | 2.41 | 2.63 |
| | C | 0.37 | 0.47 | 0.36 | 0.55 | 0.45 | 0.72 | 0.52 | 0.83 | 0.61 | 0.95 | 2.42 | 2.67 |
| **Delay** | D | 0.39 | 0.53 | 0.42 | 0.65 | 0.49 | 0.78 | 0.61 | 0.86 | 0.66 | 1.04 | 2.47 | 2.71 |
| **Time/$\Delta_{del}$** | E | 0.47 | 0.55 | 0.52 | 0.67 | 0.55 | 0.81 | 0.68 | 0.92 | 0.72 | 1.10 | 2.56 | 2.73 |
| | F | 0.41 | 0.54 | 0.45 | 0.62 | 0.52 | 0.75 | 0.65 | 0.89 | 0.70 | 1.04 | 2.50 | 2.68 |
| | G | 0.42 | 0.53 | 0.46 | 0.63 | 0.53 | 0.76 | 0.67 | 0.91 | 0.72 | 1.06 | 2.52 | 2.69 |
| | H | 0.48 | 0.59 | 0.55 | 0.69 | 0.57 | 0.83 | 0.69 | 0.97 | 0.78 | 1.12 | 2.57 | 2.75 |

Note: In this table, $\Delta_{del\_up}$ is denoted as "up", $\Delta_{del\_down}$ is denoted as "dw".

### 4.1.1. Relationship between Pedal Opening and Delay Time

The swing angle of the pedal arm or the linear displacement of the brake's master cylinder is usually used for evaluating the pedal opening [25]; the latter is adopted in this paper because of the easily measured moving displacement of the servo drive device. Thus, the pedal opening can be denoted as:

$$0 \leq \psi = \frac{l}{l_{max}} \leq 100\%, \tag{2}$$

where $l$ is the measured displacement and $l_{max}$ is the rated moving displacement, here $l_{max} = 15$ mm, and $\psi$ is the pedal opening. When analyzing the experimental results from group 1 to group 5 in Table 3, we know that the delay time of each loop was not the same due to different pedal opening though the pedal's moving speeds being approximately equal. For the full eight loops, the delay time of the rear four loops (E, F, G and H) are higher than the front four loops (A, B, C and D). For a certain loop, the higher the pedal opening, the longer the delay time. In order to explore the quantitative relationships between them, polynomial fitting is carried out for both the brake process and the brake release process using least squares method.

$$\begin{cases} \Delta_{del\_up} = K_{1\_up}\psi^2 + K_{2\_up}\psi + W_{up} \\ \Delta_{del\_down} = K_{1\_down}\psi^2 + K_{2\_down}\psi + W_{down} \end{cases} \tag{3}$$

where $K_{1\_up}$, $K_{2\_up}$, $K_{1\_down}$, $K_{2\_down}$ are coefficient matrixes and $W_{up}$, $W_{down}$ are constant matrixes. The fitting results are shown in Figure 10, and the corresponding fitting parameters are shown in Table 4. The fitting coefficient of determination $R^2$ can be calculated as follows,

$$R^2 = 1 - \frac{\sum_{a=1}^{n}\left(y_a - y_a'\right)^2}{\sum_{a=1}^{n}\left(y_a - \overline{y}\right)^2}, \tag{4}$$

where $y_a$ is the original data, $y_a'$ is the obtained fitting data, $\overline{y}$ is the average value of the original data, and $n$ is the number of original data. Each calculated $R^2$ is close to 1, indicating that the fitting is very good.

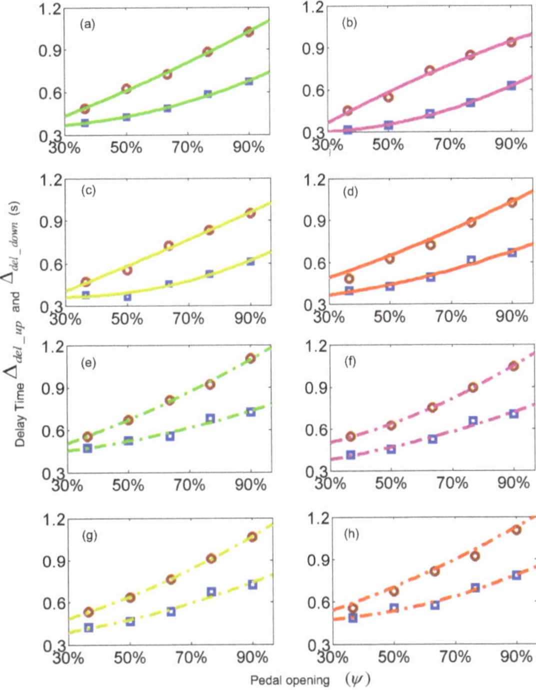

Figure 10. Relationships between pedal opening and delay time of eight loops. (**a–h**) refer to, respectively, loop A, B, C, ... , H. The curve with marker "o" is $\Delta_{del\_up}$, and with marker "□" is $\Delta_{del\_down}$.

**Table 4.** Fitting parameters.

| Loop | $K_1$ | | $K_2$ | | W | | $R_2$ | |
|---|---|---|---|---|---|---|---|---|
| | Up | Down | Up | Down | Up | Down | Up | Down |
| A | 0.597 | 0.290 | −0.238 | 0.621 | 0.378 | 0.192 | 0.997 | 0.996 |
| B | 0.759 | 0.158 | −0.423 | 0.746 | 0.356 | 0.109 | 0.998 | 0.985 |
| C | 0.754 | 0.117 | −0.520 | 0.778 | 0.447 | 0.130 | 0.980 | 0.993 |
| D | 0.393 | 0.321 | 0.026 | 0.497 | 0.308 | 0.287 | 0.975 | 0.990 |
| E | 0.353 | 0.407 | 0.026 | 0.473 | 0.401 | 0.299 | 0.958 | 0.997 |
| F | 0.355 | 0.653 | 0.115 | 0.087 | 0.299 | 0.394 | 0.975 | 0.999 |
| G | 0.397 | 0.533 | 0.081 | 0.298 | 0.316 | 0.324 | 0.971 | 0.999 |
| H | 0.601 | 0.450 | −0.243 | 0.409 | 0.488 | 0.352 | 0.977 | 0.998 |

### 4.1.2. First-Order Plus Time Delay Model

There is an obvious hysteresis of the PBS according to the above analysis. In this paper, a simplified model of the system is proposed to benefit the further development of brake controller. Considering the inherent hysteresis characteristics obtained in Section 4.1.1 and [26,27], the system can be approximately fitted to a first-order plus time delay model. This model relates the opening of the treadle valve and the pressure transient in the brake chamber. The governing equation of the model can be described like this,

$$T\frac{dP(t)}{dt} + P(t) = Ku(t - \tau),$$ 

(5)

where $T$ is the time constant of the system, $K$ is the gain, and $\tau$ is the delay time. $P(t)$ and $u(t)$ are the output and input of the system. Here, they refer to the pressure transients in the brake chamber and the opening of the brake pedal, respectively. Taking the Laplace transform of the Equation (5), we obtain the transfer function of the system.

$$G(s) = \frac{Ke^{-\tau s}}{(Ts + 1)},$$ 

(6)

Figure 11 shows a schematic description of the first-order plus time delay model of the PBS. Past research on the PBS has shown that the moving displacement of treadle valve plunger has a linear relationship with the desired braking pressure. Therefore, according to Formula (2) we assume that the relationship between pedal opening and the desired braking pressure can be expressed in Formula (7). As for the parameter $\tau$, it can be defined as Formula (8).

$$P_d = a\psi + b,$$ 

(7)

$$\begin{cases} \tau_{up} = \Delta_{del\_up} - \widetilde{T}_{up} \\ \tau_{down} = \Delta_{del\_down} - \widetilde{T}_{down} \end{cases},$$ 

(8)

where $a, b$ are constants, related to the type of the used treadle valve and brake pedal. $\tau_{up}, \tau_{down}$ are the pure time delay of the system in brake process and brake release process, respectively. $\widetilde{T}_{up}, \widetilde{T}_{down}$ are the time constants related to $T$. Where $\widetilde{T}_{up}$ is the time spent when the pressure in the chamber varies from 0 to 75% of the steady value in brake process, $\widetilde{T}_{down}$ is the time spent when the pressure in the chamber varies from the steady value to 15% in the brake release process.

With the method of system identification, the simulation model of the system can be established. In view of the deviation of hysteresis characteristics of each loop, eight loops correspond to eight different models, as shown in Figure 12, and their relevant parameters are listed in (9). We find that the models approximately coincide with the experimental results.

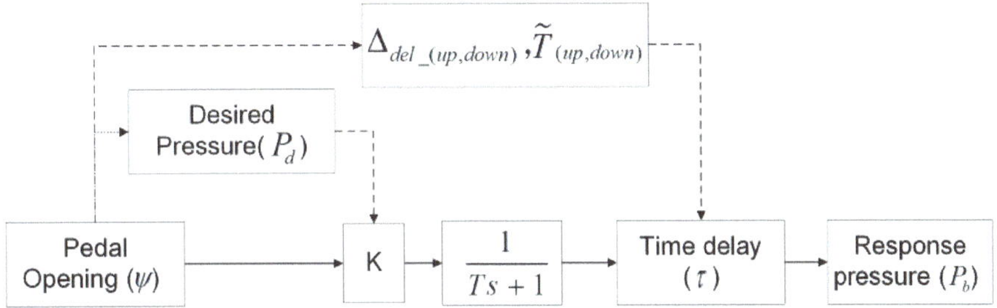

**Figure 11.** A schematic description of the first-order plus time delay model.

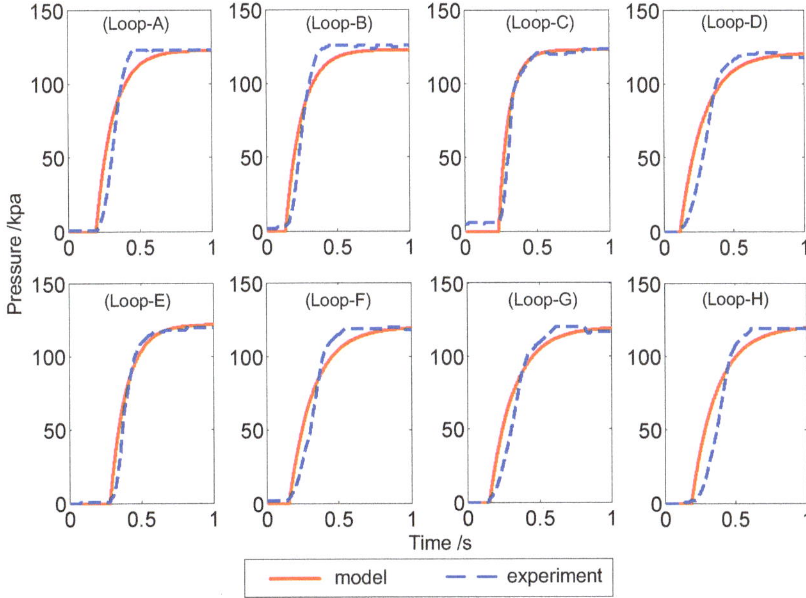


**Figure 12.** First-order plus time delay models and experimental results of the eight loops with the pedal opening 40%.

$$\begin{cases} T = [\ 0.124\ \ 0.118\ \ 0.075\ \ 0.170\ \ \ 0.114\ \ \ 0.171\ \ \ \ 0.169\ \ \ 0.184\ ] \\ \widetilde{T}_{up} = [\ \ \ 0.170\ \ 0.160\ \ \ 0.100\ \ 0.240\ \ \ \ 0.150\ \ \ 0.240\ \ \ \ 0.230\ \ \ 0.250\ ] \\ \widetilde{T}_{down} = [\ \ \ \ 0.210\ \ \ 0.200\ \ \ 0.130\ \ 0.300\ \ \ \ 0.200\ \ \ 0.310\ \ \ \ 0.300\ \ \ 0.330\ ] \end{cases}, \qquad (9)$$

Figure 13 shows the variation of the brake chamber pressure in the loop A when the brake pedal opening is set at 67% and 93%, respectively. It can be observed that the simulation results agree reasonably well with the experimental data in both the brake process and brake release process, which validate the accuracy of the model. Models of other loops are validated as well.

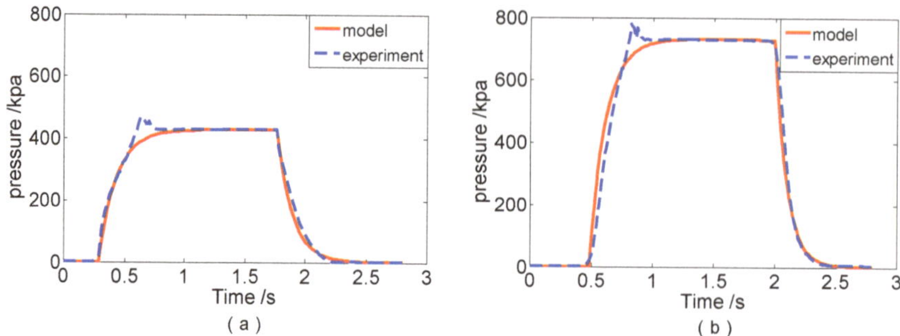

**Figure 13.** The variation of the brake chamber pressure with different pedal openings. (**a**) $\psi = 67\%$;(**b**) $\psi = 93\%$.

### 4.1.3. Relationship between Moving Speed and Delay Time

When concentrating on the results of group 5 and group 6 in Table 3, we know that when the pedal openings are the same, the faster the pedal is driven, the shorter the delay time is, and this law exists in both the brake process and brake release process. According to literature [28], the moving speed of brake pedal is easily influenced by driver's driving habits, so it is not advisable to be used as the parameter for recognizing driver's braking intention directly. Therefore, this paper does not pay much attention to the relationship between moving speed and delay time.

### 4.2. Response Time of Subassemblies in a Single Axle

Subassemblies in a single loop consist of treadle valve, control pipeline, relay valve, brake pipeline and brake chamber. Delay time of a loop is the accumulation of the response time of each subassembly. Sensors used for measuring the response time are arranged as shown in Figure 7. $t_{res\_(f,r)j}^{(up,down)}$ $(j = 1, 2, 3, 4)$ represents the response time of each subassembly in the front (C) or rear (E) loop in brake process or brake release process. Specifically, $t_{res\_(f,r)1}^{(up,down)}$, $t_{res\_(f,r)2}^{(up,down)}$, $t_{res\_(f,r)3}^{(up,down)}$, $t_{res\_(f,r)4}^{(up,down)}$ represents the response time of treadle valve, controlling pipeline, relay valve and braking pipeline, respectively. Hence, the delay time of the two loops can be described as,

$$
\begin{cases}
\Delta t_{del\_f}^{(up,down)} = \sum_{j=1}^{n} t_{res\_fj}^{(up,down)} \\
\Delta t_{del\_r}^{(up,down)} = \sum_{j=1}^{n} t_{res\_rj}^{(up,down)}
\end{cases}
, \tag{10}
$$

where $\Delta t_{del\_f}^{(up,down)}$, $\Delta t_{del\_r}^{(up,down)}$ are the delay time of the front loop C and rear loop E, respectively. The original measured data of the two loops under the testing condition of group 1 are shown in Figure 14.

Subsequently, a method for calculating the response time of each subassembly in a single loop is given as follows.

$$
\begin{cases}
\Delta t_{del\_fi}^{(up,down)} = t_{\gamma p\_fi}^{(up,down)} - t_{start\_(up,down)} \\
t_{res\_fj}^{(up,down)} = \Delta t_{del\_fi}^{(up,down)} - \Delta t_{del\_f(i-1)}^{(up,down)}
\end{cases}
, \tag{11}
$$

$$
\begin{cases}
\Delta t_{del\_ri}^{(up,down)} = t_{\gamma p\_ri}^{(up,down)} - t_{start\_(up,down)} \\
t_{res\_rj}^{(up,down)} = \Delta t_{del\_ri}^{(up,down)} - \Delta t_{del\_r(i-1)}^{(up,down)}
\end{cases}
, \tag{12}
$$

where $\Delta t_{del\_fi}^{(up,down)}$ and $\Delta t_{del\_ri}^{(up,down)}$ are the delay time of each subassembly in the front and rear loop, respectively, and they are calculated in the same way as $\Delta_{del\_(up,down)}$ mentioned in Section 4.1. $t_{\gamma p\_fi}^{(up)}$,

$t_{\gamma p\_ri}^{(up)}$ are the times when the pressure in a subassembly increase to 75% of its steady value in the brake process, and $t_{\gamma p\_fi}^{(down)}$, $t_{\gamma p\_ri}^{(down)}$ are the times when the pressure in a subassembly decreases to 15% of its steady value in the brake release process. These times are measured by sensors $f_i$ and $r_i$. Finally, the response times of each subassembly in two loops are acquired, as shown in Figures 15 and 16.

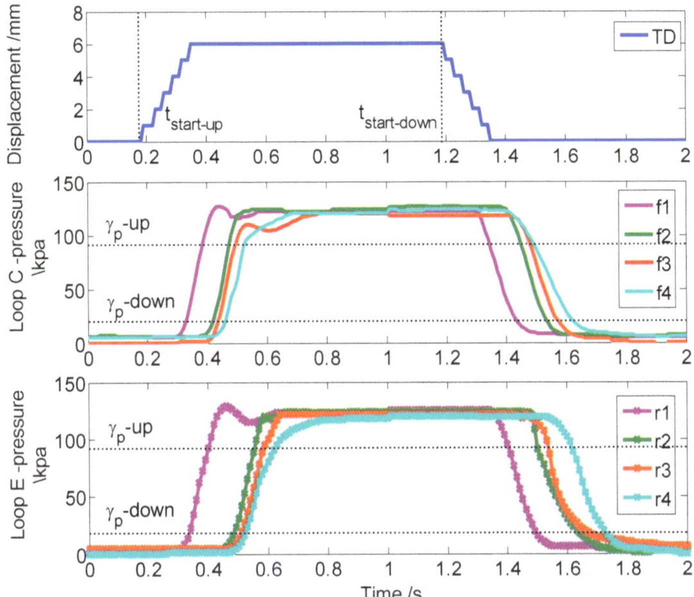


**Figure 14.** Original pressure curves of front loop C and rear loop E under the testing condition of group 1.

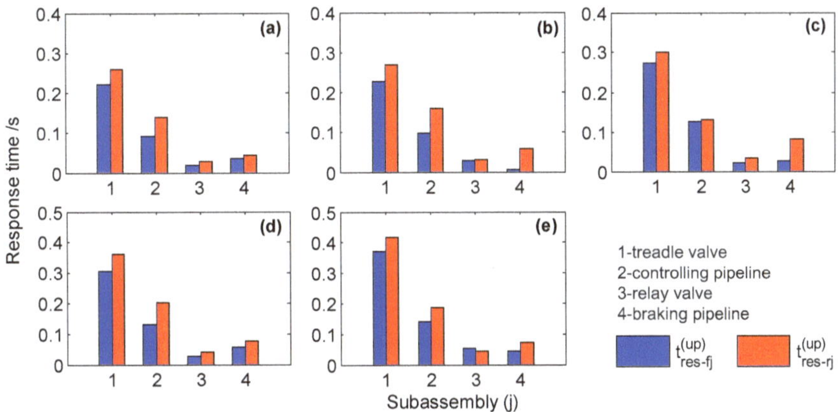


**Figure 15.** Response time of each subassembly in brake process under the testing condition from group 1 to group 5, shown in (**a**–**e**) respectively.

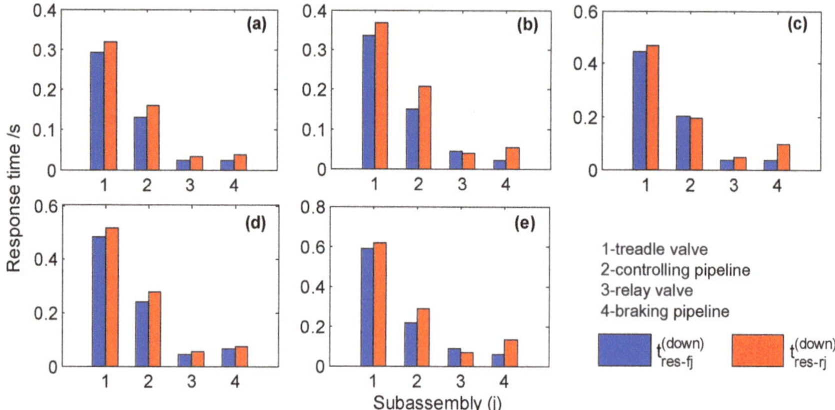

**Figure 16.** Response time of each subassembly in brake release process under the testing condition from group 1 to group 5, shown in (**a–e**), respectively. $j$ = 1, 2, 3, 4.

Finally, the proportion of response time of each subassembly can be calculated using the following formula and the corresponding results are shown in Figure 17.

$$\varphi_{res\_(f,r)j}^{(up,down)} = \frac{t_{res\_(f,r)j}^{(up,down)}}{\Delta t_{del\_(f,r)}^{(up,down)}},\tag{13}$$

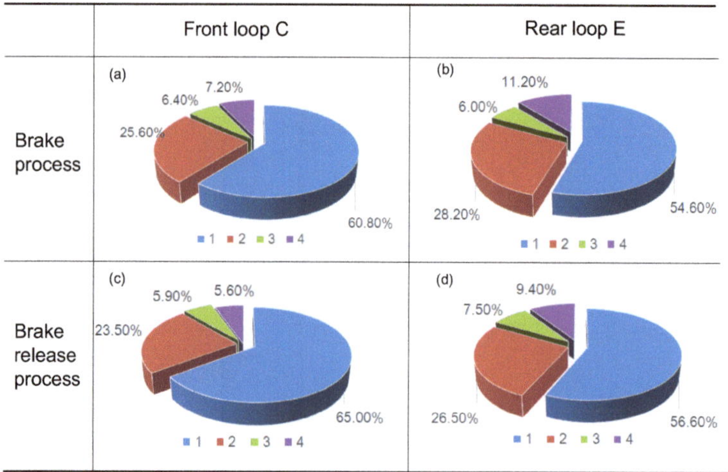

**Figure 17.** Average proportion of response time of each subassembly under the testing condition from group 1 to group 5. (**a**) Subassemblies in front loop C in brake process; (**b**) Subassemblies in rear loop E in brake process ; (**c**) Subassemblies in front loop C in brake release process; (**d**) Subassemblies in rear loop E in brake release process; (1—treadle valve; 2—controlling pipeline; 3—relay valve; 4—braking pipeline).

From Figure 17, it can be observed that the response time of the treadle valve and the controlling pipeline are much longer than that of the controlling pipeline and braking pipeline. The sum response time accounts for more than 80% of the total delay time of both the two loops in the brake process, as well as more than 85% in the brake release process. The results indicate that treadle valve and controlling pipeline are the main contributors to the hysteresis effect of the service circuit.

## 5. Conclusions

In this paper, a pneumatic braking system for an eight-axle vehicle is introduced. In order to accurately study the hysteresis effect of the system in emergency braking situation, a test bench is built. Not only the delay time of each loop corresponding to each axle but also the response time of each subassembly in a single loop is detected in real time. Moreover, the driver's different braking intensions expressed by opening and moving speed of brake pedal are accurately simulated by a servo drive device.

The acquired data are processed to search for relationships between the pedal's opening as well as moving speed and hysteresis times of eight loops in service brake circuit. The results show that under the same braking conditions, delay times of the front four loops are shorter than those of the rear four loops, and the delay times of a certain loop in the brake release process are longer than those in the brake process Under the different braking conditions, when the braking speeds are similar, the larger the pedal opening, the longer the delay time, and a quadratic curve relationship exists between the two. Given this fact, the pressure transients of each loop in the system can be fitted to a corresponding first-order plus time delay model. When the pedal openings are the same, the faster the braking speed, the shorter the delay time. In addition, response time of each subassembly in a loop is also obtained. The sum response time of treadle valve and controlling piping accounts for more than 80% of total delay time of a loop in both the brake process and brake release process, which indicates that these two subassemblies are the main contributors to the hysteresis effect of the loop.

For further study of the pneumatic braking system for MHV, the structure of the treadle valve needs to be optimized to shorten its response time, and more reasonable layout of the system should be studied for reducing the gas transmission distance. Additionally, the development of an advanced brake control strategy for the PBS in MHVs based on the established first-order plus time delay models will be a research challenge.

**Acknowledgments:** The presented work was supported by the National Natural Science Foundation of China (Grant No. 51275453).

**Author Contributions:** Zhe Wang wrote the paper. Xiaojun Zhou and Xuelei Wu conceived and designed the experiment; Zhe Wang and Zhaomeng Chen performed the experiments and analyzed the data; Chenlong Yang modified the final paper.

**Conflicts of Interest:** The authors declare no conflict of interest.

## References


1. Wu, J.L.; Zhang, H.C.; Zhang, Y.Q. Robust design of a pneumatic brake system in commercial vehicles. *SAE Int. J. Commer. Veh.* **2009**, *2*, 17–28. [CrossRef]
2. Harris, P.G.; O'Donnell, G.E.; Whelan, T. Modelling and identification of industrial pneumatic drive system. *Int. J. Adv. Manuf. Technol.* **2012**, *58*, 1075–1086. [CrossRef]
3. Pugi, L.; Malvezzi, M.; Allotta, B.; Banchi, L.; Presciani, P. Parametric library for the simulation of a Union Internationale des Chemins de Fer (UIC) pneumatic braking system. *Proc. Inst. Mech. Eng. Part F J. Rail Rapid Transit* **2004**, *218*, 117–132. [CrossRef]
4. Zheng, H.; Wang, L. Multi-objective stability control algorithm of heavy duty based on EBS. *SAE Int. J. Commer. Veh.* **2014**, *7*, 514–519. [CrossRef]
5. Miller, J.; Cebon, D. An investigation of the effects of pneumatic actuator design on slip control for heavy vehicles. *Veh. Syst. Dyn.* **2013**, *51*, 139–164. [CrossRef]
6. Kienhofer, F. Heavy Vehicle Wheel Slip Control. Ph.D. Thesis, University of Cambridge, Cambridge, UK, 2011.
7. Henderson, L.; Cebon, D. Full-scale testing of a novel slip control braking system for heavy vehicles. *Proc. Inst. Mech. Eng. Part D J. Automob. Eng.* **2016**, *230*, 1221–1238. [CrossRef]

8. Zhang, R.; Li, K.; He, Z.; Wang, H.; You, F. Advanced emergency braking control based on a nonlinear model predictive algorithm for intelligent vehicles. *Appl. Sci.* **2017**, *7*, 504. [CrossRef]

9. Bu, F.; Tan, H.S. Pneumatic brake control for precision stopping of heavy-duty vehicles. *IEEE Trans. Control Syst. Technol.* **2007**, *15*, 53–63. [CrossRef]

10. Miller, J.; Henderson, L.M.; Cebon, D. Designing and testing an advanced pneumatic braking system for heavy vehicles. *Proc. Inst. Mech. Eng. Part C J. Mech. Eng. Sci.* **2013**, *227*, 1715–1729. [CrossRef]

11. Wan, Z.; Gao, F.; Ding, J. Dynamics model and braking performance analysis of multi-axle vehicle. *China Mech. Eng.* **2008**, *19*, 365–369.

12. Jiang, F.; Gao, Z. An application of nonlinear PID control to a class of truck ABS problems. In Proceedings of the 40th IEEE Conference on Decision and Control (CDC), Orlando, FL, USA, December 2001; Volume 1, pp. 516–521.

13. Patil, J.; Palanivelu, S.; Jindal, A. Mathematical model of dual brake valve for dynamic characterization. *SAE Tech. Pap.* **2013**. [CrossRef]

14. Russomanno, A.; Gillespie, R.B.; O'Modhrain, S. Modeling pneumatic actuators for a refreshable tactile display. In *Haptics: Neuroscience, Devices, Modeling and Applications 2014*; Springer: Berlin/Heidelberg, Germany, 2014; pp. 385–393.

15. Natarajan, S.V.; Subramanian, S.C.; Darbha, S. A model of the relay valve used in an air brake system. *Nonlinear Anal. Hybrid Syst.* **2007**, *1*, 430–442. [CrossRef]

16. Gastaldi, L.; Pastorelli, S.; Sorli, M. Static and dynamic experimental investigation of a pneumatic open loop proportional valve. *Exp. Mech.* **2016**, *40*, 1377–1385. [CrossRef]

17. Selvaraj, M.; Gaikwad, S.; Suresh, A.K. Modeling and simulation of dynamic behavior of pneumatic brake system at vehicle level. *SAE Tech. Pap.* **2014**. [CrossRef]

18. He, L.; Wang, X.L. Modeling and simulation vehicle air brake system. In Proceedings of the IEEE 8th Modelica Conference, Dresden, Germany, 20–22 March 2011; pp. 430–435.

19. Pugi, L.; Malvezzi, M.; Tarasconi, A.; Palazzolo, A.; Cocci, G.; Violani, M. HIL simulation of WSP systems on MI-6 test rig. *Veh. Syst. Dyn.* **2006**, *44*, 843–852. [CrossRef]

20. Bartlett, W.D. Calculation of deceleration rates for S-Cam Air-Braked heavy trucks equipped with Anti-Lock brake systems. *SAE Tech. Pap.* **2007**. [CrossRef]

21. Lei, Y.; Liu, K.; Fu, Y. Research on driving style recognition method based on driver's dynamic demand. *Adv. Mech. Eng.* **2016**, *8*, 1–14. [CrossRef]

22. Meng, D.; Zhang, L.; Yu, Z. A dynamic model for brake pedal feel analysis in passenger cars. *Proc. Inst. Mech. Eng. Part D J. Automob. Eng.* **2015**, *230*, 955–968. [CrossRef]

23. Bowlin, C.L.; Subramanian, S.C.; Darbha, S.; Rajagopal, K.R. Pressure control scheme for air brakes in commercial vehicles. *IEE Proc. Intell. Transp. Syst.* **2006**, *153*, 21–32. [CrossRef]

24. Qin, T. Research on Delay Time Analysis and Its Control Techniques of Bus Pneumatic Braking System. Ph.D. Thesis, Wuhan University of Technology, Wuhan, China, 2012.

25. Fang, E.; Zhang, D.L.; Cheng, Z.D. Research on the method and application of driver's braking intention recognition for passenger car. *Shanghai Auto* **2016**, *1*, 49–53.

26. Kumar, Y.S.R.R.; Sonawane, D.B.; Subramanian, S.C. Application of PID control to an electro-pneumatic brake system. *Int. J. Adv. Eng. Sci. Appl. Math.* **2012**, *4*, 260–268. [CrossRef]

27. Qin, T.; Li, G.Y.; Tu, M. Bus pneumatic braking circuit delay analysis and control. *Trans. Beijing Inst. Technol.* **2012**, *32*, 470–474.

28. Wang, Y.; Ning, G.A. Parameter selection for the identification of driver's braking intention for passenger car. *Automot. Eng.* **2011**, *33*, 213–216.

*Article*

# Energy Regeneration Hydraulic System via a Relief Valve with Energy Regeneration Unit

Tianliang Lin [1], Qiang Chen [1], Haoling Ren [1,2,*], Yi Zhao [1], Cheng Miao [1], Shengjie Fu [1] and Qihuai Chen [1,2]

[1] College of Mechanical Engineering and Automation, Huaqiao University, Xiamen 361021, China; ltlkxl@163.com (T.L.); 1511303034@hqu.edu.cn (Q.C.); 1411112045@hqu.edu.cn (Y.Z.); 1400403007@hqu.edu.cn (C.M.); xmshengjie@163.com (S.F.); chen.qihuai@163.com (Q.C.)
[2] State Key Laboratory of Fluid Power and Mechatronic Systems, Zhejiang University, Hangzhou 310027, China
* Correspondence: happyrhlly@126.com; Tel.: +86-592-616-2598

Academic Editor: Yunhua Li
Received: 21 April 2017; Accepted: 9 June 2017; Published: 13 June 2017

**Featured Application: This research aims to reduce the energy loss through the relief valve, especially in the systems where the system pressure is high or the overflow flow rate is large.**

**Abstract:** Relief valves are widely used in industrial machinery. Due to the outlet of the relief valve being connected to the tank, the pressure drop of the relief valve is frequently equal to the inlet pressure. Accordingly, the energy loss of the relief valve is very high in some cases and this will worsen with an increase in the rated pressure of the hydraulic system. In order to overcome the disadvantage of overflow energy loss in a relief valve, a hydraulic energy regeneration unit (HERU) is connected to the outlet of the relief valve to decrease the pressure drop between the inlet and outlet of the relief valve. The overflow loss, which is characterized by the pressure drop, can be reduced accordingly. The approach is to convert the overflow energy loss in hydraulic form and allow for release when needed. The configuration and working principle of the relief valve with HERU is introduced in this present study. The mathematical model is established to obtain the factors influencing the stability of the relief valve. The working pressure of the hydraulic accumulator (HA) is explored. Furthermore, the control process of the operating state of the HA is scheduled to decide whether to regenerate the energy via the HERU. The software AMESim is utilized to analyze the performance and characteristics of the relief valve with HERU. Following this, the test rig is built and used to verify the effectiveness of the proposed relief valve with HERU. The experimental results show that the relief valve with the HERU connected to its outlet can still achieve better pressure-regulating characteristics. The energy regeneration efficiency saved by the HA is up to 83.6%, with a higher pre-charge pressure of the HA. This indicates that the proposed structure of the relief valve with HERU can achieve a better performance and higher regeneration efficiency.

**Keywords:** pilot relief valve; energy regeneration unit; pressure fluctuation; regeneration efficiency

---

## 1. Introduction

The energy crisis and new emissions rules have motivated the development of new proposals for energy efficient vehicles and non-road mobile machines. Construction machinery, especially the hydraulic excavator (HE), is widely used in transport and infrastructure, which can output a larger amount of power and consume a larger proportion of energy. Therefore, the energy saving and regeneration become the research hotspot to reduce the energy consumption and emissions [1–5]. The energy loss of the hydraulic system can be divided into two types: throttling loss and overflow loss.

Research examining energy saving in fluid power systems has mainly concentrated on reducing the throttling loss. There are few studies focusing on reducing the energy loss in relief valve. The objective of systems, such as the positive/negative flow control system [6,7], load sensing control [8] and separate control of actuator ports [9], is to match the pump supply to the demands of load. Hence, the throttle energy loss can be reduced to some extent. Furthermore, the pump controlled system has been widely applied [10–12]. One of the most significant merits of the pump controlled system is that the energy regeneration system (ERS) becomes efficient. However, the overflow loss through the relief valve due to the frequent start-stop function and rotary, such as the swing of a hydraulic excavator, still exists and consumes a large amount of energy [13]. Furthermore, there are few studies examining this loss. The recent fast digital valve development provided the groundwork for a wide variety of solutions and the digital valve can be used as the flow control valve to reduce the throttle energy loss [14]. However, the high speed on-off valve control system can avoid the overflow loss in the relief valve. Shi et al provided a hydropneumatic transformer to supply high-pressure oil to drive the vehicles, the efficiency of which can be improved by 10% [15].

Furthermore, there are studies on energy regeneration to save and reutilize the throttle energy produced by the potential energy of the boom. Wang et al. also presented an electric ERS, which integrates the regeneration device with the throttle valve using a different controller [16,17]. Using the pressure compensation principle, the electromagnetic torque of the generator was adapted to the loaded pressure. Therefore, the pressure drop across the throttle was guaranteed to remain a small constant pressure drop and consequently, the velocity of the boom cylinder can be regulated effectively by controlling the throttle valve opening. Furthermore, most of the throttle energy can be regenerated by the hydraulic motor and an electric motor.

Relief valves, the most important component in hydraulic systems, are widely used in industrial machinery. In hydraulic systems, the outlet of the relief valve is frequently connected to the tank and thus, the pressure loss of the relief valve, which is characterized by the inlet and outlet pressure, is equal to the inlet pressure. With an increase in the inlet pressure of the relief valve, the pressure loss increases accordingly. Therefore, the energy loss of the relief valve, which is the product of the flow rate and the inlet pressure, is very high and it will worsen with an increase in the rated pressure of the hydraulic system. If the energy through the relief valve can be saved and reutilized, the efficiency of the hydraulic system will be greatly improved.

This paper focuses on exploring the methods of regenerating the energy through the relief valve and the performance of the relief valve with the energy regeneration unit. The following part is organized as follows: The structure and working principle of the pilot relief valve (PRV) with the hydraulic energy regeneration unit (HERU) is proposed in Section 2. The mathematical model is developed and the control flow for the HA is presented in Section 3. The key parameters and control for the best working ability of the hydraulic accumulator (HA) to achieve the maximum energy saving are discussed in Section 4. After this, the experiment is carried out to verify the feasibility and the influence of the HERU on the PRV in Section 5. Following this, the concluding remarks are summarized in Section 6.

## 2. Configuration of the PRV with HERU

### 2.1. Configuration of HERU of Energy Loss in Relief Valve

To reduce the energy loss of the relief valve, a hydraulic energy regeneration unit (HERU) is connected to the outlet of the relief valve, which can increase the outlet pressure of the relief valve and thus reduce the pressure drop between the inlet and outlet [18], as shown in Figure 1. As there are many types of relief valves, the PRV is chosen as an example in this paper. The advantages are as follows:

(1) The overflow energy loss, which backs into the tank directly in the traditional PRV, is converted into hydraulic energy to be stored in the HA and released when needed. During this process, the outlet pressure of the PRV is far higher than that of the tank, which means that the pressure drop between the inlet and outlet is decreased. Therefore, the pressure drop through the PRV can be reduced. The pressure drop of the PRV is defined as

$$\Delta p = p_1 - p_2 \tag{1}$$

where $p_1$ is the inlet pressure of the PRV; $p_2$ is the outlet pressure of the PRV; and $\Delta p$ is the pressure drop between the inlet and outlet.

When the traditional PRV is connected to the tank, the pressure $p_2$ is approximately equal to zero. Therefore, the pressure drop can be simplified as

$$\Delta p = p_1 \tag{2}$$

When the HERU is connected to the outlet of the PRV, the oil will flow into the HERU and cause an increase in the outlet pressure $p_2$. Therefore, the pressure $p_2$ is far larger than zero, while the pressure drop through the PPRV is smaller than that without HERS.

The overflow energy loss in PRV is featured by the product of the flow rate and the pressure drop

$$P = \Delta p \cdot Q \tag{3}$$

where $P$ is the power lost through the PRV and $Q$ is the flow rate through the PRV.

The flow rate $Q$ through the PRV remains mostly constant. Hence, the energy loss through the PRV with HERU is less than the traditional PRV. Moreover, the energy stored in the HERU can be reutilized and the efficiency of the hydraulic system can be improved. In general, a higher outlet pressure $p_2$ results in a lower overflow energy loss of the PRV, which is characterized by the pressure drop $\Delta p$. To ensure the normal operation of the PRV, the outlet pressure $p_2$, which is caused by the HERU, must be smaller than the target relief pressure of the PRV.

(2) The proposed PRV is different to the traditional PRV in that the outlet of the pilot valve is connected to the tank alone. In comparison, the outlet of the pilot is connected to the outlet of the main valve in the traditional PRV. This difference can ensure that the pilot valve of the proposed PRV is not affected by the pressure fluctuations in the outlet of the PRV.

(3) As seen in Figure 1, the PRV has a main valve and a pilot valve. The inlet pressure $p_1$ acts on the main valve spool and the pilot valve spool at the same time. Meanwhile, the pressure $p_3$ of the spring chamber is also connected to the pilot valve spool. The outlet port of the PRV is connected to the HERU through a directional valve, which is used to determine whether to recover the overflow energy or not. The pilot oil circuit is connected to the tank alone to avoid an increase in the relief pressure as the PRV and the HERU are in series.

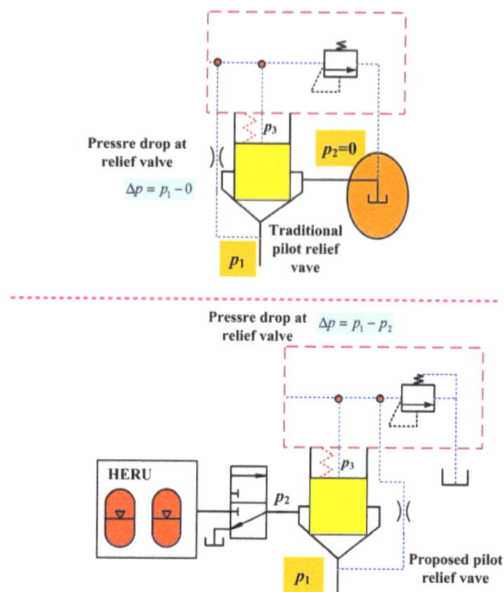


**Figure 1.** Comparison of the pilot relief valve (PRV) with and without hydraulic energy regeneration unit (HERU).

### 2.2. Working Principles of the PRV

Figure 2 is the working principle diagram of a PRV. The rated pressure of the PRV is set by the pilot operated proportional valve. A and B denote the inlet and outlet of the PRV, respectively. Back pressure in port B is produced by the HERU 5, which is connected to port B of the PRV.

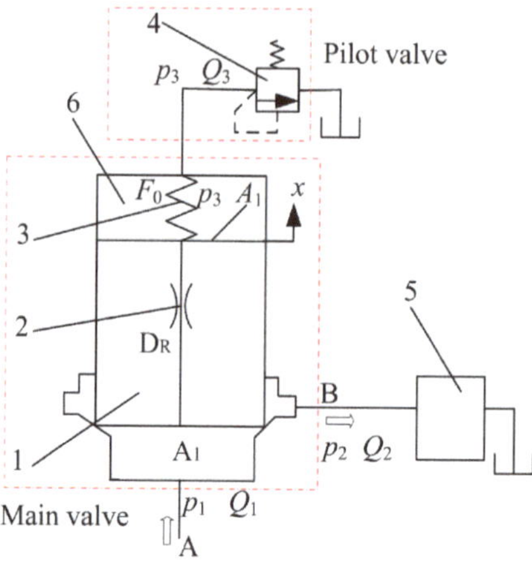


**Figure 2.** Working principles of the PRV (1—Main valve spool; 2—Orifice; 3—Reset spring; 4—Pilot valve; 5—HERU; 6—Spring chamber).

When the adjusting spring of the pilot valve is pre-compressed to some extent, the relief pressure of the PRV is set. The system pressure is equal to the relief pressure of the PRV. The oil supplied to the PRV is divided into three parts: one acts on the bottom of the main valve spool, the second on the upper of the main valve spool and the third acts on the pilot valve. The latter two are supplied through orifice 2 in the main valve spool. The pilot spool remains closed and no oil flows through it when the force produced by the pressure $p_3$ in the spring chamber 6 is less than the electromagnetic force. The main spool is also closed because the inlet pressure $p_1$, which is equal to the pressure $p_3$ in this case, is smaller than the resultant of the spring force and the pressure $p_3$. The forces on the main spool satisfy the equation

$$p_3 A_1 + F_0 = p_1 A_1 + F_0 > p_1 A_1 \tag{4}$$

where $F_0$ is the pre-tightening force of the reset spring; $A_1$ is the effective cross-sectional area of the main spool; and $p_3$ is the pressure of the spring chamber.

The main spool continues closing until the inlet pressure is higher than the spring force of the pilot valve. At this time, the pilot valve is open and the pressure $p_3$ is smaller than the inlet pressure $p_1$ because there is a pressure drop across the orifice 2. Due to the small stiffness of the reset spring, the forces on the main spool satisfy the equation

$$p_1 A_1 > p_3 A_1 + F_0 \tag{5}$$

Therefore, the main valve opens and the main valve spool moves upwards to allow the liquid to flow through port A to port B and then through HERU to the tank. The pressure in the HA of the HERU increases, while the pressure drop between the inlet and outlet of the main valve reduces. Following this, the overflow energy loss of the PRV decreases accordingly.

When the pre-compression of the spring of the pilot valve changes, the relief pressure changes accordingly. Furthermore, when the pressure of the HA changes, the performance of the PRV may be influenced. The influence of the HERU should be discussed to ensure a stable performance of the PRV and obtain a comparable, better control characteristic.

## 3. Control System

The control system has two goals. One is to recycle as much of the overflow energy as possible using the HERU. The other is to guarantee the operation characteristics of the PRV with the HERU. In other words, the PRV can still control the pressure of the system despite having an HERU, which is connected to the outlet of the PPRV. Therefore, how to choose the parameters of the HERU and how to control these parameters during the energy regeneration process are essential considerations for the control system.

### 3.1. Working Pressure of the HA

The HA is the main component of the HERU and must be properly designed to achieve the two goals of the control system. The key parameters of the HA include the rated volume, the pre-charge pressure, as well as the minimum and maximum working pressure. The appropriate parameters of the HA are beneficial in controlling how much energy can be regenerated. For example, if the working pressure of the HA is high, which means that the outlet pressure of the PRV is high, the pressure drop of PRV is consequently small and the energy lost in the orifice is low. However, the amount of oil that the HA can store is less. In comparison, if the working pressure of the HA is low, which means that the outlet pressure of the PRV is low, the energy lost in the orifice of the PRV is subsequently high. However, the amount of oil that the HA can store is more.

The HA is ideal for those confronted with frequent and short start-stop cycles in adequate space. The key advantage of using HA as energy regeneration components in HERU is its seamless interface. They can be easily integrated into a hydraulic circuit of a HE. However, the major disadvantage of a HA is that the energy storage density is severely limited compared to other competing technologies, such as the battery. Therefore, the major objective of working pressure optimization for the HA is to improve the energy density of the HA.

According to the Boyle's law, the gas in the accumulator follows the ideal gas law as

$$p_{a0} V_{a0}{}^{n} = p_{a1} V_{a1}{}^{n} = p_{a2} V_{a2}{}^{n} = C \tag{6}$$

where $p_{a0}$ is the pre-charge pressure of HA; $V_{a0}$ is the rated volume of HA; $V_{a1}$ and $p_{a1}$ are the gas volume and pressure when the HA is at the minimum working pressure, respectively; $V_{a2}$ and $p_{a2}$ are the gas volume and pressure when the HA is at the maximum working pressure, respectively; $n$ is the polytrophic exponent; and C is a constant.

The HA used in the HE mainly functions during the starting and braking process, which lasts only a short period of time, while the working cycle of the HE is about 20 s. Following this, the compression process when the HA is charged can be considered as the adiabatic process. The gas index $n$ is set to 1.4.

The volume of the HA is calculated by the equation

$$V_{a0} = \Delta V_{a1} \frac{\left(\frac{p_{a1}}{p_{a0}}\right)^{\frac{1}{n}}}{1 - \left(\frac{p_{a1}}{p_{a2}}\right)^{\frac{1}{n}}} \tag{7}$$

where $\Delta V_{a1}$ is the maximum oil volume that is stored in HA when the pressure in HA increases from $p_{a1}$ to $p_{a2}$.

The energy density $E_\rho$ of the HA can be calculated as

$$E_\rho = p_2 \frac{r^{\frac{1}{n}} - r}{1 - n} \tag{8}$$

where

$$r = \frac{p_{a1}}{p_{a2}} \tag{9}$$

As seen from the above equations, increasing the volume of the HA, improving the pre-charge pressure and the maximum working pressure or reducing the minimum working pressure can improve the energy storage. The volume of the HA is constrained by the installation space of the HE. Furthermore, the pre-charge pressure, minimum and maximum pressure all are related to one another. In order to obtain the relationships, the derivative of Equation (8) is used as

$$\frac{d(E_\rho)}{dr} = 0 \tag{10}$$

Accordingly, the maximum energy density based on the optimum pressure ratio can be represented in the form

$$r = n^{\frac{n}{1-n}} \tag{11}$$

On the other hand, the HA works frequently in the process of charging and releasing. To prolong the service life of the HA, the peak volume change of the air bladder of the HA cannot be too large. Hence, the pressures should satisfy

$$0.25p_{a2} < p_{a0} < 0.9p_{a1} \tag{12}$$

For example, if the rated pressure of the PRV is 21.5 MPa, the pressure drop of the PRV should be at least 1 MPa, which is the same as the pressure drop in the throttle control valve. This is needed to guarantee the fact that the PRV can release the redundant flow of the hydraulic system. Then, according to the pressure scope of the HE, the pressures of the HA can be chosen as

$$p_{a1} = 6.2 \text{ MPa} \tag{13}$$

$$p_{a2} = 20 \text{ MPa} \tag{14}$$

$$p_{a0} = 5 \text{ MPa} \tag{15}$$

### 3.2. Control Flow

In an energy regeneration system for the energy loss of a PRV, the decision whether to regenerate the overflow energy depends on the relationships of the inlet/outlet pressure of the PRV and the maximum working pressure of the HA. Figure 3 shows the principle diagram of the tested PRV with a HERU connected to its outlet. The overload protection unit is used to protect the pump 4 when the PRV does not work. When the valve 1 is powered off, the regulating pressure of safety valve 2 is 31.5 MPa to ensure that no oil flows through valve 2 when PRV is working. In comparison, when the PRV does not work, valve 1 is powered to allow oil to flow from the pump to the tank with a low energy loss.

The flow meters 8, 12, and 13 are used to detect the flow rate in and out of the PRV. Pressure sensors 10, 11, 19, and 20 are used to gain the pressures of the pump, the inlet and outlet of the PRV and the HA. These pressures are used to determine the energy recovery process. In Figure 3, $p_s$ is the outlet pressure of the pump and pa is the pressure of the HA. The detailed control process is shown in Figure 4. The control strategy is based on following logic:

- Energy recovery mode

When the pressures satisfy $p_a < p_{a2}$, the PRV works in the energy recovery mode. In this condition, the valve 1 is powered off and the lower electromagnet of the valve 14 is powered on to make the valve 14 work at the bottom position. The overflow energy through the PRV will be stored in the HA until it reaches the maximum working pressure.

- Traditional mode

When the pressures satisfy $p_a \geq p_{a2}$, the PRV works in the traditional mode. In this condition, valve 1 is powered on and both the electromagnets of valve 14 are powered off. The pump is unloaded through the safety valve 2, which is unloaded by valve 1. The valve 14 works at top position, while the HA is disconnected to the outside and the pressure is kept constant.

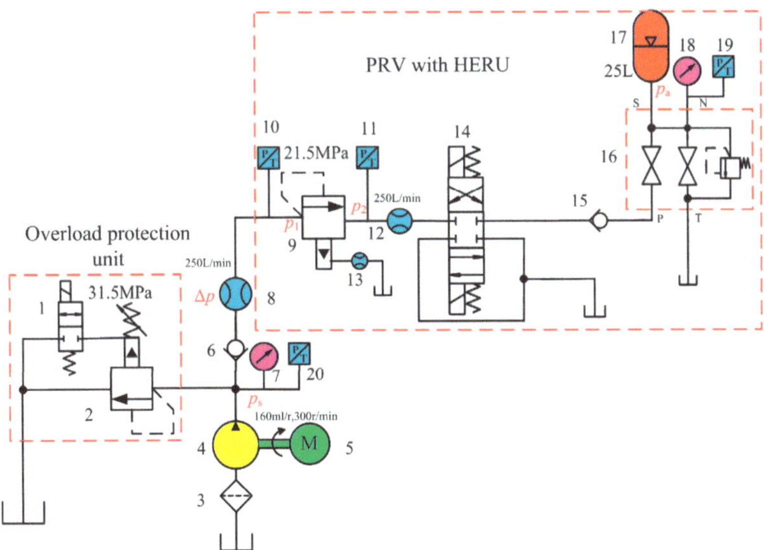


**Figure 3.** Principle diagram of the test for the PRV with HERU (1—2/2 Solenoid directional valve; 2—Safety valve; 3—Filter; 4—Fixed displacement pump; 5—Electric variable frequency motor; 6 and 15—Check valve; 7 and 18—Pressure gauge; 8, 12, and 13—Flow meter; 9—PRV; 10, 11, 19, and 20—Pressure sensor; 14—3/4 Solenoid directional valve; 16—Safety shutoff valve; 17—HA (HERU)).

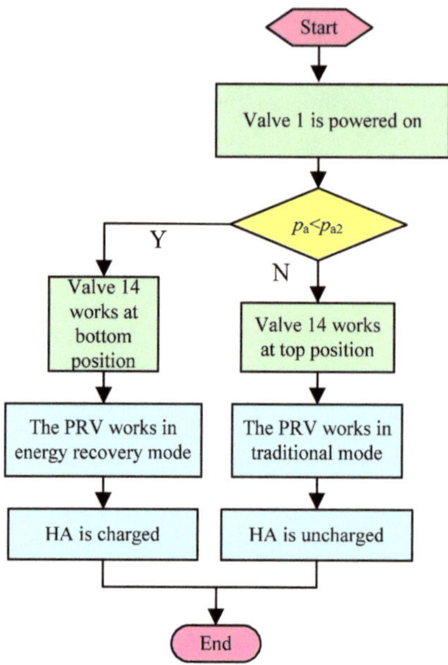



**Figure 4.** Control flow of the HERU.

## 4. Simulation Results

Figure 5 is the simulation model built in AMESim based on the principle diagram and control flow shown in Figures 3 and 4. The simulation model is used to test the influence of the HERU on the characteristics of the PRV and the influence of the pre-charge pressure on the energy regenerating efficiency of the HERU.

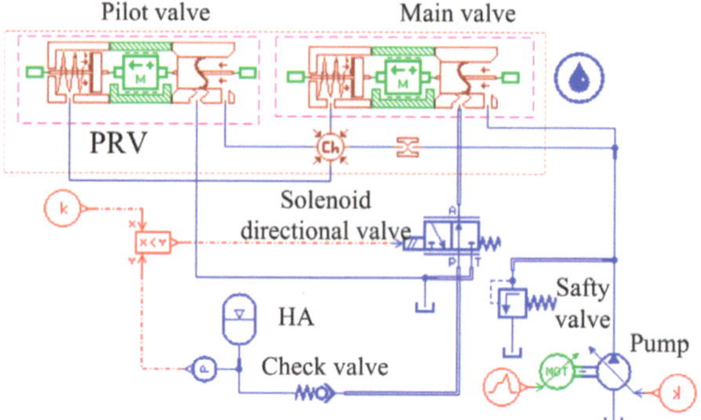



**Figure 5.** Simulation model in AMESim.

Figure 6 shows the inlet pressure of the PRV and the inlet pressure of the HA during the energy regenerating process. The pre-charge pressure of the HA is 5 MPa and the regulating pressure of the PRV is 20 MPa. As seen in Figure 6, with an increase in the regenerated energy in the HA, the pressure of the HA increases until it reaches its target value of 20 MPa. At the same time, the inlet pressure of the PRV decreases marginally from 21 MPa to 20.6 MPa. As the inlet of the HA is connected to the outlet of PRV, the pressure drop across the orifice of the PRV is 0.6 MPa. This indicates that the PRV can still maintain good performance under a lower pressure drop. It can also be deduced that the

HERU that is connected to the outlet of the PRV has little influence on the performance of the PRV. The PRV can still control the system pressure steadily.

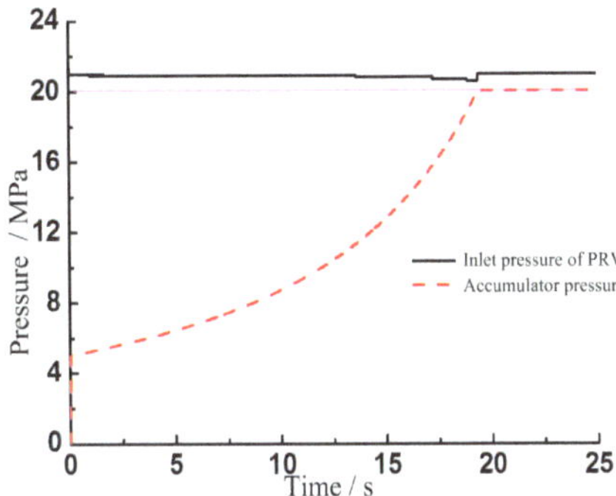

**Figure 6.** Pressures during the energy regeneration process.

During the energy regeneration, the outlet pressure $p_2$ of the PRV is nearly equal to the HA pressure $p_a$. It is shown in Figures 6 and 7 that with an increase in the outlet pressure $p_2$ of the PRV, the displacement of the main valve spool increases to keep the flow rate constant. Hence, the flow rate of the PPV is maintained at a certain value is important so that the PPV can release the redundant flow of hydraulic system. The reason is that the displacement of the main valve spool will change according to the outlet pressure of the PRV, which is shown in Figure 7.

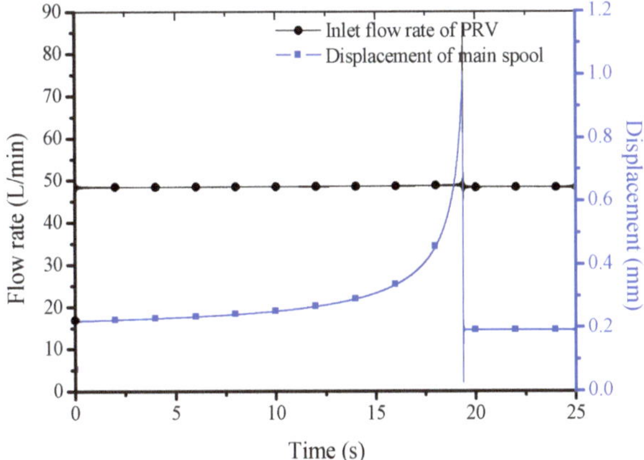

**Figure 7.** Flow rate and displacement of main valve of the PRV during the energy regeneration process.

Figure 8 shows the compression of the pre-charge pressure on the charging process of the HA. The pre-charge pressure is 5, 10, and 15 MPa, respectively. When the energy regeneration starts, the pressure of the HA increases from the pre-charge pressure to its target value. During this process, the volume of HA remains the same. It can be seen that a higher pre-charge pressure results in a shorter time needed to reach the target pressure, which also means that less energy can be stored.

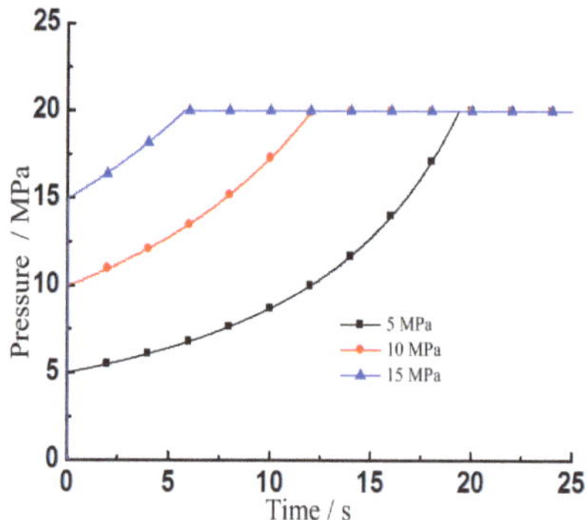

**Figure 8.** Compression of hydraulic accumulator (HA) pressure with different pre-charge pressures.

Figure 9 shows the influence of the outlet pressure of the PRV on the inlet pressure of the PRV. The regulation pressure of the PRV is set to 21.5 MPa. Before the PRV is powered on at the time of 1 s, the inlet pressure is slightly higher than the outlet pressure. After the PRV is powered on, the inlet pressure decreases a little with an increase in the outlet pressure, but the inlet pressure remains around 21.5 MPa, which is the regulating pressure of the PRV. The maximum pressure difference is 0.7 MPa, which is under 3.3% of the regulating pressure of the PRV. This indicates that the HERU connected to the outlet of the PRV has little influence on the pressure regulation characteristics.

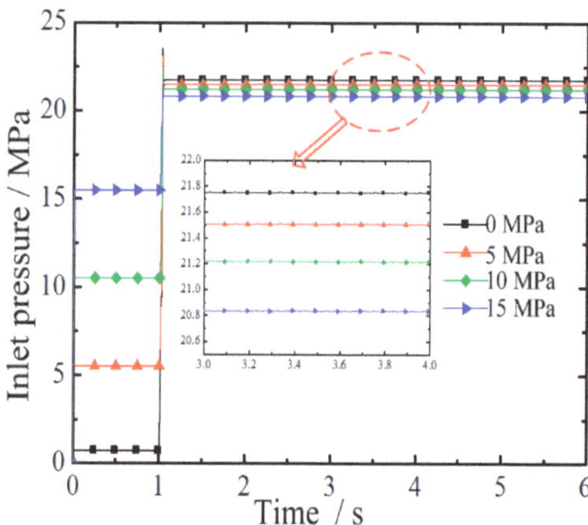

**Figure 9.** Influence of the outlet pressure of the PRV on the characteristics of the PRV.

## 5. Experimental Study and Discussion

### 5.1. Test Rig

The test rig was built to verify the characteristics of PRV with HERU. Figure 10 is the layout of the tested unit, according to the working diagram shown in Figure 3. The hydraulic pump can maximally supply a flow rate of 250 L/min under a maximum pressure of 31.5 MPa. The HERU includes a few HAs with different volumes, as shown in Figure 10. The PRV is the cartridge valve. The rated flow of the check valve and the directional valve is sufficient at 200 L/min, which can minimize the pressure drop during the test.

**Figure 10.** Layout of the tested unit.

## 5.2. Results and Discussion

### 5.2.1. Control Performance

Figure 11 shows the pressure of the HA during the energy regeneration process under different pre-charge pressures. When the valve 14 is working at the bottom position at a time of 5 s, the flow rate out of the PRV flows into the HA through the valve 14, which leads to an increase in the pressure in the HA. When the pressure in HA reaches the pre-set pressure of 20 MPa, the valve 14 works at the top position and allows flow from the PRV into the tank. The HA stops regenerating energy and there is a little decrease of the pressure when the valve 14 is powered off. This is because the connecting line between the PRV and the HA is very long in addition to the hydraulic fluid leakage occurring in check valve 15 and safety shut-off valve 16 in the test rig. The trend of the HA pressure in Figure 11 matches that in Figure 8. This indicates that the simulation model and the test rig match well, with well-chosen simulation parameters.

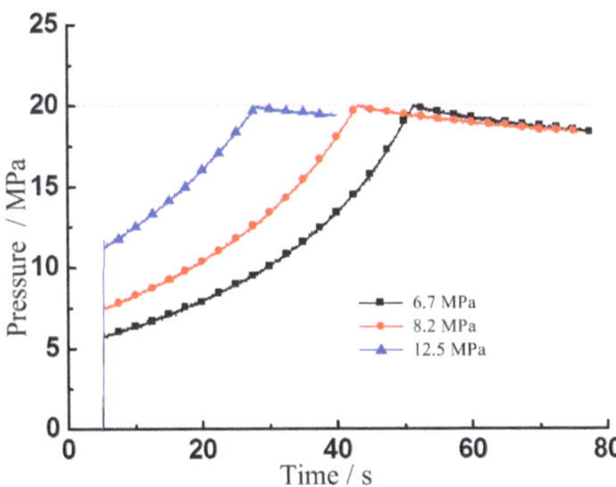

**Figure 11.** HA pressure with different pre-charge pressures during the energy regeneration process.

When the pre-charge pressure of HA is 6.7 MPa and the regulation pressure of the PRV is 21.5 MPa, the PRV is powered on at the time of 5 s and the valve 14 is powered on at the time of 7.4 s to charge the HA. It can be seen from Figure 12 that, before the valve 14 is powered on, both the pressure of the inlet pressure and the pressure in the spring chamber are kept at a constant level. When the HERU is connected to the outlet of the PRV, both the pressures decrease due to a decrease in the fluid force on the main spool until the pressure in HA is almost equal to the pressure in the spring chamber.

Following this, the three pressures increase until the pressure in HA reaches the set value and the valve switches to the top position. There is a pressure fluctuation when the valve 14 is switched from the bottom position to the top position. During this process, the port of the valve 14 and the outlet of the PRV first closes. After this, the oil flowing into the PRV has no channel to flow into and holds up in the inlet, which increase the pressure of the inlet and the pressure in the spring chamber in addition to the enlargement of the orifice of the main valve. After valve 14 begins to work at the top position, the outlet of the PRV is connected to the tank. Due to the larger orifice, the pressure in the inlet and the spring chamber decrease, which leads to the smaller orifice and an increase in the pressure. After a short period of regulation, both the inlet pressure and the pressure in the spring chamber return to their initial values.

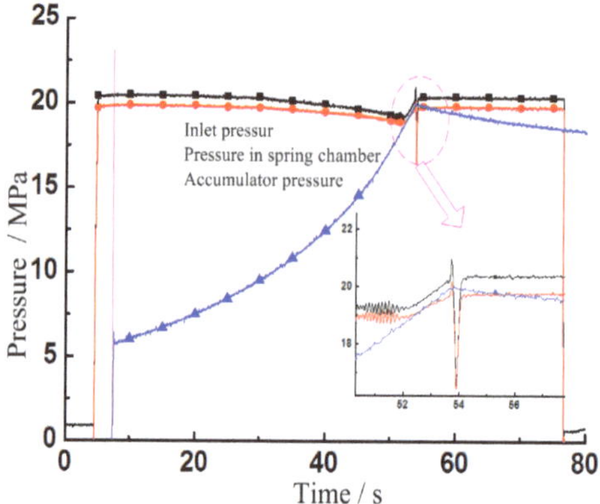


**Figure 12.** Pressures during the energy regeneration process.

Figure 12 illustrates that independent of whether HERU is connected or disconnected to the outlet of the PRV, the inlet pressure of the PRV can be still maintained around the set pressure of 21.5 MPa.

Figure 13 shows the flow rate out of the PRV through the outlet. It can be seen that independent of whether the outlet of the PRV is connected to the HERU, the flow rate out of the PRV is almost the same after the self-regulation.

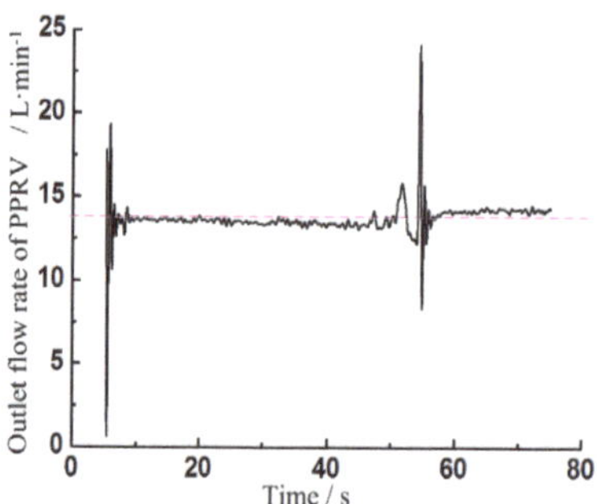

**Figure 13.** Flow rate out of the PRV.

Figure 14 shows the inlet pressure of the PRV under the different maximum pressures of HA. It can be seen that when the PRV is powered on at the time of 1 s, the inlet pressure of the PRV is almost the same as the regulating pressure, which is 21.5 MPa. The inlet pressure slightly decreases with an increase in the maximum pressure of HA, with the maximum error of 0.5 MPa being nearly 2.3% compared to the regulating pressure.

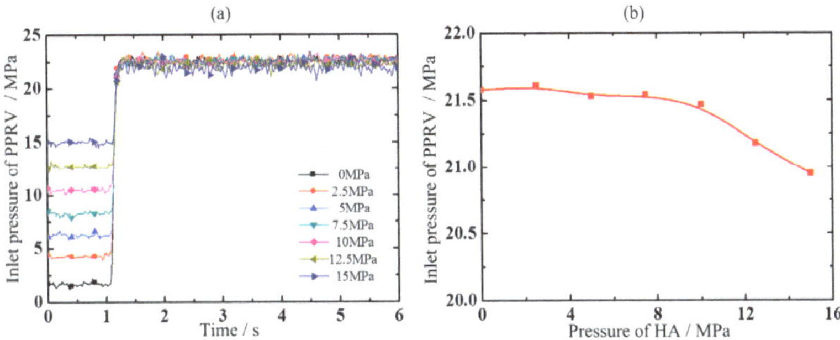

**Figure 14.** Experimental results with different maximum pressures of HA. (**a**) Inlet pressure of PRV; (**b**) Inlet pressure vs. maximum pressure of HA.

### 5.2.2. Regeneration Efficiency

One objective of the experiment is to determine how much energy can be stored by the HERU. As seen in Figure 12, the area between the pressure of the HA and inlet pressure of the PRV is the energy loss through the PRV. When combined with Figure 11, a higher pre-charge pressure results in a smaller area denoting the energy loss.

The energy loss through PRV without HERU is given by

$$E_{\mathrm{r}} = \int p_1 Q dt \tag{16}$$

where $E_{\mathrm{r}}$ is the energy loss through PRV without HERU.

The regenerated energy in HA can be obtained as

$$E_{\mathrm{a}} = \frac{p_{\mathrm{a}0} V_{\mathrm{a}0}}{n-1} \left[ \left( \frac{p_{\mathrm{a}2}}{p_{\mathrm{a}0}} \right)^{\frac{n-1}{n}} - \left( \frac{p_{\mathrm{a}1}}{p_{\mathrm{a}0}} \right)^{\frac{n-1}{n}} \right] \tag{17}$$

where $p_0$, $p_1$, and $p_2$ are the pre-charge pressure, minimum pressure, and maximum pressure of the HA, respectively; and $V_0$ is the volume of the HA. The charging process is short and the compression of the gas can be considered as the adiabatic process. The gas index of the ideal gas $n$ equals to 1.4.

The regeneration efficiency of the HERU is calculated as:

$$\eta_{\mathrm{r}} = \frac{E_{\mathrm{a}}}{E_{\mathrm{r}}} \tag{18}$$

The energy loss and the regenerated energy by HA are listed in Table 1 when the pre-charge pressure is 6.7, 8.2 and 12.5 MPa, respectively. The target pressure of the HA is set to 20 MPa for the three conditions. It can be seen from Table 1 and Figure 10 that when the pre-charge pressure is 6.7 MPa, there is maximum energy loss through PRV without HERU and maximum regenerated energy in HERU. However, the lowest regeneration efficiency happens at this time, being only 61.2%. With an increase in the pre-charge pressure, the energy loss through PRV without HERU and the regenerated energy in HERU decrease, but there is an increase in regeneration efficiency. The reason is that the volume of the HA is the same in the test rig. When the pre-charge pressure of the HA is high, the time for regenerating the energy loss of the PRV via the HA is short, which is seen in Figure 10. As the inlet pressure and the flow rate of PRV are nearly the same, a higher pre-charge pressure results in less energy loss through PRV without HERU. Furthermore, when the pre-charge pressure is high, the outlet pressure of PRV is high. Following this, the pressure drop between the inlet and outlet reduces, which means that the regeneration efficiency will increase. The volume of the HA is chosen according to the installation space. In general, a larger volume means more energy that can be stored.

The choice of pre-charge pressure depends on the requirements. If the regeneration efficiency is the main consideration, a higher pre-charge pressure should be chosen. However, if there is need to regenerate as much energy as possible, a smaller pre-charge pressure should be chosen or multiple HA should be utilized.

**Table 1.** Energy regenerated with different pre-charge pressures.

| Pre-charge pressure/MPa | 6.7 | 8.2 | 12.5 |
|---|---|---|---|
| Energy loss through PRV without HERU/kJ | 206.3 | 166.0 | 94.8 |
| Regenerated energy in HA/kJ | 126.3 | 114.8 | 79.2 |
| Regeneration efficiency/% | 61.2 | 69.2 | 83.6 |

## 6. Conclusions

The proposed configuration of the PRV with the HERU connected to its outlet was discussed through the simulation and experimental analysis. These useful conclusions are obtained:

(1)  To reduce the energy loss of the relief valve, the HERU is connected to the outlet of the PRV, which can increase the outlet pressure and thus reduce the pressure drop between the inlet and outlet. The overflow energy loss that backs into the tank directly in the traditional working conditions is converted into hydraulic energy, which is stored in the hydraulic accumulator and can be released when needed.

(2)  The PRV with HERU connected to the outlet can still have a better function in regulating pressure. The flow rate of the PRV is almost constant independent of whether the HERU is connected or disconnected to the PRV after a short adjusting time.

(3)  A higher pre-charge pressure can achieve higher regeneration efficiency, while the lower pre-charge pressure can regenerate more energy. The chosen pressure of the HA should consider the regeneration efficiency, the actual working style, and the installation space.

(4)  Further research will concentrate on the multiple HA used in the proposed HERU and the control of them to regenerate as much energy as possible with a higher regeneration efficiency.

**Acknowledgments:** The authors gratefully acknowledge the financial support of National Natural Science Foundation of China (Grant No. 51505160), Promotion Program for Young and Middle-aged Teachers in Science and Technology Research of Huaqiao University (ZQN-YX201), Industry Cooperation of Major Science and Technology Project of Fujian Province (2017H6101), Open Foundation of the State Key Laboratory of Fluid Power and Mechatronic Systems (GZKF-201517 & GZKF-201611), and Natural Science Foundation of Fujian Province (2017J01087).

**Author Contributions:** Tianliang Lin proposed the idea of using the energy regeneration unit to recovery the energy loss through the relief valve. Qiang Chen. and Haoling Ren designed the experiments. Qiang Chen, Yi Zhao and Cheng Miao performed the experiments and analyzed the data. Shengjie Fu and Qihuai Chen contributed to the control of the test rig. Tianliang Lin and Haoling Ren wrote the paper.

**Conflicts of Interest:** The authors declare no conflict of interest.

### Nomenclatures and Symbols

$p_1$ is the inlet pressure of the PRV; $p_2$ is the outlet pressure of the PRV; $\Delta p$ is the pressure drop between the inlet and outlet; $p_3$ is the pressure of the spring chamber; $p_{a0}$ is the pre-charge pressure of HA; $p_s$ is the outlet pressure of the pump; $p_a$ is the pressure of the HA; $F_0$ is the pre-tightening force of the reset spring; $A_1$ is the effective cross-sectional area of the main spool; $V_{a0}$ is the rated volume of HA; $V_{a1}$ and $p_{a1}$ are the gas volume and pressure when the HA is at the minimum working pressure; $V_{a2}$ and $p_{a2}$ are the gas volume and pressure when the HA is at the maximum working pressure; $\Delta V_{a1}$ is the maximum oil volume that is stored in HA when the pressure in HA increases from $p_{a1}$ to $p_{a2}$; $n$ is the polytrophic exponent; C is a constant; $E_\rho$ is the energy density of the HA; $E_r$ is the energy loss through PRV without HERU; $E_a$ is the regenerated energy in HA; $Q$ is the flow rate of the PRV; $\eta_r$ is the regeneration efficiency of the HERU.

*Appl. Sci.* **2017**, *7*, 613

## References


1. Shen, W.; Jiang, J.; Su, X.; Karimi, H.R. Control strategy analysis of the hydraulic hybrid excavator. *J. Frankl. Inst.* **2015**, *352*, 541–561. [CrossRef]
2. Chen, Q.; Wang, Q.; Wang, T. A novel hybrid control strategy for potential energy regeneration systems of hybrid hydraulic excavators. *Proc. Inst. Mech. Eng. Part I J. Syst. Control Eng.* **2016**, *230*, 375–384. [CrossRef]
3. Lin, T.; Wang, Q.; Hu, B.; Gong, W. Development of hybrid powered hydraulic construction machinery. *Autom. Constr.* **2010**, *19*, 11–19. [CrossRef]
4. Shen, G.; Zhu, Z.C.; Li, X.; Tang, Y.; Hou, D.; Teng, W. Real-time electro-hydraulic hybrid system for structural testing subjected to vibration and force loading. *Mechatronics* **2016**, *33*, 49–70. [CrossRef]
5. Shi, Y.; Li, F.; Cai, M.; Yu, Q. Literature review: Present state and future trends of air-powered vehicles. *J. Renew. Sustain. Energy* **2016**, *8*, 325–342. [CrossRef]
6. Chen, Q.; Wu, W.; Liu, W.; Zhang, X. Research on pump control signal of excavator positive control system. *J. Hefei Univ. Technol. (Nat. Sci.)* **2014**, *37*, 645–649.
7. Cheng, M.; Zhang, J.; Ding, R.; Xu, B. Pump-based compensation for dynamic improvement of the electrohydraulic flow matching system. *IEEE Trans. Ind. Electron.* **2016**, *64*, 2903–2913. [CrossRef]
8. Sugimura, K.; Murrenhoff, H. Hybrid Load Sensing–Displacement Controlled Archtecture for Excavators. In Proceedings of the 14th Scandinavian International Conference on Fluid Power (SICFP'15), Tampere, Finland, 20–22 May 2015.
9. Xu, B.; Liu, Y.J.; Yang, H.Y. Research on electro-hydraulic proportional load sense separate meter in and separate meter out control system. *Hunan Daxue Xuebao/J. Hunan Univ. Natl. Sci.* **2010**, *37*, 190–196.
10. Quan, Z.; Quan, L.; Zhang, J. Review of energy efficient direct pump controlled cylinder electro-hydraulic technology. *Renew. Sustain. Energy Rev.* **2014**, *35*, 336–346. [CrossRef]
11. Shen, W.; Jiang, J. Analysis and development of the hydraulic secondary regulation system based on the CPR. In Proceedings of the International Conference on Fluid Power and Mechatronics, Beijing, China, 17–20 August 2011; pp. 117–122.
12. Hippalgaonkar, R.; Ivantysynova, M.; Zimmerman, J. Fuel savings of a mini-excavator through a hydraulic hybrid displacement controlled system. In Proceedings of the 8th International Fluid Power Conference, Dresden, Germany, 26–28 March 2012; pp. 139–153.
13. Ren, H.; Lin, T.; Huang, W.; Fu, S.; Chen, Q. Characteristics of the energy regeneration and reutilization system during the acceleration stage of the swing process of a hydraulic excavator. *Proc. Inst. Mech. Eng. Part D J. Automob. Eng.* **2017**, *231*, 842–856. [CrossRef]
14. Wang, B. High Speed On-Off Valve Self-adapting Clamping System. *J. Appl. Sci.* **2014**, *14*, 279–284. [CrossRef]
15. Shi, Y.; Wu, T.; Cai, M.; Wang, Y.; Xu, W. Energy conversion characteristics of a hydropneumatic transformer in a sustainable-energy vehicle. *Appl. Energy* **2016**, *171*, 77–85. [CrossRef]
16. Wang, T.; Wang, Q. Design and analysis of compound potential energy regeneration system for hybrid hydraulic excavator. *Proc. Inst. Mech. Eng. Part I J. Syst. Control Eng.* **2012**, *226*, 1323–1334. [CrossRef]
17. Wang, T.; Wang, Q.; Lin, T. Improvement of boom control performance for hybrid hydraulic excavator with potential energy recovery. *Autom. Constr.* **2013**, *30*, 161–169. [CrossRef]
18. Lin, T.; Chen, Q.; Ren, H.; Miao, C.; Chen, Q.; Fu, S. Influence of the energy regeneration unit on pressure characteristics for a proportional relief valve. *Proc. Inst. Mech. Eng. Part I J. Syst. Control Eng.* **2017**, *231*, 189–198. [CrossRef]

*Review*

# Review on Inlet/Outlet Oil Coordinated Control for Electro-Hydraulic Power Mechanism under Sustained Negative Load

**Wei Liu [1], Yunhua Li [1,*] and Dong Li [2]**

[1]  School of Automation Science and Electrical Engineering, Beihang University, Beijing 100191, China; weiliu_buaa@buaa.edu.cn
[2]  School of Aeronautic Science and Engineering, Beihang University, Beijing 100191, China; ld890107@126.com
*   Correspondence: yhli@buaa.edu.cn; Tel.: +86-10-82339038

Received: 26 April 2018; Accepted: 23 May 2018; Published: 28 May 2018

**Abstract:** Speed control and smooth regulation in an electro-hydraulic motion control system under negative load (over-running load) are crucial to mobile machineries, vehicles, and motion simulation equipment. Problems, such as bad natural stability, bad dynamic performance for small adjusting signal, serious coupling, and difficulty to coordinate between speed smoothness and speed regulation, exist in traditional valve-controlled and pump-controlled electro-hydraulic power mechanism. The coordinated motion control scheme, which is based on independent regulation of inlet/outlet oil for electro-hydraulic power mechanism has attracted the attention of many scholars. In the last two decades, many progresses had been made in coordinated control technology, employing in inlet/outlet independent metering and pump/valves independent regulation. Moreover, the technology has been widely used in the electro-hydraulic operating system in hydraulic excavator and the speed smoothness control system in heavy transport vehicles. In this study, recent advancements and upcoming trends in coordinated control and inlet/outlet independent metering for electro-hydraulic power mechanism under sustained negative load are reviewed. Firstly, related research advancements are summarized, including flow rate regulation mechanism based on inlet/outlet coordinated control, coordinated control strategy and nonlinear control method of electro-hydraulic control system. Then, nonlinear modeling of inlet/outlet independent metering and pump/valves independent regulation in electro-hydraulic control system is presented. In addition, the electro-hydraulic speed smoothness control and energy recovery for heavy engineering vehicles under long down-slope is discussed and reviewed. Finally, existing problems and future trends of inlet/outlet coordinated control for an electro-hydraulic power mechanism under sustained negative load are presented.

**Keywords:** sustained negative load; electro-hydraulic power mechanism; independent regulation; coordinated control; speed smooth and limitation control

---

## 1. Introduction

Various types of heavy construction machinery and engineering vehicles [1] are playing an increasingly important role in the construction of large-scale projects, such as high-speed railway, highways, large bridges, as well as the construction of underground tunnels for coal mining. The working condition under sustained negative load is common in electro-hydraulic loading control system [2]. Then, the traveling drive system, hoist system and swing system of construction machineries and engineering vehicles are inseparable from the electro-hydraulic power mechanism under sustained negative load. As the crucial technology of above systems, the control technology of electro-hydraulic power mechanism has greatly influenced the control performance for construction machinery and engineering vehicles. Therefore, systematic study on advanced regulation mechanism and control strategy for electro-hydraulic power mechanism with negative load is of great value, which enhances integrated control performance for overall electro-hydraulic mechanism.

Sustained negative load, also known as over-running load, often occurs in long down-slope condition in traveling system of mobile machines, descending condition under heavy load in electro-hydraulic hoist systems of hoisting cranes, swinging and boom sticks conditions in excavators, and negative load condition of electro-hydraulic loading systems. Negative load operation is inevitable and standard working condition to a completed task cycle. The characteristics of negative load in electro-hydraulic power system is that the load force's direction is in accordance with the orientation of load movement, and the function of load force is changed from opposing motion to driving motion. In fact, the running speed of moving parts will continuously increase due to the increase of driving force incorporating the force from negative load. But the speed may even exceed the actuating components' maximum running speed which can be provided by flow rate of the hydraulic pump. Moreover, the dithering phenomenon in the process of system running can easily cause safety accidents.

Aiming at the method of traditional flow rate regulation for electro-hydraulic control system with sustained negative load, two control modes are usually used. One mode is the valve-control mode, which uses dual-port mechanical linkage control valve to realize the throttling regulation. The other mode is the pump-control mode, which uses proportional or servo variable displacement pump to achieve the volume regulation. The former adopts inlet/outlet mechanical coupling metering, and a large valve-restriction pressure difference is needed for balancing the negative load. And the latter commonly adopts the balance valve in return-oil line to form the back-pressure for speed smoothness control. In the former case, because of the throttling and coupling control with dual-port mechanical linkage control valve, it is difficult to achieve superior control effect on speed control and smooth regulation. For the latter, since the balance valve can't be actively regulated, the satisfactory control effect also can't be obtained.

Under sustained negative load, the operating points of electro-hydraulic control valve for motion control system are in quadrants II and IV, and pressure flow coefficient of the valve is reduced, then control characteristics are deteriorated when the orifice is small. The coupling control with flow rate and load pressure in traditional electro-hydraulic mechanism results in a large valve-restriction pressure difference in inlet/outlet oil loop, then a good coordination between speed control and speed-smooth regulation cannot be realized. In references [3,4], the hydraulic balance method and the stability of balance valve under negative load are studied, and relevant hydraulic balance scheme is proposed. Aiming at the problem of down dithering and secondary sliding for hoisting system under negative load, Liu [5] designed a new balance valve cover with damping network, which makes the pressure into pilot control oil chamber of the balance valve steadier. Moreover, the balance valve and brake get a good matching for the time of opening and closing. In view of function exchange for motor and pump under sustained negative load, Xian [6] presented a negative torque control scheme for hoisting mechanism type in closed volume hydraulic system. In the control scheme, by reducing the displacement volume of pump or increasing the motor's displacement volume, the speed of motor and engine keeps within the limit. Nie [7] proposed the constant power control for load input power, and it guarantees the optimal load rate of the engine. Moreover, the security and energy-saving for the crane under negative load is achieved.

At present, for inlet/outlet independent metering control system of electro-hydraulic power mechanism, the research mainly focuses on the multi-layer and coordinated control strategy and energy-saving control technology. Aiming at the existing problems of electro-hydraulic power mechanism under sustained negative load, a thorough and systematic study, such as, theoretical analysis of regulation mechanism, motion-force hierarchical control, inlet/outlet coordinated control strategy, unknown disturbance estimation of power mechanism, dynamic feature recognition for continuously running system with long time and strong robust control strategy are comparatively scarce. The coordinated motion control strategy based on inlet/outlet independent metering can be introduced for electro-hydraulic motion control system, through which controllability and practicability of control system is enhanced. Especially, the coupling problem of speed control and smooth regulation can be effectively solved. Moreover, the two independent control subsystems can be coordinately adjusted, which greatly improves the control potential for electro-hydraulic power control system.

The remainder of this paper is organized as follows: for the electro-hydraulic power control system, flow rate regulation mechanism based on inlet/outlet independent metering, coordinated control strategy, and nonlinear control algorithm are introduced in Section 2. Nonlinear modeling of inlet/outlet independent metering for pump/valves control system is elaborated Section 3. Section 4 provides electro-hydraulic speed smoothness control and energy recovery of heavy vehicles under long down-slope, followed by the main conclusions are summarized, and the future research works are prospected.

## 2. Inlet/Outlet Independent Regulation and Coordinated Control

Inlet/outlet independent regulation and coordinated control with hierarchical structure have been the focus of intensive research, resulting in a rich literature. In this section, the research fields, including flow-rate regulation mechanism of inlet/outlet independent metering control, coordinated control strategy, and nonlinear control algorithm are reviewed. Moreover, the current research statuses are summarized, and necessary analysis and evaluation is raised.

### 2.1. Flow-Rate Regulation Mechanism of Inlet/Outlet Independent Metering

Inlet/outlet independent metering system usually uses two electro-hydraulic control valves to determine the motion state of an actuator, which opens the mechanical connection between the inlet and outlet in ordinary electro-hydraulic control valve. The key question is how to coordinate the opening displacement of two spools in control valves, so that the coordinate control of pressure and flow rate can be realized. Because two throttle valves can be respectively responsible for flow rate control in inlet lines and back-pressure control in outlet lines, to reduce the throttling loss in inlet line at least is possible.

In 1987, Professor Backe [8] presented the conception of separate control of actuator ports with cartridge valve control theory, the control principle is shown in Figure 1. Through the input liquid resistance and output liquid resistance, pressure and flow rate of hydraulic cylinder chambers are independently controlled. Liquid resistances are realized in four cartridge valves. With the development of electronic technology, Backe applied electronic feedback control technology to hydraulic control system, further optimized the performance of the control system. The feedback signals, which are in the form of pressure feedback, flow rate feedback, speed feedback and acceleration feedback, realize the system feedback control function.

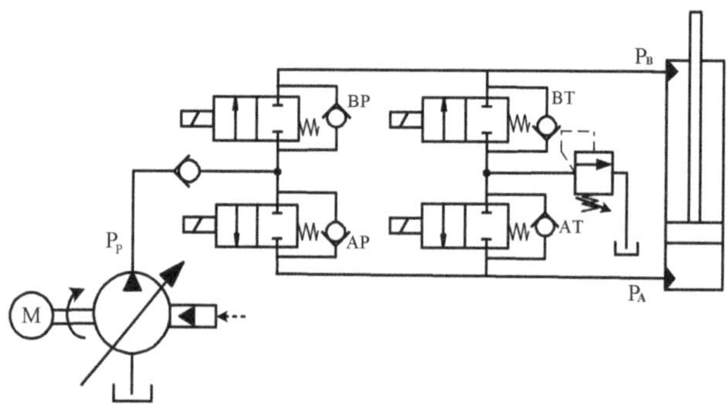

**Figure 1.** Independent metering control system based on poppet valve.

Palmberg [9] formally proposed the concept of inlet/outlet independent metering control. Four electro-hydraulic proportional poppet valves are used to realize the control of hydraulic actuators, and a constant pressure pump is selected for oil supply. In addition, the value and direction of the load are obtained by the measured pressures from two pressure sensors in two chambers, then different control strategies can be selected for optimal control according to the load and speed characteristics.

Book et al. [10] put forward the controllability and possibility of energy recovery under negative load, and the scheme adopts the inlet/outlet oil decoupling control through four-valve independent regulation. Moreover, the mathematical analysis and demonstration is derived. Compared to the traditional valve control system, the saving energy effect of the independent metering mechanism with four-valve can is better by controlling the back pressure. The electro-hydraulic system mechanism of four-valve independent regulation is shown in Figure 2.

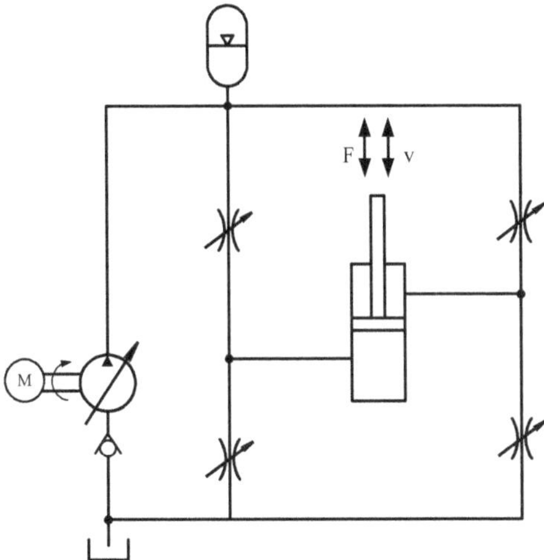


**Figure 2.** Electro-hydraulic system mechanism of four-valve independent regulation.

Sphenoid et al. [11] further studied this kind of electro-hydraulic system controlled by four valves. Since the four valves can be independently controlled, the configuration scheme can be operated with a variety of double valves in independent control modes. In addition, the regulation modes of five different working conditions are summarized, which includes power extension, high side regeneration extension, low side regeneration retraction, power retraction, and low side regeneration extension.

Elfving M. et al. [12] studied the physical decoupling control of double chamber pressure for independent metering control system, where the two chambers of hydraulic cylinder respectively use an electro-hydraulic proportional valve for flow-rate or pressure control. In this control system, an auxiliary controller and two pressure controllers are designed. The pressure decoupling control of two chambers is achieved with two pressure controllers, and the speed control of actuator is realized by auxiliary controller.

Patrick et al. [13] proposed a new control method of automatic calibration state trajectory, and applied it to the flow rate control in independent metering system. The system uses five electric hydraulic lifting valves to form the independent metering control structure. The proposed control method is used to adjust four of these valves, with which the flow rate control of hydraulic actuator's inlet/outlet oil is realized. Moreover, supply pressure is controlled by the fifth valve with open-loop operation. In view of the fact that the excavator system based on inlet/outlet independent metering control is a mechanical-hydraulic coupling system with many unknown parameters, Ding et al. [14] proposed the multi-objective optimization method based on nonlinear programming by quadratic

Lagrangian (NLPQL). The optimal value of characteristic parameters, which are difficult to estimate, are found by the method, and the speed and accuracy of parameter identification are improved.

Zhang and Hu [15,16] designed an independent metering control system consisted of five independent proportional cartridge valves, which can be programmed to achieve flexible control of the system. Changing the control logics of the valves can achieve different median performance, and the system can realize system energy saving and regeneration without changing the hardware structure by proper programming. For inlet/outlet independent metering system, Professor Andersen [17,18] carried out the study on the control strategy and the influence of control performance by valve structural parameters. The electronic control handle transmits the speed signal instruction to system controller, and the system controller sends control signals to control the proportional directional valve on both sides of the actuator, then the speed control and pressure control of actuator are realized. Meanwhile, the performance of system energy saving is enhanced. In addition, because the throttle loss of check valve is much smaller than balance valve, the efficiency of system is improved by changing components.

In the middle of 1990s, Wang et al. [19,20] carried out a full study on the impact problems during the acceleration and deceleration, the starting and brake of the large inertia load in the hydraulic excavator. A pressure difference sensing technology is put forward to reduce the pressure impact, which is produced during the acceleration and deceleration of actuators. The control method of calculation flow rate feedback is adopted to control the flow rate of throttle valve, so that the speed smooth control is achieved for large inertia load. The core idea of control is to use a piecewise control method while the large inertia load is in the process of acceleration, deceleration and braking, and the method is based on actuator speed observation and parameters online estimation. Moreover, it takes account of its control accuracy of steady state as well as fast stability of dynamic process.

In references [21,22], the speed control characteristic of independent metering control system with multiple actuators is studied when actuators run at the same time. The synchronization model of the dual-actuator for independent metering system is established. Moreover, the controller is developed, which compares the speed control characteristics under different working conditions. Speed control method for piecewise adjustment according to the delay time of the balance valve is proposed, which improves the stability of speed control in balance valve circuit.

For independent metering control system, Quan et al. [23] proposed a control scheme for differential hydraulic cylinder, and this method can reduce power loss of electric motor by about 20%, which improves the efficiency of pump-controlled differential cylinder system. In addition, an inlet/outlet independent metering control system based on flow rate matching with pump and valve was put forward. The system ensures that the output flow rate of power source always matches the required flow rate by all the actuators, and it not only achieves the implementation of the individual executor, but also reduces throttle loss of the system to a minimum.

Yuan [24] studied the control technology of inlet/outlet independent metering, which uses a double servo valve to control the cylinder system. Furthermore, an adaptive robust control method based on correction of quiescent operation point is proposed. Aiming at the pump/valves control strategy for an electro-hydraulic flow rate matching system, Cheng et al. [25] analyzed the requirements with speed control system under overrunning and impedance conditions, and an improved method of dynamic performance based on compensation with proportional valve is proposed. The proportional valve controller, which considers the compensation of dynamic and static performance at the same time, is developed to improve the performance of speed control system. A typical electro-hydraulic flow rate matching system mechanism of pump/valves control is shown in Figure 3.

Dong [26] designed a flow-rate and pressure coupling controller to control the piston speed and pressure in each chamber, and a double closed loop control in independent metering control system is used to monitor and modify the pressure and flow rate in real time. Moreover, four types of threads are designed to arrange the task. In reference [27], a programmable valve with inlet/outlet independent metering control is applied to replace the multi-way directional valve in construction machinery, which integrates the advantages of digital signal and inlet/outlet independent metering technology. Due to the increase of control freedom of electro-hydraulic system in construction machinery, the controllability of the control system is improved, and the energy consumption of system cut down by reducing the back pressure of hydraulic actuator.

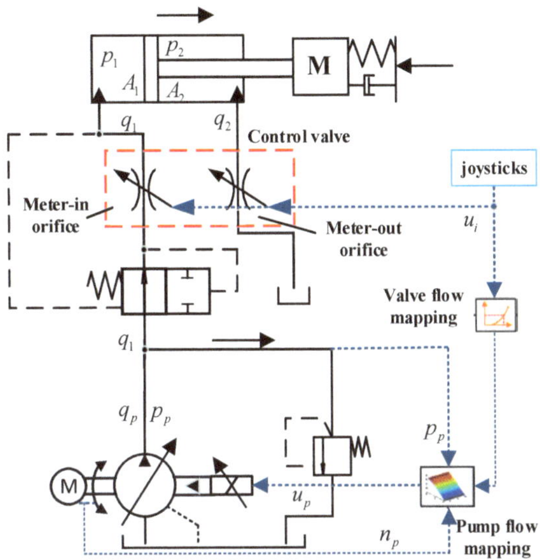


**Figure 3.** Electro-hydraulic flow rate matching system mechanism of pump/valves control.

Li [28] put forward a novel flow rate amplifier valve with digital pilot, where the digital valve is the pilot stage and the Valvistor valve is the main stage. Furthermore, applying it to independent metering control system, not only realizes the requirements of control performance and energy saving of valve control system, but also lays the foundation for realizing modern intelligent control of electro-hydraulic system. Liu et al. [29] used a flow-rate feedback calculation control and pressure calculation control for independent metering system under negative load condition to improve the speed stability. Liu [30] presented a direct proportional flow rate control by pressure compensation with load control valve. In this valve, the flow rate through load control valve is proportional to the pilot pressure in the control stroke, and load pressure compensation function means that when the load pressure is too high, the flow rate of load control valve can be limited to the vicinity of the maximum rated flow. Addressing the control for the speed system of pump controlled parallel variable displacement motor with the flow rate adaptive distribution and multiplying nonlinear link characteristics, Zheng et al. [31] proposed a compound control architecture, which realizes pressure control by variable displacement pump and speed control of drive shafts by variable displacement motors. The second order linearization models of pressure control system and speed regulating system are derived, and the expected pressure planning method is proposed to realize the dynamic pressure regulation of the system. In addition, by introducing a disturbance observer to the speed controller of variable displacement motor, the nonlinear disturbance and unknown load disturbance are suppressed, which improves the steady state accuracy with fluctuating pressure.

From above-surveyed on flow rate regulation mechanism for inlet/outlet independent metering control system, the most important research aspects are summarized, and listed below.

1. The inlet/outlet independent metering control has become an important research direction for electro-hydraulic power mechanism. It not only improves the accuracy and stability of the motion control system, but also possibly realizes the energy saving and regeneration. While maintaining

the excellent dynamic performance of the electro-hydraulic control system, how to increase the energy efficiency of system has become an important research topic.

2. With the relatively large amount of attention devoted to valve-controlled cylinder systems, references are comparatively scarce on independent and coordinated control of pump/valves for pump-controlled motor system. However, the study on inlet/outlet independent metering control for pump-controlled motor system also has important practical value in engineering application.

3. For the inlet/outlet independent metering control of electro-hydraulic system, while adopting different control strategies and structural parameters of the valve, the control performance and energy-saving efficiency are affected to a certain extent, which has been the focus of intensive research.

4. As new hydraulic components, the digital programmable valve and novel flow rate amplifier valve with digital pilot are applied to the independent metering control system, which promotes the modern intelligent control for electro-hydraulic system. However, most of the digital valves currently studied are only as high-speed on-off valves. Because of the limitation with their structure, it can only be applied to pilot control and small flow rate control applications.

5. The key issue of the study on inlet/outlet independent metering control system is to coordinate control variables of double electro-hydraulic control elements, which usually includes the flow-rate and the pressure regulation. Therefore, considering the system under different working modes, independent and coordinated regulation mechanism based on motion-force hierarchical control architecture is always an important research topic.

6. The function exchange of motor and pump under sustained negative load seriously affects the dynamic stability performance for motion control system. Aiming at the problem, development of speed smoothness control strategy with motion-force hierarchical structure have been an urgent problem to be solved.

7. With sustained negative load in engineering applications, the study of regulation mechanism for suppressing negative load and integrated energy management scheme are of great practical value for electro-hydraulic motion control system.

### 2.2. Control Strategy for Electro-Hydraulic Control System

The nonlinear characteristics and modeling uncertainty of electro-hydraulic system seriously restrict the improvement of control performance for electro-hydraulic system, which cause the classical control method based on linear theory gradually unable to meet the requirements for high performance system. Nonlinear control method and active disturbance rejection control technology based on uncertainty disturbance have become an urgent need. In addition, in view of electro-hydraulic motion control under sustained negative load, speed smooth control, and energy saving control have become new and important research topics. The coordinated control based on inlet/outlet independent metering system and motion-force hierarchical architecture also have become the effective solutions, and an intensive study in this field will play an important role in improving the performance and energy saving of electro-hydraulic motion control system. This section overviews the nonlinear control algorithm and hierarchical coordination control strategy for electro-hydraulic control system, and necessary analysis is carried out.

### 2.2.1. Nonlinear Control Algorithm of Electro-Hydraulic Control System

There are many nonlinear characteristics in the electro-hydraulic control system, such as the pressure and flow rate nonlinearity in servo valve, the structure nonlinearity of differential equation, the friction nonlinearity of actuators, and so on. And these nonlinear factors and questions in the most electro-hydraulic servo systems are inevitable to solve. Therefore, more advanced nonlinear control methods need be studied for the electro-hydraulic control system with nonlinear factors.

Addressing the uncertainties in nonlinear control systems, Astrom and Utkin et al. developed various nonlinear control methods, such as adaptive control and variable structure sliding mode control. In the 1990s, Kokotovic et al. proposed the idea of backstepping control for the mismatched and uncertainty problem in electro-hydraulic control, where the core is how to overcome all kinds of uncertainties in nonlinear models. Niksefat and Sepehri et al. [32] established the nonlinear models for the electro-hydraulic actuator (EHA), and the robust controller of EHA based on the quantitative feedback theory (QFT) is designed. Meanwhile, the LuGre friction model is introduced to EHA model, the adaptive observer is used to estimate the actuating cylinder acceleration and the friction state, which cannot be observed directly. Moreover, the adaptive control law is designed to compensate the parameters perturbation and load disturbance of the system. For the nonlinear model of EHA, Alleyne and Kaddissi [33,34] designed the backstepping controller, which meets the stability of Lyapunov, obtaining a good control effect.

Bjorn [35,36] applied the existing nonlinear control strategy to inlet/outlet independent metering control system, and the method with linear quadratic Gaussian (LQG) optimal control is proposed. In addition, the speed and pressure controller of the independent metering control system is also designed, which is applied to electro-hydraulic system of the crane. In order to test the actuator controlled by direct drive valve (DDV), Nam et al. [37,38] developed the dynamic load simulator, and the robust controller for loading system is designed, which is based on the quantitative feedback control technology (QFT). For electro-hydraulic loading system, Ahn et al. [39,40] proposed a fuzzy proportional-integral-derivative (PID) intelligent control algorithm, which combines with the fuzzy control algorithm, grey scale prediction, and PID control.

In order to solve the issue that fluid bulk elastic modulus is difficult to quantify, Bora [41] developed a control design process, where additional sensors are not needed. Moreover, the design process makes the variables in bulk modulus have robustness. Garett et al. [42] presented the derivation of nonlinear tracking control law for a hydraulic servo system. An analysis of the nonlinear system equations is used in the derivation of a Lyapunov function that provides for exponentially stable force trajectory tracking. Furthermore, this control law is then extended to provide position tracking. Addressing the nonlinear systems with unknown input dead-zones, Hu et al. [43] proposed the direct/indirect adaptive robust control combining with the desired compensation strategy to synthesize practical high performance motion controllers for precision electrical drive systems. Even if the overall system is subjected to parametric uncertainties, time varying disturbances, and other uncertain nonlinearities, certain guaranteed robust transient performance and steady-state tracking accuracy can be achieved. For multi-degree of freedom system, Takahiro et al. [44] expanded the diagonalization method based on the modal space disturbance observers, which suppresses the interferences between the position control and force control of the system. As a result, the bilateral control is realized, and that performance is better than the coordinated control. Aiming at the problem on system state and interference estimation for slotted hydraulic, Pillosu et al. [45] linearized the nonlinear dynamics system and designed an estimation algorithm based on sliding mode control and nonlinear disturbance observer. For the high nonlinearity and non-differentiable features of electro-hydraulic control system, Claude et al. [46] used indirect backstepping adaptive control strategy to achieve the real-time position control.

On the basis of the improved state observer, Yang et al. [47] proposed a new model reference adaptive control algorithm for nonlinear systems with significant uncertainties. Moreover, the algorithm adopts neural networks to improve transient and steady-state performance, and the stable tracking performance is met. For a class of linear systems with uncertain external disturbances, Zhang et al. [48] designed a memory-based adaptive output feedback sliding mode controller, which improves the transient performance for linear systems. Chiang et al. [49] proposed an adaptive fuzzy controller with self-tuning fuzzy sliding mode compensation for electro-hydraulic displacement-controlled system. Furthermore, the controller has online adjustment capability by controlling the rule parameters, which can deal with the time-varying and non-linear uncertainty behavior of the system. In reference [50], aiming at the system nonlinearities and uncertain parameters for high performance force control of hydraulic load simulator, a discontinuous projection-based nonlinear adaptive robust force controller is presented, which makes the system asymptotically stable. Moreover, it realizes the transient performance and final tracking accuracy. Considering the nonlinear input and input dead zone, Triet et al. [51] proposed an adaptive fuzzy sliding mode controller for speed control system, which is driven by hydraulic pressure coupling. And the controller combines direct adaptive fuzzy and fuzzy sliding mode, which reduces the tracking error and jitter in new structure.

Aiming at multi-input and multi-output nonlinear control system with inlet/outlet oil independent metering control, Liu [52] raised a force calculation controller based on the feedback linearization, where the controller can calculate the reference pressure in the two chamber of hydraulic cylinder. Furthermore, PID controller is used to achieve the closed loop tracking control of two chambers' pressure. During the operation of the excavator's boom lowering, the balance valve is usually used to suppress the sustained negative load, which causes the hydraulic system to generate a large amount of heat. In view of the serious heating problem, Jin and Wang [53] proposed proportional throttle valve to balance the negative load. And an adaptive backstepping controller is designed, which maintains the inlet oil pressure of single-acting cylinder at a small value, then the goal of system energy saving is achieved.

In view of the uncertain nonlinear systems with parametric uncertainties and uncertain nonlinearities, Yao et al. [54] proposed an active disturbance rejection adaptive controller for tracking control, and the controller effectively combines adaptive control with extended state observer by backstepping method. Moreover, system uncertainties are estimated by extended state observer and compensated in a feedforward way. Aiming at the trajectory tracking problem for a Delta robot with uncertain dynamical model, Luis et al. [55] raised the output-based adaptive control based on the active disturbance rejection control (ADRC) technique. Furthermore, the simultaneous observer-identifier scheme is designed. The experimental results show that the adaptive control based ADRC has better control performance than the conventional PID controller with feedforward actions and regular ADRC. Addressing the vibration attenuation problem for vehicle suspension system, Li et al. [56] applied the ADRC technique to the industrial applications of active vehicle suspension system, the controller exhibits good robustness and easy implementation. Moreover, the total disturbance of vehicle suspension system is estimated and compensated in the controller via extended state observer. Among all of the experiments, the vehicle suspension system with ADRC shows a best performance under different road profile.

### 2.2.2. Hierarchical Coordination Control Strategy of Electro-Hydraulic Control System

With the increase of control performance requirement and complexity of control system, the decoupling control for multivariable and complex systems is no longer able to achieve the design requirements. However, for many electro-hydraulic control systems, it is not required to decouple the variables, but to maintain a certain coordination of multiple variables in control process. Therefore, the hierarchical and coordinated control of electro-hydraulic control system becomes an important research field.

Multivariable decoupling control is widely used in multivariable and complex electro-hydraulic control system, and the working principle is that the coupling between controllers is used to remove the coupling effect of controlled plant. Moreover, the multivariable control system with coupling effect is decomposed into several single variable control subsystems to regulate separately. In 1960, the first international federation of automatic control and Chinese scholar Tu Xuyan first proposed the multivariable coordinated control theory. It is not only different from the non-interactional control put forward by American scholars, but also different from the theory of self-regulation by Soviet scholars.

At present, the idea of hierarchical coordination control has been widely used in different professional fields. Moghadam et al. [57] developed a hierarchical optimal controller, which is simultaneous to control both the machining force and the axis position. The whole milling process is divided into two stages, and the two level targets are adjusted in the feasible range by weight. However, during the transient process, a series of simulations show that the control method can only make one target performance better. In order to enhance the vehicle dynamics stability and handling performance for electric vehicle, an optimal two-layer control scheme with hierarchical structure is proposed in the reference [58]. In fact, the upper controller based on sliding-mode control is in charge of motion regulation, where the desired longitudinal and lateral forces and yaw moment is calculated for keeping vehicle dynamics stability. The lower layer assigns the driving/braking torques to each in-wheel motor with an optimization algorithm. Moreover, considering motor power capability and tire workload, a cost function with adjustable weight coefficients is designed. The optimal control scheme with hierarchical structure for electric vehicle is shown in Figure 4.

Wu et al. [59] developed a hierarchical control architecture based on the Takagi-Sugeno (T-S) fuzzy model. Even if there are serious behavior changes of boiler steam turbines, the hierarchical control can achieve the optimal control. Yao [60,61] developed the controller for an independent metering control system with five cone valves, and the control characteristics is studied deeply. The two-layer controller included the nonlinear robust adaptive controller and PID are designed, which achieves trajectory tracking control and energy saving control. Moreover, the control system is composed of five cone valves, where four valves are used for independently regulating the flow rate and pressure in two chambers of actuator, and one valve is directly connected with the two cylinders for energy regeneration. In addition, the two-layer controller is used for mode selection and flow-pressure control with proportional valve, which includes the upper level controller with task switching and the lower layer controller with flow-pressure control. Furthermore, both the pressure controller and flow rate controller adopt robust adaptive control based on the nonlinear system model.

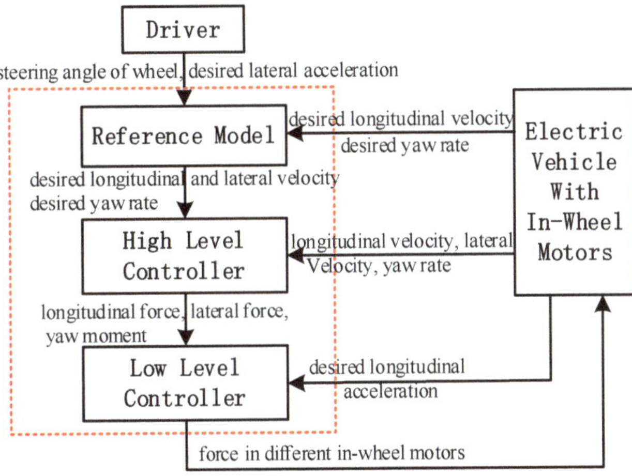


**Figure 4.** Optimal control scheme with hierarchical structure for electric vehicle.

Aiming at power distribution and coordination problem for the earth moving vehicle, Zhang et al. [62] developed a generalized model for the multi-input and multi-output electro-hydraulic transmission system, which adopts the method with linearization of nominal work points. Moreover, the controller based on $H_2$ and $H_\infty$ is designed for the control system. For the synchronous control strategy of multi mechatronic control systems, since the coupling compensation rules is difficult to be determined, and the amount of online computing is very large, Zhang et al. [63] put forward a synchronous control idea with the minimum number of related axes. Moreover, the synchronization control algorithm, which is based on the theory of adjacent coupling error and sliding mode control, is developed. For the system model with large model uncertainties and external disturbances,

Fang et al. [64] proposed a cross-coupling control (CCC) algorithm, which is based on Lyapunov stability criterion and recursive updating technology. Indeed, the contour performance is obviously enhanced by coordinating the motion of multiple axes.

In order to solve the motion synchronization problem for double-cylinder electro-hydraulic elevator system, Sun and Qiu [65] presented a nonlinear control algorithm including two-layer coordination controller. The synchronization controller of outer loop motion is designed, which is based on robust control technique on linear multi input and multiple output (MIMO) control system. Furthermore, the inner loop pressure controller for lifting oil cylinder is developed on the basis of disturbance observation for nonlinear single input and single output (SISO) control system.

Aiming at the cylinder synchronous control system based on independent metering control system, Zhu et al. [66] proposed a coordinated control strategy to realize the synchronous motion control of two cylinders. In case of the slight fluctuation of friction force, an adaptive robust pressure controller is designed to keep the pressure stable in the cylinder cavity, which is beneficial to the accurate modeling of friction force. In addition, an adaptive robust controller is designed to improve motion tracking accuracy of the cylinder, and the model compensation via the on-line assessment with flow rate coefficient is enhanced. For inlet/outlet independently metering control of electro-hydraulic system, Xu et al. [67] proposed a three-level controller to improve the energy utilization efficiency of system. The upper level controller selects the appropriate operating mode by the requirement of load and command speed. The lower level controller not only includes speed controller with the calculation flow rate feedback and pressure controller with back-pressure feedback, but also includes the flow rate controller by variable displacement pump. The middle level achieves the coordinated control between pump-control and valve-control, so that the opening of the inlet valve can be kept as large as possible. Due to the regulation of backpressure control and flow rate control, the system pressure loss is reduced to a minimum, and the energy saving control of the system is realized. The control scheme of independent metering valves-pump system for energy saving is shown in Figure 5.

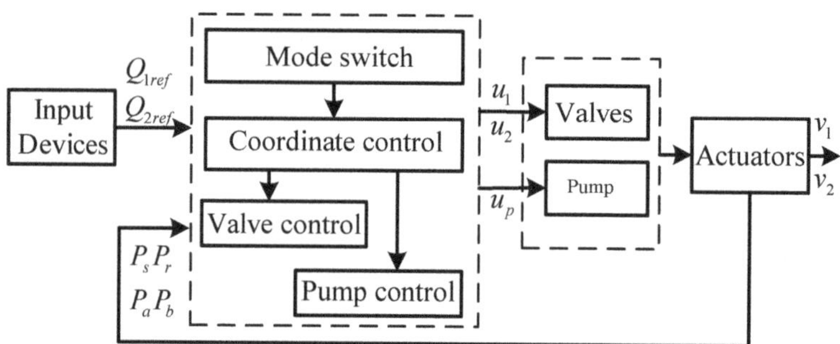


**Figure 5.** Control scheme of independent metering valves-pump system for energy saving.

In order to realize coordinated control of multi-hydraulic-motor traveling system, the influence of different characteristics amount on every hydraulic motor driving subsystem is considered. Zheng et al. [68] put forward the adjacent cross-coupling strategy on the basis of the basic control law with the tracking error, which introduces the tracking errors of the adjacent two axes into the controller of local axis. The structure of multi-axis coordinated motion system based on adjacent cross-coupling strategy is shown in Figure 6.

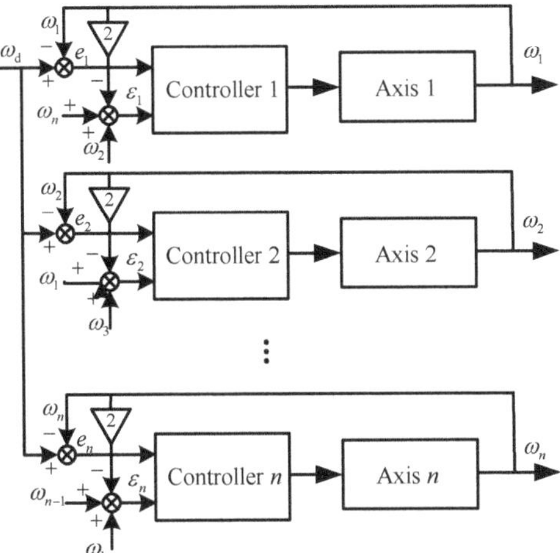


**Figure 6.** Structure of multi-axis coordinated motion system based on adjacent cross-coupling strategy.

From above-surveyed on the control strategy for electro-hydraulic control system, the most important research aspects are summarized, and listed below.

1. Because of nonlinear characteristics in electro-hydraulic control system, such as the nonlinearity of pressure and flow rate in servo valve, the structure nonlinearity of differential equation, traditional linear control theory has been unable to meet the requirements of high precision control system. Therefore, advanced nonlinear control method based on nonlinear model are always an important research topic.

2. With the rapid development of intelligent control algorithm, the compound control method incorporated intelligent control algorithm and modern control is the trend of future development.

3. With the improvement of control performance requirements and complexity for electro-hydraulic motion control system, the motion-force hierarchical and coordinated control based on independent metering control has applied in operating control system of hydraulic excavator and speed smoothness control system of heavy transport vehicles. However, electro-hydraulic speed smoothness control and energy recovery for construction machineries and engineering vehicles have been an urgent problem to be solved.

4. Aiming at the influence of various kinds of uncertainties and unknown disturbances in electro-hydraulic motion control system, the active disturbance rejection control (ADRC) technique assigns all uncertainties to the total disturbance of system and gives estimation and compensation. Moreover, because of the good robustness and easy implementation, the ADRC will be very useful in industrial applications. Then, ADRC and compound control strategy based on ADRC are new research focus for electro-hydraulic motion control system.

5. Based on independent metering control and hierarchical control for electro-hydraulic motion control system, motion-force control algorithm incorporated intelligent control algorithm become a new research topic.

## 3. Nonlinear Modeling of Independent Metering Control for Electro-Hydraulic Mechanism

### 3.1. Non-Linear Modeling of Dual-Valve Independent Metering Control System

A dual-valve independent metering control system of electro-hydraulic power mechanism is shown in Figure 7. The pressure and flow rate in inlet/outlet oil chambers of actuator are controlled by a servo valve or proportional valve separately. On the basis of force balance equation and flow rate continuum equation with two working chambers, the nonlinear tracking control law for electro-hydraulic motion control system is designed.

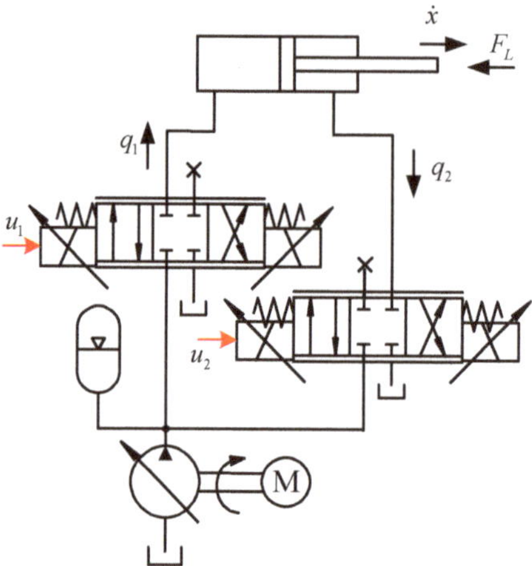

**Figure 7.** A dual-valve independent metering control system of electro-hydraulic power mechanism.

The nonlinear tracking control law mainly includes speed regulation loop and speed smoothness control loop. In the former control loop, the reference pressure of actuator inlet chamber is obtained from the desired running speed. And the reference backpressure of actuator return chamber is obtained by calculating the balanced negative load. Through the coordinated control with pressure in inlet/outlet oil chamber of actuator, the expected driving force can be tracked by hydraulic driving force, then the tracking control of the motion control system can be realized.

Dynamic equation of servo valve can be approximated by the first-order, and it is described as:

$$\dot{x}_v = -\frac{1}{\tau_v}x_v + \frac{k_{xu}}{\tau_v}u \tag{1}$$

where $x_v$, $\tau_v$, $u$, and $k_{xu}$ are the spool displacement, servo valve time constant, the control input, and the gain of the spool displacement to the input of the servo valve, respectively.

Flow rate equation of servo valve port can be described as:

$$q_1 = \begin{cases} C_d W x_{v1}\sqrt{\frac{2}{\rho}(p_s - p_1)}, & u_1 \geq 0 \\ C_d W x_{v1}\sqrt{\frac{2}{\rho}(p_1 - p_t)}, & u_1 < 0 \end{cases} \tag{2}$$

$$q_2 = \begin{cases} C_d W x_{v2}\sqrt{\frac{2}{\rho}(p_2 - p_t)}, & u_2 \geq 0 \\ C_d W x_{v2}\sqrt{\frac{2}{\rho}(p_s - p_2)}, & u_2 < 0 \end{cases} \tag{3}$$

where $C_d$, $W$, $x_v$, and $\rho$ are respectively throttle valve orifice flow coefficient, servo valve orifice area gradient, servo valve spool displacement, and hydraulic oil density, respectively. $p_s$, $p_1$, and $p_2$ are respectively the outlet pressure of the oil pump, the pressure of rodless chamber and the pressure of rod chamber of the hydraulic cylinder.

Continuous flow rate equation of hydraulic cylinder inlet oil chamber and return oil chamber:

$$q_1 = A_1\frac{dx}{dt} + \frac{V_1}{\beta_e}\frac{dp_1}{dt} \tag{4}$$

$$q_2 = A_2 \frac{dx}{dt} - \frac{V_2}{\beta_e} \frac{dp_2}{dt} \tag{5}$$

where $V_1 = V_{01} + A_1 x$, $V_2 = V_{02} - A_2 x$, $\beta_e$, $V_1$, $V_2$, $V_{01}$, and $V_{02}$ are respectively the elastic modulus of hydraulic oil, initial volume of inlet oil chamber and return oil chamber of hydraulic cylinder, respectively. In addition, the impact of system oil leakage is ignored in above equation.

The following equation represents the dynamic behavior of motion control system:

$$A_1 p_1 - A_2 p_2 = m\ddot{x} + B_m \dot{x} + F_L \tag{6}$$

where, $A_2$, $B_m$, $m$, $x$, and $F_L$ are the effective piston area of the two chambers of cylinder, effective viscous damping coefficient, mass of load, output displacement, and external load interference, respectively. In addition, the impact of system oil leakage is also ignored in above equation.

### 3.2. Non-Linear Modeling of Pump/Valves Independent Metering Control System

A pump-valve independent metering control of volumetric hydraulic power mechanism is shown in Figure 8. The flow rate and pressure in the inlet oil loop are regulated by variable displacement pump, and electro-hydraulic proportional orifice valve in return oil loop is responsible for speed smoothness control and back pressure regulation.

For positive load case, the proportional orifice valve is settled with damp operation mode. Based on the system dynamics equation, flow rate continuum equation, and motion-force hierarchical control theory, the reasonable control law design for variable displacement pump and proportional orifice valve may make the rotational speed of the motor be smoothly controlled under sustained negative load. However, the gravitational potential energy is consumed by the loss of throttling.

The torque balance equation of hydraulic motor can be written as:

$$p_L D_m = (p_1 - p_2)D_m = J_m \dot{\omega}_m + B_m \omega_m + T_L \tag{7}$$

where $J_m$ is total moment of inertia for the motor and load, $B_m$ is viscous damping coefficient, and $\omega_m$ is actual speed of motor. $T_L$, $D_m$ are external load torque and displacement of the motor, respectively. $p_1$, $p_2$ are the pressure in cylinder chambers, and $p_L$ is the load pressure of the motor.

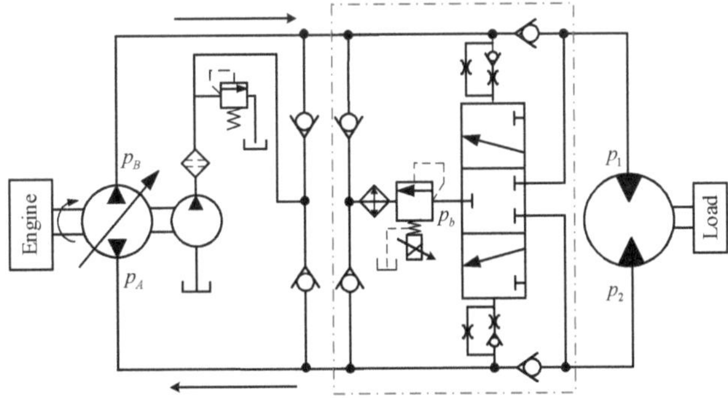


**Figure 8.** A pump-valve independent metering control of volumetric hydraulic power mechanism.

The flow rate continuous equation for pump-control-motor system is as:

$$\omega_p D_p = \frac{V_t}{\beta_e} \Delta\dot{p} + \omega_m D_m + C_L \Delta p \tag{8}$$

where $V_t$ is the volume of a chamber, $\beta_e$ is the effective volume elastic modulus, and $C_L$ is the leakage coefficient.

The displacement of variable displacement pump is regulated by the electro-hydraulic proportional driver. The controller converts the input voltage signal into the corresponding current signal, which drives the proportional actuator of variable displacement pump to control the displacement. The displacement of variable displacement pump could be shown by:

$$D_p = K_{pp}K_{pi}u_p \tag{9}$$

where $K_{pp}$ is the gain of electro-hydraulic proportional valve, with which regulate the displacement of variable pump, $K_{pi}$ is the gain of electro-hydraulic proportional driver of walking pump.

The response of proportional relief valve can be simplified to first order system, and the output pressure $p_b$ of proportional relief valve is proportional to input drive current $I$. Indeed, the proportional amplifier converts the input control voltage $u$ into the drive current $I$. The first order system for proportional relief valve can be written as:

$$p_b = \frac{K_{pv}}{T_{pv}s + 1}I = \frac{K_{pv}K_{di}}{T_{pv}s + 1}u \tag{10}$$

where $K_{pv}$ and $K_{di}$ are the amplifier gain factor of relief valve and proportional amplifier, respectively.

## 4. Electro-Hydraulic Speed Smoothness Control and Energy Recovery for Heavy Vehicles

Under the flat road and climbing conditions, the displacement pump of the walking system for heavy-duty vehicles is driven by the engine. The speed regulation of hydraulic motor is mainly controlled by displacement regulation for variable displacement pump. However, if there are no necessary retarding measures under long down-slope, the speed of vehicle will continue to increase under the influence of gravity. Traditional braking methods use the travelling brake and parking brake on each motor to maintain the speed of vehicle.

However, when the transport vehicle is downhill for a long distance, using a long-lasting brake to control the speed will cause overheating on the brake pads, which results in brake failure and safety accidents. Therefore, in order to ensure the function of speed regulation when going downhill and at a constant speed, an auxiliary continuous braking scheme is needed. Current major solutions of retarded braking for heavy vehicle include hydraulic auxiliary brake, eddy-current auxiliary brake, engine auxiliary brake, etc.

In view of electro-hydraulic retarding control strategy and energy recovery and reutilization for heavy vehicles, Amir et al. [69] proposed a nonlinear control method based on adaptive sliding mode control, which solves the problem of engine torque control in regenerative mode. Experimental results show that the controller has better performance, such as robustness, convergence of tracking error, and anti-interference characteristics.

Aiming at the problem of high idle speed and high fuel consumption with the engine in the long-distance downhill condition, Jin et al. [70] designed an anti-drag feedback electric drive system and raised a three-layer speed smooth control strategy. The upper control algorithm determines the driving state of vehicle in accordance to the influence factors, such as drag state confirmation value, vehicle speed and engine speed. The middle controller is in charge of starting and running control, and simultaneously realizes the tracking of rotor frequency and voltage to reduce the impact of starting current. The lower controller deals with the driver's brake pedal and operational intent, where normal exit and emergency braking intention are judged according to the brake pedal opening and its change rate. Moreover, the output anti drag speed is calculated by brake pedal opening and bus voltage. The control scheme of proposed three-layer speed smooth control is shown in Figure 9.

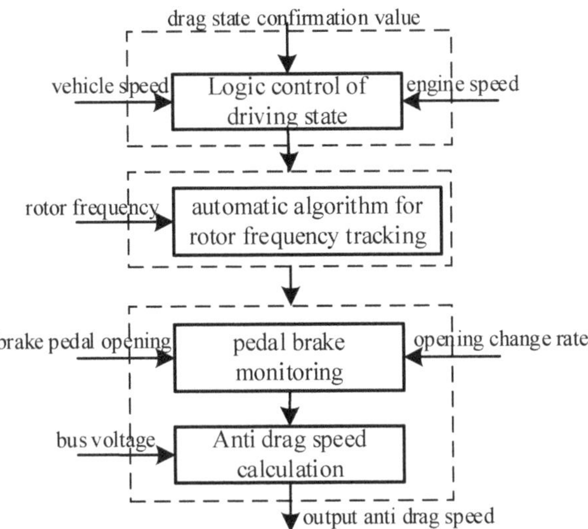



**Figure 9.** Control scheme of three-layer speed smooth control.

Aiming at the problem of speed regulation and energy recovery in hydraulic drive system for construction machinery, for instance hydraulic excavator under intermittent negative load, Wang et al. [71,72] proposed a control scheme of energy regeneration based on a hydraulic motor-generator, which achieves satisfactory control effect in the experiment. In the reference [73], an engine brake mechanism design for diesel engine application with decompression effect is presented, with which the emissions are drastically reduced, even eliminated. Considering the problem of insufficient braking torque with eddy current brakes (ECB) at low speeds, Karakoc et al. [74] studied new methods of improving the generation capacity of braking torque with ECB. Aiming at the shortcomings of heavy duty vehicle in retardation braking, Zhang et al. [75] proposed a design scheme of hydraulic auxiliary retarder, which mainly consists of hydraulic pump/motor element, accumulator and relief valve, etc.

Hong et al. [76] proposed a speed smooth control system of vehicle, which composes of the braking monitor, wheel slip controller, and target slip assignment algorithm. The extended Kalman filter is used to estimate tire braking force at each wheel, the brake disk-pad friction coefficient, and the lateral tire forces. Sliding mode control method is used to design the wheel slip controller, and the control input is calculated by the estimated braking forces and brake disk-pad friction coefficient. The performance and effectiveness of proposed vehicle speed smooth control system are proved in hardware in-the-loop simulator (HILS) experiments.

For the hydrostatic driving system of closed-type pump-control motor applied in the tunnel segment truck, Li and He [77] designed a hydraulic retarder continuous braking device embedded into the hydraulic drive system. The devices are constituted of constant pressure pump, proportional relief valve, and other electro-hydraulic components. And it makes up for the problem of brake overheating caused by braking for a long time, which easily leads to brake failure and safety accident. The architecture schematic of close pump-controlled motor hydraulic system is shown in Figure 10.

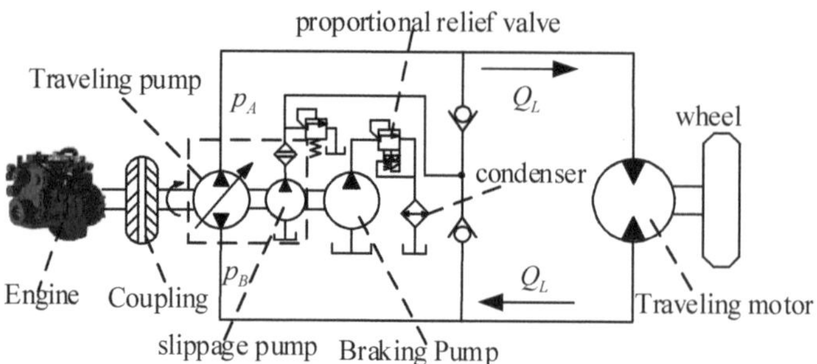


**Figure 10.** Architecture schematic of close pump-controlled motor hydraulic system.

Shen et al. [78] analyzed the influence of displacement with a hydraulic pump on braking performance for hydraulic driving vehicles under long down-slope, and proposed the principle that hydraulic system and anti-drag braking ability of engine should be exerted reasonably. For the research on eddy current retarder of heavy-duty vehicle, Ni et al. [79] optimized the design of eddy current retarder based on the robust optimization theory. Wang [80] proposed a new system structure scheme, which combines the hydraulic motor-generator energy recovery unit and series throttling valve. Furthermore, a stepwise optimization design method with structural parameters of stator and rotor is proposed.

Energy recovery of heavy engineering vehicles under sustained negative load can not only improve energy efficiency, but also extend the life of vehicle components. Moreover, it is of great significance for the requirements for energy saving and emission reduction. In order to regenerate vehicle vibration into electric power and adjust the required damping according to the road condition, Xie et al. [81] proposed an active control approach to recover energy from damped and regenerative dampers. In addition, an electromagnetic energy harvesting damper with 12 independently controlled slot winding sensors (SWT) is designed. Ho et al. [82] presented a hydraulic energy regeneration system based on closed-loop hydraulic transmission device, and a hydraulic accumulator is used as an energy storage system to recover kinetic energy without any reversal of fluid flow. Moreover, a hierarchical control system is established, and adaptive fuzzy sliding mode control is designed for speed control of secondary device.

At present, the technology has been applied to the hydraulic system of excavators. Energy recovery usually uses two types: no energy storage devices and energy storage devices. The former mainly reuses the cycle energy through energy regeneration systems, but the energy utilization is limited, which restricts its application [83]. The latter mainly includes mechanical, hydraulic recovery, electrical recovery, and hybrid.

Through mechanical recycling technology, the recovery energy is converted to kinetic energy of flywheel. And the energy is released when needed, which plays the role of auxiliary oil supply. Although the technology has been applied in hydraulic system, the controllable performance is not ideal. Hydraulic recovery technology mainly uses the accumulator to realize the recovery of system pressure and potential energy. The technology is relatively mature, and is more suitable for occasions with changeable working conditions. Caterpillar et al. [84] used accumulator to recover the large chamber pressure of hydraulic brake and redundant power of pump, and the effect energy saving is improved well. Domestic universities are also actively studying hydraulic energy recovery technology, such as Zhejiang University, Central South University, Jilin University, etc. The energy recovery of boom and swing system in hydraulic excavator had also seen great progress [85].

With electric recycling technology, the recovered energy is converted into electrical energy storage or direct output. This form has good energy saving effect and good controllability, but the cost of hydraulic system is high [86]. Hybrid technology used two or more power sources to provide power. The technology is widely used in hybrid-electric mixed applications, which is composed of engine and electric generator in construction machinery. Hybrid power technology can effectively exert the advantages of each power source to improve overall system efficiency.

Through a lot of literature analysis and actual investigation, we can know that the application of hydraulic auxiliary braking and eddy current braking is limited in the actual construction, which is restricted by the installation structure and safety requirements of heavy transport vehicles. In addition, since braking force provided by the engine is limited, it cannot be accurately controlled.

Aiming at the study on speed smooth control for heavy vehicle under long down-slope, it is mainly focuses on the following aspects: design of hydraulic retarder devices and system, multi-layer speed smooth control strategy and motion-force control algorithm, coordination control for energy saving, and integrated energy management scheme for energy recovery and utilization is also a research hotspot. However, research progresses on design of retarder system and speed smoothness control strategy with motion-force hierarchical control architecture for heavy vehicles are comparatively scarce, especially in practical engineering applications. But that is indeed a technical problem urgently needed to be solved in practical engineering.

Motion-force hierarchical and coordinated control scheme, which is based on the inlet/outlet independent metering control, is applies to the study of speed smoothness control for a heavy vehicle, and it is improved an effective solution to promote the research. Moreover, introducing active disturbance rejection control (ADRC) in the study on speed smoothness control strategy for heavy vehicles is of great practical value for engineering applications. The influence of uncertainty disturbances, such as the uncertainty of road condition, the descent force and system parameters of heavy vehicle is unified as the total disturbance of the system. Then the disturbance is estimated and compensated online by the extended state observer. Since the ADRC does not require accurate mathematical models, and has strong robustness and engineering practicality, it has become an effective solution for the design on speed smoothness control for heavy vehicles in practical applications.

The speed smoothness control for heavy vehicles means that the travelling speed can be controlled and maintain the speed smooth under sustained negative load, which needs to solve the coupling influence between speed regulation and load variation suppression. However, some disadvantages existed in current electrohydraulic retarding technology solutions and are listed below:

1. Through continuous braking of the brake, sustained negative loads may be fully suppressed, but that will cause the serious braking heat as well as decline of braking performance, especially under long down-slope. Although intermittent repeated braking can improve this problem to a certain degree, the speed smoothness is not good.

2. The solution, employing pump-controlled speed regulation and electro-hydraulic proportional balancing valve for back pressure regulation, can decouple speed control and negative load suppression, and brake slow speed is not required for long downhill slope. Moreover, the balancing valve is equipped on return line, which facilitates cooler arrangement for forced cooling, and response speed due to valve-controlled back pressure regulation. However, the disadvantage of this scheme is that it cannot recover downhill potential energy.

3. Given that the solution of speed regulation with pump-control and speed smoothness control with retarder pump retardation solution, downhill potential energy is absorbed by the engine load and other hydraulic circuits, and the remaining energy is further utilized by the retarder system. Although this scheme recovers part of the downhill potential energy, system structure is complex and limited by installation space, especially while the additional weight and installation space of the recovery device is strictly required in practical engineering application. In addition, response speed is slower compared with the above scheme.

4. Dual-valve independent metering scheme can achieve coupling influence between speed regulation and negative load suppression, and it has a simple structure and fast response. This solution, combined with the accumulator, can supply the cooling circuit with oil, and partially recover of downhill energy. But the disadvantages are that the throttling loss is serious, and the energy efficiency is lower compared with pump-control regulation.

The electro-hydraulic retarder control strategy for heavy vehicles, which is studied in this paper, is based on the inlet/outlet independent metering control of electro-hydraulic power mechanism. Due to the increase of system control variables, the flexibility of system control system is greatly improved and the energy efficiency is also raised. However, so far, the research achievements in this respect are limited, and the research group conducted little research in this aspect.

## 5. Conclusions

Inlet/outlet oil independent and coordinated control for electro-hydraulic power mechanism improve the control flexibility compared to ordinary electro-hydraulic control valve, and effectively solve the serious coupling influence between speed regulation and load variation suppression, especially sustained negative load. Combined with digital intelligent control and advanced displacement control technology, the inlet/outlet oil independent and coordinated control scheme reduces metering losses of valve-controlled and pump-controlled electro-hydraulic system, for example, reduction of throttling loss, energy regeneration and recuperation, as well as maintaining the excellent dynamic performance. However, it not only increases the control system complexity and reduces the reliability, but also increases the production and maintenance costs, which greatly restricts the application in a practical engineering project. Moreover, the inlet/outlet oil independent and coordinated control faces challenges, such as smooth switching between different control modes, combination with advanced pump control technology, and better maintainability and redundancy strategies [87].

With the development of computer control technology and advanced power electronics equipment, for instance, various high precision sensor, the new electro-hydraulic control components with digital control have appeared, and it usually include the digital programmable valve and flow rate amplifier valve with digital pilot, etc. Moreover, the new control components apply to independent metering control system, which promotes the modern intelligent control for electro-hydraulic system. However, most of the digital valves currently studied are only as high-speed on-off valves. Because of the limitation with their structure, it can only be applied to pilot control and small flow rate control applications.

For the electro-hydraulic power mechanism under sustained negative load, the inlet/outlet oil dependent metering control system with motion-force hierarchical control architecture mainly includes the following research aspects: flow rate and pressure independent regulation, the decoupling mechanism and method between speed control and speed smoothness control, the motion-force hierarchical and coordinated control, the robust control for variable parameter and soft parameter, dynamic feature recognition method for long time and continuous running system, and integrated energy management scheme for energy recovery and utilization.

Meanwhile, with the influence of various kinds of uncertainties and unknown disturbances in electro-hydraulic motion control system, the active disturbance rejection control (ADRC) technique assigns all uncertainties to the total disturbance of the system and gives estimation and compensation. In addition, because of the good robustness and easy implementation, it will be very useful in industrial applications. Consequently, ADRC and compound control strategy based on ADRC will become a hot research focus for an electro-hydraulic motion control system.

At present, the inlet/outlet oil independent metering regulation and the motion-force coordinate control have become a hot topic in the field of electro-hydraulic control. And it has been successfully applied in the electro-hydraulic swing and execution system of hydraulic excavators and the speed smooth control system of heavy transport vehicles. However, the systematic achievement and research has not appeared in the independent adjusting mechanism of hydraulic power mechanism under the sustained negative load. Especially, in the speed smooth control strategy and energy saving control for moving body machine power mechanism under sustained negative load. Energy recovery schemes in many cases are limited in case of considering the addition mass of energy recovery component (e.g., for the electro-hydraulic walking system on the long downhill, the volume and additional weight of the accumulator for storing gravity potential energy are too large to be applied in practice) and the safety-request (e.g., the grid feedback is not allowed due to the spark of the pantograph) in a construction site. In addition, it is also necessary to be considered the cooling problem caused by the braking.

In summary, for electro-hydraulic motion control system under sustained negative load, the research fields, including comprehensive analysis, the independent and coordinate control strategy based on motion-force hierarchical control architecture, and compound control strategy based on ADRC are the future trends. Especially, the research is of the important theoretical significance and engineering application value for promoting the overall control performance in electro-hydraulic motion control system. At present, there are still some systematic and thorough problems in coordinated control and energy recovery method with low cost. With the developments of cybernetic physical system, field bus, intelligent control, and big data and applications in hydraulic control system, more and more coordinated control and energy recovery method will be proposed and applied in hydraulic system under continuous negative load.

**Author Contributions:** Design of the research scheme and research supervision: Y.L.; performing of the research and writing of the manuscript: W.L. and D.L.

**Acknowledgments:** The authors would like to acknowledge the support of National Natural Science Foundation of China (Grant No. 51475019) and National Key Basic Research Program of China (Grant No. 2014CB046403).

**Conflicts of Interest:** The authors declare no conflict of interest with other researchers.

## References

1. Li, Y.H.; Wang, Y.; Geoffrey Chase, J.; Mattila, J.; Myung, H.; Sawodny, O. Survey and Introduction to Focused Section on Mechatronics for Sustainable and Resilient Civil Infrastructure. *IEEE/ASME Trans. Mechatron.* **2013**, *18*, 1637–1645. [CrossRef]
2. Sheng, Z.; Li, Y.H. Hybrid Robust Control Law with Disturbance Observer for High-Frequency Response Electro-Hydraulic Servo Loading System. *Appl. Sci.* **2016**, *6*, 98. [CrossRef]
3. Chen, M.M. *The Research on Stability of Counterbalance Valve in Over-Running Load Condition*; Jilin University: Changchun, China, 2015.
4. Liu, X.M.; Lang, X.J. Research on Speed Stability of Separate Direction Valve Control System under Over-running Load. *Chin. J. Mech. Res. Appl.* **2015**, *2*, 176–179.
5. Liu, B.C. *A Study of Winch System for Wheel Crane in the Over-Running Load Condition*; Jilin University: Changchun, China, 2011.
6. Xian, Y.P. *Research on Closed Hydraulic System of Crawler Crane in Over-Running Load Condition*; Dalian University of Technology: Dalian, China, 2012.
7. Nie, D.W. *Research on Control Method of Load for Truck Crane Hoisting Mechanism in Over-Running Load Condition*; Hunan University: Changsha, China, 2013.
8. Backe, W. Design systematics and performance of cartridge valve controls. In Proceedings of the International Conference on Fluid Power, Tampere, Finland, 1–4 June 1987; pp. 1–48.
9. Jansson, A.; Palmberg, J.O. *Separate Controls of Meter-In and Meter-Out Orifices in Mobile Hydraulic Systems*; SAE Technical Paper; SAE International: Warrendale, PA, USA, 1990.
10. Shenouda, A.; Book, W. *Energy Saving Analysis Using a Four-Valve Independent Metering Configuration Controlling a Hydraulic Cylinder*; SAE Technical Paper; SAE International: Warrendale, PA, USA, 2005.
11. Shenouda, A.; Wayne, B. Optimal Mode Switching for a Hydraulic Actuator Controlled with Four-Valve Independent Metering Configuration. *Int. J. Fluid Power* **2008**, *9*, 35–43. [CrossRef]
12. Elfving, M.; Palmberg, J.O.; Jansson, A. Distributed control of fluid power actuators-a load sensing application of a cylinder with decoupled chamber pressure control. In Proceedings of the Fifth Scandinavian International Conference on Fluid Power (SICFP'97), Linkoping, Sweden, 28–30 May 1997.
13. Opdenbosch, P.; Sadegh, N.; Enes, A.R.; Book, W.J. Auto-calibration based control for independent metering of hydraulic actuators. In Proceedings of the IEEE International Conference on Robotics and Automation (ICRA 2011), Shanghai, China, 9–13 May 2011; Volume 43, pp. 153–158.

14. Ding, R.Q.; Xu, B.; Zhang, J.; Li, G. Mechanical-hydraulic Coupling Model of Independent Metering Control Excavator and Its Test Verification. *Chin. J. Trans. Chin. Soc. Agric. Mach.* **2016**, *47*, 309–409.

15. Hu, H.; Zhang, Q. Realization of Programmable Control using a Set of Individually Controlled Electrohydraulic Valves. *Int. J. Fluid Power* **2002**, *3*, 9–34. [CrossRef]

16. Zhang, Q.; Hu, H. Multi-Function Realization Using an Integrated Programmable E/H Control Valve. *Appl. Eng. Agric.* **2003**, *19*, 283–290.

17. Pedersen, H.C.; Andersen, T.O.; Hansen, R.H.; Stubkier, S. Investigation of Separate Meter-In Separate Meter-Out Control Strategies for Systems with Over Centre Valves. In Proceedings of the Bath/ASME Symposium on Fluid Power & Motion Control (FPMC 2010), Bath, UK, 12–14 September 2010.

18. Andersen, T.O.; Mnzer, M.E.; Hanse, M.R. Evaluation of control strategies for separate meter-in separate meter-out hydraulic boom actuation in mobile applications. In Proceedings of the 17th International Conference on Hydraulic and Pneumatics, Ostrava, Czech Republic, 5–7 June 2001; pp. 450–475.

19. Wang, Q.F.; Gu, L.Y.; Lu, Y.X. Research on Principle and Performance of Meter-in, Meter-out Independent Regulate based on Pressure Decrease Sensing. *Chin. J. Mech. Eng.* **2001**, *37*, 21–24. [CrossRef]

20. Wang, Q.F. Research on Inlet/Outlet Independent Regulation of Electro-hydraulic Throttling System with Large Inertia Load. *Chin. J. Mech. Eng.* **1999**, *10*, 853–855.

21. Liu, Y.J.; Xu, B.; Yang, H.; Zeng, D. Strategy for Flow and Pressure Control of Electro-hydraulic Proportional Separate Meter-In and Meter-Out Control System. *Appl. Eng. Agric.* **2010**, *41*, 182–187.

22. Xu, B.; Zeng, D.R.; Ge, Y.Z.; Liu, Y.J. Research on Characteristic of Mode Switch of Separate Meter in and Meter out Load Sensing Control System. *J. Zhejiang Univ.* **2011**, *45*, 858–863.

23. Quan, L.; Lian, Z.S. Improving the efficiency of pump Controlled Differential Cylinder System with Inlet and Outlet Separately Controlled Principle. *Chin. J. Mech. Eng.* **2005**, *41*, 123–127. [CrossRef]

24. Yuan, M.L. *Research on Load Port Independently Controlled Double Servo Valves Cylinder System*; Beijing Institute of Technology: Beijing, China, 2015.

25. Cheng, M.; Xu, B.; Zhang, J.; Ding, R. Valve-based compensation for controllability improvement of the energy-saving electrohydraulic flow matching system. *Zhejiang Univ. Sci. A* **2017**, *18*, 430–442. [CrossRef]

26. Dong, W.; Jiao, Z.; Wu, S. Compound control strategy for independent metering directional valve of engineering machinery by simulation. In Proceedings of the 2014 IEEE International Conference on Mechatronics and Automation, Tianjin, China, 3–6 August 2014.

27. Wang, S. *Independent Metering Control Programmable Valve and Its Application in Intelligent Hydraulic Excavator*; Zhejiang University of Technology: Hangzhou, China, 2017.

28. Li, Z.Z.; Huang, H.J.; Quan, L.; Wang, S.G. The Independent Metering System Based on Digital Flow Valve. *Chin. Hydraul. Pneum.* **2016**, *2*, 17–23.

29. Liu, Y. *Research on Key Techniques of Independent Metering Directional Valve Control System*; Zhejiang University: Hangzhou, China, 2011.

30. Liu, J.; Xie, H.; Hu, L.; Fu, X. Realization of direct flow control with load pressure compensation on a load control valve applied in overrunning load hydraulic systems. *Flow Meas. Instrum.* **2017**, *53*, 261–268. [CrossRef]

31. Zheng, Q.; Li, Y.H.; Yang, L.M. Compound control strategy for the pump controlled parallel variable displacement motor speed system. *J. Beijing Univ. Aeronaut. Astron.* **2012**, *38*, 692–696.

32. Niksefat, N.; Sepehri, N. Robust Force Controller Design for an Electro-Hydraulic Actuator Based on Nonlinear Model. In Proceedings of the 1999 IEEE International Conference on Robotics & Automation, Detroit, MI, USA, 10–15 May 1999; pp. 200–206.

33. Liu, R.; Alleyne, A. Nonlinear Force/Pressure Tracking of an Electro-Hydraulic Actuator. *J. Dyn. Syst. Meas. Control* **2000**, *122*, 232–237. [CrossRef]

34. Kaddissi, C.; Kenne, J.P.; Saad, M. Identification and Real-Time Control of an Electrohydraulic Servo System Based on Nonlinear Backstepping. *IEEE/ASME Trans. Mechatron.* **2007**, *12*, 12–22. [CrossRef]

35. Eriksson, B. Mobile Fluid Power Systems Design: With a Focus on Energy Efficiency. Ph.D. Thesis, Linköping University, Linköping, Sweden, 2010.

36. Eriksson, B.; Palmberg, J.O. Individual Metering Fluid Power Systems: Challenges and Opportunities. Proceedings of the Institution of Mechanical Engineers. *Part I J. Syst. Control Eng.* **2011**, *225*, 196–211.

37. Nam, Y. QFT force loop design for the aerodynamic load simulator. *IEEE Trans. Aerosp. Electron. Syst.* **2001**, *37*, 1384–1392.

38. Nam, Y.; Hong, S.K. Force control system design for aerodynamic load simulator. *Control Eng. Pract.* **2002**, *10*, 549–558. [CrossRef]

39. Truong, D.Q.; Ahn, K.K. Force control for hydraulic load simulator using self-tuning grey predictor-fuzzy PID. *Mechatronics* **2009**, *19*, 233–246. [CrossRef]

40. Truong, D.Q.; Ahn, K.K.; Soo, K.J.; Soo, Y.H. Application of fuzzy-PID controller in hydraulic load simulator. In Proceedings of the IEEE International Conference on ICMA 2007, Harbin, China, 5–8 August 2007; pp. 3338–3343.

41. Eryilmaz, B.; Wilson, B.H. Improved tracking control of hydraulic systems. *J. Dyn. Syst. Meas. Control* **2001**, *23*, 457–462. [CrossRef]

42. Sohl, G.A.; Bobrow, J.E. Experiments and simulations on the nonlinear control of a hydraulic servo system. *IEEE Trans. Control Syst. Technol.* **1999**, *7*, 238–247. [CrossRef]

43. Hu, C.; Yao, B.; Wang, Q. Adaptive Robust Precision Motion Control of Systems with Unknown Input Dead-Zones: A Case Study with Comparative Experiments. *IEEE Trans. Ind. Electron.* **2011**, *58*, 2454–2464. [CrossRef]

44. Nozaki, T.; Mizoguchi, T.; Ohnishi, K. Decoupling Strategy for Position and Force Control Based on Modal Space Disturbance Observer. *IEEE Trans. Ind. Electron.* **2014**, *61*, 1022–1032. [CrossRef]

45. Pillosu, S.; Pisano, A.; Usai, E. Unknown-input observation techniques for infiltration and water flow estimation in open-channel hydraulic systems. *Control Eng. Pract.* **2012**, *20*, 1374–1384. [CrossRef]

46. Kaddissi, C.; Kenne, J.P.; Saad, M. Indirect Adaptive Control of an Electro-hydraulic Servo System Based on Nonlinear Backstepping. *IEEE/ASME Trans. Mechatron.* **2011**, *16*, 1171–1177. [CrossRef]

47. Yang, Y.; Balakrishnan, S.N.; Tang, L.; Landers, R.G. Electro-hydraulic Control Using Neural MRAC Based on a Modified State Observer. *IEEE/ASME Trans. Mechatron.* **2013**, *18*, 867–877. [CrossRef]

48. Zhang, J.H.; Zheng, W.X. Design of Adaptive Sliding Mode Controllers for Linear Systems via Output Feedback. *IEEE Trans. Ind. Electron.* **2014**, *61*, 3553–3562. [CrossRef]

49. Chiang, M.H.; Lee, L.W.; Liu, H.H. Adaptive fuzzy controller with self-tuning fuzzy sliding-mode compensation for position control of an electro-hydraulic displacement-controlled system. *J. Intell. Fuzzy Syst.* **2014**, *26*, 815–830.

50. Yao, J.Y.; Jiao, Z.X.; Yao, B.; Shang, Y.; Dong, W. Nonlinear Adaptive Robust Force Control of Hydraulic Load Simulator. *Chin. J. Aeronaut.* **2012**, *25*, 766–775. [CrossRef]

51. Ho, T.H.; Ahn, K.K. Speed Control of a Hydraulic Pressure Coupling Drive Using an Adaptive Fuzzy Sliding-Mode Control. *IEEE/ASME Trans. Mechatron.* **2012**, *17*, 976–986. [CrossRef]

52. Liu, Y.J.; Xu, B. Calculation Force Control Based on Linearization Feedback Model of Separate Meter in and Separate Meter Out Control System. In Proceedings of the IEEE International Conference on Automation and Logistics, Shenyang, China, 18–21 August 2009; pp. 1464–1468.

53. Jin, M.; Wang, Q. An Energy-Saving Way to Balance Variable Negative Load Based on Back-Stepping Control with Load Observer. In Proceedings of the BATH/ASME 2016 Symposium on Fluid Power and Motion Control, Bath, UK, 7–9 September 2016.

54. Yao, J.; Deng, W. Active disturbance rejection adaptive control of uncertain nonlinear systems: Theory and application. *Nonlinear Dyn.* **2017**, *89*, 1–14. [CrossRef]

55. Castañeda, L.A.; Luviano-Juárez, A.; Chairez, I. Robust Trajectory Tracking of a Delta Robot Through Adaptive Active Disturbance Rejection Control. *IEEE Trans. Control Syst. Technol.* **2015**, *23*, 1387–1398. [CrossRef]

56. Li, P.; Lam, J.; Chun, K.C. Experimental Investigation of Active Disturbance Rejection Control for Vehicle Suspension Design. *Int. J. Theor. Appl. Mech.* **2016**, *1*, 89–96.

57. Moghadam, H.Z.; Landers, R.G.; Balakrishnan, S.N. Hierarchical optimal force-position control of complex manufacturing processes. *IEEE Trans. Control Syst. Technol.* **2005**, *13*, 321–327.

58. Wang, Z.P.; Wang, Y.C.; Zhang, L.; Liu, M.C. Vehicle Stability Enhancement through Hierarchical Control for a Four-Wheel-Independently-Actuated Electric Vehicle. *Energies* **2017**, *10*, 947. [CrossRef]

59. Wu, X.; Shen, J.; Li, Y.G.; Lee, K.Y. HGLMA. Hierarchical optimization of boiler–turbine unit using fuzzy stable model predictive control. *Control Eng. Pract.* **2014**, *30*, 112–123. [CrossRef]

60. Yao, B.; Hu, C.H.; Wang, Q.F. An Orthogonal Global Task Coordinate Frame for Contouring Control of Biaxial Systems. *IEEE/ASME Trans. Mechatron.* **2012**, *17*, 622–634.

61. Liu, S.; Yao, B. Coordinate Control of Energy Saving Programmable Valves. *IEEE Trans. Control Syst. Technol.* **2007**, *16*, 34–45. [CrossRef]

62. Zhang, R.; Alleyne, A.; Prasetiawan, E. Modeling and H2/H∞ MIMO control of an Earthmoving vehicle powertrain. *J. Dyn. Syst. Meas. Control.* **2002**, *124*, 625–636. [CrossRef]

63. Zhang, C.H.; Shi, Q.; Cheng, J. Synchronization Control Strategy in Multi-motor Systems Based on the Adjacent Coupling Error. *Proc. CSEE* **2007**, *15*, 59–63.

64. Fang, R.W.; Chen, J.S. Cross-coupling control for a direct-drive robot. *JSME Int. J. Ser. C* **2002**, *45*, 749–757. [CrossRef]

65. Sun, H.; Chiu, T.C. Motion synchronization for dual-cylinder electro-hydraulic lift systems. *IEEE/ASME Trans. Mechatron.* **2002**, *7*, 171–181.

66. Zhu, X.; Cao, J.; Tao, G.; Yao, B. Synchronization Strategy Research of Pneumatic Servo System Based on Separate Control of Meter-in and Meter-out. In Proceedings of the 2009 IEEE/ASME International Conference on Advanced Intelligent Mechatronics, Singapore, 14–17 July 2009; pp. 14–17.

67. Xu, B.; Ding, R.Q.; Zhang, J.H.; Cheng, M.; Sun, T. Pump/valves coordinate control of the independent metering system for mobile machinery. *Autom. Constr.* **2015**, *57*, 98–111. [CrossRef]

68. Li, Y.H.; Zheng, Q.; Yang, L.M. Design of robust sliding mode control with disturbance observer for multi-axis coordinated traveling system. *Comput. Math. Appl.* **2012**, *64*, 759–765. [CrossRef]

69. Fazeli, A.; Zeinali, M.; Khajepour, A. Application of Adaptive Sliding Mode Control for Regenerative Braking Torque Control. *IEEE/ASME Trans. Mechatron.* **2012**, *17*, 745–755. [CrossRef]

70. Jin, C.; Zheng, S.Y.; Tian, H.Y. Anti-drive System Design and Hierarchy Control Strategy of Electric Drive Vehicle. *Trans. Chin. Soc. Agric. Mach.* **2014**, *45*, 22–27.

71. Chen, Q.; Wang, Q.; Wang, T. A novel hybrid control strategy for potential energy regeneration systems of hybrid hydraulic excavators. *Proc. Inst. Mech. Eng. Part I J. Syst. Control Eng.* **2016**, *230*, 375–384. [CrossRef]

72. Wang, T.; Wang, Q.; Lin, T. Improvement of boom control performance for hybrid hydraulic excavator with potential energy recovery. *Autom. Constr.* **2013**, *30*, 161–169. [CrossRef]

73. Barbieri, F.A.A.; Andreatta, E.C.; Argachoy, C.; Brandao, H. Decompression engine brake modeling and design for diesel engine application. *SAE Int. J. Engines* **2010**, *3*, 92–102. [CrossRef]

74. Karakoc, K.; Park, E.J.; Suleman, A. Improved braking torque generation capacity of an eddy current brake with time varying magnetic field: A numerical study. *Finite Elements Anal. Des.* **2012**, *59*, 66–75. [CrossRef]

75. Zhang, H.; Guo, K.; Wang, C.Y. Design and analysis of hydraulic retarder device for heavy vehicle. *Sci. Technol. Eng.* **2017**, *17*, 25–29.

76. Hong, D.; Wang, I.; Yoon, P.; Huh, K. Development of a vehicle stability control system using brake-by-wire actuators. *J. Dyn. Syst. Meas. Control* **2008**, *130*, 011008. [CrossRef]

77. Li, Y.H.; He, L.Y. Counterbalancing Speed Control for Hydrostatic Drive Heavy Vehicle under Long Down-Slope. *IEEE/ASME Trans. Mechatron.* **2015**, *20*, 1533–1542. [CrossRef]

78. Shen, J.J.; Liu, B.X.; Zhang, Z.F. Discussion on Brake Performance of Hydraulic Driving Vehicle at Downhill. *J. Zhengzhou Univ. (Eng. Sci.)* **2007**, *28*, 113–116.

79. Ni, J.S. Research of the Braking Performance and Optimization Design of Heavy Duty Construction Vehicle Eddy Current Retarder. *Chin. J. Mech. Transm.* **2015**, *9*, 81–83.

80. Wang, T. *Study on Boom Energy Recovery Element and System of Hybrid Excavator*; Zhejiang University: Hangzhou, China, 2013.

81. Xie, L.H.; Li, J.H.; Cai, S.; Li, X. Electromagnetic Energy-Harvesting Damper with Multiple Independently Controlled Transducers: On-Demand Damping and Optimal Energy Regeneration. *IEEE/ASME Trans. Mechatron.* **2017**, *22*, 2705–2713. [CrossRef]

82. Ho, T.H.; Ahn, K.K. Design and control of a closed-loop hydraulic energy-regenerative system. *Autom. Constr.* **2012**, *22*, 444–458. [CrossRef]

83. Truong, B.N.M.; Truong, D.Q.; Young, L.S. Study on energy regeneration system for hybrid hydraulic excavator. In Proceedings of the 2015 International Conference on Fluid Power and Mechatronics (FPM), Harbin, China, 25–26 December 2015; pp. 1349–1354.

84. Wang, Y.J. *Research on Working Device Joint Simulation and Static & Dynamic Characteristics of Large Hydraulic Excavators*; Taiyuan University of Technology: Taiyuan, China, 2014.

85. Lin, T.L. *Research on the Potential Energy Regeneration System based on the Hydraulic Excavators*; Zhejiang University: Hangzhou, China, 2011.

86. Zheng, T.; He, Y.C. Simulation of hydraulic excavators slewing system based on hydraulic accumulator. In Proceedings of the 2015 International Conference on Fluid Power and Mechatronics (FPM), Harbin, China, 25–26 December 2015; pp. 1063–1066.

87. Xu, B.; Cheng, M. Motion control of multi-actuator hydraudlic systems for mobile machineries: Recent advancements and future trends. *Front. Mech. Eng.* **2018**, *13*, 151–166. [CrossRef]

*Review*

# Research Progress of Related Technologies of Electric-Pneumatic Pressure Proportional Valves

**Fangwei Ning [1,†], Yan Shi [1,2,*,†], Maolin Cai [1], Yixuan Wang [1] and Weiqing Xu [1,*]**

[1]  School of Automation Science and Electrical Engineering, Beihang University, Beijing 100191, China; nfangwei@163.com (F.N.); caimaolin@buaa.edu.cn (M.C.); 15652928692@163.com (Y.W.)

[2]  The State Key Laboratory of Fluid Power and Mechatronic Systems, Zhejiang University, Hangzhou 10058, China

*  Correspondence: yesoyou@gmail.com (Y.S.); weiqing.xu@buaa.edu.cn (W.X.); Tel.: +86-10-8233-5821 (Y.S.)

†  These authors contribute equally to this paper.

Received: 1 September 2017; Accepted: 27 September 2017; Published: 17 October 2017

**Abstract:** Because of its cleanness, safety, explosion proof, and other characteristics, pneumatic technologies have been applied in numerous industrial automation fields. As a key controlling element of a pneumatic system, electric-pneumatic pressure proportional valves have attracted the attention of many scholars in recent years. In this paper, in order to illustrate the research status and the development trend of electric-pneumatic pressure proportional valves, firstly, several related technologies will be introduced, for example, simulation methods and experimental modes. In addition, controlling methods, structural styles, and feedback forms are also compared in several types of pressure proportional valves. Moreover, the controlling strategy, as a significant relevant factor affecting the efficiency of valves, will be discussed in this paper. At the end, the conclusion and worksof electric-pneumatic pressure proportional valves in the future will bediscussed to achieve the electrical integration.

**Keywords:** pneumatic system; proportional pressure valve; proportional electromagnet; compressed air; regulator

---

## 1. Introduction

Pneumatic technology was first developed in the military field to realize the stable control of the space vehicle and the missile altitude. Shearer [1] has developed the pneumatic technology in the control system in 1956. Burrows [2,3] has designed a switch servo control technology, and applied it to the pneumatic servo mechanism so that a point-to-point control can be achieved although with a low accuracy. Pneumatic technology is an important method of automatic production. Pneumatic systems are widely used in various fields, such as semiconductor/microelectronic manufacturing, biological engineering, medicine industry, and so on. Proportional valves play an important role in pneumatic servo control system. The first person who developed pneumatic servo valves is W. Professor [4] back in 1979, who promoted the rapid development of servo control technology. In the late 1980s, the pneumatic control system developed into an accurate proportional servo control based on nozzle flapper pneumatic servo valves and pneumatic proportional solenoid valves, which was replaced by the position and speed of the programmable controller [5–7]. The electric proportional pressure valve was born and started to develop rapidly since then.

Electric-pneumatic pressure proportional valves are introduced in control mode, control method, internal structure, and other aspects in this paper. In addition, the research methods, research contents, and developing trend of the electric-pneumatic pressure proportional valves will also be introduced.

*Appl. Sci.* **2017**, *7*, 1074; doi:10.3390/app7101074　　　　　　　　　www.mdpi.com/journal/applsci

## 2. Research Status of Electric-Pneumatic Pressure Proportional Valve

Pneumatic high-speed switch solenoid valve contains many advantages, such as small size, convenience in integrating, anti-pollution, low production costs, and high speed. It is more common for solenoid valves to be the pilot level proportional pressure valves, which are composed of a main valve, a pilot control valve, a digital controller, and a pressure sensor [8–11]. The pressure sensor is used to detect the pressure of the outlet; it is a closed-loop control, as shown in Figures 1 and 2.

Figure 1 is a modified polyphenylene oxide (MPPE) type electric proportional valve of FESTO Company; Figure 2 is the ITV series valve of SMC Company. There are differences between them, but the principles of both are the same. The square wave signal is controlled by the digital controller output of the electromagnetic pneumatic pressure proportional valve in Figure 3, pilot inlet valve $a$ and exhaust valve $b$ are controlled by square wave signal. When the intake valve $a$ is opened and the exhaust valve $b$ is closed, the output pressure of the pilot valve can be increased, at this time, the active valve core is pushed, and the airflow is increased due to the opening of the opened throttle, the intake port $P$ is connected to the output port $Pa$.

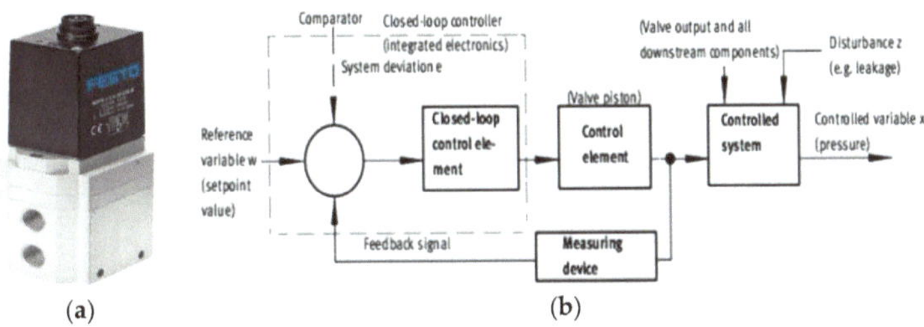

Figure 1. FESTO proportional valve. (a) FESTO pneumatic pressure proportional valve modified polyphenylene oxide (MPPE); (b) FESTO MPPE proportional valve working principle diagram.

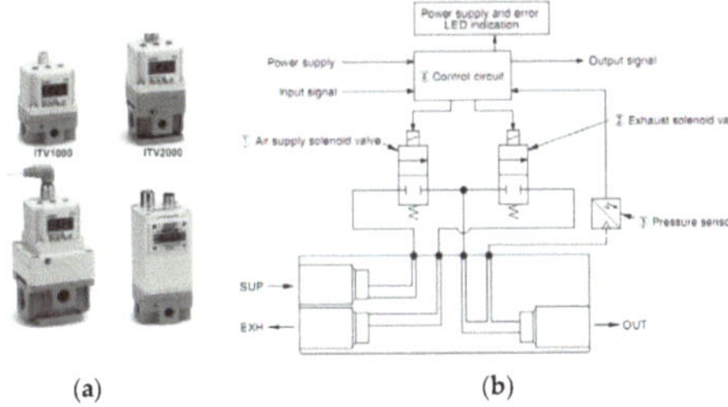

Figure 2. SMC pneumatic pressure proportional valve. (a) SMC ITV series electric proportional valve; (b) SMC ITV electric proportional valve working principle diagram.

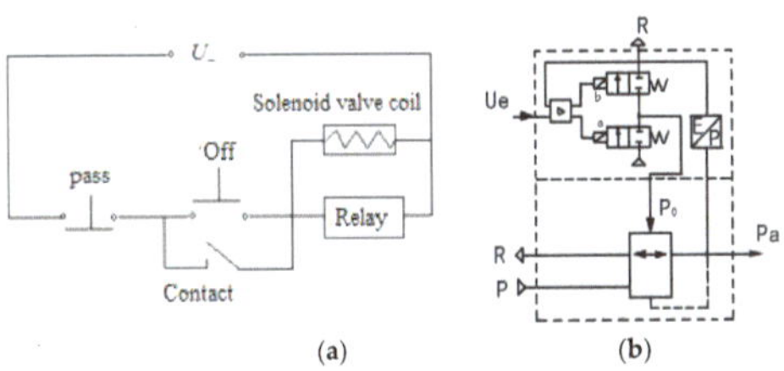

Figure 3. Switch solenoid pneumatic pressure proportional valve. (a) Solenoid valve control circuit diagram; (b) Schematic diagram of switch solenoid valve.

Switching solenoid pneumatic pressure proportional valve has been studied by a large number of scholars both at home and abroad [12–15], among whom the most representative research is the Parker P3P-R pneumatic electromagnetic proportional valve conducted by Massimo Sorli [16–18]. Parker P3P-R pneumatic solenoid proportional pressure relief valve is a common market in the typical switch solenoid pneumatic pressure proportional valve, as shown in Figure 4a,b. Based on the relationship among the internal parts of the valve, the nonlinear mathematical model of the valve is established, and the dynamic behavior of the valve in the time domain and the frequency domain is studied with numerical analysis and experiments.

Massimo Sorli discovered that when the output pressure of the pressure-reducing valve reaches the preset pressure, the two switches in the control chamber will remain closed, hence there is no excessive energy consumption. However, when the preset pressure changes, it needs to adjust the control valve in the cavity to control the pressure in the valve cavity, frequent opening and closing, and the slow response of the valve, which causes the output frequency response of low pressure valve.

At the same time, the correctness of the mathematical model of Massimo Sorli has been verified, as shown in Figure 4c. Figure 4d is the frequency response of the valve; the results show that the cutoff frequency of the pressure relief valve is 5 Hz. When the valve reaches a frequency of 5 Hz, the pressure chamber will show instantaneous fluctuating pressure and flow and the valve precision will be affected due to the unavoidable limited movement position that is caused by valve position.

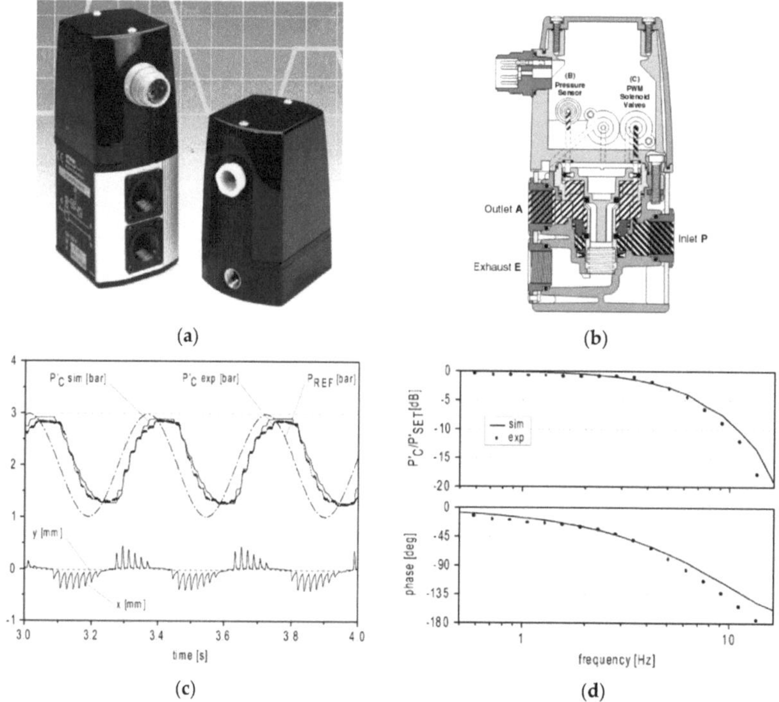


**Figure 4.** ParkerP3P-R pneumatic solenoid pneumatic pressure proportional valve. (**a**) P3P-R pneumatic solenoid pneumatic pressure proportional valve; (**b**) Schematic of P arkerP3P-R pneumatic solenoid; (**c**) Sinusoidal tracking signal; (**d**) Frequency response.

Switch electromagnetic pneumatic pressure proportional valve has the advantages of consuming low energy, having a simple structure, low costs, and so on; it has obvious advantages in static pressure regulation. It is insensitive to the rapidly changing pressure signal because it has a low frequency response. At the same time, in the process of pressure regulation, the pressure loss in the control chamber is large, which seriously affects the service life and the reliability of the pressure reducing valve and also restricts the application of switch electromagnetic pneumatic pressure proportional valve.

At present, the most effective way to improve the lifespan of switch electromagnetic pneumatic pressure proportional valve is to increase the frequency of the switch [16–19].

## 3. Related Technology Research Progresses

### 3.1. Numerical Simulation Researches

Pneumatic pressure proportional valve is composed of more complex pneumatic components, including a variety of gas, pneumatic actuators, controllers, and electrical mechanical converters. When analyzing the relationships among various components of the valve, the numerical analysis method is often used, which can predict the performance of the valve, optimize the structure design of the valve, and improve the design efficiency.

Electric mechanical converter will be the proportion of the input current or voltage converted into force or displacement output through the armature of special design. To establish the mathematical model is the key for modeling proportional solenoid valve. Vaughan, N.D., Gamble, J.B. [20] proposed a kind of nonlinear dynamic model of high speed solenoid valve that comprises of a proportional solenoid and the spool, and the model can accurately predict for a given input voltage steady state and dynamic response. The dynamic analysis of the proportional pressure valve is carried out by Dasgupta, K. and Karmakar, R. [21,22]. The dynamic mathematical model of the valve is established, and the flow characteristics of the valve pressure are taken into account to improve the dynamic response characteristics of the pressure valve.

Excepting the direct relationship between the input signal and the output force, the magnetic circuit analysis is a common method thatcan clearly identify each physical quantity and can be used to analyze the static performance of electromagnet. However, the magnetic circuit analysis method cannot describe the dynamic of the electromagnet. In the current study, the researchers tend to limit the method to establish the mathematical model of the electromagnet [23].

For example, Kawase, Y. and Ohachi, Y. [24] analyzed various characteristics of electromagnets by using the finite element method, including resistivity, response speed, and material properties. The finite element model has the advantages of high precision, but it is difficult to use it to find out effective results because there is a large amount of calculation and complex algorithm.

Furthermore, except for the mathematical models, there are semi-empirical modeling methods in which the experimental data are used to fit the nonlinear part of the system [9,20,25–27]. In the research of proportional electromagnet circuit, there is the nonlinear induced voltage of the electromagnet motion and in the modeling of a proportional electromagnet, it is hard to find the reasonable approximation of induced voltage.

Many scholars estimated the induced voltage indirectly, that is to say, that they obtained the relationship between the induced voltages by fitting the experimental data. The semi-empirical model is used to study the approximate the induced voltage in XU [25], which is based on the relationship between the output force of the electromagnet and the armature. The linear least square method is used to fit the static experimental results of electromagnetic force, and further obtain the approximate nonlinear induction voltage.

Vaughan [20] used a semi-empirical modeling method to estimate the electromagnet as a pure combination of resistance and nonlinear inductance. The coil current and magnetic flux are calculated by experiment. In Elmer's Research [26], which is based on the linear relationship between the reciprocal of the equivalent inductance of the coil and the coil current, the linear coefficients are fitted by the experimental data but are less effective. The reason is the ignorance of the influence of the armature displacement on the output force of the proportional electromagnet.

A pneumatic pressure proportional valve for ankle foot orthopedic instruments was studied by Lescano, C.N. [28]. Based on the linear model of the voltage signal and the output pressure, the dynamic response of the pneumatic pressure proportional valve is studied and the open-loop transfer function model of the system is obtained, as shown in Figure 5. Finally, the model is verified by experiments. It can guide the development of driving system of medical instrument gas.

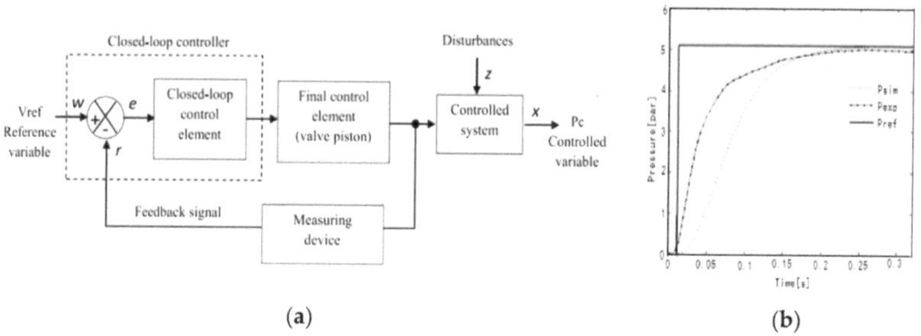


(a)      (b)

**Figure 5.** Linear model of Lescano, C.N. voltage signal and output pressure. (**a**) Schematic of pneumatic pressure proportional valve circuit; (**b**) Comparison of model validation.

When compared with the finite element method, the semi-empirical modeling method can solve the problem in proportional electromagnet design. However, the non-human factors often affect the accuracy of calculation. Therefore, the researchers often combine finite element methods such as the magnetic circuit structure, material properties, and other physical quantities obtained by the finite element method, thus improving the accuracy of the calculation done by the semi-empirical modeling method.

Massimo Sorli [29] used the finite element method and the semi-empirical model to build the mathematical model of the proportional electromagnet, as shown in Figure 6. The data of Speedance and differential inductance are calculated by finite element method. The semi-empirical modeling method, combined with the finite element method, has the advantages of a high accuracy in calculation, simplified calculation, and high efficiency, and is widely used now [27].

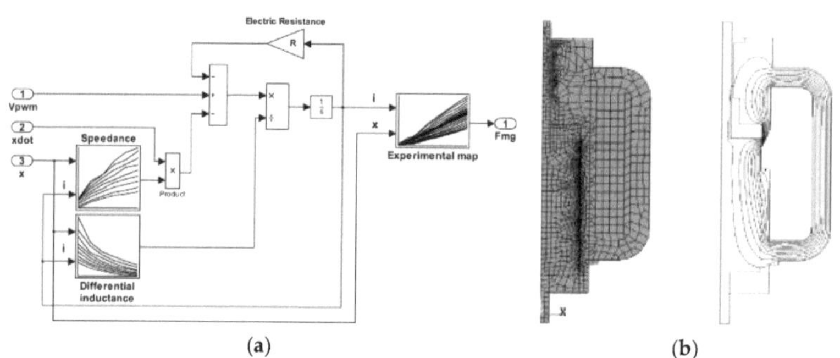


(a)      (b)

**Figure 6.** Electromagnet model of Massimo Sorli. (**a**) Electromagnet model; (**b**) Finite element model.

### 3.2. Modeling of Mechanical Structure

The key of structural modeling of pneumatic pressure proportional valve is the modeling of the valve core, in which the force balance is the essence of pneumatic pressure proportional valve. The forces acting on the valve core include friction force, driving force, aerodynamic force, and contact force, and in this paper, various forces will be introduced.

The contact forces also include the spool stroke limit force and the contact force between the valve cores, while the spool stroke limit force is the key to the valve core modeling. The earliest research on the spool stroke limit force is done by Kukulka [15]. They used a simple method to make it easy to travel limit force to a special spring, which is related to the amount of compression. Kukulka determined the relationship between the stiffness and the amount of compression.

Elmer [26] continued the study of Kukulka and found that there was no relationship between the stiffness of the special stiffness and the amount of compression. Based on this, the limit force is simplified as an ordinary spring with an elastic coefficient. It has been pointed out that the contact force between the valve core and the contact material is the hard point in valve core modeling. Massimo Sorli [16] studied this problem and came up with the assumption that the core and contact material are simplified as a spring damping system, and the results are verified by numerical simulation.

The friction force is generated by the contact between the sealing ring and the metal wall of the valve core, and it has been studied by a large number of scholars [30,31]. At present, the most widely used friction model is "Stribeck + viscous friction + Kulun + static friction":

$$F_f = \begin{cases} \mathrm{sag}(\dot{x})\left[F_c + (F_s - F_c)e - |\dot{x}/\dot{x}_s|\right]^{\delta_s} + b\dot{x} & \dot{x} \neq 0 \\ F_e & \text{else} \end{cases} \tag{1}$$

In the formula, $F_c$ refers to the dynamic friction force of Kulun, $F_s$ refers to the maximum static friction force, $b$ is a viscous friction coefficient, and $\dot{x}$ is Stribeck velocity. The input multiple superimposed flutter signal allows the spool to do a small reciprocating motion, which can effectively eliminates the static friction, thereby eliminating the influence of static friction.

Therefore, the model can be considered without static friction. The viscous friction force is often used for modeling the viscous friction; the absolute value of the spool speed is in proportion to the size of the viscous friction, while the direction of the force is opposite to the direction of the spool [14,19,25]. The viscous friction coefficient is determined by the trial and error method.

Aerodynamic force is the key factor that must be taken into account during the modeling process of the valve core. This is because the changing rate of the aerodynamic force with the displacement of the spool is large, which makes the aerodynamic force enable to produce considerable equivalent stiffness [32]. For this problem, the method of gas resistance is often used, [33]. According to the law of gas flow to the general conversion formula, Piao [34] proposed the concept of air resistance characteristic in series and obtained the conclusion that all of the nonlinear gas resistance has a common series of characteristics. Liu [35] established the model and condition of gas volume, air resistance, and gas resistance in the pneumatic stripping system, and obtained the parameters of air volume and air-flow in pneumatic stripping. In addition, some other scholars have made a preliminary study of the gas resistance and gas content [36].

*3.3. Modeling of Gas Flow Characteristics*

In terms of the analysis of gas flow characteristics of pneumatic pressure proportional valve, many scholars studied the flow characteristics of the gas passing through the valve port and the thermodynamic properties of the gas in the cavity [32].

By simplifying the complex process of flowing through the valve port of the gas, the process is considered as a constant flow of the nozzle in the research of Venant. If it is assumed that the specific heat of gas is constant, the one-dimensional constant drop flow equation of the ideal gas passing through the contraction nozzle can be presented like:

$$\dot{m}_t = \begin{cases} \dfrac{A}{\sqrt{RT_u}}\sqrt{\dfrac{2k}{k-1}}P_u\sqrt{\left(\dfrac{P_d}{P_u}\right)^{2/k} - \left(\dfrac{P_d}{P_u}\right)^{(k+1)/(k)}} & C_t < \dfrac{P_d}{P_u} \leq 1 \\[4mm] \dfrac{A}{\sqrt{RT_u}}P_u\sqrt{k\left(\dfrac{2}{k+1}\right)^{(k+1)/(k-1)}} & \dfrac{P_d}{P_u} \leq C_t \end{cases} \tag{2}$$

In the formula, $R$ is the gas constant, $T_u$ is the upper temperature of the gas, $m$ is the gas mass flow through the valve port, $P_u$ and $P_d$ are the downstream pressure of the contraction pipe, and $C_t$ is the critical pressure ratio. For air, $C_t = 0.528$. When the pressure difference between upstream and downstream is very small as $\Delta P = (P_u - P_d) \to 0$, this excessive gain will enlarge the error in numerical calculation. At the same time when $\Delta P \to 0$, the gas flow rate approaches 0, and the flow of gas is laminar flow, expressing the gas flow in the quality problems. In order to solve this problem, the following modified formula is needed:

$$
\dot{m}_t = \begin{cases}
ka \cdot A \cdot T_u^{-0.5}(P_u - P_d) & \beta_{lam} \le \frac{P_d}{P_u} < 1 \\[2ex]
A \cdot P_u \cdot \sqrt{\frac{2k}{k-1} \cdot \frac{1}{R \cdot T_u}\left[\left(\left(\frac{P_d}{P_u}\right)^{2/k} - \left(\frac{P_d}{P_u}\right)^{(k+1)/(k)}\right)\right]} & b \le \frac{P_d}{P_u} < \beta_{lam} \\[2ex]
A \cdot P_u \cdot \sqrt{\frac{k}{R \cdot T_u} \cdot b^{\frac{K+1}{K}}} & \frac{P_d}{P_u} < b
\end{cases}
\tag{3}
$$

In the formula, $\beta_{lam}$ is the threshold value which is 0.995~0.999, $b$ is $C_t$ in the formula and $ka$ is the compensation coefficient.

The actual pressure proportional valve orifice is not smooth; it causes the flow loss of the convective gas [37]. Therefore, the actual flow quality is less than the theoretical value of the formula, in order to further improve the flow quality it needs the following formula:

$$
\dot{m} = C_d \cdot \dot{m}_t
\tag{4}
$$

The experimental results showed that the flow coefficient $C_d$ in different types of orifice is different. Perry [37] further suggested that the flow coefficient $C_d$ in the orifice is related to the upstream and downstream pressure ratio, and established the corresponding relation. Reid [38] and Fleischer [39] established the relationship between the flow coefficient and the orifice geometry. The empirical formula of flow coefficient $C_d$ and upstream and downstream pressure ratio is given by Mozer [40]. In addition, some scholars further standardized the calculation of $C_d$ based on the empirical formula of Mozer.

For the study of gas in the cavity of thermodynamics, both domestic and foreign scholars all referred to the ideal gas state equation, the mass continuity equation, and the first law of thermodynamics. The differential equations of pressure and temperature are established, and then the differential equation is simplified. Massimo Sorli [14] calculated the gas in the cavity, and the variation law of pressure in the cavity is given by using the ideal gas state equation. According to the law of adiabatic change, the change of pressure and temperature of the gas in the cavity is obtained according to the change rule in research on Nabi [41]. Cai thought that the heat exchange between the gas and the inner wall of the cavity could be deduced from the energy equation. Richer [38] thought that the gas's physical changes of entering and leaving the gas chamber are different. It is suggested that when the gas enters the gas chamber, the adiabatic process can be used to estimate the gas. But when the gas flows out of the chamber, it is isothermal.

## 4. Proportional Controller

### 4.1. Research on Proportional Controller

Proportional controller and proportional solenoid actuator cooperate with each other to convert the electrical signal (voltage and current) into force proportional to the displacement of the physical output, and their performance often depends on the level of them. The proportional controller is the key of the whole conversion element, which has attracted much attention in both domestic and abroad researches.

The earliest research on proportional controller is developed in Europe and the United States, and other hydraulic components companies, including Moog, Towler, Vickers, as well as Japan's oil research company. With the development of pneumatic technology, some pneumatic components manufacturers such as ASCO, SMC, FESTO, etc., also joined the ranks of the proportional controller. Figure 7a is a Control D type proportional NUMATICS's controller that comprises a power unit, signal processing unit, and a main control unit, according to the input coil current in proportional valve control. It can also be connected to the PC through the mini USB, to support open-loop control, single closed-loop control, and double closed-loop control mode, as shown in Figure 7b.

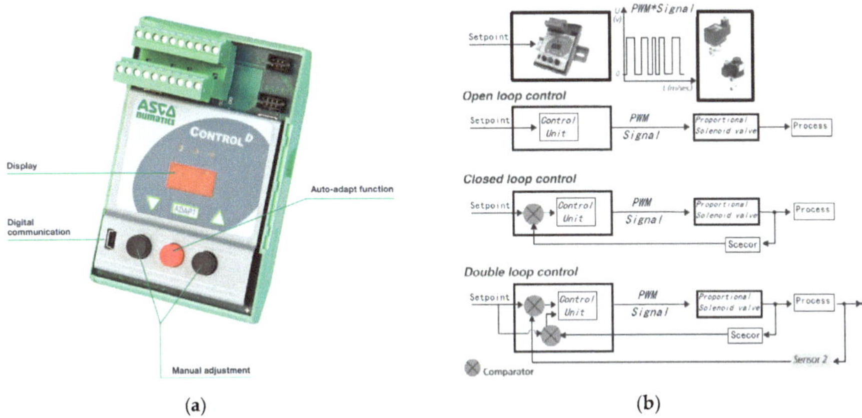


**Figure 7.** NUMATICS D Control proportional controller. (**a**) Control type proportional controller; (**b**) Schematic diagram of control mode.

In the research of component level controller, the power drive circuit is the main research object. The performance of the simulation is poor, the power is low, and it is not easy to integrate. The switch has many advantages, such as high efficiency, small size, and low power consumption.

Figure 8 is the topology of the common high side single tube drive circuit. Liu [42,43] wanted to improve the performance of a single drive circuit based on the structure of the proposed "anti-unloading structure". As shown in Figure 9, the improved structure and the current drop speed are obvious, but it consumes more power. Nie [44] further improved the anti-unloading structure, including the proposed three state modulation proportional electromagnet power anti-unloading mode, the driver, and the realization of arbitrary current control and tracking, which has made great progress. Rexroth proposed the "supercharged driven" method, in which when the current rises, it provides the voltage booster circuit, making the current rapidly rise and improving the driving circuit and the frequency of response. Amirante [45] used the method of "boost drive" in the control of proportional directional valve and put forward the method of increasing current drive, which improved the response time of the control circuit.

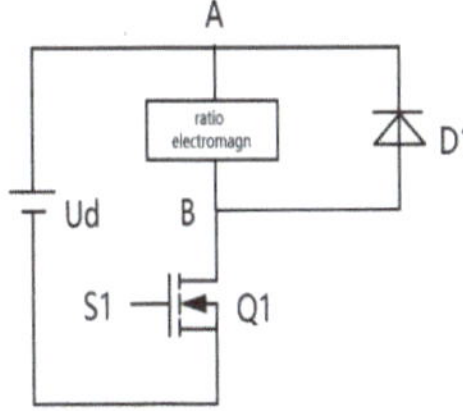


**Figure 8.** High side single tube driving circuit.

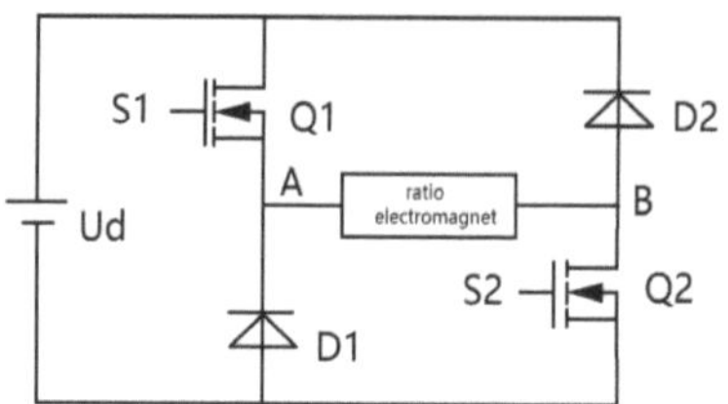

**Figure 9.** Anti unloading driving structure.

With the development of electronic technology, the integration of the proportional controller is increasing. In addition to the proportional control circuit, there are many independent modules in the control chip, with the advantages of high integration, low cost, and high performance. Figure 10 represents the design of Infineon R & D of proportional solenoid driver chip, which contains the driver transistor, the transistor, and the current recovery resistor. It is composed of a current detection circuit, power driver circuit, and chatter generation circuit with the purpose of realizing the closed-loop control of the coil current. The principle of the control is to adjust the switch of the power tube by dynamic adjustment, so that the coil current reaches the set value. It is widely used in the middle precision proportional control.

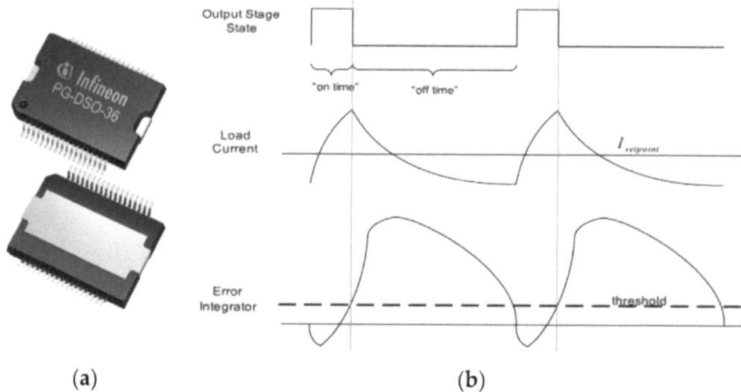

(a)        (b)

**Figure 10.** Infineon Company's proportional solenoid drive chip. (**a**) Driver chip; (**b**) Schematic diagram of current control.

When compared with the design of system level and component level proportional controller proportional controller, the former one more focuses on the overall performance of stability, reliability, and universality, and it is the component proportion controller after a certain stage of development of the new design idea. The latter satisfies the reliability level of the proportional controller. As far as the domestic research design is concerned, there are a number of universities and research institutes in China that are only in the experimental stage.

*4.2. Research on Control Strategy*

The control strategy determines the performance of the pneumatic pressure proportional valve to a great extent, and in the choice of control strategy, it needs to be combined with various components of the function to give full play to the performance of various components. Effective control strategy is the principle in designing the proportional valve. According to proportional pressure relief valve with no pressure electrical feedback, pneumatic pressure proportional valve can be divided into closed-loop control and open-loop control. While the control modes can be divided into sliding mode control and adaptive control.

The current control of solenoid coil is the key of open-loop control design. In the early stage, the coil current is controlled by the P or PI circuit, but it has many disadvantages like poor flexibility and a complicated operation process. Subsequently, the digital controller has been widely used and the coil current control is generally controlled by Proportion Integration Differentiation (PID). Many foreign and domestic scholars have conducted a research on how to improve the accuracy of current control. Ally [46] and Liu [47] studied the control system and since the effect of controller depends on the exact mathematical model of the controlled system, the adaptive parameter method, based on Lyapunov analysis, can compensate the uncertainty of system parameters. Lee, S.R. [48,49] used the least square method to identify the system in real time. Based on this, the controller is designed according to the pole placement.

Based on the embedded model control theory, the embedded model is established in Canuto's control system [50]. The utility model achieved the purpose of suppressing the disturbance of the current value and the voltage value of the coil, and realized the accurate control. The control strategies are based on the comparison between the current and the current in the coil. But in fact, the increase in the coil temperature leads to changes in coil resistance and furthers the instability of the current in the coil, therefore, some scholars think from the perspective of how to keep the coil resistance constant. Based on the iterative method derived from the fixed point theory, Jung [51,52] estimated the parameters of the coil resistance. Furthermore, the current value of the coil keeps a linear relationship with the duty cycle of the Pulse Width Modulation (PWM) drive signal. Lannone [53] proposed a hardware compensation scheme of coil resistance, the resistance increase of auxiliary circuit real-time measurement coil values, the measured resistance value, and the duty ratio of PWM signal can fix coil working state of the look-up table operation. These methods could effectively improve the control accuracy.

The nonlinear problem in the control of pneumatic pressure proportional valve is caused by the compressibility of the working medium of the gas, and is how to obtain the high precision control. The domestic and foreign famous companies are solved by the closed loop control. PID control is the main control strategy, and the conventional PID control has poor adaptability and low flexibility so it is necessary to improve the PID control. Hamdan [54] believed that the nonlinear problems in the proportional valve have hysteretic characteristics, therefore, they put forward the control strategy of conventional PID, Anti-windup, Bang-bang, and feed forward variable gain control combination, as shown in Figure 11. When compared with the conventional PID control, the control effect of this control strategy is improved obviously, but there still exist many problems such as overshoot and the lack of accuracy.

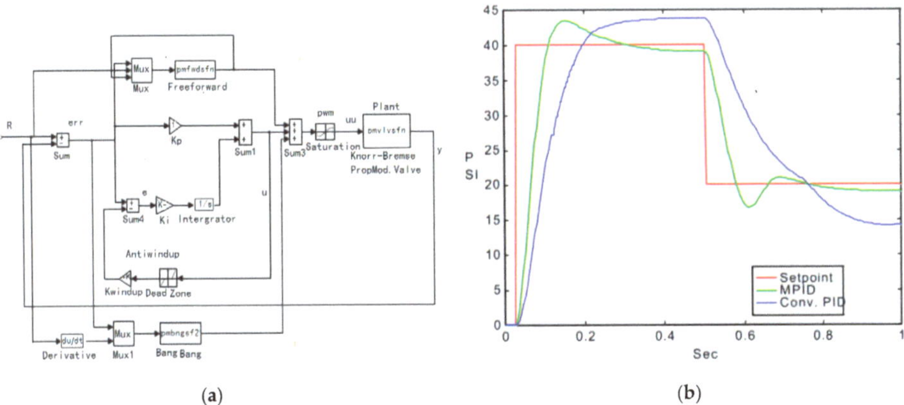

(a)        (b)

**Figure 11.** ModifiedProportion Integration Differentiation (MPID) control strategy of Hamdan. (**a**) Schematic diagram of control strategy; (**b**) Comparison of conventional PID and MPID.

Wu [55] adopted a generalized predictive control that considered the influence of the change of system stiffness and the setting value, and compensated for the high nonlinearity of the system. As compared with the PID control, the performance of the system using generalized predictive control is improved significantly and the robustness is stronger. Li [56] put forward a variable gain neuron adaptive PID controller to realize the real-time control of piston displacement and the valve controlled cylinder pneumatic position servo control system. This method does not require the accurate modeling of the system. The controller has a simple structure, convenient design, and steady performance.

PWM has been successfully applied to pneumatic switch control [57–59], and has become the most extensive and effective control method, and it can improve the regulating precision of the regulating valve and reduce the power consumption. After turning on the solenoid valve, the circuit is switched to PWM operation at once. The frequency of the PWM pulse is much larger than the response frequency of the solenoid valve, so the time of the pulse flow on the coil is averaged to form a sustainable current. The PWM drive mode also has the advantages of simple circuit structure, small size, and low energy consumption, but the problem of slow opening and closing time of the pilot valve is widespread [60–62]. In recent years, with the development of integrated circuits, insulated gate bipolar transistors (IGBT) have gradually replaced Metal-Oxide-Semiconductor Field-Effect Transistor (MOSFET) and Bipolar junction transistor (BJT) devices in PWM drive control of switching valves [63].

By measuring the input and output information, the adaptive control can adjust the air quantity according to the dynamic characteristics of the controlled object and the system error, which means to adjust the motion law of the controller so that the control performance of the system can be maintain to optimally [64]. Wang [65] controlled the pressure in the gas tank, and established the three-order controlled auto-regressive moving average model. For the state variables in the model and the parameters of the model, Wang used the Calman filtering method and the recursive least squares algorithm with a recursive factor.

In order to make the model more universal, Wang has also designed a linear two-time self-tuning pressure regulator, which has a better pressure control effect. In order to reduce the overshoot of the system and improve the precision of the system, Pipat Chaewieang [66] designed a generalized predictive control method to control the load pressure, and Wang also used a similar solution in the calculation process to achieve the desired objectives. Steven Lambeck [67] presented an exact linearization control method for the nonlinear dynamic model based on pneumatic pressure proportional valve. It can adapt great to the volume of gas in the valve chamber, which improves the output linearity of the pneumatic pressure proportional valve.

The influence of friction force on the pressure pneumatic pressure proportional valve control system is nonlinear. Regarding the study of friction, the general approach is to compensate the friction force, especially in the spool type pressure pneumatic pressure proportional valve. The influence of the uncertainty of the large friction parameters on the nonlinear robust control has been studied by Shen [68], and the corresponding control method has been proposed. In the Arkan's [69] study of friction compensation, in order to realize a better linearity in the input and output, the discrete filtering method is used and the Luenberger observer is used to realize the accurate displacement observation, so as to compensate the influence of the friction force on the control system. Tore [70] analyzed the motion of the valve core, including the amplitude, pulse width, and the fundamental frequency, and put forward a reasonable superposition of the flutter signal that can be compensated for the impact of friction on the control system. Except for the compensating friction, there are other ways to reduce friction. For example, the air bearing has the advantages such as clean, small friction, and etc. Kayihan [69] used air bearing to reduce the friction of spool movement. However, it is complex and costly to apply the air bearing pressure pneumatic pressure proportional valve structure.

## 5. Conclusions

As the people's awareness of environmental protection has enhanced, green clean energy is developing rapidly. Compressed air has the unique advantages regarding the energy storage and transmission. It has been widely favored, and pneumatic components have also been developed rapidly. For electric proportional valve pressure, to complete the precise regulation is no longer the only goal, it aims to achieve the electrical integration and integration of the development direction of the composite. "PLC + sensor + pneumatic components" has become a typical control system. The pneumatic components or mechanisms with different functions form new integrated components or mechanisms with additional functions are to make them composite and integrated. Composite integration can shorten the design cycle, reduce wiring, piping, and components, while reducing the impact and debugging time, thus improving work efficiency has become the development trend of electric proportional valves.

High-pressure can improve the dynamic characteristics, the hardness, and the response speed of the pneumatic system, and also realize the miniaturization and high speed of pneumatic components. With the maturation of compressed gases, especially energy density devices, such as new fuel cell vehicles, can effectively solve these problems, such as stringent energy storage, volume ratio, and high-pressure. In addition, the pressure reduction can also reduce the volume of the actuating element, and the corresponding auxiliary elements such as filters and electromagnetic valves. Pressure reducing valves are correspondingly reduced in size. Thus, the cost, installation space, and resources are also reduced. In the future, the development of high-pressure pneumatic components, energy saving and safety research of components and systems will be the main research field of the electrical pneumatic pressure proportional valve.

Moreover, regarding the research and application of the electric-pneumatic pressure proportional valves, it is necessary to meet the following requirements:

### 5.1. Oil-Free

The mechanical components in the pneumatic system need to be lubricated so that the oil sprayer in the system is indispensable. In the pneumatic system, there is no return pipeline for oil mist recovery, and the exhaust gas causes pollution to the environment. The development of the non-oil system is required in the development of the electric-pneumatic pressure proportional valves. With the extensive research of self-lubrication special material and gas film lubrication, its unique lubricity can be applied to an oil-free system, which provides the idea for oil-free system [71–76].

### 5.2. Miniaturization and Light-Weight

With the miniaturization of mechanical components and mechanical devices, the miniaturization of pneumatic components, pneumatic auxiliary components, and control elements is becoming much higher. The miniaturization and light-weight of pneumatic components consume less energy and cost, and can thus reduce the overall cost of the system. At present, most of the components are made of aluminum alloy and new plastic materials and they are made of ultra-thin, ultra short, and ultra-small parts. But as the demand of miniaturization keeps growing, it is necessary to explore new methods of miniaturization design.

### 5.3. Energy Saving

The manufacturing industry in all countries focuses on energy saving. While the energy saving of pneumatic technology can be carried out in two ways: reducing the consumption of gas and reducing the power of the system. Reducing power consumption as an important aspect of energy saving aims to develop a variety of small power solenoid valves and the power of less than 2 W solenoid valve is popular. Reducing the gas consumption aims to reduce the power consumption of the compressor thus to achieve the effect of saving energy.

**Acknowledgments:** The research is funded by Grants (51575020, 51675020) of the National Natural Science Foundation of China and Open Foundation of the State Key Laboratory of Fluid Power and Mechatronic Systems.

**Author Contributions:** Fangwei Ning was responsible for consulting the literature and drafting the paper. Yan Shi, Maolin Cai, Yixuan Wang, Weiqing Xu contributed to directing the thesis.

**Conflicts of Interest:** The authors declare no conflict of interest.

## References


1. Shearer, J.L. Study of Pneumatic Process in the Continuous Control of Motion with Compressed Air-I. *Trans. ASME* **1956**, *2*, 233–242.
2. Burrows, C.R.; Webb, C.R. Simulation of an On-Off Pneumatic Servomechanism. *Proc. Inst. Mech. Eng.* **1967**, *182*, 631–642. [CrossRef]
3. Burrows, C.R.; Webb, C.R. Further Study of a Low-Pressure on-off Pneumatic Servomechanism. *Proc. Inst. Mech. Eng.* **1969**, *184*, 849–858. [CrossRef]
4. Xu, H. Theory Analysis and Study of Control Method of Pneumatic-Hydraulic Combination Control Position System. Ph.D. Thesis, Harbin Institute of Technology, Harbin, China, 1 April 2001. (In Chinese)
5. Ioannidis, I.; Nguyen, T. Microcomputer-Controlled Servo-Pneumatic Drives. In Proceedings of the 7th International Fluid Power Symposium, Bath, UK, 16–18 September 1986; pp. 155–164.
6. Liu, S.; Bobrow, J.E. An Analysis of a Pneumatic Servo System and Its Application to a Computer-Controlled Robot. *J. Dyn. Syst. Meas. Control* **1988**, *110*, 228–235. [CrossRef]
7. Lai, J.Y.; Menq, C.H.; Singh, R. Accurate Position Control of a Pneumatic Actuator. *ASME J. Dyn. Syst. Meas. Control* **1990**, *112*, 734–739. [CrossRef]
8. Numatics. *Proportional Technology Precise Control of Pressure and Flow*; Numatics: Norwich, MI, USA, 2012; Available online: www.numatics.com (accessed on 8 July 2017).
9. Matrix. Panel Mounting Electronic Pressure Regulator. 2011. Available online: http://www.matrix.to.it/matrix2005/business3.aspid=14&id2=7&idbarea2=17 (accessed on 8 July 2017).
10. Proportion Air. QPV&MPV Proportional Pressure Control Valves. 2011. Available online: http://www.proportionair.com (accessed on 8 July 2017).
11. Festo. *Proportional Pressure Regulators VPPE without Display*; Festo: Shanghai, China, 2010; Available online: http://www.festo.com.cn (accessed on 8 July 2017).
12. Belforte, G. Design of a new pressure regulator with electronic control: Friction force analysis. In Proceedings of the International Conference on Functional Programming (ICFP'2005), Hangzhou, China, 5–8 April 2005; Available online: http://porto.polito.it/1419478/ (accessed on 17 October 2017).
13. Szente, V.; Vad, J.; Lorant, G.; Fries, A. Computational and experimental investigation on dynamics of electric braking systems. In Proceedings of the Scandinavian. International Conference on Fluid Power, Linkoping, Sweden, 30 May–1 June 2001.
14. Sorli, M.; Figliolini, G.; Pastorelli, S. Dynamic model and experimental investigation of a pneumatic proportional pressure valve. *IEEE ASME Trans. Mechatron.* **2004**, *9*, 78–86. [CrossRef]
15. Sorli, M.; Figliolini, G.; Almondo, A. Mechatronic Model and Experimental Validation of a Pneumatic Servo-Solenoid Valve. *J. Dyn. Syst. Meas. Control.* **2010**, *132*, 626–634. [CrossRef]
16. Ahn, K.; Yokota, S. Intelligent switching control of pneumatic actuator using on/off solenoid valves. *Mechatronics* **2005**, *15*, 683–702. [CrossRef]
17. Passarini, L.C.; Nakajima, R.P. Development of a high-speed solenoid valve: An investigation of the importance of the armature mass on the dynamic response. *J. Braz. Soc. Mech. Sci. Eng.* **2003**, *25*, 1143–1147. [CrossRef]
18. Kajima, T.; Kawamura, Y. Development of a high-speed solenoid valve: Investigation of solenoids. *IEEE Trans. Ind. Electron.* **2002**, *42*, 1–8. [CrossRef]
19. SMC. *E-P Hyregvyl*; SMC: Changzhou, China, 2010; Available online: http://www.smcworld.com/ (accessed on 8 July 2017).
20. Vaughan, N.D.; Gamble, J.B. The Modeling and Simulation of a Proportional Solenoid Valve. *J. Dyn. Syst. Meas. Control* **1996**, *118*, 120–125. [CrossRef]
21. Dasgupta, K.; Karmakar, R. Modelling and dynamics of single-stage pressure relief valve with directional damping. *Simul. Model. Pract. Theory* **2002**, *10*, 51–67. [CrossRef]
22. Dasgupta, K.; Watton, J. Dynamic analysis of proportional solenoid controlled piloted relief valve by bondgraph. *Simul. Model. Pract. Theory* **2005**, *13*, 21–38. [CrossRef]

23. Lequesne, B.P. Finite-element analysis of a constant-force solenoid for fluid flow control. *IEEE Trans. Ind. Appl.* **1988**, *24*, 574–581. [CrossRef]

24. Kawase, Y.; Ohachi, Y. Dynamic analysis of automotive solenoid valve using finite element method. *IEEE Trans. Magn.* **1991**, *27*, 39–42. [CrossRef]

25. Xu, Y.; Jones, B. A simple means of predicting the dynamic response of electromagnetic actuators. *Mechatronics* **1997**, *7*, 589–598. [CrossRef]

26. Elmer, K.F.; Gentle, C.R. A parsimonious model for the proportional control valve. *J. Mech. Eng. Sci.* **2001**, *215*, 1357–1363. [CrossRef]

27. Gaeta, A.D.; Glielmo, L.; Diglio, V. Modeling of an Electromechanical Engine Valve Actuator Based on a Hybrid Analytical-FEM Approach. *IEEE ASME Trans. Mechatron.* **2008**, *13*, 625–637. [CrossRef]

28. Lescano, C.N.; Rodrigo, S.E.; Herrera, C.V. Dynamic Response of a Pneumatic Pressure Valve Applied to the Design of an Actuation System in Assistive Robotics. *IFMBE Proc.* **2015**, *49*, 952–955.

29. Kukulka, D.J.; Benzoni, A.; Mollendorf, J.C. Digital Simulation of a Pneumatic Pressure Regulator. *Soc. Model. Simul. Int.* **1994**, *63*, 252–266. [CrossRef]

30. Armstrong, H.B.; Dupont, E.; Canudas, W.C. A Survey of Models, analysis tools and Compensation Methods for the Control of Machines with Friction. *Automatica* **1994**, *307*, 1083–1138. [CrossRef]

31. Canudas, D.W.C.; Olsson, H.; Astron, K.J. A New Model for Control of Systems with Friction. *IEEE Trans. Autom. Control* **1995**, *40*, 419–425. [CrossRef]

32. Beater, P. *Pneumatic Drives: System Design, Modelling and Control*; Springer: Berlin, Germany, 2007. [CrossRef]

33. Idelchik, I.E.; Fried, E. *Handbook of Hydraulic Resistance*; Hemisphere Publishing: New York, NY, USA, 1986.

34. Andersen, B.W.; Binder, R.C. The Analysis and Design of Pneumatic Systems. *J. Appl. Mech.* **1967**, *34*, 1055. [CrossRef]

35. Liu, M.W.; Zhang, L.F. Application of Pneumatic Capacity Pneumatic Inductance and Pneumatic Resistance on Pneumatic Peeling-off. *Chin. Hydraul. Pneum.* **2006**, *2006*, 14–15. (In Chinese)

36. Xu, Z.P.; Wang, X.Y.; Pi, Y.J. Numerical Simulation of PPRV Based on Pneumatic Bridge and Control Networks. In Proceedings of the Fifth Fluid Power Transmission and control, Hangzhou, China, 3–5 April 2007; pp. 961–964.

37. Perry, J.A. Critical Flow through Sharp-edged Orifices. *Trans. ASME* **1949**, *71*, 757–764.

38. Reid, J.; Stewart, C.D. A review of critical flow nozzles for the mass flow measurement of gases. In Proceedings of the 2nd International Symposium on Fluid Control Measurement Mechanics and Flow Visualization, Sheffield, UK, 5–9 September 1988; pp. 454–457.

39. Fleischer, H. *Manual of Pneumatic System Operation*; McGraw-Hill: New York, NY, USA, 1995.

40. Szente, V.; Mózer, Z.; Ákos, T. Experimental investigation on pneumatic components. In Proceedings of the 12th International Conference on Modelling Fluid Flow, Budapest, Hungary, 3–6 September 2003.

41. Nabi, A.; Wacholder, E.; Dayan, J. Dynamic Model for a Dome-Loaded Pressure Regulator. *J. Dyn. Syst. Meas. Control* **2000**, *122*, 290–297. [CrossRef]

42. High, A.; Riche, E.; Hurmuzlu, Y. A High Performance Pneumatic Force Actuator System Part 1—Nonlinear Mathematical Model. *J. Dyn. Syst. Meas. Control* **2001**, *122*, 416–425.

43. Liu, X.H. Optimization of Static Characteristics of Proportional Control Amplifier and Proportional electro Magnet. *Hangzhou Zhejiang Univ.* **1987**, *17*, 126–225.

44. Nie, Y. Research on the Key Technology of A New Programmable Electro Hydraulic Proportional Controller. Ph.D. Thesis, Zhejiang University, Hangzhou, China, 2017. Available online: http://cdmd.cnki.com.cn/Article/CDMD-10335-1011068969.htm (accessed on 8 July 2017).

45. Amirante, R.; Innone, A.; Catalano, L.A. Boosted PWM Open Loop Control of Hydraulic Proportional Valves. *Energy Convers. Manag.* **2008**, *49*, 2225–2236. [CrossRef]

46. Liu, R.; Alleyne, A. Nonlinear Force/Pressure Tracking of an Electro-Hydraulic Actuator. *J. Dyn. Syst. Meas. Control* **2000**, *122*, 232–237. [CrossRef]

47. Alleyne, A.; Liu, R. A simplified approach to force control for electro-hydraulic systems. *Control Eng. Pract.* **2000**, *8*, 1347–1356. [CrossRef]

48. Lee, S.R.; Srinivasan, K. On-Line Identification of Process Models in Closed Loop Material Testing. In Proceedings of the American Control Conference, Atlanta, GA, USA, 15–17 June 1988; pp. 1909–1916.

49. Lee, S.R.; Srinivasan, K. Self-Tuning Control Application to Closed-Loop Servohydraulic Material Testing. *J. Dyn. Syst. Meas. Control* **1990**, *112*, 680–689. [CrossRef]

50. Canuto, E.; Acuria-Bravo, W.; Agostani, M.; Bonadei, M. Digital current regulator for proportional electro-hydraulic valves with unknown disturbance rejection. *ISA Trans.* **2014**, *53*, 909–919. [CrossRef] [PubMed]

51. Jung, H.; Hwang, J.; Yoon, P.; Kim, J. Robust Solenoid Current Control for EHB. *Math. Oper. Res.* **2005**, *28*, 11–14.

52. Jung, H.G.; Hwang, J.Y.; Yoon, P.J.; Kim, J. Resistance estimation of a pwm-driven solenoid. *Int. J. Automot. Technol.* **2007**, *8*, 249–258.

53. Iannone, C.A.; Turner, K.W. Apparatus and Method for Monitoring and Compensating for Variation in Solenoid Resistance during Use. U.S. Patent 7,054,772, 30 May 2006.

54. Hamdan, M.; Gao, Z. A novel PID Controller for Pneumatic Proportional Valves with Hysteresis. In Proceedings of the 2000 Conference Record of the IEEE Industry Applications Conference, Rome, Italy, 8–12 October 2000; Volume 2, pp. 1198–1201.

55. Wu, G.; Sepehri, N.; Ziaei, K. Design of hydraulic force control system using a generalized predictive control algorithm. *IEE Proc. Control Theory Appl.* **1998**, *145*, 428–436. [CrossRef]

56. Li, B.; Wu, J.B.; Du, J.M. Research on Control Strategy of High-pressure Pneumatic Servo Position System. *Hydraul. Pneum. Seals* **2002**, *2*, 5–7.

57. Zhou, H. The technology of pneumatic proportional control and its application. *Chin. Hydraul. Pneum.* **1999**, *3*, 1–3.

58. Liu, X.; Jia, Q.; Liu, G.B. The Study on Controlling the Position of Air Cylinder by Electropneumatic Proportional Valve. *Mech. Eng.* **2002**, *6*, 19–25.

59. Han, J.H.; Zhang, H.X. The pneumatic proportional servo control technology and its application. *Mach. Tools Hydraul.* **2001**, *1*, 3–7.

60. Li, J.X.; Yuan, G.Q. Study of PWM controller based on AT 89C2051 single-chip micro-controller. *J. Zhejiang Univ. Technol.* **2000**, *5*, 419–425.

61. Zhou, X.; Shan, X.H.; Chen, L. The study of the digital proportional valve control system. *Mech. Electr. Eng. Technol.* **1997**, *2*, 51–52.

62. Yu, L. Methods to generate pulse-width modulation wave with single-chip microprocessor. *J. Fujian Agric. Univ.* **2001**, *17*, 332–347.

63. Tian, J. Developing PWM Signal Genevator for High Speed On-off Valve. *J. Civ. Aviat. Univ. China* **2003**, *6*, 256–301.

64. Topçu, E.E.; Yüksel, İ.; Kamış, Z. Development of electro-pneumatic fast switching valve and investigation of its characteristics. *Mechatronics* **2006**, *16*, 365–378. [CrossRef]

65. Wang, X.S.; Cheng, Y.H.; Peng, G.Z. Modeling and self-tuning pressure regulator design for pneumatic-pressure–load systems. *Control Eng. Pract.* **2007**, *15*, 1161–1168. [CrossRef]

66. Chaewieang, P.; Sirisantisamrit, K.; Thepmanee, T. Pressure control of pneumatic-pressure-load system using generalized predictive controller. In Proceedings of the IEEE International Conference on Mechatronics and Automation, Takamatsu, Japan, 5–8 August 2008; pp. 788–791.

67. Lambeck, S.; Busch, C. Exact Linearization Control for a pneumatic proportional pressure control valve. In Proceedings of the IEEE International Conference on Control and Automation, Xiamen, China, 9–11 June 2010; pp. 22–27.

68. Shen, T.; Tamura, K.; Kaminaga, H.; Henmi, N.; Nakazawa, T. Robust Nonlinear Control of Parametric Uncertain Systems with Unknown Friction and Its Application to a Pneumatic Control Valve. *J. Dyn. Syst. Meas. Control* **2000**, *122*, 257–262. [CrossRef]

69. Kayihan, A.; Francis, J.D. Friction compensation for a process control valve. *Control Eng. Pract.* **2000**, *8*, 799–812. [CrossRef]

70. Hägglund, T. A friction compensator for pneumatic control valves. *J. Process Control* **2002**, *12*, 897–904. [CrossRef]

71. Shi, Y.; Wang, Y.; Cai, M.; Zhang, B.; Zhu, J. Study on the Aviation Oxygen Supply System Based on a Mechanical Ventilation Model. *Chin. J. Aeronaut.* **2017**. accepted.

72. Ren, S.; Shi, Y.; Cai, M.; Xu, W. Influence of secretion on airflow dynamics of mechanical ventilated respiratory system. *IEEE/ACM Trans. Comput. Biol. Bioinform.* **2017**, *99*, 1. [CrossRef] [PubMed]

73. Ren, S.; Cai, M.; Shi, Y.; Xu, W.; Zhang, X.D. Influence of Bronchial Diameter Change on the airflow dynamics Based on a Pressure-controlled Ventilation System. *Int. J. Numer. Methods Biomed. Eng.* **2017**. [CrossRef] [PubMed]

74. Shi, Y.; Zhang, B.; Cai, M.; Xu, W. Coupling Effect of Double Lungs on a VCV Ventilator with Automatic Secretion Clearance Function. *IEEE/ACM Trans. Comput. Biol. Bioinform.* **2017**, *99*, 1. [CrossRef] [PubMed]

75. Shi, Y.; Zhang, B.; Cai, M.; Zhang, D. Numerical Simulation of volume-controlled mechanical ventilated respiratory system with two different lungs. *Int. J. Numer. Methods Biomed. Eng.* **2016**, *33*, 2852. [CrossRef] [PubMed]

76. Niu, J.; Shi, Y.; Cai, M.; Cao, Z.; Wang, D.; Zhang, Z.; Zhang, D.X. Detection of Sputum by Interpreting the Time-frequency Distribution of Respiratory Sound Signal Using Image Processing Techniques. *Bioinformatics* **2017**. [CrossRef]

MDPI AG
Grosspeteranlage 5
4052 Basel
Switzerland
Tel.: +41 61 683 77 34

*Applied Sciences* Editorial Office
E-mail: applsci@mdpi.com
www.mdpi.com/journal/applsci